NONLINEAR SPECTROSCOPY FOR MOLECULAR STRUCTURE DETERMINATION

MEMBERSHIP OF THE COMMISSION

CHAIRMAN

J.E. Bertie *(Canada; 1994–97)*

SECRETARY

J.F. Sullivan *(USA; 1990–95)*
P. Klaeboe *(Norway; 1996–97)*

TITULAR MEMBERS

A.M. Heyns *(RSA; 1994–97)*
R. Janoschek *(Austria; 1994–97)*
P. Klaeboe *(Norway; 1994–97)*
S. Tsuchiya *(Japan; 1994–97)*
B.P. Winnewisser *(Germany; 1992–97)*

ASSOCIATE MEMBERS

A.M. Bradshaw *(FRG; 1992–95)*
S.M. Cabral de Menezes *(Brazil; 1994–97)*
B.G. Derendjaev *(Russia; 1992–95)*
E. Hirota *(Japan; 1994–97)*
J. Kowalewski *(Sweden; 1996–97)*
A. Oskam *(Netherlands; 1992–97)*
C. Zhang *(China; 1990–97)*

NATIONAL REPRESENTATIVES

J.E. Collin *(Belgium; 1986–97)*
M. Chowdhury *(India; 1986–95)*
S. Califano *(Italy; 1990–97)*
J.J.C. Teixeira-Dias *(Portugal; 1992–97)*
Y.S. Lee *(Republic of Korea; 1990–97)*
D. Escolar *(Spain; 1992–97)*
J. Kowalewski *(Sweden; 1994–95)*
S. Suzer *(Turkey; 1996–97)*
S. Içli *(Turkey; 1994–95)*
R.K. Harris *(UK; 1994–97)*

INTERNATIONAL UNION OF PURE AND APPLIED CHEMISTRY

Nonlinear Spectroscopy for Molecular Structure Determination

Prepared under the auspices of
the IUPAC Commission on Molecular Structure
and Spectroscopy

EDITED BY
R.W. FIELD, E. HIROTA, J.P. MAIER
AND S. TSUCHIYA (Coordinator)

Blackwell
Science

and published for them by
Blackwell Science Ltd
Editorial Offices:
Osney Mead, Oxford OX2 0EL
25 John Street, London WC1N 2BL
23 Ainslie Place, Edinburgh EH3 6AJ
350 Main Street, Malden
MA 02148 5018, USA
54 University Street, Carlton
Victoria 3053, Australia
10, rue Casimir Delavigne
75006 Paris, France

Other Editorial Offices:
Blackwell Wissenschafts-Verlag GmbH
Kurfürstendamm 57
10707 Berlin, Germany

Blackwell Science KK
MG Kodenmacho Building
7–10 Kodenmacho Nihombashi
Chuo-ku, Tokyo 104, Japan

First published 1998

Printed and bound in Great Britain
by MPG Books Ltd, Bodmin

DISTRIBUTORS

Marston Book Services Ltd
PO Box 269
Abingdon, Oxon OX14 4YN
(*Orders*: Tel: 01235 465500
Fax: 01235 465555)

USA
Blackwell Science, Inc.
Commerce Place
350 Main Street
Malden, MA 02148 5018
(*Orders*: Tel: 800 759 6102
781 388 8250
Fax: 781 388 8255)

Canada
Login Brothers Book Company
324 Saulteaux Cresent
Winnipeg, Manitoba R3J 3T2
(*Orders*: Tel: 204 224-4068)

Australia
Blackwell Science Pty Ltd
54 University Street
Carlton, Victoria 3053
(*Orders*: Tel: 3 9347 0300
Fax: 3 9347 5001)

A catalogue record for this title
is available from the British Library

ISBN 0-632-04217-6

Contents

Editors

R.W. FIELD *Department of Chemistry, Massachusetts Institute of Technology, Cambridge, MA 02139, USA*

E. HIROTA *Graduate University for Advanced Studies, Kmiyamaguchi, Hayama, Kanagawa 240-01, Japan*

J.P. MAIER *Institut für Physikalische Chemie, Universität Basel, Klingelbergstrasse 80, CH-4056 Basel, Switzerland*

S. TSUCHIYA (Coordinator) *Department of Chemical and Biological Sciences, Japan Women's University, Mejirodai, Bunkyo-ku, Tokyo, 112 Japan*

Contributors

M.N.R. ASHFOLD *School of Chemistry, University of Bristol, Bristol BS8 1TS, UK* [127]

H. BITTO *Physikalisch-Chemisches Institut der Universität Zürich, Winterthurerstrasse 190, CH-8057 Zürich, Switzerland* [223]

M.C.R. COCKETT *Department of Chemistry, The University of York, Heslington, York YO1 5DD, UK* [167]

H-L. DAI *Department of Chemistry, University of Pennsylvania, Philadelphia, PA 19104-6323, USA* [55]

T. EBATA *Department of Chemistry, Graduate School of Science, Tohoku University, Sendai 980-77, Japan* [149]

Y. ENDO *Department of Pure and Applied Sciences, The University of Tokyo, Komaba, Tokyo 153, Japan* [29]

M. FUJII *Department of Electronic Structure, Institute for Molecular Science, 38 Myodaiji-chyo, Okazaki 444, Japan* [29]

H. HAMAGUCHI *Department of Chemistry, School of Science, The University of Tokyo, 7-3-1 Hongo, Tokyo 113, Japan* [203]

J.R. HUBER *Physikalisch-Chemisches Institut der Universität Zürich, Winterthurerstrasse 190, CH-8057 Zürich, Switzerland* [223]

K. MÜLLER-DETHLEFS *Department of Chemistry, The University of York, Heslington, York YO1 5DD, UK* [167]

Y. R. SHEN *Department of Physics, University of California and Materials Sciences Division, Lawrence Berkeley National Laboratory, Berkeley, CA 94720, USA* [249]

M. TAKAMI *The Institute of Physical and Chemical Research (RIKEN), Wako-shi, Saitama 351-01, Japan* [1]

P.H. VACCARO *Department of Chemistry, Yale University, 225 Prospect Street, New Haven, CT 06520-8107, USA* [75]

Preface

Since its advent in 1960, the laser has played many very important roles in molecular spectroscopy. At present it would be almost impossible to find a laboratory that investigates molecular structure and dynamics which is not equipped with any lasers. Even in the early period of the development of laser applications, scientists realized that the most remarkable characteristics of laser light are coherence, high power density and high spectral purity. These are the ideal characteristics for the observation of nonlinear phenomena. Nonlinear spectroscopy, made feasible by lasers, has proved to be an extremely powerful tool for molecular structure studies. It allows us to examine new aspects of molecular processes in detail and to determine molecular parameters orders of magnitude more precisely and rapidly than with conventional light sources.

A large variety of nonlinear spectroscopic methods have so far been developed and applied to studies of molecular structure and dynamics. Some of these methods utilize complex techniques, making it difficult to understand the underlying principles. Furthermore, most of these methods are referred to by nicknames or acronyms, which keeps them shrouded in mystery for most non-specialists who are thereby discouraged from mastering and applying the methods to molecular systems.

Under these circumstances it was judged appropriate and timely to overview those nonlinear spectroscopic methods which are currently particularly useful for studies of molecular structure and dynamics. Thus, the Commission on Molecular Structure and Spectroscopy of the International Union of Pure and Applied Chemistry decided to publish a book, intended to provide an introduction to nonlinear spectroscopy for non-specialists .

The first chapter of the book is devoted to introduction of the basic principle of coherent interaction of the laser light field with molecule including higher-order nonlinear effects. This chapter is intended to minimize difficulties which readers might face in understanding the following chapters. Chapters 2 through 10 present nine kinds of laser spectroscopic methods which elucidate the structure and dynamics of excited molecules, chemically reactive species, molecules in liquids or on surfaces, and so on. Each chapter describes first the underlying principle of the method and its typical experimental setup in order to allow a clear understanding by readers who are not familiar with laser spectroscopy. In the remaining part of each chapter, several typical examples of the application are given together with the current status of the method.

The editors are grateful to Professor John E. Bertie, Chairperson of IUPAC Commission I.5 and its members for adopting our proposal for the publication of the present book and the encouragement they gave us during the editing procedure. We also wish to thank the authors of the respective chapters for accepting our invitation to contribute to the book and for spending the time to write their chapter and to revise it based on the comments of the editors. Last, but not least, we are grateful to Professor K. Kuchitsu, the former President of Physical Chemistry Division of IUPAC, who made invaluable comments on the present book and carefully checked the symbols, nomenclature, and units in all of the chapters for consistency with the recommendations of the Physical Chemistry Division of IUPAC.

The Editors
R.W. Field
E. Hirota
J.P. Maier
S. Tsuchiya(Coordinator)

List of Abbreviations

ASE: amplified spontaneous emission

BO: Born-Oppenheimer

BOXCARS: CARS technique implemented in a box type phase-matching scheme

CARS: coherent anti-Stokes Raman scattering

CSRS: coherent Stokes Raman scattering

CW: continuous wave

DF: dispersed fluorescence

DFG: difference frequency generation

DFWM: degenerate four-wave mixing

ESR: electron spin resonance

FID: free induction decay

FWHM: full width at half maximum

HOMO: highest occupied molecular orbital

HWHM: half width at half maximum

IDIRS: ionization-detected infrared spectroscopy

IDSR: ionization-detected stimulated Raman

IDSRS: ionization-detected stimulated Raman spectroscopy

IP: ionization potential

IR: infrared

IRS: inverse Raman spectroscopy

IVR: intramolecular vibrational redistribution

LIF: laser-induced fluorescence

LUMO: lowest unoccupied molecular orbital

MATI: mass-analyzed threshold ionization

MO: molecular orbital

MODR: microwave-optical double resonance

MOPS: microwave-optical polarization spectroscopy

MPI: multiphoton ionization

MQDT: multichannel quantum defect theory

MW: microwave

NMR: nuclear magnetic resonance

OODR: optical-optical double resonance

OPO: optical parametric oscillation

PAD: photoelectron angular distribution

PAR: photoacoustic Raman

PES: photoelectron spectroscopy

PFI: pulsed field ionization

PSCARS: polarization sensitive CARS

Re-TOF: reflectron time of flight

REMPI: resonance enhanced multiphoton ionization

REMPI-PES: REMPI photoelectron spectroscopy

REMPI-PIE: REMPI photoionization efficiency
RIKE: Raman induced Kerr effect
SEP: stimulated emission pumping
SES: stimulated emission spectroscopy
SF: sum frequency
SFG: sum frequency generation
SHG: second harmonic generation
SRG: stimulated Raman gain
SRL: stimulated Raman loss
SRS: stimulated Raman scattering
TC-LIGS: two-color laser-induced grating
TOF: time of flight
UHV: ultrahigh vacuum
UV: ultraviolet
VMP: vibrationally-mediated photodissociation
VUV: vacuum ultraviolet
XUV: extreme ultraviolet
ZEKE: zero kinetic energy
ZEKE-PES: ZEKE photoelectron spectroscopy
2-D CARS: two-dimensional multiplex CARS

1 Interaction of Molecules with Coherent Radiation

Michio Takami
The Institute of Physical and Chemical Research (RIKEN), Wako-shi, Saitama, 351-01 Japan

1.1 Introduction

Many techniques in non-linear spectroscopy have their roots in radiofrequency spectroscopy (NMR, ESR, Microwave Spectroscopy), developed in 1950-1960s. However, the potential of nonlinear spectroscopy is far greater in laser spectroscopy. Advantages of nonlinear spectroscopy with lasers originate from their shorter wavelength (higher frequency) as well as high coherence. The shorter wavelength allows a laser beam to be focused in a very small area because the minimum spot size of focused electromagnetic radiation is limited by its wavelength. Also, a higher frequency is essential for the generation of a very short pulse because the pulse width is limited by the reciprocal of the bandwidth of the gain medium. These characteristics allow the laser to produce a high frequency electric field which is orders of magnitude stronger than the fields available in the radiofrequency region. A number of new nonlinear spectroscopic techniques have been developed in laser spectroscopy based on these characteristics. The aim of this chapter is to present a theoretical basis for linear and nonlinear spectroscopic techniques and related phenomena in laser spectroscopy. We use a well established theory of coherent radiation-matter interaction. Readers are expected to be familiar with elementary quantum mechanics, statistical mechanics, and linear algebra. The discussion is restricted on basic phenomena in gas phase molecules because specific theories for various nonlinear laser spectroscopy will be given in the following chapters. Important techniques, phenomena, and concepts are shown as ***keywords***, frequently with figures as visual aids.

We first discuss a two-level system interacting with near-resonant coherent radiation by solving a conventional Schrödinger equation and present some important phenomena. On most occasions we use classical electromagnetic waves interacting with quantized molecular states. The theory based on classical electromagnetic waves is easy to understand, but has a disadvantage that emission processes (spontaneous and stimulated emission) cannot be introduced in a systematic manner. A density matrix formalism is used, in a non-conventional manner, as a method to obtain steady-state solutions for multilevel systems. A quantized electromagnetic field is discussed briefly. In all calculations, the magnetic quantum number M has been neglected to avoid excessive complexing. Algebraic manipulations for the derivation of mathematical expressions have been made by the *Maple* program in *Scientific Work Place*. The curves in the Figures are also drawn by *Maple*, with the parameters given in figure captions. Finally, I would like to thank two reviewers and Dr. Yukari Matsuo for their valuable comments on this manuscript.

1.2 Two-level system

1.2.1 *Basic formulation*

We first consider a two-level system (Fig.1.1) interacting with near-resonant coherent radiation. In general, a linearly polarized electromagnetic wave propagating to the direction $\vec{k}$ with an angular frequency ω ($= 2\pi\nu$, where ν is a frequency in units of s^{-1}) is given by

$$\vec{E} = E\,\vec{e}\cos(\omega t - \vec{k}\cdot\vec{r}), \tag{1.1}$$

where E is an amplitude of the oscillating electric field, $\vec{e}$ is a unit vector specifying the direction of linear polarization, and $\vec{r} = (x, y, z)$. In vacuum, $\vec{e}$ is perpendicular to $\vec{k}$, and $|k| = \omega/c_0$ must be satisfied for the electromagnetic wave to propagate at the velocity of light c_0. When an isolated molecule interacts with laser radiation, we can ignore the term $-\vec{k}\cdot\vec{r}$ because the wavelength λ $(= c_0/\nu)$ is much larger than the size of the molecule. The $-\vec{k}\cdot\vec{r}$ term should be included when the dispersion of co-propagating electromagnetic waves in a bulk material (including gas phase molecules) is important (see, for example, Chapter 4).

The Hamiltonian of the interacting system is expressed by

$$H = H_0 + H_{\text{int}}\,, \tag{1.2}$$

where H_0 is the Hamiltonian of a free molecule. The interaction Hamiltonian H_{int} is given by

$$H_{\text{int}} = -\vec{E}\cdot\vec{\mu}\,, \tag{1.3}$$

where only the lowest-order interaction (electric dipole interaction) is considered. Here $\vec{\mu}\,(=\sum_i e_i\vec{r}_i)$ is a dipole moment operator produced by charges e_i distributed at $\vec{r}_i$, and

$$\vec{E} = E\,\vec{e}\cos\omega t = \frac{1}{2}E\,\vec{e}\,(e^{i\omega t} + e^{-i\omega t})\,. \tag{1.4}$$

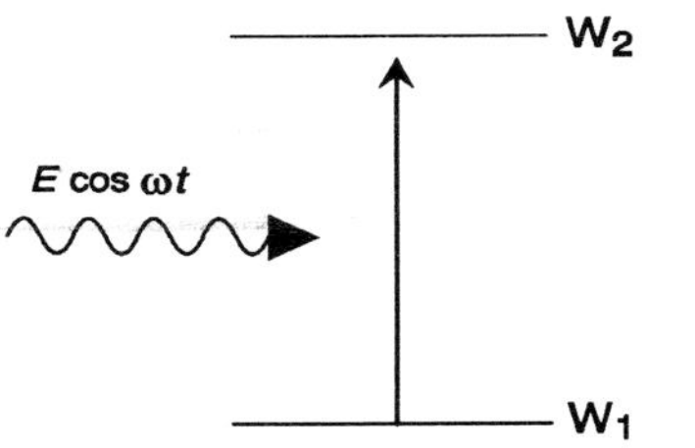

Fig. 1.1. A two-level system.

To calculate the eigenvalues of the above time-dependent Hamiltonian, we use time-dependent eigenfunctions for a free molecule, $|\psi_1\rangle$ and $|\psi_2\rangle$[1], as basis functions. They are given explicitly by,

$$|\psi_1\rangle = e^{-iW_1t/\hbar}|\phi_1\rangle\,, \tag{1.5}$$

$$|\psi_2\rangle = e^{-iW_2t/\hbar}|\phi_2\rangle\,, \tag{1.6}$$

where W_1 and W_2 are the energies of levels 1 and 2 (See Fig.1.1), and $|\phi_i\rangle$ is the time-independent molecular eigenfunction defined by

$$H_0|\phi_i\rangle = W_i|\phi_i\rangle\ (i = 1, 2). \tag{1.7}$$

For calculating a transition dipole moment from the interaction Hamiltonian, we assume that the wavefunction $|\phi_i\rangle$ is non-degenerate. For non-degenerate wavefunctions, we can define a quantum number called ***parity***. When Cartesian coordinates are inverted $(\vec{r}\rightarrow-\vec{r})$, non-degenerate molecular eigenfunctions transform either by $\phi(-\vec{r}) = \phi(\vec{r})$ or $\phi(-\vec{r}) = -\phi(\vec{r})$. The former eigenstate has even parity (+), and the latter odd parity (−). Parity is also defined for some operators. For example, the dipole moment operator $\vec{\mu} = \sum_i e_i\vec{r}_i$ has odd parity because the vector $\vec{r}$ changes sign under the inversion of the coordinate system. All measurable physical quantities should not change sign by such an artificial transformation of the coordinate system. Thus the transition dipole moment, which is a measurable physical quantity, is non-zero only between even and odd states so that the matrix element should have even parity as a whole, or in other words, the dipole moment operator μ has only off-diagonal matrix elements,[2]

$$\langle\phi_1|\mu|\phi_2\rangle = \mu_{12}\,, \tag{1.8}$$

[1] In this chapter we use Dirac's **bra** and **ket** notation, $\langle\phi| = \phi^*$, $|\phi\rangle = \phi$, and $\langle\phi|H|\phi'\rangle = \int\phi^* H\phi' dv$ for simplicity.

[2] From here on we take $\vec{\mu}$ and $\vec{E}$ as scalar quantities by taking the projection of $\vec{\mu}$ along $\vec{E}$.

and μ_{21} is the complex conjugate of μ_{12} ($\mu_{21} = \mu_{12}^*$). There is another requirement on the transition dipole moment to be non-vanishing. As will be discussed in 1.4, the absorption and emission of radiation occur by an exchange of a photon. A photon is an elementary particle defined in the quantized electromagnetic field (see 1.4), and has a discrete energy $\hbar\omega$ ($h\nu$) and one unit of angular momentum $\hbar$ ($J = 1$). Thus the conservation of total angular momentum allows only transitions that satisfy $\Delta J \leq 1$.

The interacting system is expressed in general by a linear combination of the two time-dependent basis functions,

$$|\Psi\rangle = a(t)|\psi_1\rangle + b(t)|\psi_2\rangle. \tag{1.9}$$

Then the problem is to solve the following Schrödinger equation,

$$i\hbar\frac{\partial}{\partial t}|\Psi\rangle = H|\Psi\rangle, \tag{1.10}$$

and to determine the coefficients $a(t)$ and $b(t)$ under appropriate initial conditions. By substituting Eqs.1.5, 1.6, and 1.9 into Eq.1.10, and applying $\langle\phi_1|$ and $\langle\phi_2|$ from the left, one can obtain the following differential equations for $a(t)$ and $b(t)$,

$$i\hbar\frac{d}{dt}a(t) = -\frac{1}{2}\mu_{12}Eb(t)e^{-i\omega_0 t}(e^{-i\omega t} + e^{i\omega t}), \tag{1.11}$$

$$i\hbar\frac{d}{dt}b(t) = -\frac{1}{2}\mu_{21}Ea(t)e^{i\omega_0 t}(e^{-i\omega t} + e^{i\omega t}), \tag{1.12}$$

where $\hbar\omega_0 = W_2 - W_1$. For a near-resonant case, $\omega \approx \omega_0$, the terms containing $e^{\pm i(\omega+\omega_0)}$ are neglected because, after integration over time, such terms have a very small factor $\pm 1/(\omega + \omega_0)$ (***rotating wave approximation***). Then the above equations reduce to,

$$\frac{d}{dt}a(t) = \frac{i}{2}xb(t)e^{i(\omega-\omega_0)t}, \tag{1.13}$$

$$\frac{d}{dt}b(t) = \frac{i}{2}x^*a(t)e^{-i(\omega-\omega_0)t}, \tag{1.14}$$

where $|x|$ ($x = \mu_{12}E/\hbar$) is called the ***Rabi frequency*** (in angular frequency units), which is an important quantity to represent the strength of interaction. Note that x^2 is proportional to the laser power because radiation energy is proportional to E^2 (see Eq.1.22).

These differential equations can be solved easily. With the initial condition that the molecule is in the lower level at $t = 0$ ($a(0) = 1, b(0) = 0$), the solutions are given by

$$a(t) = \{\cos\frac{\Omega}{2}t - i\frac{\omega - \omega_0}{\Omega}\sin\frac{\Omega}{2}t\}\exp\{\frac{i}{2}(\omega - \omega_0)t\}, \tag{1.15}$$

$$b(t) = \frac{ix^*}{\Omega}\sin\frac{\Omega}{2}t\exp\{-\frac{i}{2}(\omega - \omega_0)t\}, \tag{1.16}$$

where $\Omega = \sqrt{(\omega - \omega_0)^2 + x^2}$. Then the probability of finding the molecule in the upper level (level 2) at time t is given by,

$$|b(t)|^2 = \frac{x^2}{\Omega^2}\sin^2\frac{\Omega}{2}t \equiv \frac{x^2}{(\omega - \omega_0)^2 + x^2}\sin^2(\frac{1}{2}\sqrt{(\omega - \omega_0)^2 + x^2}\,t). \tag{1.17}$$

The time dependence of $|b(t)|^2$ is shown in Fig.1.2 for two cases, $\omega - \omega_0 = 0$ and $|x|$. The probability for the resonant case (i) ($\omega - \omega_0 = 0$) shows a sinusoidal oscillation between 0 and 1 with time. Thus the probability of finding the molecule in the upper level oscillates with the frequency $|x|/2\pi$. This is called ***Rabi oscillation***. If one uses a pulsed radiation field, which satisfies the condition $\Delta t = \pi/|x|$, the molecule is found in the upper level with 100 % probability. This is a well known technique called *π-pulse* excitation, developed first in NMR. If the incident radiation is off-resonant, the Rabi oscillation has a higher frequency and the maximum value is smaller than 1 as shown in Fig.1.2(ii).

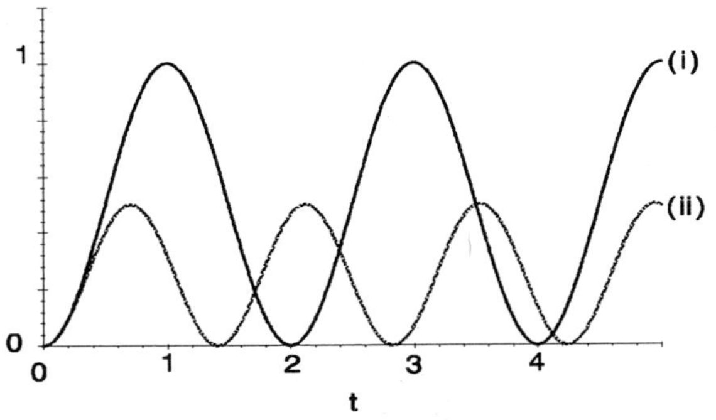

Fig. 1.2. Rabi oscillation for (i) $\omega - \omega_0 = 0$ and (ii) $\omega - \omega_0 = |x|$. The time scale is in units of $\pi/|x|$.

1.2.2 *Collisional effects*

The Rabi oscillation will be well defined for a single free molecule. When many molecules are interacting with the radiation simultaneously, coherent Rabi oscillation of the whole molecules is observable only when all molecules have the same absorption frequency and interaction time (e.g. a velocity-selected molecular beam). In a real system, the interaction time is limited by relaxation processes such as collisions, spontaneous emission, etc. In the gas phase, molecular collisions change the internal (ro-vibrational) states, limiting the interaction time of the relevant states with the radiation. Because the change of states by a collision is a stochastic process (a process in which the probability of an event to occur per unit time is constant), each molecule has a different interaction time with the radiation. Suppose the radiation is present from $t = -\infty$, and the collisional process populates and depopulates level 1 with an average rate, $\gamma = 1/\tau$. Then the probability of finding a molecule, brought into level 1 by a collision at time t and still remains in level 1 at $t = t_0$, is given by $\mathrm{e}^{-(t_0-t)/\tau}/\tau$ where $1/\tau$ is a normalization factor. The average over different interaction time is given by the integration,

$$\begin{aligned}\overline{|b(t)|^2} &= \int_{-\infty}^{t_0} |b(t_0 - t)|^2 \frac{1}{\tau} \mathrm{e}^{-(t_0-t)/\tau} \mathrm{d}t \\ &= \frac{\frac{1}{2}x^2}{(\omega - \omega_0)^2 + x^2 + 1/\tau^2}\,.\end{aligned} \tag{1.18}$$

Then the laser power absorbed by the molecules (J/s) is given by

$$P_{\text{abs}} = N_1^0 \frac{\hbar\omega}{\tau} \frac{\frac{1}{2}x^2}{(\omega - \omega_0)^2 + x^2 + \gamma^2}\,, \tag{1.19}$$

where N_1^0 is the number of molecules in level 1 at thermal equilibrium, $\hbar\omega$ is the photon energy, and $N_2^0 = 0$ is assumed. The factor $1/\tau$ is necessary because, at thermal equilibrium, the average residence time of a molecule in level 1 is τ, and therefore the actual number of molecules which absorb the radiation should be N_1^0/τ.

The frequency profile of Eq.1.19 provides several important results. The frequency dependence of $\overline{|b(t)|^2}$ is shown in Fig.1.3 for $|x| = 1/\tau$, $2/\tau$, and $5/\tau$. This line shape, which has the form of $1/\{(\omega - \omega_0)^2 + \Gamma^2\}$, is called ***Lorentzian*** and has a long tail (see Fig.1.5). The line width Γ (Half Width at Half Maximum, HWHM) is given by $\Gamma^2 = x^2 + 1/\tau^2$. When $|x| \gg 1/\tau$, the width is very close to $|x|$. Thus the line width increases in proportion to the square root of laser power. This effect is called ***saturation broadening***. The mean interaction time τ also affects the width. If $1/\tau \gg |x|$, the width is given by $1/\tau$. This is equivalent to the width derived from uncertainty principle for time and energy in quantum mechanics ($\Delta t \Delta E \geq \hbar$). The width originating from τ is called ***collisional broadening***, or ***pressure broadening*** because $1/\tau$ is proportional to gas pressure (when the pressure is not very high). The probability $\overline{|b(t)|^2}$ converges to 1/2 as x^2 (laser power) increases. At the limit of high laser power, the probabilities of finding a molecule in levels 1 and 2 are equal. This effect is called ***saturation***. A plot of $\overline{|b(t)|^2}$ against x^2/γ^2 is shown in the lower part of Fig. 1.4.

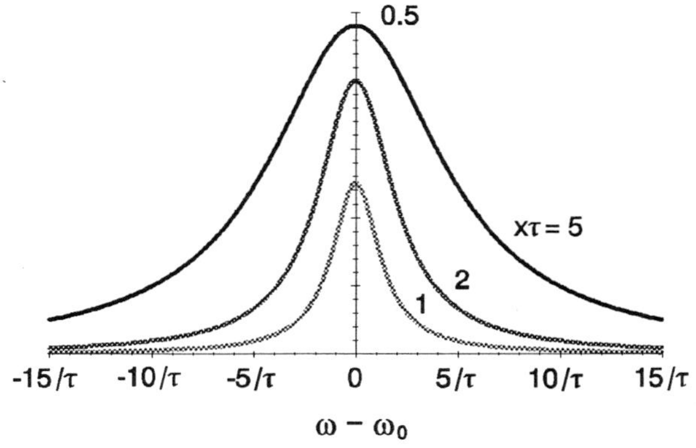

Fig. 1.3. Absorption line profiles for $|x| = 1/\tau$, $2/\tau$, and $5/\tau$ (see Eq.1.18). Note that, when $|x| > 1/\tau$, the linewidth increases with $|x|$ almost linearly while the peak value converges to 1/2 due to the saturation effect.

So far we assumed that collisions always change the internal states of molecules. However, there are collisions which are not strong enough to cause changes of internal states but still perturb the molecules. Because such perturbations shift the resonance frequency ω_0, the Rabi oscillation is reset almost to 0 due to the factor of x^2/Ω^2 in Eq.1.17. Thus such weak collisions also limit the build-up time of Rabi oscillation and should be included in the evaluation of $\overline{|b(t)|^2}$. The events of state-changing and state-non-changing collisions are independent. Thus we redefine the relaxation rate γ as $\gamma = 1/\tau + 1/\tau'$ where τ' is the mean time between the state-non-changing collisions. This relaxation constant γ is called the ***transverse relaxation rate***, and $1/\gamma$ the ***transverse relaxation time***. The latter corresponds to T_2 in magnetic resonance. Similarly, τ is called the ***longitudinal relaxation time***, which is equivalent to T_1 in magnetic resonance. It should be noted that there are other effects which influence the line width. For example, if a narrow laser beam excites molecules in low pressure gas or crosses a molecular beam perpendicularly, the interaction time is limited by the transit time of the molecule across the laser beam, causing additional line broadening. The origin of this broadening is not a stochastic process and therefore the width cannot be added as another relaxation term. Nevertheless, the influence of this finite interaction time may be included in γ by adding another term $1/\tau''$ where τ'' is the mean interaction time. This effect is important only when the line profile is measured at the resolution higher than 1 MHz, unless the average velocity is much higher than the thermal velocity at a room temperature.

A molecule in an excited state can change states by emitting a photon. This is another longitudinal relaxation process called ***spontaneous emission***. Spontaneous emission is also a stochastic process. The probability of finding a molecule in the upper level after the excitation is given by $e^{-t/\tau_r}/\tau_r$ where τ_r is the (mean) radiative lifetime given by

$$\tau_r^{-1} = \frac{\omega^3 \mu_{21}^2}{3\pi\varepsilon_0 \hbar c_0^3}, \tag{1.20}$$

and $\varepsilon_0 = 8.8542 \times 10^{-12}$ Fm^{-1} is the permittivity of vacuum. When many molecules are excited simultaneously to an upper level, the time dependence of the emission intensity shows an exponential decay with a lifetime τ_r. When a single molecule is excited, the expression $e^{-t/\tau_r}/\tau_r$ simply gives the probability of finding an emitted photon at time t after the excitation. Note that the lifetime is inversely proportional to ω^3. Therefore, spontaneous emission is usually negligible in the microwave and infrared regions relative to other relaxation processes, while it can be a dominant relaxation process in the visible and UV regions. A typical lifetime for an allowed visible/UV transition is in the range of nanoseconds.

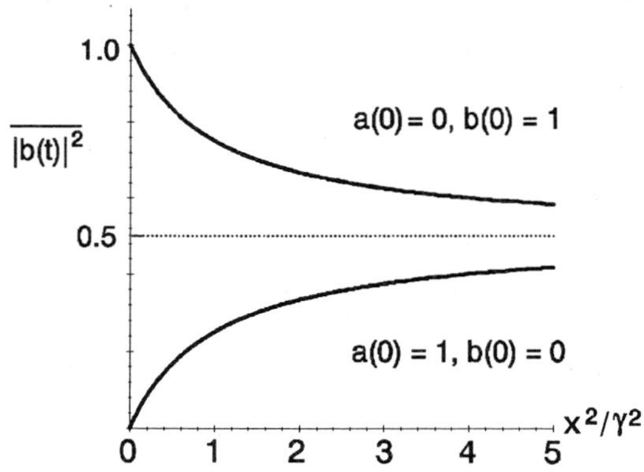

Fig. 1.4. Probability of finding a molecule in the upper level (level 2) after the average over collisional relaxation. The upper trace is for the initial conditions $a(0) = 0$, $b(0) = 1$ ($N_1^0 = 0$ and $N_2^0 = N^0$), and the lower trace for $a(0) = 1$, $b(0) = 0$ ($N_1^0 = N^0$ and $N_2^0 = 0$).

Here we discuss the case of another initial condition. If we assume that a molecule is in the upper level at $t = 0$ ($a(0) = 0$ and $b(0) = 1$), we obtain a similar expression as Eq.1.19, but N_1^0 is replaced by $-N_2^0$. The resulting negative sign of absorbed energy means that the radiation energy is increased by ***stimulated emission*** (see 1.4). Stimulated emission, also called ***induced emission***, is a phenomenon whereby a molecule in the upper level emits a photon with the same frequency ω, wave vector $\vec{k}$, and polarization $\vec{e}$ as the incident radiation. When both the lower and upper levels are populated, the loss of radiation energy by absorption from level 1 and the increase of radiation energy by stimulated emission from level 2 should be considered simultaneously. Thus the net absorbed power is given by

$$P_{\text{abs}} = (N_1^0 - N_2^0)\frac{\hbar\omega}{\tau}\frac{\frac{1}{2}x^2}{(\omega-\omega_0)^2 + x^2 + \gamma^2}\,. \tag{1.21}$$

When $N_2^0 > N_1^0$, P_{abs} is negative and $|P_{\text{abs}}|$ gives emitted power (power gain). The population distribution[3] in which an upper level has a larger number of molecules is called ***population inversion***, or ***negative temperature state*** because the inverted population distribution can be described by a Boltzmann distribution with a negative temperature. When the population is inverted, the incident radiation is amplified. In fact, the laser is an optical oscillator based on the optical feedback of radiation into the population-inverted medium, to be amplified by stimulated emission, with a pair of mirrors. (***LASER***, first demonstrated in 1960 [1], is the acronym of ***L***ight ***A***mplification by ***S***timulated ***E***mission of ***R***adiation, taken from ***MASER***, a similar oscillator in the microwave region operated first in 1954 [2].) The saturation of stimulated emission by the increase of laser power is shown in the upper part of Fig.1.4.

Evaluation of the Rabi frequency, $|x|$, is sometimes important for finding proper experimental conditions. First we have to know the optical electric field intensity E. This is derived from the following relation describing the energy of electromagnetic radiation passing through a unit area per unit time,

$$W_{\text{rad}} = \frac{1}{2}\int(\varepsilon_0\vec{E}^2 + \mu_0\vec{H}^2)dv = \int\varepsilon_0\vec{E}^2 dv = \frac{1}{2}\varepsilon_0 c_0 E^2\,, \tag{1.22}$$

[3]Number of molecules (atoms) in a specific internal state is frequently called ***population***.

where the integration is over space and the second relation is obtained from the equipartition law between the electric and magnetic energies of radiation. For a CW laser beam of 1 W/m^2 power density, the electric field intensity E is 27.4 V/m. A convenient conversion number for calculating the Rabi frequency is $|x|/2\pi = 5.035$ kHz for 1 debye transition dipole moment (10^{-18} c.g.s.e.s.u. or 3.3356×10^{-30} Cm) and 1 V/m electric field intensity. It should be noted that the optical electric field intensity is proportional to the square root of the laser power density W/m^2. Thus, if a W Watt laser beam is focused to the area of S m^2, the above electric field intensity should be multiplied by a factor of $\sqrt{W/S}$. Similarly, for a laser pulse of T s width and P Joule energy per pulse, the intensity should be multiplied by a factor of $\sqrt{P/T}$. For example, if a 1 mJ laser pulse of 10 ns width is focused to the 1mm x 1 mm area, $E = 27.4\sqrt{P/ST} = 8.66 \times 10^6$ V/m, and the Rabi frequency is given by $5.035 \times 27.4\sqrt{P/ST}$ kHz ≈ 43.7 GHz in frequency unit (≈ 1.5 cm^{-1}) for 1 D transition dipole moment. This is a value easily resolved with a conventional pulsed dye laser. If a weak second laser is used to probe the excited level without time delay, this Rabi frequency causes a 1.5 cm^{-1} splitting, or a shift of the same order of magnitude when the pump laser is slightly off-resonant, providing the spectral line profile entirely different from a single-peaked Lorentzian shape (see 1.3).

1.2.3 *Doppler effect*

If a molecule is moving with a velocity $\vec{v} = (v_x, v_y, v_z)$ and the radiation is propagating to the $+z$ direction, the molecule feels the radiation frequency ω as

$$\omega' = \frac{1 - v_z/c_0}{\sqrt{1 - v^2/c_0^2}}\,\omega\,, \tag{1.23}$$

by a Doppler effect. Because $v/c_0 \approx 10^{-6}$ at room temperature (see Eq.1.27), one can neglect the term v^2/c_0^2, and the resonance frequency in the laboratory frame changes by the first-order Doppler shift to,

$$\omega_0' = (1 + v_z/c_0)\omega_0\,. \tag{1.24}$$

In thermal equilibrium, the velocity distribution of molecules along the z direction is given by the Boltzmann distribution

$$n(v_z) = \frac{N_0}{\sqrt{\pi}u} \mathrm{e}^{-v_z^2/u^2}\,, \tag{1.25}$$

where $n(v_z)$ is the number density of molecules with the velocity component v_z, N_0 is the total number of molecules, and u is the most probable velocity defined by $k_B T = \frac{1}{2}mu^2$ where k_B is the Boltzmann constant, T is the absolute temperature of the gas, and m is the mass of the molecule. The absorption line profile of gas phase molecules is given by solving v_z in the resonance condition $\omega - \omega_0' = \omega - (1 + v_z/c_0)\omega_0 = 0$ and substituting it into Eq.1.25,

$$n(\omega) = \frac{N_0}{\sqrt{\pi}u} \exp\{-(\frac{\omega - \omega_0}{k_0 u})^2\}, \tag{1.26}$$

where $k_0 = \omega_0/c_0$ and the profile has been given as the number density of molecules as a function of ω. This line profile, which has the form of $\exp(-ax^2)$, is called ***Gaussian*** (see Fig. 1.5). The line width $\Delta\omega_D$ (HWHM) is given by

$$\Delta\omega_D/\omega_0 = \sqrt{\ln 2}u/c_0 = \sqrt{2(\ln 2)k_B T/mc_0^2} \approx 3.58\sqrt{T/\overline{m}} \times 10^{-7}\,, \tag{1.27}$$

where $\overline{m}$ is the mass of the molecule in atomic mass unit m_u. If we take $T = 300$ K and $\overline{m} = 30$, $\Delta\omega_D/\omega_0 \approx 1 \times 10^{-6}$. This value is not very sensitive to mass and temperature because it has a square-root dependence on these parameters.

In 1.2.2, we introduced the line width Γ originating from collisional relaxation and saturation broadening. Such width is called ***homogeneous width***. This name originates from the fact that all molecules have the same absorption frequency and interact with the radiation simultaneously. On the other hand, when the Doppler width is larger than the homogeneous width, each molecule has different absorption frequency depending on its velocity component along the direction of the radiation. As a consequence, only a part of the molecules absorb the radiation. Such width is called ***inhomogeneous width***. These two widths have quite different characteristics. The homogeneous width is given by an absolute value and does not depend on the frequency. In the gas phase, pressure broadening and saturation broadening are the two major origins of the homogeneous width. The former is proportional to the gas pressure and usually on the order of 10 MHz/torr for a polar molecule. The saturation broadening is given by $|x|/2\pi$, which is proportional to the square root of the laser power density as discussed previously. In the visible/UV region, spontaneous emission is another important source of homogeneous broadening, which is given by $1/2\pi\tau_r$ in frequency units. On the other hand, the Doppler width is proportional to the transition frequency. In the microwave region, the Doppler width is less than 30 kHz and usually much smaller than the homogeneous width by collisions. In the optical region, the Doppler width dominates the homogeneous widths at room temperature. For example, at 20000 cm^{-1}, the Doppler width is about 0.02 cm^{-1} (600 MHz), and can be easily resolved with a CW ring dye laser.

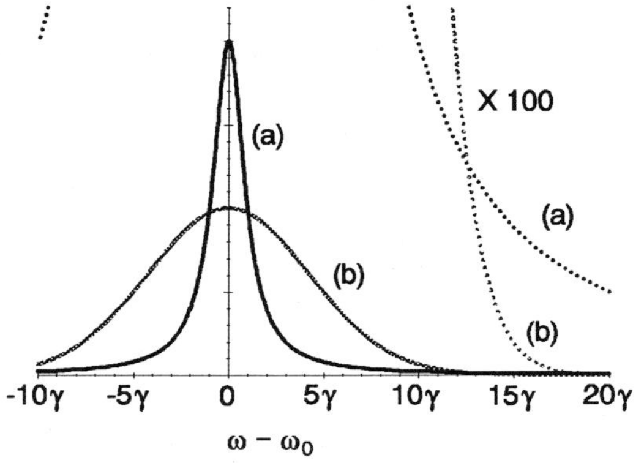

Fig. 1.5 Lorentzian (a) and Gaussian (b) line profiles for homogeneously and inhomogeneously broadened lines. For the Gaussian profile, $uk_0 = 5\gamma$ is assumed and the peak height is set to 1/2 of the Lorentzian peak. For $|\omega - \omega_0| > 15\gamma$, the Gaussian curve converges rapidly to 0 while the Lorentzian curve still has a considerable value due to the slowly decreasing factor $1/(\omega - \omega_0)^2$ at a large frequency offset.

When the Doppler width is larger than the homogeneous width, the previous calculations on the two level system should be modified because the absorption occurs only by the molecules which are Doppler-shifted close to the laser frequency. The number density of molecules in level 2 with the velocity component v_z is obtained by replacing ω_0 in Eq.1.18 with $\omega_0' = \omega_0 + k_0 v_z$ and multiplying by the Gaussian velocity distribution

$$n_2(v_z) = \frac{N_1^0}{\sqrt{\pi}u} \frac{\frac{1}{2}x^2}{(\omega - k_0 v_z - \omega_0)^2 + \Gamma^2} e^{-v_z^2/u^2}, \tag{1.28}$$

where $N_2^0 = 0$ is assumed. If $u \gg \Gamma$, the velocity profile $n_2(v_z)$ is almost proportional to the Lorentzian shape with the width $\Delta v_z = \Gamma/k_0 = c_0\Gamma/\omega_0$. Thus the laser excites only those molecules that have velocity in the range of about $v_z = (\omega - \omega_0)/k_0 \pm \Gamma/k_0$. The population distributions for the lower and upper levels under the presence of laser excitation are shown in Figs. 1.6(a) and (b), respectively.

Figure 1.6(a) shows that the laser excitation produces a population dip in the lower level around $v_z = (\omega - \omega_0)/k_0$. This effect is called ***hole burning***. The absorbed laser power is obtained by replacing N_1^0 with $N_1^0 - N_2^0$ in Eq.1.28 to include stimulation emission and integrating over v_z,

$$P'_{\text{abs}} = \frac{(N_1^0 - N_2^0)\gamma\hbar\omega}{\sqrt{\pi}u} \int_{-\infty}^{\infty} \frac{\frac{1}{2}x^2}{(\omega - k_0 v_z - \omega_0)^2 + \Gamma^2} e^{-v_z^2/u^2}\, dv_z \,. \tag{1.29}$$

This integration cannot be expressed by an analytic function. When $\Delta\omega_D \gg \Gamma$, one can ignore the variation of the Gaussian function over the Lorentzian profile, and obtain the following expression,

$$P'_{\text{abs}}(\omega) = \frac{\sqrt{\pi}(N_1^0 - N_2^0)\gamma\hbar\omega x^2}{2\Gamma k_0 u} \exp\{-(\frac{\omega - \omega_0}{k_0 u})^2\} \,. \tag{1.30}$$

The peak value of this function is about $\sqrt{\pi}\Gamma/k_0 u$ times the peak value of Eq.1.19, reflecting the ratio of two absorption linewidths with and without Doppler broadening. For efficient pumping of a Doppler-broadened line, one has to use a broad band laser or strong laser power so that the Rabi frequency $|x|/2\pi$ is comparable with or larger than the Doppler width.

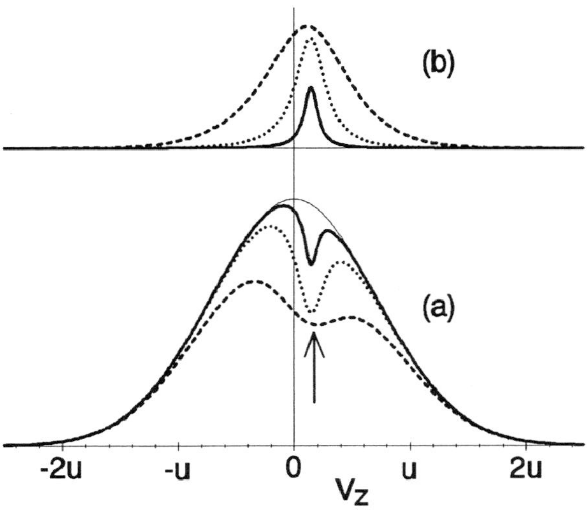

Fig. 1.6 A hole burning effect for a Doppler-broadened line. Population distributions for the (a) lower and (b) upper levels are shown as a function of v_z. The parameters used are $k_0 u = 20\gamma$, $\omega - \omega_0 = 3\gamma$, and $|x| = \gamma$ (solid line), 3γ (dotted line), and 10γ (broken line), respectively. The arrow indicates the location of the pump laser frequency. The thin line in (a) indicates the Doppler profile without laser irradiation. As the laser power increases, the hole becomes deeper and also wider due to the saturation effect. The ratio of homogeneous to Doppler width, $\gamma/k_0 u$, is set much larger than the value usually expected in the optical region.

When $|\omega - \omega_0| \gg k_0 u$ (the laser frequency is far outside the Doppler profile), the main contribution in the above integral arises from the region of $v_z = 0$ because, when the resonance offset is large, the Gaussian function converges exponentially to 0 while the Lorentzian function decreases much more slowly as $1/(\omega - \omega_0)^2$. By neglecting higher-order terms, the integration gives

$$P''_{\text{abs}} \approx (N_1^0 - N_2^0)\gamma\hbar\omega x^2/2(\omega - \omega_0)^2 \,. \tag{1.31}$$

This is equivalent to Eq.1.21 at a large resonance offset. Thus the absorption at far off-resonance is due to the tail of the Lorentzian, and the Doppler effect does not appear explicitly in the lowest-order approximation.

1.3 Three-level system

1.3.1 *Density matrix formalism*

The Schrödinger equation used in the two-level system can be extended to a three-level system [3]. However, the interaction of a multi-level system with coherent radiation can be handled much more easily by a density matrix method [4]. In contrast to Schrödinger equation where the variables are the coefficients of basis functions ($a(t)$ and $b(t)$ of the two level system in 1.2.1), variables (or elements) in the density matrix $[\rho]$ are the quadratic forms of these coefficients. In the two-level system, aa^*, ab^*, a^*b, and bb^* correspond to ρ_{11}, ρ_{12}, ρ_{21}, and ρ_{22}, respectively. To introduce relaxation processes, we take a statistical average (ensemble average) of these variables and redefine them as the density matrix. The equation of motion for the density matrix ρ is then given by,[4]

$$i\hbar\dot{\rho} = H\rho - \rho H - i\hbar\Gamma_r\rho\,, \tag{1.32}$$

where H is a 2×2 matrix with the element $H_{ij} = \langle\varphi_i|H|\varphi_j\rangle$, Γ_r is a matrix representing relaxation rates[5], and $\rho_{ji} = \rho^*_{ij}$ and $H_{ji} = H^*_{ij}$ by definition. One of the advantages of the density matrix method is that all variables are related to well-defined physical quantities. For example, $\rho_{11} = aa^*$ gives the probability of finding a molecule in level 1, and the number of molecules in level i is given by $N_i = N^0\rho_{ii}$ where N^0 is the total number of molecules. As a consequence, phenomenological relaxation constants can be used in Γ_r. Also, multiphoton transitions are introduced in a systematic manner as will be shown later.

First we apply the density matrix method to the two level system discussed in 1.2.1. The relaxation matrix has the elements, $\Gamma_{r,ii} = 1/\tau_i$ and $\Gamma_{r,ij} = \Gamma_{r,ji} = \gamma$. With the Hamiltonian in 1.2.1, one can obtain,

$$\begin{aligned}
\dot{\rho}_{11} &= -\frac{i}{\hbar}(H_{11}\rho_{11} + H_{12}\rho_{21} - \rho_{11}H_{11} - \rho_{12}H_{21}) - (\rho_{11} - \rho^0_{11})/\tau_1 \\
&= \frac{i}{2}(x\rho_{21} - x^*\rho_{12})(\mathrm{e}^{-i\omega t} + \mathrm{e}^{i\omega t}) - (\rho_{11} - \rho^0_{11})/\tau_1\,, && (1.33)\\
\dot{\rho}_{22} &= -\frac{i}{2}(x\rho_{21} - x^*\rho_{12})(\mathrm{e}^{-i\omega t} + \mathrm{e}^{i\omega t}) - (\rho_{22} - \rho^0_{22})/\tau_2\,, && (1.34)\\
\dot{\rho}_{12} &= -\frac{i}{\hbar}(H_{11}\rho_{12} + H_{12}\rho_{22} - \rho_{11}H_{12} - \rho_{12}H_{22}) - \gamma\rho_{12} \\
&= i\omega_0\rho_{12} + \frac{i}{2}x(\mathrm{e}^{i\omega t} + \mathrm{e}^{-i\omega t})(\rho_{22} - \rho_{11}) - \gamma\rho_{12}\,, && (1.35)
\end{aligned}$$

where ρ^0_{ii} is the value of ρ_{ii} at thermal equilibrium, and $\dot{\rho}_{21}$ is given as a complex conjugate of $\dot{\rho}_{12}$. The relaxation terms, $-(\rho_{ii} - \rho^0_{ii})/\tau_i$'s and $-\rho\gamma$'s, are added so that $\rho_{ii} = \rho^0_{ii}$ and $\rho_{ij} = 0\,(i \neq j)$ when there is no radiation (at thermal equilibrium).

Equation 1.35 indicates that, when $|x| \ll \omega$, ρ_{12} is roughly proportional to $\mathrm{e}^{i\omega_0 t}$. In order to define slowly varying variables for introducing the rotating wave approximation, we define the following symmetrized variables,

$$\begin{aligned}
\rho^- &= x\rho_{21}\mathrm{e}^{i\omega t} - x^*\rho_{12}\mathrm{e}^{-i\omega t}\,, && (1.36)\\
\rho^+ &= x\rho_{21}\mathrm{e}^{i\omega t} + x^*\rho_{12}\mathrm{e}^{-i\omega t}\,. && (1.37)
\end{aligned}$$

Then, with the rotating wave approximation, the equations of motion become

$$\begin{aligned}
\dot{\rho}_{11} &= \frac{i}{2}\rho^- - (\rho_{11} - \rho^0_{11})/\tau_1\,, && (1.38)\\
\dot{\rho}_{22} &= -\frac{i}{2}\rho^- - (\rho_{22} - \rho^0_{22})/\tau_2\,, && (1.39)\\
\dot{\rho}^- &= i(\omega - \omega_0)\rho^+ - ix^2(\rho_{22} - \rho_{11}) - \gamma\rho^-\,, && (1.40)\\
\dot{\rho}^+ &= i(\omega - \omega_0)\rho^- - \gamma\rho^+\,. && (1.41)
\end{aligned}$$

These equations may also be derived by using the definition of $\rho_{11} = aa^*$ etc. and the differential equations for a and b in 1.2.1.

---

[4]Hereafter we use the notation $\dot{F}$ instead of $\mathrm{d}F/\mathrm{d}t$.

[5]The relaxation (damping) term $\Gamma_r\rho$ should be read as a 2×2 matrix with elements $\Gamma_{r,ij}\rho_{ij}$ and not as a multiplication of two 2×2 matrices.

The time evolution of the population in levels 1 and 2 is obtained by solving the above complex linear differential equations with a given initial condition. If the radiation is continuous, the system reaches a steady state after an interaction time much longer than the relaxation times τ_i and $1/\gamma$. The solution for the steady state is obtained by setting $\dot{\rho} = 0$ and solving the linear homogeneous equations. For example, the increase of the population density in the upper level after reaching at equilibrium is calculated to be

$$\rho_{22} - \rho_{22}^0 = \frac{\frac{1}{2}\gamma\tau x^2(\rho_{11}^0 - \rho_{22}^0)}{(\omega - \omega_0)^2 + \gamma^2 + \gamma\tau x^2}, \tag{1.42}$$

where $\tau_1 = \tau_2 = \tau$ is assumed. Note that if we assume $\gamma = 1/\tau$, $\rho_{22}^0 = 0$, $\rho_{11}^0 = 1$, and multiply the factor of $\hbar\omega/\tau$, the above expression becomes equivalent to Eq.1.21 derived by solving a Schrödinger equation.

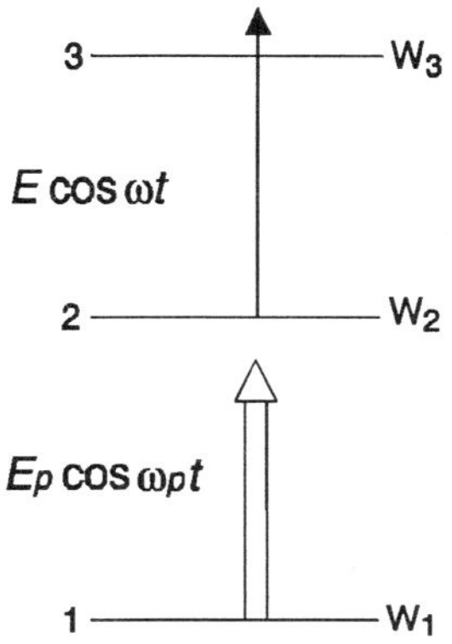

Fig. 1.7 A three level system.

1.3.2 *Coherent effects in a three-level system*

One of the most popular techniques in nonlinear laser spectroscopy is two-step excitation. Here we discuss typical nonlinear effects in a three level system interacting with two near-resonant coherent radiation fields (see Fig.1.7). Note that the following discussion applies only when the two lasers are continuously oscillating or the two laser pulses are overlapped in time. The equation of motion for a density matrix is again,

$$i\hbar\dot{\rho} = H\rho - \rho H - i\hbar\Gamma_r\rho, \tag{1.43}$$

where ρ, H, and $\Gamma_r\rho$ are now 3×3 matrices. For simplicity, we use only one relaxation rate γ for the nine components of Γ_r. The interaction Hamiltonian is given by

$$H_{\text{int}} = -\frac{1}{2}\vec{E}_p \cdot \vec{\mu}(\mathrm{e}^{i\omega_p t} - \mathrm{e}^{-i\omega_p t}) - \frac{1}{2}\vec{E} \cdot \vec{\mu}(\mathrm{e}^{i\omega t} + \mathrm{e}^{-i\omega t}), \tag{1.44}$$

where E_p, E, ω_p, and ω are the amplitudes and angular frequencies of the incident pump ($1 \longrightarrow 2$) and probe ($2 \longrightarrow 3$) radiation, and $\vec{\mu}$ is the electric dipole moment. Elements of the Hamiltonian matrix H are $H_{ii} = W_i$, $H_{13} = H_{31} = 0$, and

$$H_{12} = -\frac{1}{2}\hbar x_p(\mathrm{e}^{i\omega_p t} + \mathrm{e}^{-i\omega_p t}), \tag{1.45}$$

$$H_{23} = -\frac{1}{2}\hbar x(\mathrm{e}^{i\omega t} + \mathrm{e}^{-i\omega t}), \tag{1.46}$$

where $x_p = E_p\mu_{12}/\hbar$ and $x = E\mu_{23}/\hbar$ are the Rabi frequencies for the pump and probe transitions, and $H_{ji} = H_{ij}^*$ ($i \neq j$). It should be noted that ρ_{13} is non-zero despite $H_{13} = H_{31} = 0$ and is given by $i\hbar\dot{\rho}_{13} = H_{11}\rho_{13} + H_{12}\rho_{23} - \rho_{11}H_{13} - \rho_{12}H_{23}$.

As in the two level system, we introduce the following variables,

$$\rho_p^- = x_p\rho_{21}e^{i\omega_p t} - x_p^*\rho_{12}e^{-i\omega_p t} \tag{1.47}$$
$$\rho_p^+ = x_p\rho_{21}e^{i\omega_p t} + x_p^*\rho_{12}e^{-i\omega_p t} \tag{1.48}$$
$$\rho^- = x\rho_{32}e^{i\omega t} - x^*\rho_{23}e^{-i\omega t} \tag{1.49}$$
$$\rho^+ = x\rho_{32}e^{i\omega t} + x^*\rho_{23}e^{-i\omega t} \tag{1.50}$$
$$\rho_{13}^- = x_p x\rho_{31}e^{i(\omega_p+\omega)t} - x_p^* x^*\rho_{13}e^{-i(\omega_p+\omega)t} \tag{1.51}$$
$$\rho_{13}^+ = x_p x\rho_{31}e^{i(\omega_p+\omega)t} + x_p^* x^*\rho_{13}e^{-i(\omega_p+\omega)t}. \tag{1.52}$$

With the rotating wave approximation, the equations of motion are,

$$\dot\rho_{11} = \frac{i}{2}\rho_p^- - \gamma(\rho_{11} - \rho_{11}^0), \tag{1.53}$$
$$\dot\rho_{22} = -\frac{i}{2}\rho_p^- + \frac{i}{2}\rho^- - \gamma(\rho_{22} - \rho_{22}^0), \tag{1.54}$$
$$\dot\rho_{33} = -\frac{i}{2}\rho^- - \gamma(\rho_{33} - \rho_{33}^0), \tag{1.55}$$
$$\dot\rho_p^- = i(\omega_p - \omega_{p0})\rho_p^+ + \frac{i}{2}\rho_{13}^+ - ix_p^2(\rho_{22} - \rho_{11}) - \gamma\rho_p^-, \tag{1.56}$$
$$\dot\rho_p^+ = i(\omega_p - \omega_{p0})\rho_p^- + \frac{i}{2}\rho_{13}^- - \gamma\rho_p^+, \tag{1.57}$$
$$\dot\rho^- = i(\omega - \omega_0)\rho^+ - \frac{i}{2}\rho_{13}^+ - ix^2(\rho_{33} - \rho_{22}) - \gamma\rho^-, \tag{1.58}$$
$$\dot\rho^+ = i(\omega - \omega_0)\rho^- - \frac{i}{2}\rho_{13}^- - \gamma\rho^+, \tag{1.59}$$
$$\dot\rho_{13}^- = i(\omega_p + \omega - \omega_{p0} - \omega_0)\rho_{13}^+ - \frac{i}{2}x_p^2\rho^+ + \frac{i}{2}x^2\rho_p^+ - \gamma\rho_{13}^-, \tag{1.60}$$
$$\dot\rho_{13}^+ = i(\omega_p + \omega - \omega_{p0} - \omega_0)\rho_{13}^- - \frac{i}{2}x_p^2\rho^- + \frac{i}{2}x^2\rho_p^- - \gamma\rho_{13}^+. \tag{1.61}$$

A steady-state solution is obtained by setting $\dot\rho = 0$ and solving the linear homogeneous equations. In principle, this method provides exact steady-state solutions for any number of levels as far as the rotating wave approximation is valid. However, solving linear homogeneous equations with nine variables is not a simple task, and even when explicit expressions are obtained, it is not easy to derive clear physical interpretation for the final results. Thus we calculate the spectral profile in the lowest order of probe laser power, x^2, while leaving the pump laser power arbitrary. Then, after lengthy algebraic manipulation, one can obtain the following expression for ρ_{33},

$$\rho_{33} = \frac{x_p^2 x^2(\Delta_p^2 + \Delta^2 + \Delta_p\Delta + 3\gamma^2 + \frac{3}{4}x_p^2)}{8(\Delta_p^2 + \gamma^2 + x_p^2)(\Delta^2 + \gamma^2)[\{\Delta_p + \Delta(1 - \frac{x_p^2}{4(\Delta^2+\gamma^2)})\}^2 + \gamma^2(1 + \frac{x_p^2}{4(\Delta^2+\gamma^2)})^2]}\rho_{11}^0, \tag{1.62}$$

where $\Delta_p = \omega_p - \omega_{p0}$, $\Delta = \omega - \omega_0$, and $\rho_{22}^0 = \rho_{33}^0 = 0$ is assumed. The absorbed photon energy for the probe laser is calculated by replacing $|b(t)|^2$ with ρ_{33} in Eq.1.18 and multiplying by $N_1^0\gamma\hbar\omega$. Because the above expression is still too complex to understand the physical process, we discuss two extreme cases, $\Delta_p = 0$ and $\Delta_p \gg \gamma$.

When the pump laser is resonant ($\Delta_p = 0$), the above expression is simplified to

$$\begin{aligned}\rho_{33} &= \frac{x_p^2 x^2}{8(\gamma^2 + x_p^2)}\frac{\Delta^2 + 3\gamma^2 + \frac{3}{4}x_p^2}{\{(\Delta + \frac{1}{2}x_p)^2 + \gamma^2\}\{(\Delta - \frac{1}{2}x_p)^2 + \gamma^2\}}\rho_{11}^0 \\ &= \frac{x_p^2 x^2}{16(\gamma^2 + x_p^2)}\left[\frac{1}{(\Delta + \frac{1}{2}x_p)^2 + \gamma^2} + \frac{1}{(\Delta - \frac{1}{2}x_p)^2 + \gamma^2}\right. \\ &\quad \left. + \frac{4\gamma^2 + x_p^2}{\{(\Delta + \frac{1}{2}x_p)^2 + \gamma^2\}\{(\Delta - \frac{1}{2}x_p)^2 + \gamma^2\}}\right]\rho_{11}^0. \end{aligned} \tag{1.63}$$

In the weak pump laser limit $(x_p \ll \gamma)^6$, we obtain

$$\rho_{33} = \frac{x_p^2 x^2}{8\gamma^2}\{\frac{1}{\Delta^2+\gamma^2} + \frac{2\gamma^2}{(\Delta^2+\gamma^2)^2}\}\rho_{11}^0 . \tag{1.64}$$

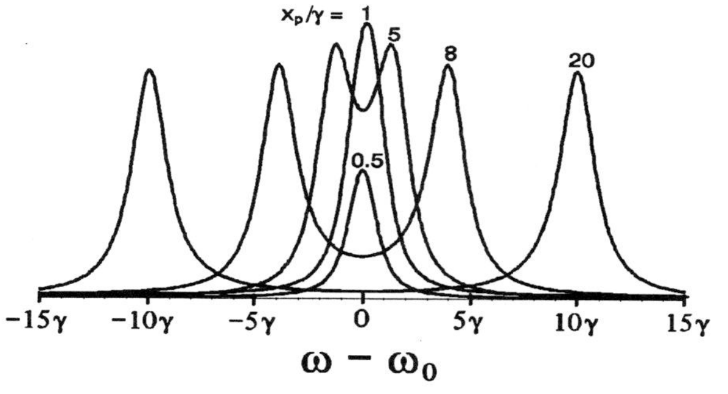

Fig. 1.8 The Rabi splitting for various values of x_p/γ. $x \ll \gamma$ is assumed.

Thus the signal has a single peak at $\omega = \omega_0$ while the linewidth is narrower than γ and the lineshape deviates from a Lorentzian profile due to the second term in curly brackets. When x_p is larger than γ, the lineshape shows a totally different feature. The expression in Eq.1.63 indicates that two resonance peaks are expected at around $\Delta = \pm x_p/2$ (or $\omega = \omega_0 \pm x_p/2$). Thus the probe transition line splits into a doublet centered at $\omega = \omega_0$ with a spacing of $\approx x_p$. The absorption line profiles calculated from Eq.1.63 are shown in Fig.1.8 for various values of x_p/γ. This splitting is called ***Rabi splitting*** or ***resonance splitting***. The origin of this splitting can be found in the two-level system discussed previously. If we assume $\omega = \omega_0$, $a(t)$ and $b(t)$ in Eqs.1.15 and 1.16 become

$$a(t) = \cos\frac{x}{2}t = \frac{1}{2}(\mathrm{e}^{ixt/2} + \mathrm{e}^{-ixt/2}), \tag{1.65}$$

$$b(t) = i\sin\frac{x}{2}t = \frac{1}{2}(\mathrm{e}^{ixt/2} - \mathrm{e}^{-ixt/2}). \tag{1.66}$$

With these coefficients, the time-dependent wave functions in Eq.1.9 may be rewritten as

$$a(t)|\psi_1\rangle = \frac{1}{2}(\mathrm{e}^{-i(W_1+\hbar x/2)t/\hbar} + \mathrm{e}^{-i(W_1-\hbar x/2)t/\hbar})|\phi_1\rangle, \tag{1.67}$$

$$b(t)|\psi_2\rangle = \frac{1}{2}(\mathrm{e}^{-i(W_2-\hbar x/2)t/\hbar} - \mathrm{e}^{-i(W_2+\hbar x/2)t/\hbar})|\phi_2\rangle. \tag{1.68}$$

Thus the energy levels W_1 and W_2 look as if they are split into doublets at $W_1 \pm \hbar x/2$ and $W_2 \pm \hbar x/2$, respectively. These levels are called ***virtual states*** because they are not molecular eigenstates but artificial states produced by the interaction with the coherent radiation. These doublets are observed only when they are probed with a second weak probe laser.

When $|\Delta_p| \gg \gamma$, we rewrite Eq.1.62 while dropping all terms containing γ except the one in the denominator to avoid divergence, and obtain the following expression.

$$\rho_{33} \approx \frac{x_p^2 x^2(\Delta_p\Delta + \Delta_p^2 + \Delta^2 + \frac{3}{4}x_p^2)}{8(x_p^2+\Delta_p^2)\{(\Delta+\Delta_p+x_p^2/4\Delta_p)^2(\Delta - x_p^2/4\Delta_p)^2 + \gamma^2(\Delta_p^2+x_p^2)\}}\rho_{11}^0 . \tag{1.69}$$

The calculated absorption line profiles are shown in Fig.1.9 for three different values of x_p/γ. The spectrum has two peaks corresponding to two minima in the denominator at $\Delta + \Delta_p + x_p^2/4\Delta_p = 0$

[6]Hereafter we assume x_p and x are real parameters because they always appear in a quadratic form in physical quantities

and $\Delta - x_p^2/4\Delta_p = 0$, or more explicitly at $\omega + \omega_p = \omega_0 + \omega_{p0} - x_p^2/4\Delta_p$ and $\omega = \omega_0 + x_p^2/4\Delta_p$. If we ignore the term $x_p^2/4\Delta_p$, which is small in the present approximation, one can easily find that the former peak corresponds to a two-photon transition and the latter to a two-step transition. However, the actual line positions are shifted from these conditions by $\pm x_p^2/4\Delta_p$. This effect is understood again by going back to the two level system. When $\Delta_p \gg x_p$, Eqs.1.15 and 1.16 again show that the energy levels 1 and 2 are split asymmetrically as

$$W_1' \approx \begin{cases} W_1 + \Delta_p + x_p^2/4\Delta_p \\ W_1 - x_p^2/4 \end{cases} \tag{1.70}$$

$$W_2' \approx \begin{cases} W_2 - \Delta_p - x_p^2/4\Delta_p \\ W_2 + x_p^2/4\,. \end{cases} \tag{1.71}$$

What is observed as two peaks are the resonant transitions from these split levels to the unperturbed level 3. The splitting and shifts discussed in this section are sometimes called ***coherent effects*** because they are observable only when coherent radiation is interacting with molecules.

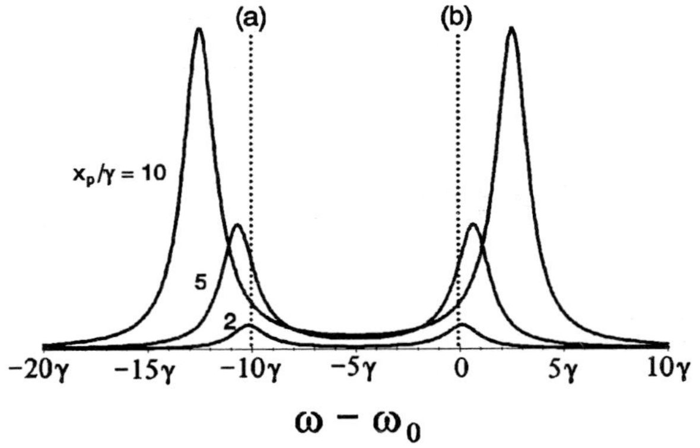

Fig. 1.9 Two-photon transition and optical Stark shift. Line profiles are given for $x_p = 2\gamma$, 5γ, and 10γ. Dotted lines show the line positions for (a)two-photon and (b)two-step transitions without optical Stark shift. Note that the peak intensities and the amount of shifts increase quadratically with x_p.

We showed already how a resonant optical field splits a level into a symmetric doublet. Here it is shown that a near-resonant optical field also splits the level but does so asymmetrically. The modification of the energy levels by an optical field is called the ***optical Stark effect*** in analogy with the Stark effect in a static electric field. The splitting due to resonant radiation, the Rabi splitting, corresponds to the first-order Stark effect, and the shift of the level by a non-resonant radiation, $x_p^2/4\Delta_p$, is called the ***optical Stark shift*** or ***high frequency Stark shift***, corresponds to the second-order Stark effect. The splitting and shifts are interpreted as resonant two-step transitions through virtual states produced by the optical Stark effects as shown in Fig.1.10.

When $|\Delta_p| \gg \gamma$ and x_p, all optical Stark shifts are negligible and the two-photon transition occurs at $\Delta_p + \Delta = 0$. Under this condition, Eq.1.69 is simplified to

$$\rho_{33} = \frac{x_p^2 x^2}{8\gamma^2\Delta_p^2}\rho_{11}^0. \tag{1.72}$$

If we compare this rate of excitation for ρ_{33} with that of the resonant transition in the two level system, $x^2/2\gamma^2$ derived from Eq.1.42 by assuming $\gamma \gg x$ and $\gamma\tau = 1$, the above rate has an extra factor of $x_p^2/4\Delta_p^2$. This factor decreases rather slowly by increasing the off-resonance frequency $|\Delta_p|$, in particular when x_p^2 is large. Thus a two-photon transition is easy to observe even when level 2 is far from resonance. Although the incident radiation fields are off-resonant, the two-photon transition dipole moment is proportional to the product of the moments of transitions $2 \leftarrow 1$ and $3 \leftarrow 2$.

Level 2 which provides the two-photon transition dipole moment is called the ***intermediate state***. The two-photon transition is also described as a resonant two-step transition through a ***virtual state*** at $W = W_1 + \hbar\omega_p$ as shown in Fig.1.10. The variation of line shapes, as Δ_p increases from 0 to 20γ, is shown in Fig.1.11. As the frequency offset Δ_p increases, the left peak shifts smoothly from one of the resonance doublets to the two-photon transition. Note that each doublet has the same peak height, with the center of doublet at Δ_p.

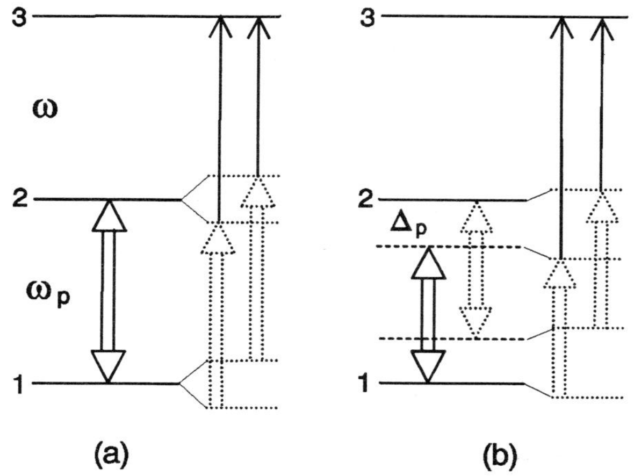

Fig. 1.10 Optical Stark effects observed as resonant transitions through virtual states. Thick arrows indicate the incident pump radiation, and dotted arrows the resonant transitions between virtual states. (a) The resonant pump laser splits the lower and upper levels. The Rabi splitting is observed as resonant two-step transitions through virtual states. Thin arrows indicate the probe transitions. (b) Off-resonant pump laser shifts the energy levels as indicated by dotted lines. The pump laser also produces virtual states shown by broken lines. The Stark-shifted probe transitions are interpreted as resonant transitions among Stark-shifted virtual states.

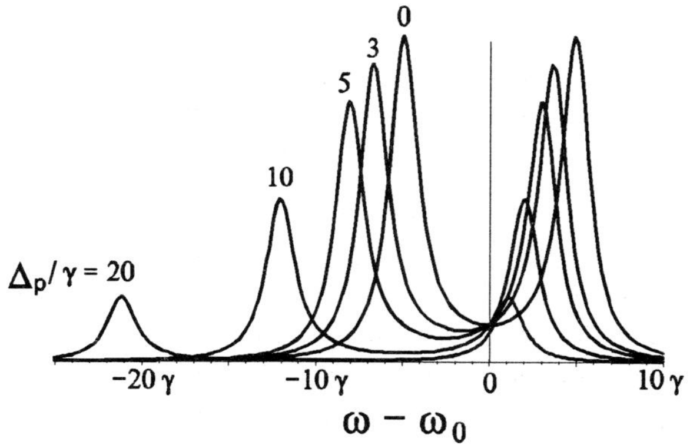

Fig. 1.11 Variation of the probe transition line shapes as Δ_p increases from 0 to 20γ. $x_p = 10\gamma$ is assumed.

1.3.3 *Rate equations*

There is another technique called ***rate equations*** to handle multi-level systems interacting with radiation. In rate equations, the variables are the numbers of molecules in all relevant levels, $N_i = N^0 \rho_{ii}$. For example, the equations of motion for a two level system may be written as

$$\dot{N}_1 = (N_2 - N_1)S_p - (N_1 - N_1^0)/\tau\,, \tag{1.73}$$

$$\dot{N}_2 = -(N_2 - N_1)S_p - (N_2 - N_2^0)/\tau\,, \tag{1.74}$$

where S_p is a pumping rate and N_i^0 is the population at thermal equilibrium, and τ is the relaxation time to restore thermal equilibrium. Assuming $S_p = x^2\gamma/2\{(\omega - \omega_0)^2 + \gamma^2\}$, a steady-state solution is given by

$$N_2 - N_2^0 = \frac{\frac{1}{2}\gamma\tau x^2(N_1^0 - N_2^0)}{(\omega-\omega_0)^2 + \gamma^2 + \gamma\tau x^2}, \tag{1.75}$$

which is equivalent to Eq.1.42. However, a similar calculation on a three level system provides solutions entirely different from those derived by the density matrix method because coherent effects are entirely absent in the rate equation method.

The rate equation is convenient to discuss the dynamics of multi-level systems interacting with radiation. However, the number of variables for an n-level system is n in rate equations while it is n^2 in the density matrix method. This difference is the reason why the rate equation method cannot describe coherent effects correctly. Actually, what the rate equations provide is the integrated line intensity over the entire spectral profile derived by the density matrix method. The rate equation method will be applicable if a laser linewidth is larger than the Rabi frequency, $x/2\pi$, because all fine structures originating from coherent effects are averaged out by the laser linewidth.

1.4 Quantized electromagnetic field

Although the classical treatment of electromagnetic radiation is quite successful for describing molecule-radiation interaction, there are a number of phenomena that can be understood much better by introducing the particle nature of radiation. A trivial example is photon counting. Another example is the instantaneous nature of optical transitions. Because there are no molecular states which can absorb a fraction of the transition energy, the change of the states by the absorption or emission of electromagnetic radiation should be instantaneous (***quantum jump***) and involve the exchange of a discrete energy $\hbar\omega$ carried by a photon. In this section, we will present the outline of a quantization procedure [5], and briefly discuss characteristics of the quantized electromagnetic field.

Classical electromagnetic waves, as in Eq.1.1, are obtained by solving Maxwell's equations. Alternatively, when there is no charge, $\vec{E}$ and $\vec{H}$ are derived from the vector potential $\vec{A}$ by the following relations

$$\vec{H} = \mathrm{curl}\vec{A}, \tag{1.76}$$

$$\vec{E} = -\frac{\partial}{\partial t}\vec{A}, \tag{1.77}$$

where $\vec{A}$ is given as a solution of the following homogeneous wave equation,

$$\nabla^2\vec{A} - \frac{1}{c^2}\frac{\partial^2}{\partial t^2}\vec{A} = 0. \tag{1.78}$$

The solution of this equation is given in a general form by,

$$\vec{A} = b(t)\vec{u}(\vec{r}) + b^*(t)\vec{u}^*(\vec{r}), \tag{1.79}$$

where $\vec{u}(\vec{r}) = \vec{e}u(\vec{r})$ ($\vec{e}$ is the direction of polarization), and $b(t)$ and $u(\vec{r})$ are given as solutions of the following differential equations

$$\nabla^2 u(\vec{r}) + k^2 u(\vec{r}) = 0, \tag{1.80}$$

$$\frac{d^2}{dt^2}b(t) + \omega^2 b(t) = 0, \tag{1.81}$$

where $k = \omega/c_0$. It is customary to solve the first differential equation with periodic boundary conditions in a L^3 cubic space, i.e. $u(\xi) = u(\xi + L)$ and $\dot{u}(\xi) = \dot{u}(\xi + L)$ where $\xi = x, y, z$. The solutions are given by

$$u(\vec{r}) \propto \mathrm{e}^{i\vec{k}\cdot\vec{r}}. \tag{1.82}$$

Each component of the k vector, k_ξ, spans an infinite number of discrete values, $k_\xi = 2\pi n_\xi / L$ ($n_\xi = 0, 1, 2 \cdots$), depending on the number of periodic cycles, n_ξ, in the unit length L. The complete set of vectors (k_x, k_y, k_z) with the direction of polarization, $\vec{e}$, provide basis functions to describe an arbitrary electric field distribution in the L^3 space. The electric field distribution associated with each basis function is called a ***mode***. For each mode, $\omega = |k|c_0$ is defined uniquely. We specify a particular mode with a suffix λ when necessary. The second differential equation, Eq.1.81, is solved with ω_λ derived from the eigenvalue k_λ,

$$b_\lambda(t) \propto \mathrm{e}^{-i\omega_\lambda t}, \tag{1.83}$$

where the normalization factor is omitted. Thus each mode is specified by $\vec{k}_\lambda$, ω_λ, and $\vec{e}_\lambda$ as discussed in 1.2.1. Although these eigenvalues have discrete values, they can be regarded as continuous by assuming $L \longrightarrow \infty$.

Next we introduce the following variables,

$$Q_\lambda = b_\lambda + b_\lambda^*, \tag{1.84}$$
$$P_\lambda = -i\omega_\lambda(b_\lambda - b_\lambda^*). \tag{1.85}$$

From Eq.1.22 and the relations between $(\vec{E}, \vec{H})$ and (b, b^*), the energy of radiation is calculated to be[7]

$$H_\lambda = 2\omega_\lambda^2 b_\lambda b_\lambda^* = \frac{1}{2}(P_\lambda^2 + \omega_\lambda^2 Q_\lambda^2). \tag{1.86}$$

This Hamiltonian is equivalent to that of a harmonic oscillator because P_λ and Q_λ satisfy the conditions for canonical variables, $\partial H_\lambda/\partial Q_\lambda = -\partial P_\lambda/\partial t$ and $\partial H_\lambda/\partial P_\lambda = \partial Q_\lambda/\partial t$. In analogy with the harmonic oscillator, quantization of the radiation field is achieved by introducing the following commutation relation,

$$P_\lambda Q_{\lambda'} - Q_\lambda P_{\lambda'} = -i\hbar\delta_{\lambda\lambda'}, \tag{1.87}$$

and the total energy of the radiation is given as well by

$$E_\lambda = (n_\lambda + \frac{1}{2})\hbar\omega_\lambda, \tag{1.88}$$

where n_λ (0, 1, 2$\cdots$) is the number of photons in mode λ. The photon state (radiation field) is specified by describing the number of photons in each mode, $|\cdots, n_\lambda, n_{\lambda'}, \cdots\rangle$.

Now we introduce the quantized radiation field in the interaction Hamiltonian. The electric field is given by

$$\vec{E} = -\frac{\partial}{\partial t}\vec{A} \propto i\omega(b\vec{u}(\vec{r}) - b^*\vec{u}^*(\vec{r})), \tag{1.89}$$

where the suffix λ and the variable t are omitted for simplicity. Since b and b^* are now operators, we rewrite b^* as b^+ to make its physical meaning clearer, as we will find below. From the matrix elements of Q and P in a harmonic oscillator $\langle n|Q|n+1\rangle = \langle n+1|Q^*|n\rangle = \sqrt{\hbar(n+1)/2\omega}$ and $\langle n|Q|n'\rangle = 0$ ($n' \neq n \pm 1$), we obtain the following important properties of the operators b and b^+ with the eigenfunctions of the photon states,

$$b|n\rangle = \sqrt{n}|n-1\rangle \quad (b|0\rangle = 0),$$
$$b^+|n\rangle = \sqrt{n+1}|n+1\rangle. \tag{1.90}$$

The operators b and b^+, which satisfy the following commutation relation requirement for a Bose particle,

$$b_\lambda b_{\lambda'} + b_\lambda^+ b_{\lambda'}^+ = \delta_{\lambda\lambda'}, \tag{1.91}$$

[7]This expression is derived with appropriate normalization factors for b and u which we omitted in the present discussion. See [5].

are called the ***annihilation operator*** and the ***creation operator*** because they decrease and increase the number of photons by one. From Eq.1.90, one can obtain

$$b^+b|n\rangle = n|n\rangle\,. \tag{1.92}$$

The operator b^+b is called the ***number operator*** because its eigenvalue is the number of photons in the relevant mode. The total energy of radiation is given by the following Hamiltonian,

$$H = \hbar\omega(b^+b+\frac{1}{2})\,, \tag{1.93}$$

$$H|n\rangle = \hbar\omega(n+\frac{1}{2})|n\rangle\,. \tag{1.94}$$

Now we are ready to calculate the transition dipole moment. We describe the basis functions by specifying both atomic and photon states simultaneously, $|\phi\,;\cdots,n_\lambda,n'_\lambda,\cdots\rangle$. The interaction Hamiltonian is given by

$$H_{\text{int}} \propto \vec{e}\cdot\vec{\mu}(b\,u(\vec{r}) - b^+u^*(\vec{r}))\,, \tag{1.95}$$

where we consider only one mode and omit λ. With the basis functions $|\phi;n\rangle \equiv |\phi\rangle|n\rangle$, the matrix element between the initial (0) and the final (f) states is given by

$$\begin{aligned}\langle\phi_f;n_f|\vec{e}\cdot\vec{\mu}\{bu(\vec{r})-b^+u^*(\vec{r})\}|\phi_0;n_0\rangle &= \langle\phi_f|\vec{e}\cdot\vec{\mu}u(\vec{r})|\phi_0\rangle\langle n_f|b|n_0\rangle - \langle\phi_f|\vec{e}\cdot\vec{\mu}u^*(\vec{r})|\phi_0\rangle\langle n_f|b^+|n_0\rangle \\ &\propto x_{f0}\langle n_f|b|n_0\rangle - x^*_{f0}\langle n'_f|b^+|n_0\rangle \\ &= \sqrt{n_0}x_{f0} - \sqrt{n_0+1}x^*_{f0}\,,\end{aligned} \tag{1.96}$$

where $u(\vec{r})$ and $u^*(\vec{r})$ have been omitted because the size of a molecule is usually much smaller than the wavelength of the radiation as discussed in 1.2.1. The properties of b and b^+ in Eq.1.90 allow only $n_f = n_0 - 1$ and $n'_f = n_0 + 1$ for the final states, corresponding to the absorption and emission of one photon, respectively. Because transition probability is proportional to the square of a transition dipole moment, the absorbed and emitted photon energy have the following relations with the incident photon number n_0.

$$P_{\text{abs}} \propto n_0 x^2_{f0}\,, \tag{1.97}$$

$$P_{\text{em}} \propto (n_0+1)x^2_{f0}. \tag{1.98}$$

The first relation represents linear absorption, because the absorbed power is proportional to the numbers of incident photons (laser power). In P_{em}, the term proportional to n_0 represents stimulated emission where the emission occurs in the same mode as that of the incident photon. The additional term originating from 1 in (n_0+1) corresponds to the spontaneous emission because the emission occurs even when $n_0 = 0$ (no incident photon). However, this also implies that the spontaneous emission is allowed in all modes because the incident photon (radiation field) is not necessary. Thus the spontaneous emission is incoherent because it occurs in any direction $\vec{k}$ and polarization $\vec{e}$ though the frequency ω is limited by the energy conservation law, with the width allowed by the energy uncertainty principle (finite interaction time) and the Doppler effect. A similar process occurs in multiphoton transitions involving emission processes.

1.5 Simple examples

1.5.1 *Higher-order transitions and interference effects in two-photon transitions*

So far we discussed the two-photon absorption process on the basis of rigorous solutions for a three level system. However, second- and higher-order transitions are frequently discussed using a time-dependent perturbation theory. The theory gives the effective transition dipole moment (corresponding to $\hbar x$) between the initial and final states as

$$\overline{H'}_{f0} = \langle f|H'|0\rangle + \sum_{i1}\frac{\langle f|H'|i1\rangle\langle i1|H'|0\rangle}{W'_0 - W'_{i1}} + \sum_{i1}\sum_{i2}\frac{\langle f|H'|i2\rangle\langle i2|H'|i1\rangle\langle i1|H'|0\rangle}{(W'_0-W'_{i1})(W'_0-W'_{i2})} + \cdots\,, \tag{1.99}$$

where H' is the interaction Hamiltonian, and 0, f, ij are the subscripts specifying the initial, final, and i-th intermediate states. All the denominators are assumed to be non-zero because the perturbation method is no longer valid when there is a resonance. The terms on the right side correspond to one-, two-, and three-photon transitions, as is easily recognized from the number of transition matrix elements.

The definition of W' is somewhat complicated because it includes the photon energy. For example, if a molecule in the W_0 state absorbs a photon $\hbar\omega$ and changes state to W_f ($W_0 < W_f$), the transition may be described as $|W_f; (n-1)\hbar\omega\rangle \leftarrow |W_0; n\hbar\omega\rangle$. Suppose we take the two-photon term (the second term) and use the notation in Fig.1.7 for the incident radiation. Then the denominator becomes $W_1' - W_2' = W_1 + \hbar\omega_p - W_2 = \hbar\Delta_p$. With the relations $\langle f|H|i1\rangle = \hbar x/2$ and $\langle i1|H|0\rangle = \hbar x_p/2$, the effective Rabi frequency for the two-photon transition, x_{tp}, is given by

$$x_{\mathrm{tp}} \equiv 2\overline{H_{f0}^{(2)}}/\hbar = \frac{x x_p}{2\Delta_p}. \tag{1.100}$$

If we substitute x_{tp} into x in Eq.1.18 and assume $\omega - \omega_0 = 0$ $(\Delta + \Delta_p = 0)$ and $x \ll \gamma$, we obtain

$$\overline{|b'(t)|^2} = \frac{x_p^2 x^2}{8\gamma^2\Delta_p^2}. \tag{1.101}$$

Because $\overline{|b'(t)|^2} = \rho_{33}$ and $\rho_{11}^0 \approx 1$ when $\rho_{33} \ll 1$, the above expression is equivalent to Eq.1.72. Thus the second-order perturbation describes two-photon processes without coherent effects.

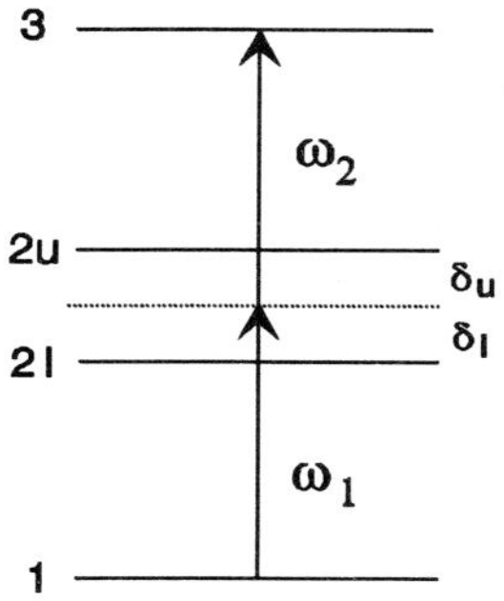

Fig. 1.12 An interference effect in a two-photon transition.

In general, the summation in Eq.1.99 should include all allowed intermediate states. When there are two or more intermediate states near resonance, matrix elements originating from all intermediate states should be added. This results in interference among the transition dipole moments from different intermediate states. An example is shown in Fig.1.12 where the $W_1 + \hbar\omega_1$ virtual state lies between the two intermediate states, 2ℓ and 2u. If we define the absolute energy differences between $\hbar\omega_1$ and the two intermediate states as δ_ℓ and δ_{u}, the total matrix element is given by

$$\begin{aligned}\overline{H_{f0}^{(2)}} &= \frac{\langle 3|H'|2u\rangle\langle 2u|H'|1\rangle}{W_0 - W_{2u} + \hbar\omega_1} + \frac{\langle 3|H'|2\ell\rangle\langle 2\ell|H'|1\rangle}{W_0 - W_{2\ell} + \hbar\omega_1} \\ &= -\frac{\langle 3|H'|2u\rangle\langle 2u|H'|1\rangle}{\delta_u} + \frac{\langle 3|H'|2\ell\rangle\langle 2\ell|H'|1\rangle}{\delta_\ell}.\end{aligned} \tag{1.102}$$

Because the above two terms have opposite signs, they will cancel each other at a certain value of ω_1. Such an interference effect is not common in two-photon spectroscopy, but one should remain alerted to the possibility of this effect when the virtual state exists among a dense energy level manifold. A similar interference may occur also in Raman transitions.

In a quantized electromagnetic field, the matrix element for a two-photon transition (numerator in the above expression) is given by

$$\langle f|H'|i\rangle\langle i|H'|0\rangle \propto \langle f|\vec{e}\cdot\vec{\mu}\{bu(\vec{r}) - b^{+}u^{*}(\vec{r})\}|i\rangle\langle i|\vec{e}_p\cdot\vec{\mu}\{b_p u_p(\vec{r}) - b_p^{+}u_p^{*}(\vec{r})\}|0\rangle\,, \tag{1.103}$$

where the suffix p has been used to specify the first step transition. This matrix element contains four different terms corresponding to four different combinations of (b, b^+) and (b_p, b_p^+). By defining the initial photon state as $|n_p, n\rangle$, the non-vanishing matrix elements are

$$\langle n_p - 1, n-1|b_p b|n_p, n\rangle = \sqrt{n_p n}\,, \tag{1.104}$$
$$\langle n_p - 1, n+1|b_p b^+|n_p, n\rangle = \sqrt{n_p(n+1)}\,, \tag{1.105}$$
$$\langle n_p + 1, n-1|b_p^+ b|n_p, n\rangle = \sqrt{(n_p+1)n}\,, \tag{1.106}$$
$$\langle n_p + 1, n+1|b_p^+ b^+|n_p, n\rangle = \sqrt{(n_p+1)(n+1)}\,. \tag{1.107}$$

These matrix elements correspond respectively to two-photon absorption, Raman transition, V-shaped two-photon transition, and two-photon emission as shown in Fig.1.13 (a)-(d). As discussed previously, all emission processes contain contributions from spontaneous and stimulated emission. For example, there are four different processes in two-photon emission corresponding to four different combinations of stimulated and spontaneous emission for ω_p and ω.

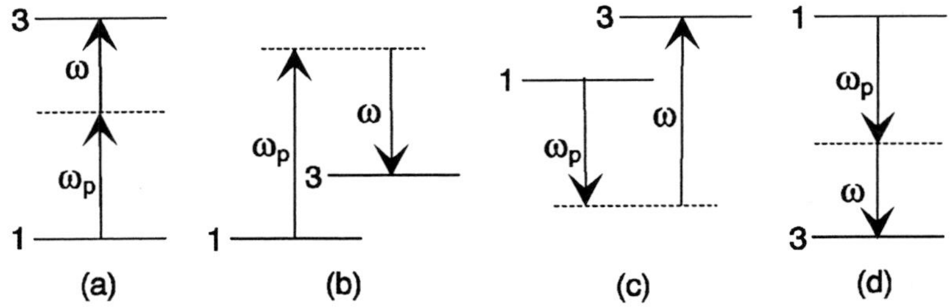

Fig. 1.13 Four different types of two-photon transition. (a)Two-photon absorption. (b)Raman transition. (c)V-shaped two-photon transition. (d)Two-photon emission. Broken lines indicate virtual states.

1.5.2 *Raman transition*

In 1.3, we discussed a three-level system in which the energy level scheme allows only absorption. When level 3 lies below level 2 (see Fig.1.14), the second step becomes an emission process. In the density matrix formalism, ρ_{33} in a Raman transition can be obtained simply by changing the sign of Δ in all expressions for the two-photon transition. When $|\Delta_p| \gg \gamma$ and x_p, resonance peaks appear at $\Delta = \Delta_p - x_p^2/4\Delta_p$ and $\Delta = -x_p^2/4\Delta_p$. The former resonance corresponds to a ***Raman transition***. This process amplifies the incident probe laser beam by stimulated emission. Thus by scanning the probe laser, one can access to the energy levels that cannot be probed from level 1 by one-photon transition. This method is called ***Raman gain spectroscopy***, reported first in 1977 [6]. As Δ_p approaches 0, the intensity of Raman scattering increases in proportion to $1/\Delta_p^2$. Raman spectroscopy with a near resonant condition is called ***resonance Raman spectroscopy***. When $\Delta_p = 0$, we see the same Rabi splitting as shown in Fig.1.8 where the vertical axis corresponds to the increase in probe laser power by stimulated emission.

From Eq.1.105, the photon energy released in a Raman transition has the following relation,

$$P_{\text{Raman}} \propto n_p(n+1)\,, \tag{1.108}$$

where n_p is the number of photons in the pump mode and n is the number in the probe mode. The term proportional to $n_p n$ represents stimulated emission by the probe laser beam as discussed above. When the probe laser is absent, one can still observe emission at around ω due to the term $n_p(\times 1)$. This is the ordinary ***Raman scattering***, also called ***spontaneous Raman scattering*** which was discovered first by Raman [7] in 1928. Although the spontaneously emitted radiation is incoherent, this is a pure two-photon process and the time profile of the scattered radiation follows that of the pump radiation. It should be noted that spontaneous Raman scattering has a linear dependence on the power of the pump radiation. As a consequence, spontaneous Raman scattering is much easier to observe compared with the two-photon absorption, which is a true nonlinear process. This is the reason why two-photon absorption was observed only with a laser in 1961 [8] while spontaneous Raman scattering was observed much earlier using incoherent pump radiation. When the pump laser is strong, the scattered light is amplified and produces coherent emission called ***stimulated Raman scattering***, observed first in 1962 [9]. This technique has been used for frequency conversion of a strong laser pulse, by producing emission in many Stokes and anti-Stokes lines.

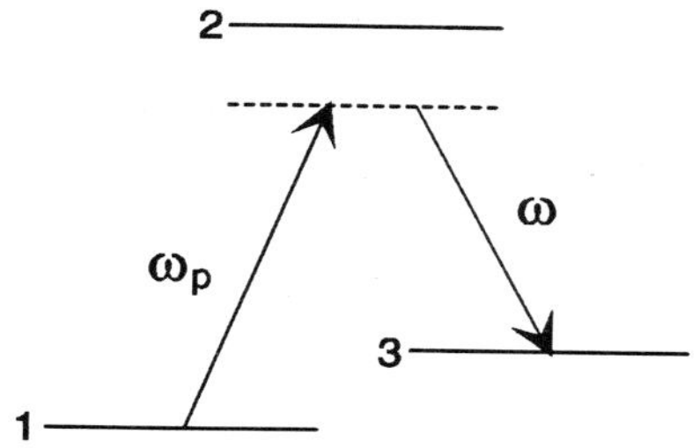

Fig. 1.14 Raman transition.

1.5.3 *Double resonance and related techniques*

In ***double resonance***, a three level system interacts with two (near) resonant radiation fields (see chapter 2). Double resonance signals show quite different features depending on the relative magnitude of the Doppler and homogeneous line widths. The results described in 1.3 are double resonance without Doppler broadening and are exemplified by ***microwave/radiofrequency double resonance***. (The bandwidth of the radiation is assumed to be narrower than the homogeneous and Doppler widths.) When one of the radiation fields is in the infrared or visible region, the Doppler width for gas phase molecules is substantially larger than the homogeneous width. This is the situation for ***microwave-optical double resonance (MODR)*** or ***infrared-microwave double resonance***. For the latter, a theory including the influence of Doppler broadening has been developed on the basis of the density matrix formalism [10]. A main feature in this case is that most of the coherent effects are smeared out by averaging over velocity distribution unless the saturation broadening is comparable with or larger than the Doppler width. A significant complication arises when both transitions are associated with Doppler broadening (***optical-optical double resonance***). This case will be discussed in the following subsection.

When two laser frequencies are resonant with a three-level system, the observed effect is a mixture of two-photon and two-step processes, though the separation of these two processes is not always straightforward. They are characterized by the fact that the latter is influenced by relaxation in the intermediate state while the former is not. A well defined two-step excitation is the case when two laser pulses excite a three-level system with a time delay longer than the pulse width for the second laser. When the laser line width is comparable or larger than the Doppler and saturation widths, all of the coherent effects are smeared out and rate equations are applicable to describe the behavior of the system.

There are a number of variants in double resonance, in particular with two pulsed lasers. In Fig.1.15, an excitation with ω_p results in spontaneous emission from level 2 to many low-lying levels.

Spectroscopy of such emission, ***dispersed fluorescence spectroscopy***, provides important information on the structure of the low-lying energy levels which cannot be accessed from the ground state by a one photon process. The excitation also produces population inversions between level 2 and many other low-lying levels. If the pump laser is strong, the spontaneous emission is amplified, and ***amplified spontaneous emission, ASE***, is observed. By setting up a proper optical cavity (a pair of mirrors) for one of the emission lines, one can obtain laser oscillation, which is known as ***optically pumped laser***.

When the dump laser ω_{dump} is on, the intensity of the fluorescence from level 2 to 3 decreases because the molecules initially excited to level 2 are transferred to level 4 by the dump laser. By scanning the dump laser while monitoring the intensity of the fluorescence, one can probe low-lying energy level structure with much higher accuracy and sensitivity (***stimulated emission pumping spectroscopy, SEP***) [11]. Higher sensitivity is obtained because a strong dump laser beam can stimulate emission even in transitions with small transition dipole moments. When a third laser (or the same pump laser) ionizes the molecule from level 2 by two-step excitation, the dump laser decreases the ion signal intensity. This is ***ion-dip spectroscopy*** [12], and has extremely high sensitivity because the dump transition is probed by detecting ions.

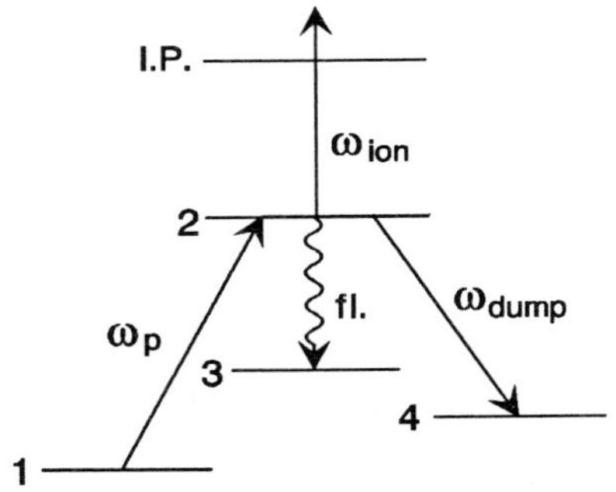

Fig. 1.15 Variants of double resonance spectroscopy.

1.5.4 *Doppler-free spectroscopy*

In the early days of laser spectroscopy, nonlinear spectroscopic techniques were developed to measure normally Doppler-broadened optical transitions with the resolution of the homogeneous linewidth. The techniques are based on the fact that, in the hole burning effect discussed in 1.2.3, the population dip (or peak in the upper level) produced by laser irradiation has a homogeneous line width. Thus, by probing the population dip in level 1 or the peak in level 2 with a second weak probe laser beam, the dip or peak can be measured with the resolution of the homogeneous linewidth.

In Fig.1.16, the pump-probe scheme and experimental setup are shown for three different arrangements of two laser beams; co-propagating, counter-propagating, and perpendicular. The pump laser beam is assumed to be propagating to the $+v_z$ direction. Suppose a strong pump laser excites molecules from level 1 to 2, with the laser frequency slightly higher than the Doppler-free transition frequency ω_0 (Fig.1.17(a)). This produces a narrow population dip in level 1 and a peak in level 2 at $v_z = c_0\Delta\omega_p/\omega_0$ where $\Delta\omega_p$ is the frequency offset for the pump laser. Then a second weak laser probes the absorption for the same transition, detecting only the population change produced by the pump laser. The frequency profiles of the absorption spectra for the three different laser alignments are shown in Fig.1.17(b). When the two laser beams are co-propagating, the absorption signal is observed on the high frequency side of the Doppler profile (on the same frequency side as the pump laser). When the pump and probe laser beams are counter-propagating, the absorption signal appears on the low frequency side (on the opposite side of the pump frequency) because the

Doppler shift for the probe beam, which is counter-propagating against the pump laser beam, has the opposite sign. When the probe beam is perpendicular to the pump beam, only a weak signal with a full Doppler profile is observed because there is no velocity-selective excitation in the direction of the probe laser beam.

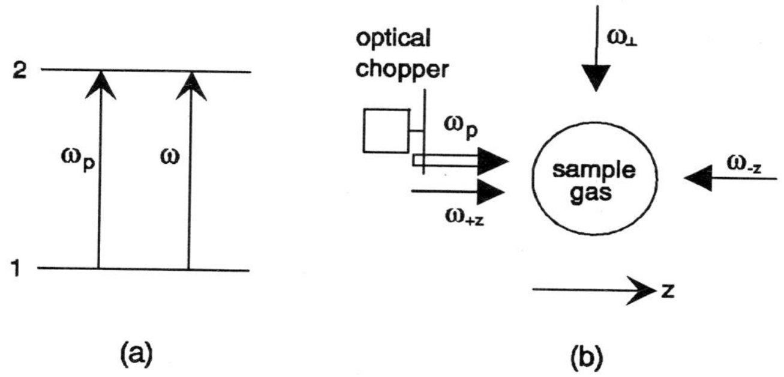

Fig. 1.16 Experimental arrangements for Doppler free spectroscopy with two independent laser beams. For detecting only the hole and dip produced by the pump beam, the pump laser is amplitude-modulated (100 %) while the probe laser detects only the modulated signal by phase sensitive detection method (lock-in detection). (a) an energy level diagram and (b) three different arrangements of the pump and probe laser beams; co-propagating ω_{+z}, counter-propagating ω_{-z}, and perpendicular $\omega_{\perp}$.

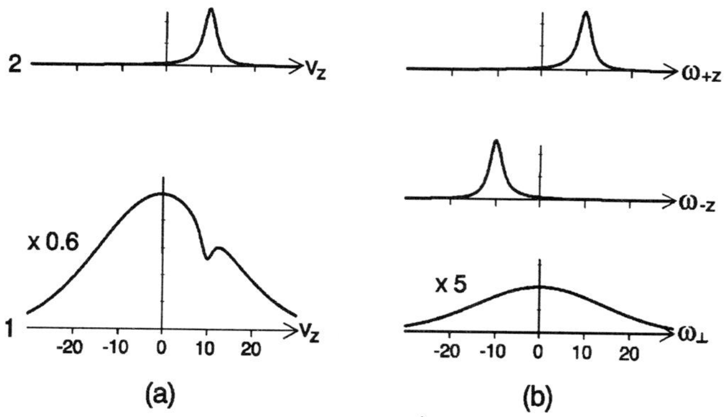

Fig. 1.17 Spectral profiles measured with the experimental arrangements in Fig.1.16. (a) shows the population hole and peak in levels 1 and 2 produced by a pump laser with offset $\Delta\omega_p = 10\gamma$. The pump laser power is assumed to satisfy the condition $x^2 = 2\gamma^2$. The scale is in units of γ/k ($k = \omega/c_0$). The Doppler width corresponds to $u = 20\gamma$, which is more than one order of magnitude smaller than the value expected in the visible region. The spectra in (b) show profiles of the absorption signals produced by the pump laser for co-propagating (ω_{+z}), counter-propagating (ω_{-z}), and perpendicular ($\omega_{\perp}$) probe beams. The horizontal scale is in units of γ. The vertical scale is not equal to that of (a). Note that a narrow peak is observed only when the two laser beams are co- or counter-propagating.

Although this technique allows the measurement of absorption lines with resolution limited only by the homogeneous linewidth, there is a disadvantage that the line position depends on the pump laser frequency and does not necessarily reflect the Doppler-free absorption frequency ω_0. This problem is solved by using two counter-propagating laser beams, a strong pump beam and a weak probe beam, from the same laser (Fig.1.18(a)). When the laser frequency ω is off-resonant from ω_0, the pump laser from the left excites molecules with $v_z = c_0\Delta\omega/\omega_0$, while the probe laser from the right monitors the absorption by molecules with $-v_z$ (see Fig.1.18 (b)). Thus the probe laser simply detects Doppler-broadened absorption profile but does not see the influence of the narrow hole burnt by the pump laser. As ω approaches ω_0, the probe laser starts detecting the presence of the hole, and as the laser frequency crosses over ω_0, one can observe a Doppler-free absorption profile as shown in Fig.1.18 (c), with the minimum of the dip located at $\omega = \omega_0$. To increase the sensitivity, one can modulate the pump laser beam using an optical chopper and detect only the modulated absorption signal with a lock-in amplifier as shown in Fig.1.18 (d). It should be noted that the widths of the signals in (c) and (d) are one half of the width of the hole in (b) because the pump and probe laser frequencies are scanned in opposite directions simultaneously.

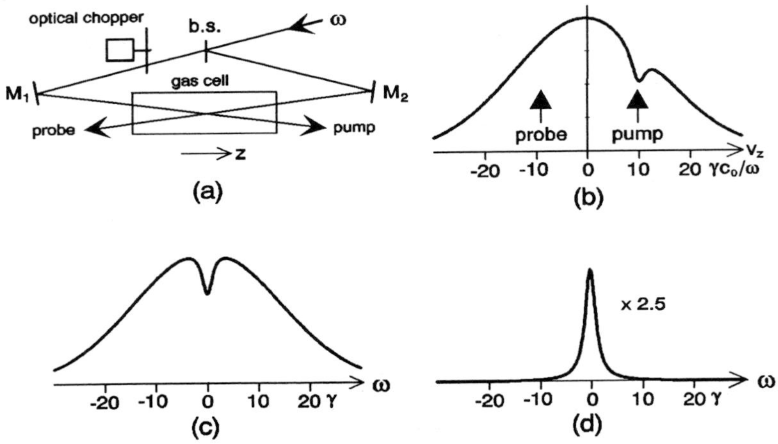

Fig. 1.18 Doppler free spectroscopy with strong pump and weak probe laser beams at the same frequency. The horizontal scales and the parameters used for calculating the spectral profile are the same as those in Fig.1.17. (a)Experimental arrangement. (b)Velocity components interacting with the pump and probe laser beams. $\omega > \omega_0$ is assumed. A hole is burnt only by the pump laser beam. The scale is in units of $\gamma c_0/\omega$. (c) Absorption spectrum recorded by the probe laser. A hole burnt by the pump laser is observed at line-center. (d)Lock-in detection signal (sign is reversed) when the pump laser is modulated.

One can also observe Doppler-free absorption lines when two strong counter-propagating laser beams are used (see the upper inset of Fig.1.19). Suppose we detect the total fluorescence from the upper level while scanning the laser frequency. When the laser frequency is off-resonant from ω_0, the two laser beams burn two holes as shown in Fig.1.19(a). The fluorescence intensity is proportional to the total area of the two burnt holes. When the laser frequency is at ω_0, molecules with $v_z = 0$ interact with the two laser beams simultaneously, with twice as strong laser intensity. If the absorption is linear, twice as strong laser intensity implies twice as much absorption (and therefore twice as much fluorescence intensity). Thus there is no difference in net fluorescence intensity whether the laser frequency is at the line-center or not. If the absorption is partially saturated, however, the absorption at twice the laser intensity is smaller than twice the absorption with a single laser beam (see Fig.1.19 (a) and (b)). As a consequence, the fluorescence shows a narrow dip at the resonance frequency, thus allowing the measurement of the Doppler-free line position with the accuracy of the homogeneous linewidth.

For quantitative discussion, we start with Eq.1.28 and neglect the exponential factor because we only consider the molecules very close to $v_z = 0$. When the resonance detuning is larger than the homogeneous width, the number density of molecules in the upper level is given by

$$n_2(v_z) = \frac{N_1^0}{\sqrt{\pi}u} \frac{\frac{1}{2}x^2}{(\omega \pm k_0 v_z - \omega_0)^2 + \Gamma^2}, \tag{1.109}$$

where $\pm$ corresponds to the two counter-propagating laser beams and $N_2^0 = 0$ is assumed. By integrating over v_z, the total number of excited molecules is

$$N_2 = \frac{N_1^0 \sqrt{\pi} x^2}{k_0 u \Gamma}, \tag{1.110}$$

where the value has been multiplied by a factor of two corresponding to two independent absorption peaks at $v_z = \pm(\omega - \omega_0)/k_0$. When the laser is at the center of the absorption line, the laser power becomes twice as strong. By replacing x^2 with $2x^2$ and integrating over v_z, we obtain

$$N_2' = \frac{N_1^0\sqrt{\pi}x^2}{k_0 u\Gamma'}, \tag{1.111}$$

where $\Gamma^2 = x^2 + \gamma^2$ and $\Gamma'^2 = 2x^2 + \gamma^2$. The difference between these two cases is

$$N_2 - N_2' = \frac{N_1^0\sqrt{\pi}x^2}{k_0 u}\left(\frac{1}{\Gamma} - \frac{1}{\Gamma'}\right). \tag{1.112}$$

If we assume $|x| \gg \gamma$, $N_2 - N_2'$ is equal to $N_2/2$. In other words, at a strong saturation limit, the absorption at $\omega - \omega_0 = 0$ is 1/2 of the absorption with two off-resonant beams. Because the signal is observed as a very narrow dip in the fluorescence profile, relatively small saturation is sufficient to detect the center-dip.

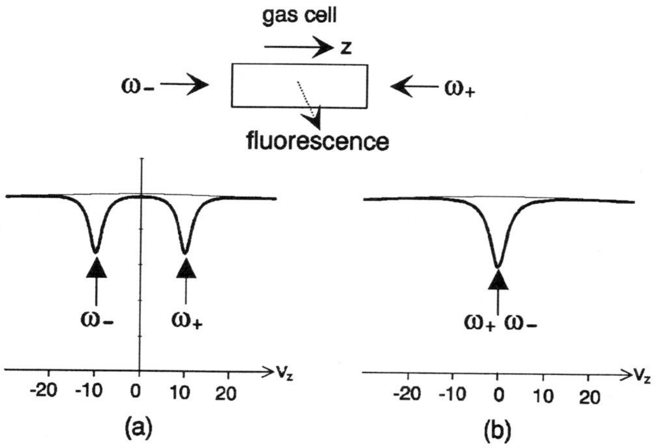

Fig. 1.19 Doppler-free spectroscopy with two strong laser beams. Here we use a more realistic Doppler width, $u = 200\gamma$. For saturation, $x^2 = 2\gamma^2$ is assumed. The scale is in units of γ. Note that the area in (b) ($\omega = \omega_0$) is smaller than the total area of the two dips in (a), where two off-resonant laser beams burn two holes.

Historically this effect was observed first in He-Ne gas laser oscillation at 1.152 μm [13]. This laser consists of a plasma tube containing a mixture of He and Ne gas and two mirrors which form an optical cavity. In general, a two mirror cavity with mirror spacing L has narrow longitudinal resonance frequencies at $\omega = n\pi c_0/L$ ($n = 1, 2\cdots$), with a regular interval of 150 MHz when $L = 1$ m. The resonant optical field is a standing wave, which is described as a superposition of two counter-propagating traveling waves at the same frequency. Because the Doppler width of Ne gas at 1 μm is the same order of magnitude as this interval, the laser oscillation occurs at only one cavity resonance frequency, and the oscillation frequency can be tuned by changing the length of the cavity. By monitoring the laser power while scanning the cavity length, a dip was observed on top of a broad Doppler-broadened laser power profile. This dip is caused exactly by the same mechanism as discussed above, and was named ***Lamb dip*** because the dip was predicted first by Lamb in his work on the theory of laser oscillation [14].

Doppler-free absorption lines have been observed in many molecular transitions. The first experiment was done on CH_4 by inserting an absorption cell in the optical cavity of 3.39 μm He-Ne laser. The CH_4 $\nu_3 F_1^{(2)} - \text{g.s.} F_2^{(2)}$ molecular transition at 2947.906 cm^{-1} is in accidental coincidence with the laser line. When the laser frequency was scanned by changing the cavity length, a very narrow peak was observed at the center of the absorption line [15]. In contrast to the dip in LIF, the decrease of intracavity absorption by the saturation effect causes the laser power to increase, resulting in an increase in laser output power when the laser frequency is at the absorption line-center. This is called the ***inverted Lamb dip***. This technique was extended later to ultrahigh resolution spectroscopy to measure ***recoil doublet*** of the same transition [16]. The recoil doublet originates from the momentum of a photon, as given by $|\vec{p}_{\text{photon}}| = \hbar\omega/c_0$. A slight difference in the velocity of a molecule before and after the absorption of a photon causes a very small splitting in the absorption line. The 2.2 kHz splitting was observed by using a frequency-stabilized laser with a 200 Hz line width.

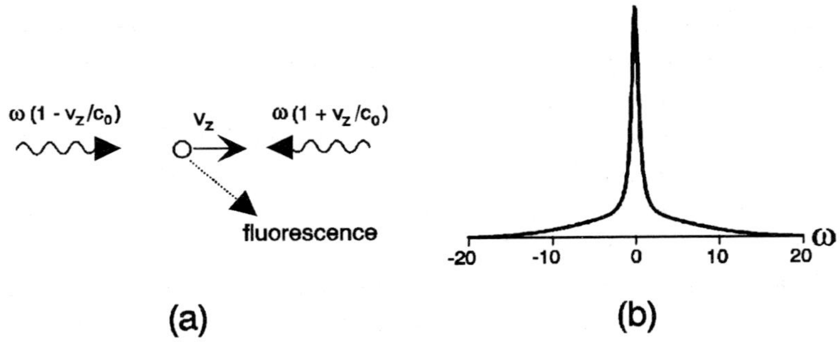

Fig. 1.20 The Doppler-free two-photon spectroscopy. (a) shows a schematic diagram for the experiment. The molecule moving with the velocity v_z feels the incident radiation (ω) as $\omega(1+v_z/c_0)$ for the right laser beam and $\omega(1-v_z/c_0)$ for the left laser beam by the Doppler effect. (b) shows the superposition of a narrow Doppler-free two photon absorption and a Doppler-broadened two-photon absorption (pedestal). $x_{\text{tp}} \ll \gamma$ is assumed. A very narrow Doppler width, $k_0u = 20\gamma$, has been used to show the Doppler-free and Doppler-broadened components simultaneously. In a more realistic system, the Doppler-broadened pedestal is almost invisible in the above scale for transitions in the visible wavelength region.

Nonlinear Doppler-free spectroscopy has also been demonstrated in two-photon transitions. Suppose we irradiate molecules with two counter-propagating beams from the same laser (with frequency ω) as shown in Fig.1.20, and total fluorescence intensity is monitored. A molecule with a velocity component v_z finds the frequency of photon from the $+z$ beam as $\omega(1+v_z/c_0)$, and a photon from the $-z$ direction as $\omega(1-v_z/c_0)$. By absorbing one photon each from the right and left laser beams, the molecule absorbs $2\hbar\omega$ as a total energy, which is Doppler-free. Therefore, when the laser frequency is scanned, all molecules with different velocity components can absorb two photons resonantly at $2\hbar\omega = W_2 - W_1$, which is observed as a narrow emission line. However, there are two-photon absorption processes involving only the right beam and only the left beam, and these produce a Doppler-broadened two-photon absorption profile as a background absorption (pedestal).

For calculating the spectral profile, we use the effective two-photon Rabi frequency in Eq.1.100 by assuming that there are no near-resonant intermediate states. With the incident laser frequency, ω, and effective two-photon Rabi frequency, x_{tp}, the number density of molecules excited by the two-photon absorption is given by

$$n_2(v_z) = \frac{N_1^0 x_{\text{tp}}^2}{2\sqrt{\pi}u}\{\frac{1}{(2\omega - 2k_0v_z - \omega_0)^2 + \Gamma^2} + \frac{1}{(2\omega + 2k_0v_z - \omega_0)^2 + \Gamma^2} + \frac{1}{(2\omega - \omega_0)^2 + \Gamma^2}\}e^{-v_z^2/u^2}, \tag{1.113}$$

where the three terms in curly brackets correspond to the two-photon absorption from the left beam, from the right beam, and from both beams. The integration over v_z, by assuming $u \gg \Gamma$, gives the absorbed photon energy,

$$P'_{\text{abs}}(\omega) = N_1^0 \gamma\hbar\omega x_{tp}^2[\frac{\sqrt{\pi}}{\Gamma k_0 u}\exp\{-\frac{c^2}{u^2}(\frac{2\omega - \omega_0}{\omega_0})^2\} + \frac{1}{(2\omega - \omega_0)^2 + \Gamma^2}]. \tag{1.114}$$

Thus the signal is a superposition of a Doppler-broadened Gaussian absorption profile and a narrow Doppler-free Lorentzian absorption profile. The peak height of the Doppler-free two-photon absorption is $k_0u/\sqrt{\pi}\Gamma$ times larger than the peak height of the Doppler-broadened line. This ratio, which is about the ratio of Doppler and homogeneous widths, implies that the Doppler-free two-photon signal is about two orders of magnitude stronger at the line center. The simulated spectrum is shown in Fig.1.20(b). The ***Doppler-free two photon spectroscopy*** was first proposed by Vasilenko *et al.* [17] in 1970, and demonstrated experimentally by Biraben *et al.* [18], Levenson and Bloembergen [19], and Pritchard *et al.* [20] in 1974.

1.6 *Summary*

In 1.2.1, we discussed a two-level system interacting with near-resonant coherent radiation by solving a Schrödinger equation. The ***Rabi frequency*** was defined as a parameter to represent the strength of interaction. With ***rotating wave approximation***, the probability of finding a molecule in the upper or lower level was found to show sinusoidal oscillation with time (***Rabi oscillation***). In 1.2.2, the influence of collisions among gas phase molecules was discussed. A stochastic change of molecular states by collisions was shown to produce a ***Lorentzian*** line shape. Under the presence of collisions, the energy absorbed by molecules shows a saturation effect as the laser power increases. Two origins for the line width, ***collisional broadening*** (***pressure broadening***) and ***saturation broadening***, were defined. Two collisional relaxation processes, ***longitudinal relaxation*** and ***transverse relaxation***, were introduced as the relaxation only by state-changing collisions for the former, and as the relaxation by both state-changing and state-nonchanging processes for the latter. Spontaneous and stimulated emission were defined as fundamental processes in an interacting two level system. In 1.2.3. Doppler effect in the gas phase molecule was discussed. A Doppler-broadened transition line was shown to have ***Gaussian*** line profile. Two origin of the line width, ***homogeneous*** and ***inhomogeneous widths***, were defined. A ***hole burning effect*** in an inhomogeneously-broadened absorption line was demonstrated.

In 1.3, coherent effects in a three-level system interacting with near-resonant coherent radiation were discussed. A density matrix formalism was introduced as a general method to obtain steady-state solutions for multi-level systems. In a three level system, a strong resonant or off-resonant pump radiation was shown to produce ***Rabi splitting*** or ***optical Stark shifts***. These effects (***coherent effects***) were interpreted as resonant transitions through ***virtual states*** produced by a strong pump radiation. A ***rate equation*** was introduced as a method to solve multilevel systems without coherent effects.

In 1.4, the classical electromagnetic radiation was quantized in analogy with the quantization of a harmonic oscillator. By quantization, the radiation field was converted to a ***photon state*** specified by the number of photons in each mode which has unique frequency ω, wave vector $\vec{k}$, and direction of polarization $\vec{e}$. The interaction Hamiltonian was transformed to the operator containing ***annihilation*** and ***creation operators*** for photons. From the matrix element of the creation operator, ***stimulated*** and ***spontaneous emission*** were defined.

Section 1.5 describes simple examples of nonlinear spectroscopy. In 1.5.1 a higher-order effective transition dipole moment obtained by a perturbation method was compared with the solution obtained by the density matrix method. An interference effect in two-photon transitions, when there are more than one near-resonant intermediate states, was discussed. In 1.5.2, ***Raman transition*** was discussed as a variant of ***two-photon transition***. Characteristics of spontaneous and stimulated Raman transitions were described. Various types of ***double resonance spectroscopy*** were introduced briefly in 1.5.3. Finally, ***Doppler-free spectroscopy*** in Doppler-broadened optical transitions was demonstrated for one- and two-photon transitions.

References

1. T. H. Maiman, *Nature* **187**, 493 (1960).
2. J. P. Gordon, H. J. Zeiger, and C. H. Townes. *Phys. Rev.* **95**, 282 (1954).
3. A. Javan, *Phys. Rev.* **107**, 1579 (1957).
4. See, for example, N. Bloembergen, *Non-Linear Optics*, Benjamin,(1965).
5. W. Heitler, *The Quantum Theory of Radiation*, Oxford, (1954).
6. A. Owyoung and E. D. Jones, *Opt. Lett.* **1**, 152 (1977).
7. C. V. Raman and K. S. Krishnan, *Nature* **121**, 501 (1928).
8. W. Kaiser and C. G. B. Garret, *Phys. Rev. Lett.* **7**, 229 (1961).

9. E. J. Woodbury and W. K. Ng, *Proc. IRE* **50**, 2367 (1962).
10. M. Takami, *Jpn. J. Appl. Phys.* **15**, 1063 (1976); ibid. **15**, 1889 (1976).
11. C. Kittrell, *et.al.* *J. Chem. Phys.* **75**, 2056 (1981). See also C. E. Hamilton, J. L. Kinsey, and R. W. Field, *Ann. Rev. Phys. Chem.* **37**, 493 (1986).
12. D. E. Cooper, C. M. Klimcak, and J. E. Wessel, *Phys. Rev. Lett.* **46**, 324 (1981).
13. R. A. McFarlane, W. R. Bennett, Jr., and W. E. Lamb, Jr. *Appl. Phys. Lett.* **2**, 189 (1963).
14. W. E. Lamb, Jr. *Phys. Rev.* **134**, A1429 (1964).
15. R. L. Barger and J. L. Hall, *Phys. Rev. Lett.* **22**, 4 (1969).
16. J. L. Hall, C. J. Bordé, and K. Uehara, *Phys. Rev. Lett.* **37**, 1339 (1976).
17. L. S. Vasilenko, V. P. Chebotaev, and A. V. Shishaev, *JETP Lett.* **12**, 113 (1970).
18. F. Biraben, B. Cagnac, and G. Grynberg, *Phys. Rev. Lett.* **32**, 643 (1974).
19. M. D. Levenson and N. Bloembergen, *Phys. Rev. Lett.* **32**, 645 (1974).
20. D. Pritchard, J. Apt, and T. W. Ducas, *Phys. Rev. Lett.* **32**, 641 (1974).

2 Double Resonance (MODR, OODR) Spectroscopy

Yasuki Endo* and Masaaki Fujii**

**Department of Pure and Applied Sciences, The University of Tokyo, Komaba, Tokyo 153, Japan*

***Department of Electronic Structure, Institute for Molecular Science, 38 Myodaiji-chyo, Okazaki 444, Japan*

2.1 Basic principle of double resonance spectroscopy

Among various nonlinear spectroscopic techniques used currently, double resonance spectroscopy, especially microwave-optical double resonance (MODR) spectroscopy, is one of those used from an early stage of laser spectroscopy. The idea of double resonance goes back before the advent of lasers. When CW lasers became available to molecular spectroscopists, MODR was tried and successfully applied to molecular systems in early 1970s. The main motivation to use the MODR method is to obtain very high resolution, usually attained by conventional microwave spectroscopy, in the optical region, where resolution is limited by the Doppler width when ordinary linear spectroscopy is applied for bulk samples. In addition to very high resolution, MODR spectroscopy has another advantage over single color linear spectroscopy. It is possible to select specific lines among congested spectrum observed in one spectral region, and obtain definite rotational assignments for these lines by labeling the transitions with another radiation field. This advantage is fully utilized also in optical-optical double resonance spectroscopy, although sub-Doppler resolution is not necessarily obtained. Since MODR spectroscopy requires a combination of quite different spectroscopic techniques, microwave spectroscopy and optical spectroscopy, and a tunable CW laser with sufficiently large output power is required in most MODR experiments, investigations of MODR spectroscopy are rather limited to a small number of laboratories. On the other hand, since only optical laser techniques are required and more versatile pulsed lasers can be used, optical-optical double resonance spectroscopy has been used more extensively in recent years, where various schemes of double resonance and variety of detection systems have been utilized.

In this chapter, we will explain the basic principles of double resonance spectroscopy in the first section and review MODR spectroscopy in the next section. Various techniques used in optical-optical double resonance spectroscopy will be described in the last section. As will be shown in later sections, double resonance spectroscopy has been providing various types of valuable information on molecular systems, which are often quite difficult to obtain by other spectroscopic methods.

2.1.1 Saturation of population

Double resonance spectroscopy involves three levels and two photons usually of different frequencies which are resonant with two of the energy separations among the three levels. Such a situation is shown in Fig. 2.1 for the cases of microwave optical double resonance, where either a cascaded or a folded scheme is possible. In the case of OODR, as shown in Fig. 2.2 (a), the cascaded configuration is usually called OODR, and the folded scheme is further divided into two, as shown in Fig. 2.2 (b) and (c), where the former is called depletion spectroscopy and the latter SEP (stimulated emission pumping) spectroscopy. These schemes will be discussed separately in different chapters. In the case of MODR, one of the sources of radiation is in the microwave region and connects rotational levels within one vibronic state. There are thus two possibilities, whether the microwave radiation is in resonance with a transition in the ground state or in the excited state, as illustrated in Fig. 2.1 (a) and (b). These cases are called ground state or excited state double resonance. It is not important whether the microwave transition takes place in either the cascaded or folded configuration.

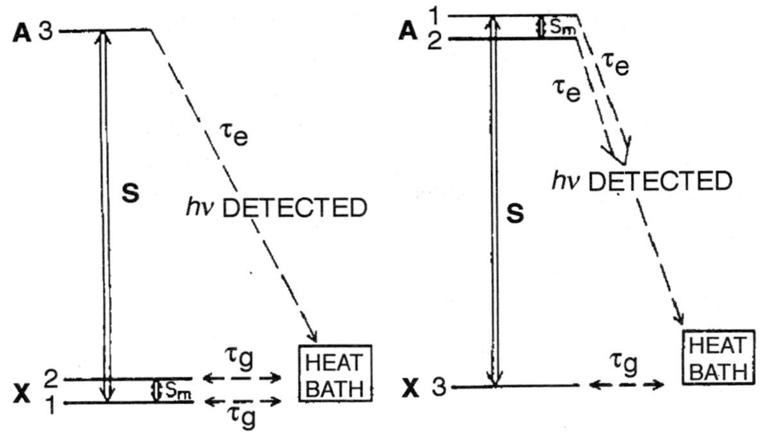

Fig. 2.1 Energy level diagrams of microwave-optical double resonance (from Ref. [1]). (a) corresponds to ground state MODR and (b) to excited state MODR. S and S_m are the laser and microwave transition rate constants. Levels in the lower state are in equilibrium with the heat bath.

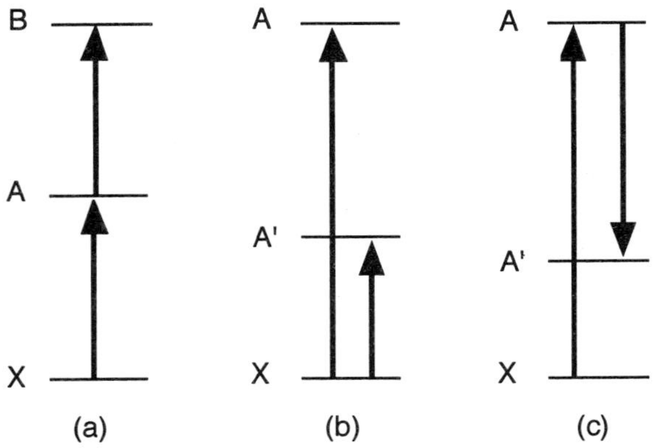

Fig. 2.2 Energy level diagrams of optical-optical double resonance, where (a) is a cascaded scheme corresponding to OODR, (b) a folded scheme corresponding to depletion spectroscopy, and (c) another folded scheme corresponding to stimulated emission pumping (SEP) spectroscopy.

The double resonance signal is observed as a change in the absorption rate of one of the radiation fields, which is kept on resonance (the pump), while the other radiation frequency is swept around the resonance frequency (the probe). The double resonance signal is observed as either an increase or decrease of the absorption when the second radiation frequency is tuned into resonance. The double resonance signal corresponds to a change in population of the level shared by the two radiations. Therefore, the double resonance signal is efficiently observed when at least one of the transitions is saturated by the radiation. Field *et al.* [1] have discussed the conditions for the double resonance signals to be observed, in which they distinguished two cases: the nonlinear optical pumping case, where the saturation effect is dominant, and the linear optical pumping case, where a polarization effect is more important. Here, we will reproduce their discussion.

As will be discussed later, the most popular detection scheme is to monitor the total fluorescence signal in the optical frequency region, and to detect the change in the fluorescence signal when the microwave transition is on resonance. If the optical pumping is strong enough to saturate the transition, the ground state population is depleted by the optical pumping and the excited state has a considerable population. In the case of a ground state double resonance, the microwave radiation partially replenishes the population of the depleted level in the ground state, and the total fluorescence rate increases. On the other hand, in the case of excited state double resonance, the microwave radiation depletes the excited state population prepared by the optical pumping, with the consequence that the optical pumping rate and the observed fluorescence rate increases further when the microwave radiation is on resonance, when fluorescence signals from both of the rotational states in the excited state are observed simultaneously.

These situations can be understood quantitatively by using a steady state rate equation approach. The rotational levels of the ground state are assumed to be in equilibrium with a heat bath with a characteristic relaxation time τ_g. The excited level population relaxes by spontaneous emission with a relaxation time τ_e. Usually, collision broadening and power broadening are both larger than the Doppler width in the microwave region, while the linewidth of the laser and the homogeneous broadening are both much smaller than the Doppler width in the optical region. Therefore, we have to consider the velocity dependent optical transition rate constant S and the velocity independent microwave transition rate constant S_m. They are

$$\begin{aligned} S &= (X^2/2)(\tau/\{1+[\omega(1-v/c)-\omega_{31}]^2\tau^2\}), \\ S_m &= (X_m^2/2)(\tau_m/\{1+(\omega_m-\omega_{21})^2\tau_m^2\}), \end{aligned} \tag{2.1}$$

where

$$\begin{aligned} X &= |\mu_{31}|E/\hbar, \\ X_m &= |\mu_{21}|E_m/\hbar, \\ \tau &\approx \tau_e, \end{aligned} \tag{2.2}$$

where μ is the electric dipole moment matrix element, $2E$ and $2E_m$ the peak to peak electric field strengths of the optical and microwave fields, v the component of velocity along the propagation direction of the laser, ω and ω_m the angular frequencies of the laser and microwave radiations, and ω_{31} and ω_{21} are the angular frequencies corresponding to the molecular energy level separations.

The rate equations corresponding to the ground state double resonance are

$$\begin{aligned} dn_1/dt &= S(n_3-n_1)+S_m(n_2-n_1)+(n_1^0-n_1)/\tau_g, \\ dn_2/dt &= S_m(n_1-n_2)+(n_2^0-n_2)/\tau_g, \\ dn_3/dt &= S(n_1-n_3)-(n_3/\tau_e), \end{aligned} \tag{2.3}$$

where n_i is the population per unit velocity of the ith level with velocity v and n_i^0 is the value of n_i in the absence of both of the radiation fields. The fluorescence intensity, which is used to monitor the occurrence of double resonance, is proportional to the total population of level 3 integrated over all velocities. If a Boltzmann distribution at a temperature T is assumed (in the case of an ordinary static cell experiment) and consequently the optical transition has a Doppler line profile, the steady state population N_3 is calculated to be

$$\begin{aligned} N_3 &= \frac{\pi^{1/2}X^2\tau_e}{2\omega(u/c)}\exp\left[-\left(\frac{\omega-\omega_{31}}{\omega\, u/c}\right)\right]\frac{N_1^0+(N_1^0+N_2^0)S_m\tau_g}{2S_m\tau_g+1} \\ &\quad\times\left[1+\tau X^2\frac{S_m\tau_g(\tau_g+2\tau_e)+\tau_g+\tau_e}{4S_m\tau_g+2}\right]^{-1/2}, \end{aligned} \tag{2.4}$$

where $u = [2k_BT/M]^{1/2}$. The value of the fractional change in fluorescence at resonance, $\Delta I/I = [I(\text{on}) - I(\text{off})]/I(\text{off})$, is derived from the above equation as

$$\Delta I/I = \left[\frac{(2x_m+1)(1+p)}{2x_m+1+p\{x_m(\tau_e+2\tau_g)/(\tau_e+\tau_g)+1\}}\right]^{1/2} - 1, \tag{2.5}$$

where we defined two dimensionless parameters, $x_m = S_m\tau_g$ and $p = \tau(\tau_g + \tau_e)X^2/2$, which are measures of saturation for microwave and optical transitions, respectively. In the strong microwave and optical pumping limit ($x_m >> 1$ and $p >> 1$), the fractional change, $\Delta I/I$, approaches a limiting value of about 40% when $\tau_g >> \tau_e$. On the other hand, it is only 15% when $\tau_g = \tau_e$. Since the ground state relaxation time is governed by collisions, it can be controlled by the sample pressure. It is thus desirable to reduce the sample pressure to satisfy the condition, $\tau_g > \tau_e$. However, it is not always possible in practice to attain the strong pumping limit for both microwave and optical transitions. Furthermore, if the microwave transition is strongly saturated, the linewidth of the double resonance signal becomes so broad that the high resolution attainable by MODR cannot be realized. Figure 2.3 shows plots of the fractional change as function of the microwave saturation parameter, x_m, for the cases $p = 0.1$, 0.3, and 1, with $\tau_g >> \tau_e$ and $\tau_g = \tau_e$. It is not difficult to realize the condition $p = 1$ for most electronic transitions when the laser output of about 100 mW is available, and a 1% change of the fluorescence signal is easily detectable. It is much easier to increase microwave power density to attain the near saturation condition, $x_m = 1$, since it is not difficult to obtain a microwave source with high enough power. Use of a microwave cavity would further enhance the power density if necessary. In fact, as will be shown later, even magnetic dipole transitions of an open-shell species have been observed by MODR, where the line strength is about 10^{-4} times smaller than the usual electric dipole allowed transitions.

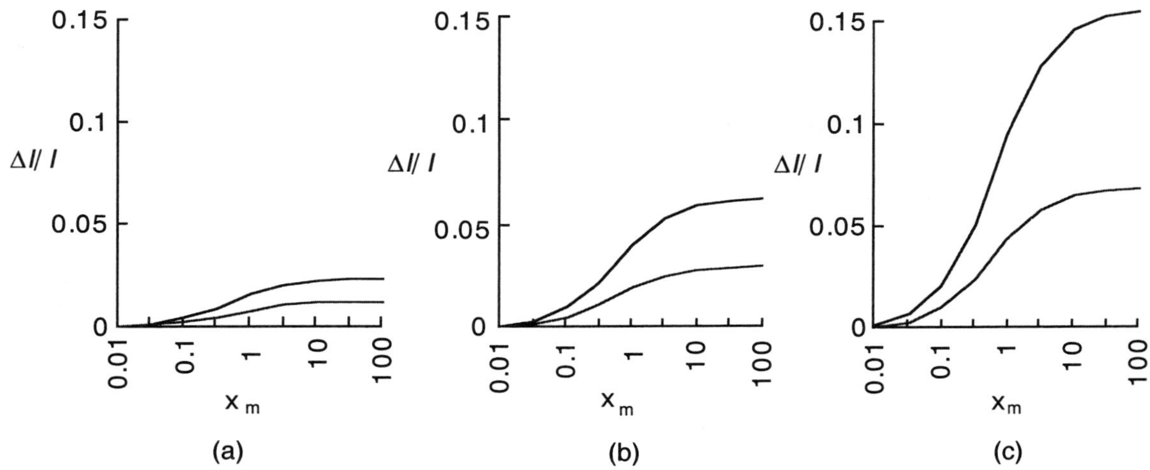

Fig. 2.3 Fractional change of population plotted against the microwave pumping parameter, x_m, for the cases, (a) optical pumping parameter $p = 0.1$, (b) $p = 0.3$, and (c) $p = 1$. In each plot, the upper trace corresponds to the case $\tau_g >> \tau_e$, and the lower $\tau_g = \tau_e$.

The estimate made so far for the MODR signal assumes a static cell condition at temperature high enough to satisfy the condition, $k_BT >> \hbar\omega_{12}$, with the ground state relaxation time constant of order of μs. If the MODR setup is combined with a supersonic molecular beam technique, situations very different from the one given above can be realized. In the molecular beam, since there are essentially no collisions, the optical pumping can deplete the ground state population almost completely, and a stronger MODR signal is obtained. A detailed explanation for this configuration will be given in a following subsection.

Conditions for excited state double resonance can be obtained similarly, where the corresponding rate equations are

$$\begin{aligned} dn_1/dt &= S(n_3 - n_1) + S_m(n_2 - n_1) - (n_1/\tau_e), \\ dn_2/dt &= S_m(n_1 - n_2) - (n_2/\tau_e), \\ dn_3/dt &= S(n_1 - n_3) + (n_3^0 - n_3)/\tau_g. \end{aligned} \tag{2.6}$$

The optical radiation directly pumps population into level 1, while level 2 is populated by microwave transitions from level 1. The fluorescence signal is proportional to the sum of the total populations of levels 1 and 2. The fractional change in the fluorescence when microwave is on and off is

$$\begin{aligned}\Delta I/I &= \frac{N_1 + \alpha N_2}{N_1(S_m = 0)} - 1 \\ &= \frac{1+(1+\alpha)S_m\tau_e}{(2S_m\tau_e+1)^{1/2}}\left[\frac{1+\dfrac{\tau X^2}{2}(\tau_e+\tau_g)}{2S_m\tau_e+1+\dfrac{\tau X^2}{2}\{S_m\tau_e(\tau_e+2\tau_g)+\tau_e+\tau_g\}}\right]^{1/2} - 1,\end{aligned} \tag{2.7}$$

where N_1 ($S_m = 0$) is the population of level 1 in the absence of the microwave radiation, and N_1 and N_2 are the populations in the presence of microwave radiation. The parameter α is the detection efficiency of fluorescence from level 2 relative to that from level 1. Thus, for example, if one employs a rotationally resolved fluorescence signal detecting only fluorescence from level 1 to monitor the double resonance signal, $\alpha = 0$. If, on the other hand, rotationally-unresolved total fluorescence is used, α is approximately 1. By using the saturation parameters for microwave, $x_{me} = S_m\tau_e$, and optical pumping, p, as have been given above, the fractional change in fluorescence on resonance can be expressed as

$$\Delta I/I = \frac{1+(1+\alpha)x_{me}}{(2x_{me}+1)^{1/2}}\left[\frac{1+p}{2x_{me}+1+p\{x_{me}(\tau_e+2\tau_g)/(\tau_e+\tau_g)+1\}}\right]^{1/2} - 1. \tag{2.8}$$

If $\tau_g >> \tau_e$, the above expression is simplified to

$$\Delta I/I = \frac{1+(1+\alpha)x_{me}}{2x_{me}+1} - 1, \tag{2.9}$$

so that there is no MODR signal when $\alpha = 1$, which is quite different from the situation for the ground state MODR. This is because the $\tau_g >> \tau_e$ assumption requires that the rotational relaxation in the ground state cannot keep up with the optical depletion. If $\tau_g = \tau_e$, the expression becomes

$$\Delta I/I = \frac{1+(1+\alpha)x_{me}}{(2x_{me}+1)^{1/2}}\left[\frac{1+p}{2x_{me}+1+p\{3x_{me}/2+1\}}\right]^{1/2} - 1. \tag{2.10}$$

In this case, a double resonance signal can be observed even when $\alpha = 1$. At the strong microwave and optical pumping limit, a fractional change of about 15% is obtained for $\alpha = 1$. At the saturation point, where $x_{me} = 1$ and $p = 1$, this is reduced to about 4.4%. In the case of excited state double resonance, microwave saturation is attained when $x_{me} = S_m\tau_e = 1$, where τ_e is the excited state lifetime, which is much shorter than the ground state relaxation time τ_g under normal experimental conditions, so that higher microwave power is required to saturate the microwave transition.

2.1.2 Polarization effect

Even in the case of the linear optical pumping limit, where the saturation effect of optical pumping is negligible, excited state double resonance could be observed due to a polarization effect. Molecular fluorescence emitted with laboratory-fixed Z polarization propagating in the XY plane is proportional to the time-averaged squared projection of the molecular transition moment along the Z-axis, $\langle\mu_Z^2\rangle$. The present polarization effect is caused by the change of $\langle\mu_Z^2\rangle$ due to the microwave transitions in the excited state.

Here, we consider only $^1\Sigma$ - $^1\Sigma$ electronic transitions of a linear molecule for simplicity. In this case the rotational angular momentum $\boldsymbol{J}$ is perpendicular to the molecular axis, and the projection of $\boldsymbol{J}$ on the space-fixed Z axis is M. If we define the angle θ to be that between $\boldsymbol{J}$ and Z and ϕ to be the angle of rotation about $\boldsymbol{J}$ by which μ is rotated out of the plane defined by Z and $\boldsymbol{J}$, then the projection of μ on Z is

$$\mu_Z = \mu\cos[(\pi/2) - \theta]\cos\phi$$
$$\phi = \Omega t \tag{2.11}$$

where Ω is the molecular rotation frequency. If we average over ϕ, as the rotation period is much shorter than the radiative lifetime, it follows that

$$<\mu_Z{}^2(\theta)> = (1/2\pi)\int_0^{2\pi} \mu^2 \sin^2\theta \cos^2\phi d\phi. \tag{2.12}$$

Because

$$\cos\theta = M/[J(J+1)]^{1/2}, \tag{2.13}$$

the expression may be further simplified to

$$<\mu_Z^2(M)> = \frac{\mu^2}{2}[1 - M^2/J(J+1)]. \tag{2.14}$$

If the molecule with J has a distribution of M levels denoted as $f(M)$, the average squared transition dipole moment is

$$<\mu_Z^2> = \frac{\mu^2}{2}\sum_{M=-J}^{J} f(M)[1 - M^2/J(J+1)]. \tag{2.15}$$

When there is a strong microwave field with polarization along Z, which saturates the (J, M) - (J+1, M) microwave (MW) transitions for the excited molecules, the average squared transition dipole moment becomes

$$\begin{aligned} <\mu_Z^2>_{\mathrm{MW}} &= \frac{\mu^2}{2}\sum_M f(M)\left[1 - \frac{1}{2}\frac{M^2}{J(J+1)} - \frac{1}{2}\frac{M^2}{(J+1)(J+2)}\right] \\ &= \frac{\mu^2}{2}\sum_M f(M)\left[1 - \frac{M^2}{J(J+2)}\right]. \end{aligned} \tag{2.16}$$

Thus the difference of the average squared transition dipole moment when microwave field is on and off is

$$\begin{aligned} \Delta <\mu_Z^2> &= <\mu_Z^2>_{\mathrm{MW}} - <\mu_Z^2> \\ &= \frac{\mu^2}{2}\sum_M f(M)\frac{M^2}{J(J+1)(J+2)} \end{aligned} \tag{2.17}$$

There is a net increase of $<\mu_Z{}^2>$ when a strong microwave field is applied connecting J, J+1 levels. The microwave-induced change of $<\mu_Z{}^2>$, *viz* $\Delta<\mu_Z{}^2>$, is compensated by changes of $<\mu_X{}^2>$ and $<\mu_Y{}^2>$ such that

$$\Delta <\mu_Z^2> = -\Delta <\mu_X^2> - \Delta <\mu_Y^2>. \tag{2.18}$$

There is no net change in $<\mu^2>$. If fluorescence is observed in a direction in the XY-plane without polarizers, the net change in fluorescence is $\Delta<\mu_Z{}^2>/2$. If the J - (J-1) transition is saturated, the resultant change is opposite in sign to that for the J - (J+1) transition.

In the simplest case, if we put $f(M) = (2J+1)^{-1}$ the fractional change at microwave resonance is

$$\begin{aligned} \Delta <\mu_Z^2>/<\mu_Z^2> &= \sum_M M^2(J+2)^{-1}/\sum_M[J(J+1) - M^2] \\ &= 1/[2(J+2)]. \end{aligned} \tag{2.19}$$

When a microwave transition with $J' = 2 - 3$ is saturated, the net change of fluorescence observed in a direction in the XY-plane is about 6%.

As has been shown above, a large change of fluorescence yield is observed for ground state double resonance when both microwave and optical transitions are near saturation. On the other hand, in the case of linear optical pumping, the polarization effect is much smaller for ground state double resonance than for excited state double resonance since the change of population is very small when $\hbar\omega << k_B T$. It is thus quite difficult to detect ground state double resonance using the polarization effect and detecting a change of total fluorescence. However, the polarization effect for ground state double resonance can be detected as a polarization rotation effect of transmitted optical light. It is possible to construct quite a sensitive detection system by using crossed polarizers to monitor the polarization effect due to ground state microwave resonance. Such experiments have been done by Ernst and Törring [2] as MOPS (Microwave Optical Polarization Spectroscopy).

2.1.3 Detection schemes

LIF detection The first successful MODR experiment was done by using a fixed frequency Ar ion laser for the optical pumping [3]. However, all subsequent MODR experiments have been performed by using tunable lasers, especially CW dye lasers. The experimental configuration for an early double resonance experiment performed by Field *et al.* [1], who observed MODR signals of BaO, is shown in Fig. 2.4, where the fluorescence signal was detected at a right angle to the incident laser beam. The microwave beam irradiated the region where fluorescence signal was detected. In the case of the example shown in the figure, a microwave horn pointing toward the fluorescence region was used. For production of BaO in the cell, Ba metal was evaporated in an oven and the vapor was reacted with oxygen. They observed double resonance signals both in the ground state and in the excited state as changes of the total

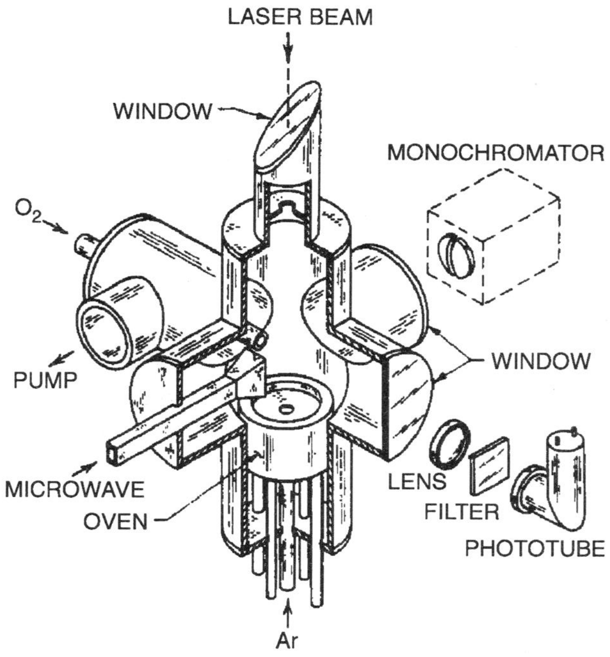

Fig. 2.4 The double resonance cell used for the study of BaO (from Ref. [1]).

fluorescence signals. Up to 5% change of the fluorescence signal was observed when the microwave frequency was on resonance, suggesting that sufficient saturation for the optical transition was obtained in the experiment with about 50 mW of CW dye laser output. They also observed the polarization effect for the excited state double resonance.

Most of the subsequent MODR experiments followed this basic configuration, using tunable CW dye lasers and monitoring a change of fluorescence intensity. Except for the experiments on NO_2, which were done shortly after the successful MODR experiment on BaO, most of the experiments investigated unstable molecules that should be produced in a fast flow system just before the double resonance cell, because CW lasers used for the MODR experiments oscillates only in the visible region, and molecules having an electronic transition in the visible region are mostly unstable molecules. Therefore, various different designs of double resonance cells have been proposed, in which the fluorescence observation system was combined with efficient production systems of the reactive molecules to be studied. For example, Takagi *et al.* used a crossed waveguide for the double resonance cell in their study of HNO [4]. Usually, metal surfaces should be avoided since most reactive species are destroyed on metal surfaces. However, HNO was not destroyed by the metal surfaces while undesirable chemiluminescence was quenched on the metal surfaces. Takagi *et al.* also made a Stark effect measurement in the excited electronic state by inserting a Stark plate in the double resonance cell [5]. The same system was used to observe MODR spectra of H_2CS by Suzuki *et al.* [6].

If a more intense microwave field is required to observe the double resonance signals, it is possible to incorporate a microwave cavity into the double resonance cell. Curl and coworkers [7-14] used a microwave cavity for their MODR experiments on NH_2. The MODR cell they used is shown in Fig. 2.5, where a cavity in the X-band region was used to obtain a strong microwave field [10]. With this configuration, they were able to observe very weak magnetic dipole transitions in the microwave region for both the ground and excited states. Suzuki *et al.* used a Fabry-Perot cavity for the observation of the double resonance signals of CCN in the mm-wave region [15]. There are two advantages in using the Fabry-Perot cavity. One is that it is easier to saturate the microwave transition due to the large microwave field in the cavity. The other advantage is that it is rather easy to incorporate a high speed flow system by utilizing a large volume of the cavity in order to efficiently produce short lived reactive species.

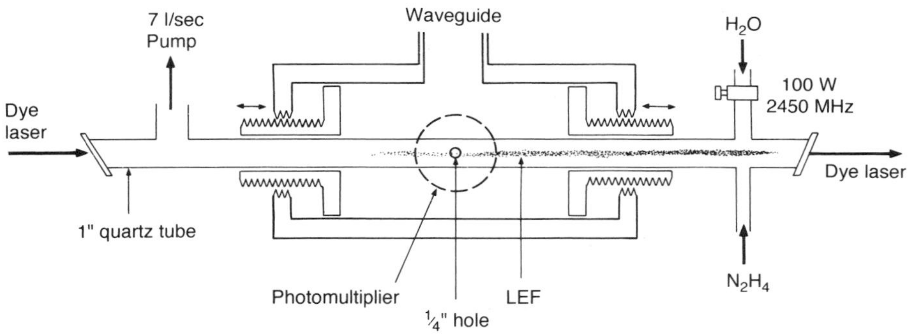

Fig. 2.5 The microwave cavity and the flow system used for the MODR experiment of NH_2 (from Ref. [10]).

In all the examples mentioned above, an intensity change of the fluorescence signal, when the microwave radiation is on and off, was detected. Usually, the laser frequency is fixed at a resonant position and the frequency of the microwave radiation is swept while monitoring whether the double resonance signal is observed or not. In order to increase the sensitivity, the microwave radiation is either amplitude or frequency modulated, and the fluorescence signal is detected by a phase-sensitive detector using the modulation frequency as a reference. It is not difficult to detect a change of fluorescence intensity of less than 1%.

Microwave-optical polarization spectroscopy Ernst and Törring proposed microwave-optical polarization spectroscopy (MOPS) [2]. This method detects the transmitted light, instead of the fluorescence signal, by using a pair of crossed polarizers; one polarizer is inserted before the absorption cell and another after the absorption cell, as shown in Fig. 2.6. When there is no resonant microwave radiation, the transmitted light is almost zero due to the crossed polarizers. When there is a resonant microwave radiation, which is circularly polarized or linearly polarized with its polarization rotated 45° with respect to the optical beam polarization, a small change of polarization of the transmitted beam results, and an increase in the detected signal is observed. They demonstrated the applicability of the method by observing double resonance signals of BaO and claimed that the method was more sensitive than conventional MODR. Later, they also demonstrated that even excited state double resonance could be observed by this technique [16].

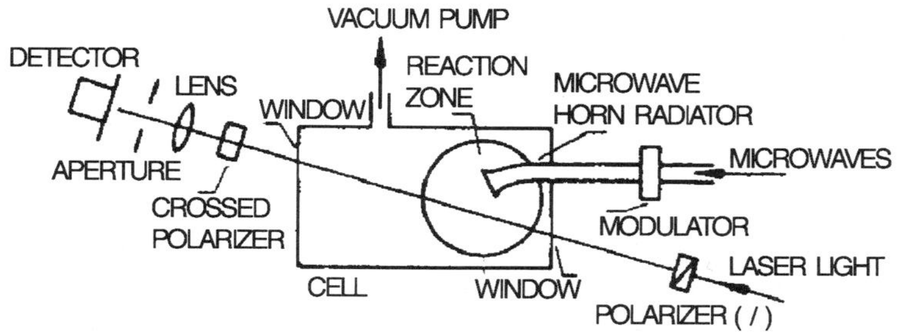

Fig. 2.6 The experimental setup for MOPS. Transmitted optical light after two crossed polarizers was detected (from Ref. [2]).

Molecular beam MODR Another modification of the conventional MODR setup is to combine the method with the molecular beam technique [17,18]. This method is similar to molecular beam electric resonance in the radio-frequency and microwave regions, where laser beams are used instead of the focusing fields. A typical experimental setup is shown in Fig. 2.7 [19]. The optical radiation is divided into two; one for pump and the other for probe. The pump beam irradiates the molecular beam at a point

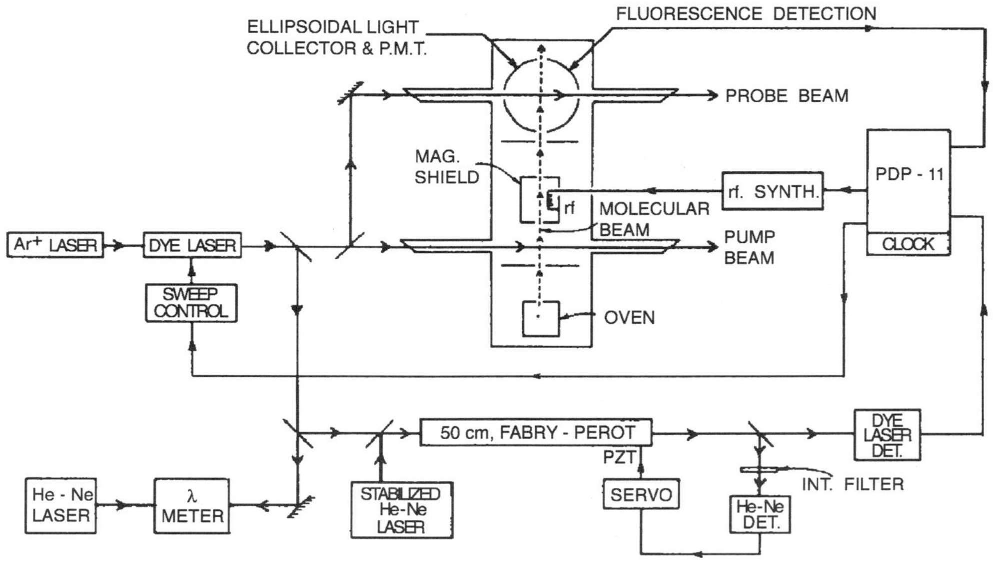

Fig. 2.7 The experimental setup of the molecular beam mw (rf) optical double resonance spectrometer (from Ref. [19]).

closer to the beam source. The pump beam depletes the ground state population of a level which is resonant with the optical radiation. The probe beam then samples the same level at a point further from the beam source, where the fluorescence signal excited by the probe beam is observed. There is practically no fluorescence signal when the pump beam completely depletes the ground state population. The pump and probe beams are separated sufficiently so that the region between the two beams can be used to partially restore the depleted population, by irradiating with microwave or radio-frequency radiation to induce a transition terminating on the depleted level. When the microwave transition is on resonance, the depleted ground state population is partially restored by the microwave transition, resulting in an increase in the fluorescence signal excited by the probe beam.

One advantage of this method is that, when the saturation (depopulation) due to the pump beam is complete, there is no background fluorescence. In addition, since the method is combined with a molecular beam, it is rather easy to saturate optical transitions by a pump beam with a relatively small power, because there are no significant collisions in the molecular beam. The method also provides extremely high resolution, which is even higher than that of conventional microwave spectroscopy. As will be described in examples in a later section, the typical spectral resolution attained by this method is on the order of 10 kHz. On the other hand, the typical linewidth is as much as a few MHz in the case of an ordinary MODR configuration, which is much wider than that of microwave spectroscopy. One serious disadvantage of this method is that it is not possible to observe excited state double resonance, since excited state population relaxes by fluorescence completely during the flight time between the pump and probe beams. The extremely high resolution obtained by this method is thus limited to ground state rotational transitions. However, it is possible to utilize the method to select specific transitions among complicated and overlapped optical transitions by labeling the ground state levels when the microwave frequency is fixed to a transition and the optical frequency is scanned. Furthermore, this method gives Doppler free spectra for the optical transitions, because a collimated molecular beam is used to excite optical transitions. Another advantage is that it is easier to observe microwave transitions with a very small transition moment, when sufficient microwave power is supplied to the repopulation region. Even magnetic dipole transitions can be observed by this method.

The method has been extensively applied to double resonance studies of metal containing molecules, because the beam method is easily combined with vaporization sources such as high temperature ovens to evaporate metals or metal containing molecules. In a recent paper, Fletcher *et al.* have incorporated a laser ablation source to obtain a molecular beam of refractory species [20].

Microwave detection using a pulsed optical laser Lehmann and Coy introduced a completely different MODR technique, in which changes of microwave absorption are observed when the microwave frequency is fixed at a rotational transition and an optical laser is scanned to observe double resonance signals [21, 22]. In their experiment, a pulsed dye laser was used as the optical pumping source. When the frequency of the pulsed dye laser is on resonance with an optical transition involving a level monitored by microwave absorption, a transient population change is induced, causing a transient microwave nutation signal. The transient power change is either absorptive or emissive depending on whether the optical transition pumps either the upper or lower level of the microwave transition. Since the change of microwave absorption is observed, the method requires a double resonance cell with a relatively long path length to obtain sufficiently high sensitivity. Furthermore, the optical laser should efficiently pump the whole active volume of the microwave absorption cell. In contrast to other MODR techniques, the microwave frequency is held fixed at a known transition frequency while the optical frequency is scanned to observe the MODR signal. This method is thus mainly used to simplify complicated optical spectra by using microwave transitions to select specific levels. This technique can be applied to the studies of molecules which have no fluorescence, since a double resonance signal is detected by a change in the microwave absorption. This method was thus first applied to the observation of high overtone bands of ammonia in the optical region [21]. A block diagram of the experimental setup is shown in Fig. 2.8, where the double resonance cell was arranged in a microwave bridge to detect a small change in microwave intensity, where relatively strong microwave radiation was fed to the double resonance cell. Later, it was shown that excited state double resonance signals could be observed by scanning the microwave frequency while the laser frequency was held fixed at an optical transition [23] where, although an experimental setup similar to that used by Lehmann and Coy was used, the microwave bridge was not used.

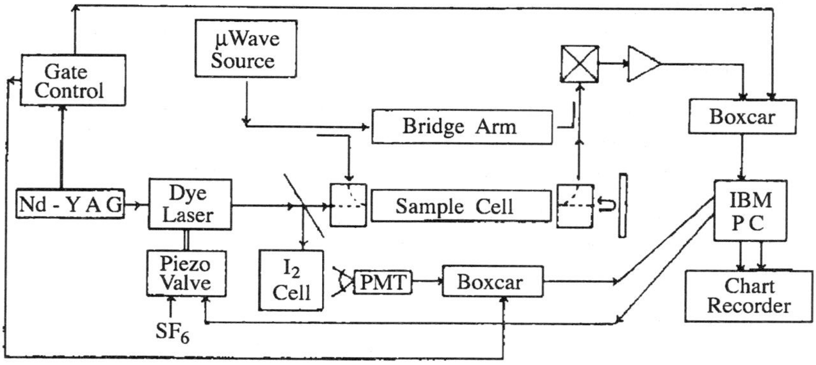

Fig. 2.8 The experimental setup for microwave-detected microwave-optical double resonance (from Ref. [22]).

2.2 Microwave-optical double resonance (MODR) spectroscopy

In this section we will present typical examples of MODR spectra recorded using optical lasers. Here we will not include MW-IR double resonance experiments, which were also done fairly extensively in the early stages of double resonance studies with a primary goal to elucidate the detailed mechanism of double resonance. However, the spectroscopic information obtained by MW-IR double resonance is rather limited unless specialized techniques are used. MW-IR double resonance is now mainly used to obtain spectroscopic data with extremely high resolution on the order of a few kHz by using such specialized techniques.

As has been mentioned, the first MODR experiment applied to a molecule by monitoring an electronic transition in the optical region was done with a CW Ar ion laser [3]. One of the Ar ion laser lines was accidentally on resonance with a BaO A - X transition. Soon after the first experiment using a fixed frequency laser, another experiment using a tunable dye laser was successfully performed [1]. This study proved that MODR was a powerful experimental technique for obtaining detailed spectroscopic information on various molecules. Since the frequency coverage of CW dye lasers with sufficiently high power to saturate the optical transitions is limited to the visible region, molecules subjected to the double resonance experiments must have electronic transitions in the visible region. Thus, molecules to be studied by MODR are limited. In this respect, diatomic molecules containing a metal atom are good candidates for MODR studies, and this is a reason why the first successful experiment was done on BaO, which has strong absorption in the visible region. Another molecule studied rather extensively in the early stages of MODR experiments is NO_2, which is a stable molecule having absorption in the visible region.

2.2.1 MODR studies of free radicals

There are two motivations to study free radicals by MODR. One is that many free radicals with open shell electronic configurations have strong electronic transitions in the visible region. Another motivation is that the rotational structures of free radicals are usually more complicated than those of closed shell singlet molecules, because they have fine and hyperfine structures. Usually, it is not possible to resolve the hyperfine splittings by ordinary Doppler-limited spectroscopy. On the other hand, except for the examples of NO_2, which has been studied at an early stage of MODR experiments, most of the free radicals are quite reactive and short-lived. It is thus required to combine efficient free radical production methods with appropriate double resonance cells, compatible with both optical and microwave radiations.

The visible absorption spectrum of NO_2 is quite complicated because there are many excited vibronic states which interact with each other, preventing definite rovibronic assignments for the observed features. The first MODR experiment on NO_2 was reported by Solarz and Levy [24] by using a 488.0 nm Ar ion laser. They observed an excited state rotational transition in the 9 GHz region. However, they were unable to give a definite assignment for the observed rotational transition in the excited electronic state. Tanaka *et al.* [25] reported an observation of ground state double resonance by using a tunable CW dye laser. They fixed the microwave frequencies to known fine and hyperfine components of a ground state rotational transition, $10_{0,10}$ - $9_{1,9}$, and observed several double resonance signals near 593.6 nm. They were able to assign the rotational quantum numbers of the excited electronic state observed by double resonance. In subsequent papers [26, 27], they were able to observe the excited state double resonance, where the microwave transition $9_{0,9}$ - $8_{1,8}$ in the excited state was observed. They were also able to resolve the spin-rotation splittings and the nitrogen hyperfine splittings. From the pressure dependence of the observed double resonance signals, they found that the $8_{1,8}$ level had considerably lower fluorescence yield than $9_{0,9}$. They also observed extra transitions in the upper state microwave spectrum and assigned the state thus observed to a level which perturbs $8_{1,8}$ in the excited state, causing the low quantum yield for that level. In a later paper, Weber *et al.* measured excited state *g*-values by using optical radiofrequency double resonance [28]. Lehmann and Coy used the microwave detected double resonance technique to study the excited state character of NO_2 [29]. Details will be given in a later section.

NH_2 is one of the important free radicals in reaction chemistry. This radical is also interesting as a typical species showing a strong Renner-Teller effect, which splits the orbitally degenerate ground electronic state into two distinct electronic states and causes the equilibrium ground state structure to be bent. There exists the first excited electronic state in the visible region, which corresponds to this Renner-Teller splitting, and electronic transitions between these states have been observed. Curl and coworkers observed MODR spectra of NH_2 [7-14] utilizing this visible electronic transition. Since the NH_2 radical is a reactive species, a fast flow reaction system had to be incorporated into the MODR system. They used a microwave cavity to obtain a strong microwave field, which was tunable from 6 to 12 GHz. The NH_2 radical was produced by reacting 2450 MHz discharge products in water with anhydrous hydrazine. In their first report, they observed the ground state rotational transition, $5_{2,3}$ - $6_{1,6}$, $J = 11/2$ - $13/2$, at 9 GHz, where they were able to resolve the hyperfine splittings [7]. They reported an observation of the other spin component of the $5_{2,3}$ - $6_{1,6}$ transition in a subsequent paper. However, this $5_{2,3}$ - $6_{1,6}$ rotational transition is the only rotational transition accessible in the cm wave region because the NH_2 radical has large rotational constants. It is thus not possible to derive detailed molecular constants using only one rotational transition. In order to determine detailed molecular constants for this radical, they observed magnetic dipole transitions by the MODR technique. As they used a microwave cavity and a microwave power amplifier, they were able to obtain a sufficiently intense microwave magnetic field to observe double resonance signals by using magnetic dipole transitions. They determined accurate fine and hyperfine coupling constants both for the ground state and the excited electronic state $\tilde{A}^2A_1(0,10,0)$. They also observed weak collision induced signals and assigned them to be transitions in the $\tilde{X}^2B_1(0,12,0)$ state. In a later paper, they observed MODR signals for NH_2 $\tilde{A}(090)$ and for the NHD species [13, 14].

Takagi *et al.* studied HNO by MODR [4]. This species also has a strong electronic transition in the visible region. Although it is a short lived reactive species, its electronic states are singlet in both the $\tilde{X}$ and $\tilde{A}$ states. They produced HNO by reacting discharged products of oxygen with a mixture of propylene and NO. The reaction products were directed into a copper MODR cell made of a waveguide. They observed excited state double resonances and determined precise rotational constants for the excited states, $\tilde{A}^2A''(100)$ and (020). They found small perturbations in the excited state rotational structure. Some of the perturbations were explained by a Coriolis interaction between the (100) and (020) vibronic states. Furthermore, they were able to observe microwave transitions directly connecting these two vibrational levels. In a later paper, they inserted an electrode into the double resonance cell and observed the Stark effect to determine the dipole moment in the excited $\tilde{A}^1A''(100)$ state [5]. They also observed a small K_a dependence of the dipole moment.

H_2CS is isovalent to formaldehyde and has a vibronically induced electronic transition in the visible region. Suzuki *et al.* studied its $\tilde{a}^3A_2$ state by using MODR [6]. Different from the case of H_2CO, H_2CS

had to be produced in a flow system since it is an unstable molecule. They produced the species by thermally decomposing trimethylene sulfide in a flow system. They employed waveguide cells of various sizes for the double resonance experiment, depending on the frequency regions observed. The frequency range they covered extended from a radiofrequency region below 100 MHz to a mm-wave region up to 86 GHz. They were able to resolve fine structure splittings in the excited $\tilde{a}^3A_2$ state, as well as the proton magnetic hyperfine splittings. The MODR data were combined with extensive laser induced fluorescence data, leading to precise molecular constants for the excited state including the fine and hyperfine coupling constants. In a subsequent paper, they determined the dipole moments of the excited $\tilde{a}^3A_2$ and $\tilde{A}^1A_2$ states by using MODR [30]. Petersen and Ramsay performed an MODR study of the $4^1{}_0$ vibronic level of the $\tilde{A}^1A_2$ state and found many microwave transitions in addition to those expected for the normal rotational transitions in the 4^1_0 level [31, 32]. They attributed the unusual transitions to those between levels of the excited state and high rovibronic levels of the ground state. They found that the number of such transitions increased almost linearly with rotational quantum number *J*. They also examined the Zeeman effect for such extra transitions and found that one of the extra transitions showed an anti-crossing with a level in the $\tilde{a}^3A_2$ state [33].

An MODR experiment on the CCN radical in the $\tilde{A}^2\Delta(000)$ state was reported by Suzuki *et al.* [15], where a Fabry-Perot cavity was used to observe mm-wave optical double resonance. They also observed rf-optical double resonance below 100 MHz by using a transmission type radiofrequency cell made of a X-band waveguide. They were able to observe excited state double resonance, yielding a precise determination of the magnetic and electric quadrupole hyperfine coupling constants for the excited state. However, ground state double resonance could not be observed, although the observation was tried carefully. The result was attributed to the fact that the ground state permanent dipole moment is much smaller than that of the excited state. In fact, the pure rotational spectrum of CCN in the ground electronic state was observed only quite recently using a Fourier-transform microwave spectrometer [34]. Suzuki *et al.* also studied a similar free radical, NCO, by using the MODR and intermodulated fluorescence techniques, where the latter technique uses two counter propagating laser beams to observe Doppler-free spectra by using a saturation effect. MODR signals were observed using a similar system as used in the CCN study [35]. Although the linewidths obtained by double resonance are of the order of 10 MHz in their study, the accuracy of MODR was about 10 times better than that obtained by the intermodulated fluorescence technique.

2.2.2 Metal containing molecules

As the early successful experiments on BaO had demonstrated, metal containing molecules are interesting and suitable species studied by MODR, since there are many metal containing molecules which have relatively strong absorption in the visible region. The metal containing molecules so far studied are mostly metal halides and metal oxides. Various MODR schemes have been applied to studies of such metal containing molecules; microwave-optical polarization spectroscopy and the molecular beam method were both successfully applied. Here we will review some representative results.

Among the metal containing molecules, BaO has strong absorption in the visible region and was subjected to various optical laser spectroscopic studies. Field *et al.* reported the first successful application of MODR for the BaO $A^1\Sigma$ - $X^1\Sigma$ transition by using a fixed frequency Ar ion laser, as mentioned above [3]. This study was followed by another paper reporting an MODR study using a tunable dye laser [1]. In the latter paper, they presented a detailed mechanism for MODR by using rate equations for the three levels involved in the double resonance. They successfully observed both the ground and excited state double resonances. For excited state double resonance, they observed spectra for $v' = 0$ to 7 and obtained accurate rotational constants for these vibronic states, which were found to be subjected to strong perturbations. Accurate results obtained by MODR were of great help to analyze the perturbations.

Molecules with open shell electron configurations are more interesting for MODR studies, since there are fine structures and magnetic hyperfine splittings which give valuable information on the bonding nature of these molecules. In this respect, alkaline earth metal halides are more interesting than the oxides, since the former species have $^2\Pi$, $^2\Sigma$ - $^2\Sigma$ electronic transitions in the visible region, while the latter species have mostly $^1\Sigma$ - $^1\Sigma$ electronic transitions, as is the case for BaO. Nakagawa *et al.* observed the MODR spectrum of CaF $A^2\Pi_r$ - $X^2\Sigma$ [36]. They observed ground state double resonance for this molecule. Although they could resolve the spin doublings in the ground electronic state, they could not resolve the hyperfine splittings for the chlorine nucleus which has the nuclear spin quantum number I = 3/2, because the half-widths of the MODR signals were as large as 20 MHz. Childs and Goodman applied the laser-rf double resonance technique in a molecular beam and thereby reduced the linewidth of the double resonance signals, claiming a precision of 1 kHz or better [37, 38]. They obtained precise values for the ground state fine and hyperfine constants [39]. A similar laser-rf double resonance study in a molecular beam was reported on an analogous molecule, CaCl, by using the $B^2\Sigma$ -$X^2\Sigma$ transition [19, 40]. Because the pump-probe molecular beam method was used, only the ground state double resonance could be observed. However, double resonance was used as a means to label the hyperfine unresolved optical transitions and estimated an upper limit for the hyperfine splittings in the excited state [19]. Ernst *et al.* applied the MOPS technique to investigate the CaCl molecule, where hyperfine splittings were completely resolved with linewidth between 1 and 2 MHz, by which they claimed a resolution much higher than that obtained by the ordinary MODR scheme [41]. They also applied the molecular-beam MODR method, where they determined the ground state permanent dipole moment by applying electric fields in the double resonance [42, 43]. The next member of the Ca halides, CaBr, was studied by Childs *et al.* by the molecular beam laser-rf double resonance method, yielding a precise determination of the ground state fine and hyperfine coupling constants [44, 45]. Laser-rf double resonance spectroscopy of SrF has been reported by Childs and his coworkers [46], and three isotopomers, ^{86}Sr, ^{88}Sr, and ^{87}Sr, have also been studied [47].

After rather extensive studies for the alkaline earth halides, molecules containing transition metals were subjected to the MODR studies. In general, they have even more complicated electronic structures, often with multiplicity higher than doublet. MODR is a method best suited for investigating such systems, since it not only gives high enough resolution to resolve fine and hyperfine splittings but also helps to assign extremely complicated spectra. Gerry *et al.* successfully applied the double resonance technique to investigate the ground state of CuO, $X^2\Pi$ [48]. In addition to the accurate ground state rotational constants and fine structure constants, they obtained accurate values for the hyperfine coupling constants for ^{63}Cu, and discussed the bonding character of CuO on the basis of these results. This work was followed by Steimle *et al.*, who used a tri-plate transmission line to introduce low frequency microwave radiation to observe Λ-doubling transitions, in addition to the mm-wave pure rotational transitions [49]. They obtained an improved set of hyperfine coupling constants.

A number of other metal-containing diatomics have been studied by the double resonance techniques, mostly combined with the molecular beam method [52-56]. Thus, most of the metal containing molecules studied so far by MODR were diatomics. Triatomic species, SrOH [19] and CaOH [57], have also been observed by the molecular beam double resonance technique, where the laser ablation method was successfully applied to produce these species in the molecular beam.

2.2.3 Application of microwave detected double resonance

Microwave detected double resonance spectroscopy has been applied to molecules completely different from those studied by ordinary MODR and molecular beam double resonance, since the method can be applied to molecules which do not fluoresce or do not have electronic transitions in the visible region. In fact, Lehmann and Coy applied the method to observe overtone transitions of ammonia, $5\nu_{N\text{-}H}$ and $6\nu_{N\text{-}H}$ in the visible region by using a pulsed dye laser [21, 22], where double resonance signals were observed by the microwave frequency fixed at an inversion splitting while the laser frequency was scanned. An example of the observed double resonance signals is shown in Fig. 2.9. Definite rotational assignments for most of the features in the overtone bands were obtained by using the double resonance technique. In these overtone bands, a large number of vibrational states are coupled with each other showing strong Fermi resonances. Lehmann and Coy showed that the band structure can be qualitatively explained on the basis of a local mode description.

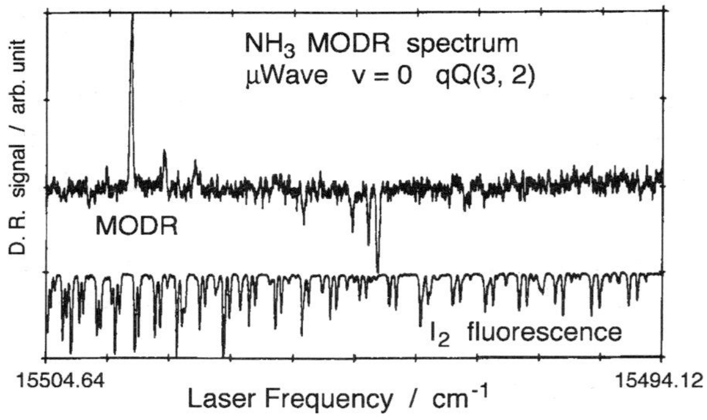

Fig. 2.9 An example of the microwave-detected MODR spectrum in $5\nu_{NH}$ of NH_3 (from Ref. [22]). The lower trace is an I_2 spectrum used for frequency calibration.

Matsuo *et al.* [23, 58] observed the excited state double resonance of the five quanta NH stretching band by using a similar experimental setup, where they determined the inversion doubling frequencies of many rotational levels in the excited state. They found that the linewidths of the double resonance signals were about 4 MHz when the mm-wave frequency was scanned. The observed linewidths were well explained by considering pressure broadening, the width due to a limited observation time (100 ns), and saturation broadening. No evidence for extra relaxation mechanisms in the five quanta excited state was observed.

As mentioned, Lehmann and Coy applied the technique to study the excited state character of NO_2 [29]. They fixed the microwave frequency at 40661.44 MHz observing a fine and hyperfine component of the $10_{0,10}$ - $9_{1,9}$ rotational transition, and scanned the frequency of a pulsed dye laser with a 0.04 cm^{-1} linewidth and 30 mJ output. Many double resonance signals were observed as changes in the microwave absorption as seen in Fig. 2.10. The double resonance signal was observed as an increase in the microwave absorption when an electronic transition from the $10_{0,10}$ level in the ground state is on resonance, or as a decrease from the $9_{1,9}$ level. They observed 324 MODR transitions in a range 16801 - 17100 cm^{-1}. The number was about 8 times larger than that of the vibronic levels reported by Smalley *et al.* [59], who used the supersonic jet technique to simplify the congested spectrum in this

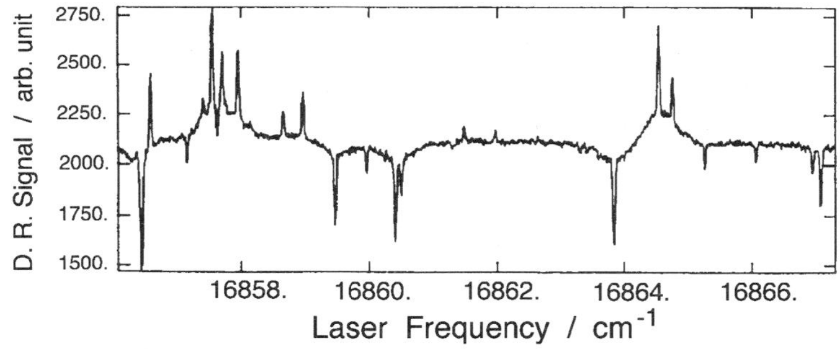

Fig. 2.10 Microwave-detected MODR signal for NO_2 (from Ref. [29]). Transitions from the $10_{0,10}$ (F_1) level go upwards and those from the $9_{1,9}$ (F_1) levels go downwards.

region. Rotational transitions with relatively low rotational quantum numbers were observed in the latter method. The double resonance spectrum, on the other hand, contains transitions with relatively high rotational quantum numbers. They attributed the difference of the numbers of the vibronic states to the fact that a symmetry selection rule was violated due to considerable mixing in the excited state, and the levels observed were approaching to a statistically defined limit of chaotic behavior.

This method can be applied to molecular systems with highly dissociative excited electronic states. Endo *et al.* observed the ammonia Ã state by this method [60]. The ammonia Ã state has been known to predissociate to NH_2 + H very rapidly, giving diffuse, structureless absorption. When the laser frequency was scanned while the microwave frequency was fixed at an inversion frequency, rotationally selected electronic transitions shown in Fig. 2.11 were observed. It was not difficult to fit the observed line shapes to a Lorentzian function, yielding accurate transition frequencies and linewidth parameters. They observed a clear *J*-dependence of the linewidths, ranging from 17 to 30 cm^{-1} for transitions up to J = 12. The rotational constants for the origin band in the Ã state have also been determined. Henck *et al.* studied NH_3, NH_2D, NHD_2, and ND_3 by the same double resonance method [61, 62]. For the observation of the double resonance signals of ND_3, which has inversion doublings in the 1.6 GHz region, they constructed a strip-line cell. They made extensive measurements for all four isotopomers for the ν_2 progression up to ν_2 = 4, providing vibrational dependences of the rotational constants and the predissociation rates. The latter information is particularly important to understand the predissociation mechanism of the ammonia Ã state.

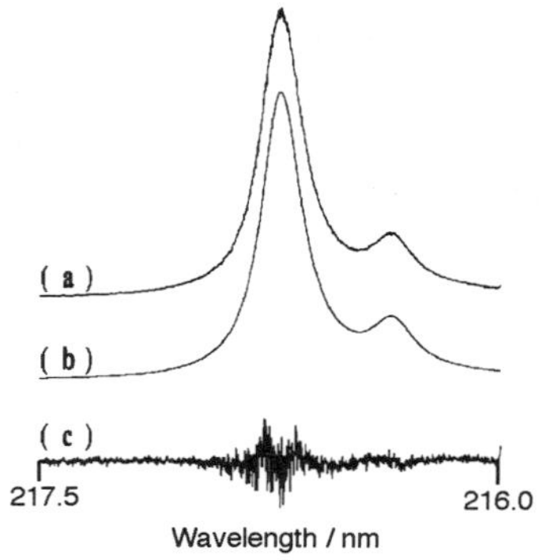

Fig. 2.11 Microwave-detected MODR signal for the Ã - X̃ transition of NH_3. Trace (a) corresponds the observed, (b) calculated, and (c) residual (scaled by 5 times). The observed line was broadened by fast predissociation (from Ref. [60]).

2.3 Optical-Optical Double Resonance Spectroscopy

When both of the radiations for the double resonance experiments are in the optical frequency region, the double resonance scheme is called optical-optical double resonance (OODR). The basic principle of the double resonance is the same as that of MODR, utilizing 3 levels with 2 radiations in resonance with the energy separations among the 3 levels. Although the accuracy of the measurements is limited in comparison with MODR because both the radiations are in the optical frequency region, OODR has several advantages over MODR and is used widely in recent years.

One of the advantages of OODR is that it is not necessary to combine two completely different experimental techniques as the cases of MODR, where techniques of optical spectroscopy should be combined with microwave spectroscopy. It is quite easy to overlap two radiations by focusing the light beams using optical lenses. Therefore, OODR signals can be observed by using various types of absorption cells; those used for ordinary single photon spectroscopy are used without essential modification. It is thus not difficult to combine the method, for example, with the supersonic molecular beam technique. Another advantage is that it is easy to scan wide frequency regions, because there are good optical lasers that are tunable in wide frequency regions. Furthermore, OODR can be observed by using pulsed lasers. In the case of MODR, microwave-detected double resonance is a rather specific method and has not been applied widely. On the other hand, OODR experiments by using pulsed lasers are commonly used, enabling wider varieties of molecules to be studied, since pulsed tunable lasers are available in much wider frequency regions than CW lasers. Although the accuracy and resolution of most of the OODR experiments are inferior to MODR, it is possible to observe Doppler-free spectra by using sufficiently narrow-linewidth laser sources, since the first laser saturates a specific velocity component among the thermally distributed molecules and double resonance signals are observed only for the molecules saturated by the first laser beam.

An early attempt to observe OODR was reported by Field *et al.* [63] for BaO, where they used an Ar ion laser and a CW dye laser for two radiation sources. Thus, their measurement was rather limited and could not obtain detailed assignment of the upper excited state they observed. However, they used two CW dye lasers in a later work and obtained detailed information on the highly excited states of BaO, which could not be reached by using only one dye laser oscillating in the visible region [64]. Since then, a number of papers have been published using the OODR technique to study highly excited states of various molecules. During these experiments many detection schemes have been developed and successfully applied. Here we will review some of the important methods currently used.

2.3.1 Two-color Fluorescence Excitation Spectroscopy

Figure 2.12 shows an excitation scheme of two-color fluorescence excitation spectroscopy. This spectroscopic method detects transitions to a highly excited state by observing fluorescence from the state. Briefly, sample molecules are excited to a specific (ro)vibronic level in a low-lying electronic state A by a first laser with frequency ν_1. The molecules excited to the A state are further excited to a highly excited state R by a second laser with ν_2. When the frequency of the second laser is resonant to state R, a fluorescence signal originating from R appears. Usually, since the wavelength of the fluorescence from R is largely different from that emitted from the intermediate state A, it is easy to distinguish the fluorescence signals from the two states by using a small monochromator or a sharp cut filter.

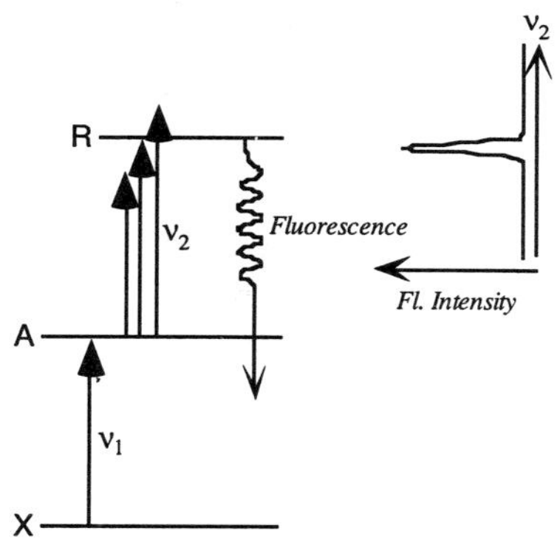

Fig. 2.12 Schematic diagram of two-color fluorescence excitation spectroscopy

Figure 2.13 shows an example of the two-color fluorescence excitation spectrum of bromine in a vapor phase. The bromine molecule Br_2 is excited to the second excited electronic state B [65]. The frequency of the second laser ν_2 was scanned while monitoring the total intensity of the fluorescence from a highly excited state. The spectrum clearly shows well-resolved rovibronic structures of the highly excited state. The upper electronic state is assigned to an ion pair state. Since the bond in the ion-pair state is significantly longer than that in the ground state X, direct observation from the ground state is almost impossible due to unfavorable Franck-Condon factors. This disadvantage is eliminated by the two-color double resonance method, because the internuclear distance in the B state is reasonably longer than that in the ground state, where the first laser can excite the molecule to the intermediate level using the inner wall of the potential curve of the B state and the second laser then excites the molecule to the final levels using the outer wall of the potential curve of the B state. So far, a number of ion pair states of halogen diatomics have been studied by OODR [66-74].

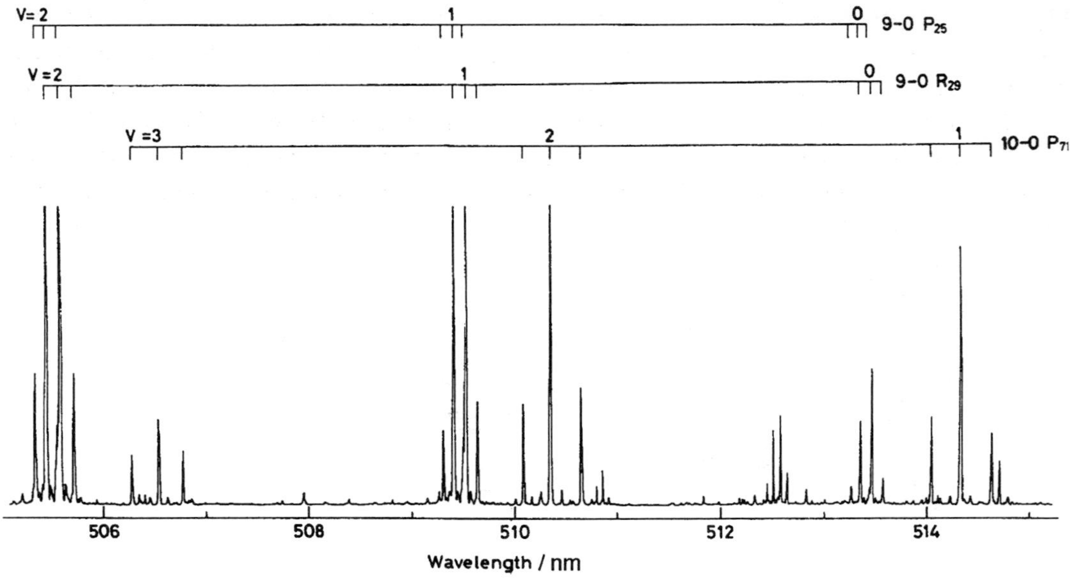

Fig. 2.13 OODR spectrum of I_2 detected by UV fluorescence (from Ref. [65])

A drawback of the two-color fluorescence excitation detection method is that highly excited states to be observed must be fluorescent. For a small molecule like iodine, it is not a serious limitation. On the other hand, for a large molecule like benzene, it is exceptional that highly excited states are fluorescent, because there exist fast nonradiative processes. Application of this detection scheme is thus practically limited to small molecules. Another point to be noted is that this detection method can be applied to both CW laser OODR and pulsed laser OODR. It is possible to obtain Doppler-free spectra when narrow linewidth CW lasers are used. Thus detailed spectroscopic studies of relatively simple molecular systems in highly excited electronic states have been extensively performed. Among such simple molecules, metal containing molecules, which have strong absorptions in the visible region and whose final states to be studied are fluorescent, have been studied extensively by this method. Homonuclear metal diatomics, such as Na_2 [75-78] and K_2 [79-81], are representative targets for the OODR studies. One point to be noted is that these species are nonpolar and it is not possible to apply MODR for these species. For the molecules whose final states are nonfluorescent, it is still possible to detect a decrease in the fluorescence from the intermediate state. However, this detection scheme has drawbacks that sensitivity is much lower because a small change in the fluorescence yield from the intermediate state is detected and undesirable background fluorescence due to the second laser has to be eliminated by carefully selecting fluorescence lines to be monitored.

Schawlow and coworkers used polarization rotation to observe double resonance [82, 83]. The basic principle is almost the same as that of microwave-optical polarization spectroscopy as explained in 2.1.3, where microwave radiation was used to polarize molecules to be studied, while the present method uses optical radiation. The polarization detection method can be applied to various energy level configurations such as SEP and depletion schemes as well as the optical-optical double resonance configurations. Schawlow *et al.* applied the method to a double resonance study of Na_2.

2.3.2 Two-color multiphoton ionization spectroscopy

Two-color multiphoton ionization spectroscopy For most of the molecules to be studied by OODR, the final states to be observed are highly excited electronic states very close to the ionization limit. Therefore, it is a natural selection to introduce the ionization detection method for the observation of the OODR signals. This is one of the sensitive detection methods because almost all the ions produced are collected by applying appropriate electric field and the collected ions can be detected with high quantum yield. Figure 2.14 illustrates a basic principle of the two-color multiphoton ionization detection scheme. This method detects a highly excited state R by observing resonance enhanced multiphoton ionization signals from an intermediate state A. The sample molecules are excited to a specific rovibronic level in the intermediate electronic state A by the first laser with ν_1, and are further excited to a highly excited state R by the second laser with ν_2. When the frequency of the second laser ν_2 coincides to an energy difference between R and A, multiphoton ionization from A is strongly enhanced. The highly excited state is thus detected by enhancement of the ion current.

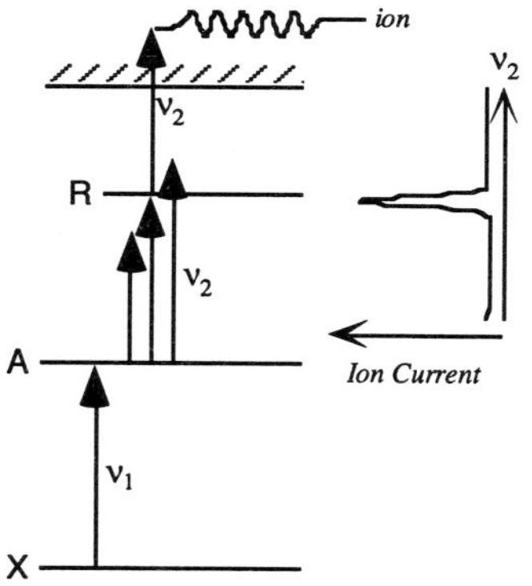

Fig. 2.14 Schematic diagram of the two-color multiphoton ionization detection method.

This detection scheme has an advantage that nonfluorescent electronic states can be detected. It is an important advantage when highly excited states of a large polyatomic molecule are studied. Figure 2.15 shows a two-color multiphoton ionization spectrum of 1,4-difluorobenzene [84]. The sample molecules

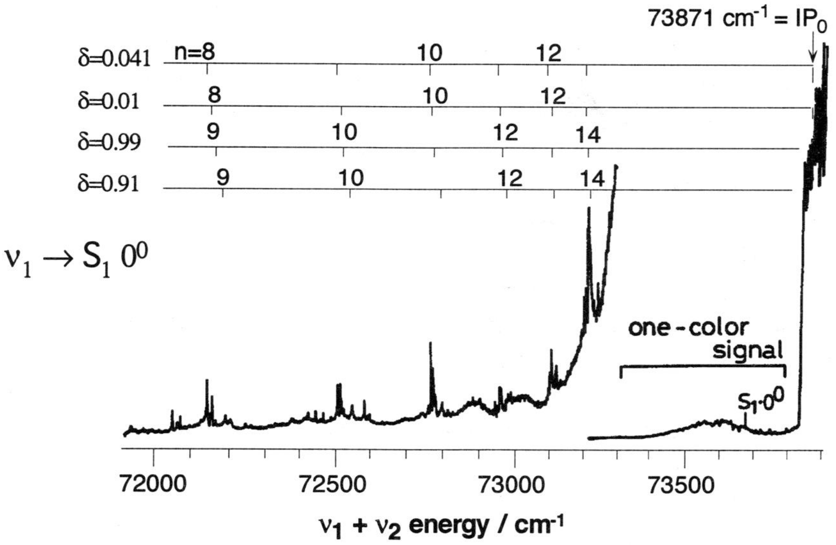

Fig. 2.15 Two-color multiphoton ionization spectrum of 1,4-difluorobenzene (from Ref. [84]).

are pumped to the S_1 state by the first laser ν_1. The S_1 molecules are further excited to high lying Rydberg states by the second laser ν_2. The spectrum in the region below 73300 cm^{-1} were observed by using a pulsed laser with rather large output power. The spectrum showed a well-resolved structure due to the Rydberg - S_1 transitions. The intense step at ~73800 cm^{-1} is the ionization threshold corresponding to the adiabatic ionization potential (the zero vibrational level E_0 in the corresponding cation). Energies of the Rydberg states E have been analyzed by a formula

$$E = E_0 - R_\infty / (n - \delta)^2, \tag{2.20}$$

where R_∞ is the Rydberg constant, n is the principle quantum number, and δ is the quantum defect. Calculated values are shown by the solid lines in the figure. As shown in the figure, the calculation well reproduced the observed positions of the Rydberg states. Four Rydberg series converging to IP_0 have been assigned in the spectrum.

As described above, the region where highly excited Rydberg series appeared was observed by using a laser with high output power. It is a disadvantage of this method. In general, highly excited states have fast relaxation paths, such as internal conversion and predissociation, and these fast relaxations compete with the ionization from the highly excited states. To overcome the competition with the relaxations, a high power laser is required to invoke multiphoton ionization. However, when combined with an appropriate ion detector, it is still possible to use CW lasers to observe double resonance signals by this detection method. Tsei *et al.* [85] observed highly excited states of Na_2 by the CW-laser OODR technique using a space-charge-limited diode ionization detector they had developed [86]. They were able to detect highly excited vibrational levels of several high lying electronic states which were impossible to observe by using the fluorescence detection scheme.

Two-color multiphoton ionization spectroscopy assisted by auto-ionization

It is known that molecules in highly excited states are known to ionize spontaneously when the states lie in the ionization continuum by a process known as *autoionization*. It is possible to use autoionization to observe highly excited states in the energy region above the ionization potential. Figure 2.16 shows an energy scheme of the two-color multiphoton ionization detection method assisted by autoionization. Similar to the two-color multiphoton ionization detection method, a highly excited state R is probed from a well defined level of an intermediate state A. In the present case the highly excited state is ionized by autoionization. No further excitation by the second laser ν_2 is required in this method. Consequently, the highly excited state can be probed by one-photon from the A state and can be observed clearly even if the output power of the second laser ν_2 is small.

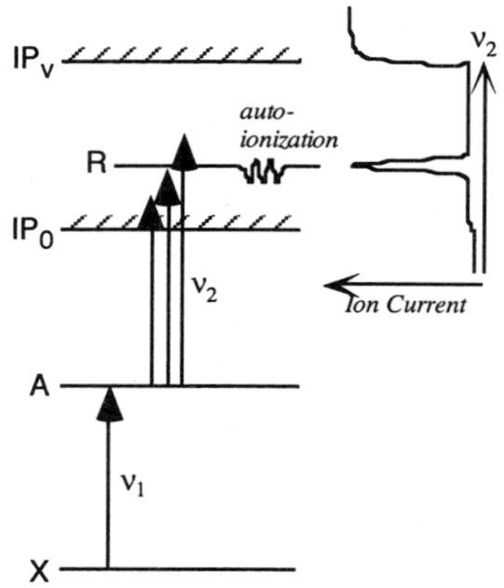

Fig. 2.16 Schematic diagram of two-color multiphoton ionization detection method assisted by autoionization.

Figure 2.17 shows an example of the two-color multiphoton ionization spectra assisted by autoionization, showing a spectrum of jet-cooled pyrazine [87]. The three traces in the figure correspond to the cases, where the first laser ν_1 was fixed to (a) the S_1 origin (30876 cm^{-1}), (b) $S_1 10a^1$ (origin + 383 cm^{-1}) and (c) $S_1 10a^2$ (origin + 823 cm^{-1}). The traces were drawn so as to align the total energy $\nu_1 + \nu_2$. The spectrum obtained after exciting the S_1 origin, (a), shows a sharp rise of the ionization threshold corresponding to the zero vibrational level of the pyrazine cation (the adiabatic ionization potential IP_0). No additional bands are seen in the spectrum. The ionization threshold also appears in the spectrum when $S_1 10a^1$ is excited by ν_1 (trace (b)). Since the position of the ionization threshold is different from that observed from the S_1 origin, it is assigned to the $10a^1$ vibrational level of the cation (the vertical ionization potential IP_v). In addition to the ionization threshold, many sharp peaks were found in this spectrum. The positions of these peaks follow the formula of Rydberg series given above (Eq. 2.20); therefore these peaks are assigned to the Rydberg series of the pyrazine. The calculated values with the assignments are shown by solid lines in the figure. Rydberg series appears only in the energy region above the ionization potential. This fact shows that the Rydberg states are ionized by autoionization. Because the intensities of these peaks depend linearly on the laser power of ν_2, the autoionization of the Rydberg states is confirmed.

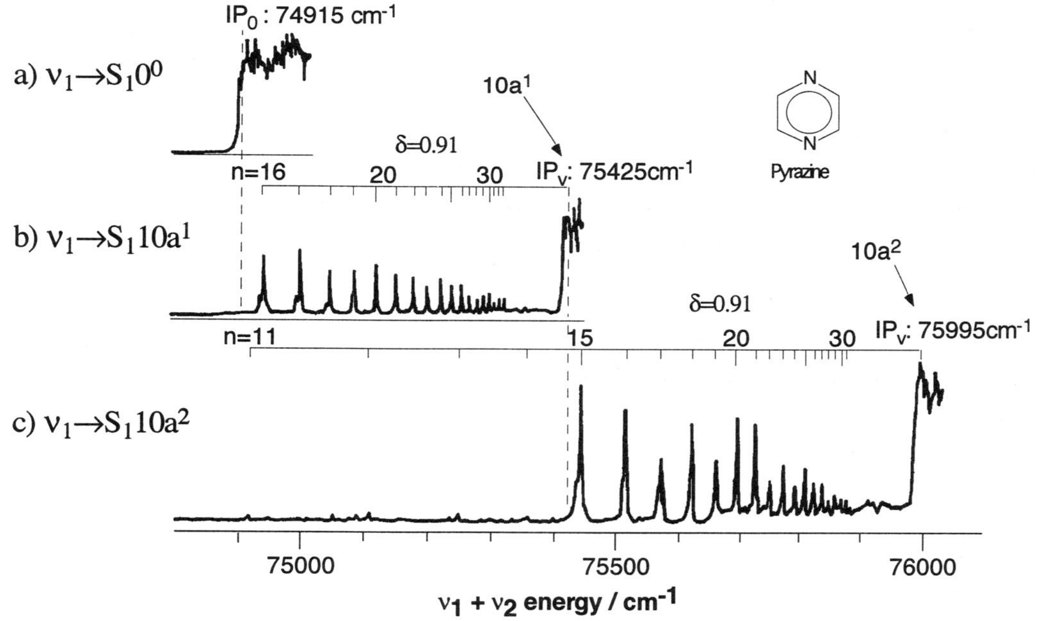

Fig. 2.17 Two-color multiphoton ionization spectrum of pyrazine assisted by autoionization (from ref. [87]). Each spectrum was obtained when the first laser ν_1 is fixed to a) S_1 origin, b) $S_1 10a^1$, and c) $S_1 10a^2$ vibronic levels.

In addition to the observation of the high lying excited states, this method provides us with information on the autoionization mechanism. Figure 2.17 (c) shows a spectrum obtained after exciting the molecule to the overtone of the 10a vibrational mode in S_1 ($S_1 10a^1$). This spectrum also shows a Rydberg series converging to the vertical ionization potential IP_v, which is the $10a^2$ vibrational level of the cation. From the convergence to the $10a^2$ level of the cation, it is concluded that the observed bands are the $10a^2$ vibrational level in the Rydberg states. Since the Rydberg series appears in the region above the adiabatic ionization potential IP_0, these bands are ionize by autoionization as well as those observed from $S_1 10a^1$. However, the intensity distribution in the Rydberg series is quite different; their intensities increase suddenly in the region above the $10a^1$ vibrational level of the cation. This strongly suggests that the Rydberg states have two channels of autoionization; autoionization to the zero vibrational level and that to the $10a_1$ vibrational level of the cation. It can be seen from the spectrum that the latter is a strongly allowed channel. Therefore, it is concluded that autoionization of the Rydberg state in pyrazine has strong propensity $\Delta v = -1$, suggesting the vibrational autoionization mechanism.

As described above, two-color multiphoton ionization assisted by autoionization is a useful tool to observe high Rydberg states lying above the ionization potential. The autoionization mechanism can also be studied from the observed spectra. Combination of the present method with high resolution photoelectron spectroscopy provides detailed information on the autoionization mechanism. However, this detection scheme has essential disadvantages. One obvious disadvantage is that states lying below the ionization potential cannot be observed. Similarly, states having very low autoionization efficiency are hard to observe by this method. Thus, this method is difficult to apply to the observation of highly excited valence states. In some cases, autoionization itself causes a trouble in observing highly excited states. When an ionization continuum appears, overlapped with highly excited states of a neutral molecule, the autoionization signal distorts the band shape by introducing wavy first-derivative shapes (Fano shape) as a result of an interference effect between them. If the S/N ratio of the observed spectrum is sufficiently high, one can still detect highly excited states having Fano shapes. In practice, however, wavy broadened band shapes make it difficult to distinguish autoionization peaks from noisy ionization continuum. Consequently, not all the molecules studied have shown peaks due to autoionization of highly excited states, although the ionization thresholds have been observed in many molecules.

Two-color multiphoton ionization spectroscopy assisted by collisional Ionization As described above, even a weak assistance for ionization such as autoionization helps the detection of highly excited states because of its high sensitivity of the ionization detection scheme. One of the most useful assistance effects for ionization may be ionization by applying a weak external electric field (field ionization), as will be discussed in Chapter 7. In this section, another external assistance effect for ionization, *i. e.* collisional ionization, is described briefly.

Figure 2.18 shows a basic principle of the two-color multiphoton ionization detection scheme assisted by collision. The sample molecules are excited to a low-lying state A by the first laser ν_1 and are further excited to a highly excited state R which lie just below the ionization potential *IP*. When a molecule in a state just below *IP* collides with other molecules, it is easily ionized and an increase in the ion current due

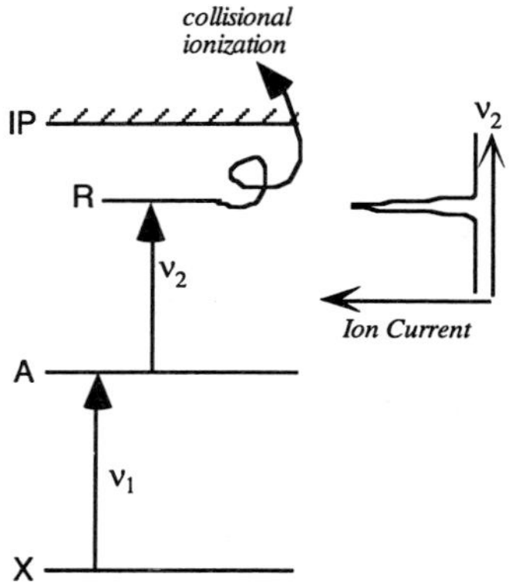

Fig. 2.18 Schematic diagram of the two-color multiphoton ionization detection scheme assisted by collision

to double resonance is detected. An example of the collisionally assisted two-color multiphoton ionization spectrum is shown in Fig. 2.19a [88]. This spectrum is obtained by exciting the 1,4-diazabicyclo [2.2.2]octane (DABCO) molecule in a static cell to the origin of the S_1 state (35783 cm^{-1}) by ν_1. Peaks due to highly excited Rydberg states appear in the region below the adiabatic ionization potential IP_0. For comparison, a spectrum in the same frequency region measured in a supersonic jet (Fig. 2.19b) is shown. All the peaks assigned to the Rydberg series disappear in the spectrum in a jet. Instead of the Rydberg series, only the ionization threshold which corresponds to IP_0 appears at ~58000 cm^{-1}. Here, the ionization threshold is shifted to red because of the field ionization of very high Rydberg states. From the comparison, it is understood that the Rydberg series observed only in the static cell spectrum is ionized by self collision. The collisional ionization is useful in the observation of highly excited states lying just below the ionization potential. However, since the collisional energy is rather small (~200 cm^{-1} at room temperature), it is difficult to observe a wide energy region.

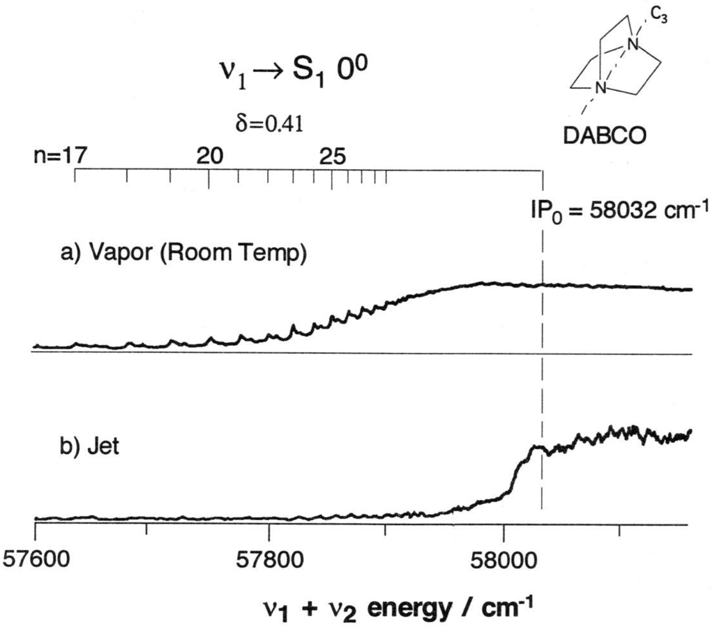

Fig. 2.19 Two-color multiphoton ionization spectrum of 1,4-diazabicyclo[2.2.2]octane measured a) in vapor at room temperature and b) in a supersonic jet (from Ref. [88]). The first laser ν_1 was fixed to S_1 origin (35783 cm^{-1}) in both of the spectra.

REFERENCES

1. R. W. Field, A. D. English, T. Tanaka, D. O. Harris, and D. A. Jennings, *J. Chem. Phys.* **59**, 2191 (1973).
2. W. E. Ernst and T. Törring, *Phys. Rev.* **A25**, 1236 (1982).
3. R. W. Field, R. S. Bradford, D. O. Harris, and H. P. Broida, *J. Chem. Phys.* **56**, 4712 (1972).
4. K. Takagi, S. Saito, M. Kakimoto, and E. Hirota, *J. Chem. Phys.* **73**, 2570 (1980).
5. K. Takagi, T. Suzuki, S. Saito, and E. Hirota, *J. Chem. Phys.* **83**, 535 (1985).
6. T. Suzuki, S. Saito, and E. Hirota, *J. Chem. Phys.* **79**, 1641 (1983).
7. J. M. Cook, G. W. Hills, and R. F. Curl, Jr., *Astrophys. J.* **207**, L139 (1976).
8. G. W. Hills, J. M. Cook, R. F. Curl, Jr., and F. K. Tittel, *J. Chem. Phys.* **65**, 823 (1976).
9. G. W. Hills and R. F. Curl, Jr., *J. Chem. Phys.* **66**, 1507 (1977).
10. J. M. Cook, G. W. Hills, and R. F. Curl, Jr., *J. Chem. Phys.* **67**, 1450 (1977).
11. G. W. Hills, R. S. Lowe, J. M. Cook, and R. F. Curl, Jr., *J. Chem. Phys.* **68**, 4073 (1978).
12. R. S. Lowe, J. V. V. Kasper, G. W. Hills, W. Dillenschneider, and R. F. Curl, Jr., *J. Chem. Phys.* **70**, 3356 (1979).
13. T. C. Steimle, J. M. Brown, and R. F. Curl, Jr., *J. Chem. Phys.* **73**, 2552 (1980).
14. G. W. Hills, C. R. Brazier, J. M. Brown, J. M. Cook, and R. F. Curl, Jr., *J. Chem. Phys.* **76**, 240 (1982).
15. T. Suzuki, S. Saito, and E. Hirota, *J. Chem. Phys.* **83**, 6154 (1985).
16. W. E. Ernst and J. O. Schroeder, *Phys. Rev.* **A30**, 665 (1984).
17. S. D. Rosner, T. D. Gaily, and R. A. Holt, *Phys. Rev. Lett.* **35**, 785 (1975).
18. W. Ertman and P. Hofer, *Z. Phys.* A **276**, 9 (1976).
19. W. J. Childs, D. R. Cok, L. S. Goodman, and O. Poulsen, *Phys. Rev. Lett.* **47**, 1389 (1981).
20. D. A. Fletcher, K. Y. Jung, C. T. Scurlock, and T. C. Steimle, *J. Chem. Phys.* **98**, 1837 (1993).
21. K. K. Lehmann and S. L. Coy, *J. Chem. Phys.* **81**, 3744 (1984).
22. S. L. Coy and K. K. Lehmann, *J. Chem. Phys.* **84**, 5239 (1986).
23. Y. Matsuo, Y. Endo, E. Hirota, and T. Shimizu, J. Chem. Phys. **87**, 4395 (1987).
24. R. Solarz and D. H. Levy, *J. Chem. Phys.* **58**, 4026 (1973).
25. T. Tanaka, A. D. English, R. W. Field, D. A. Jennings, and D. O. Harris, *J. Chem. Phys.* **59**, 5217 (1973).
26. T. Tanaka, R. W. Field, and D. O. Harris, *J. Chem. Phys.* **61**, 3401 (1974).
27. T. Tanaka, R. W. Field, and D. O. Harris, *J. Mol. Spectrosc.* **56**, 188 (1975).
28. H. G. Weber, P. J. Brucat, W. Demtröder, and R. N. Zare, *J. Mol. Spectrosc.* **75**, 58 (1979).
29. K. K. Lehmann and S. L. Coy, *J. Chem Phys.* **83**, 3290 (1985).

30. T. Suzuki, S. Saito, and E. Hirota, *J. Mol. Spectrosc.* **111**, 54 (1985).
31. J. C. Petersen and D. A. Ramsay, *Chem. Phys. Lett.* **118**, 31 (1985).
32. J. C. Petersen and D. A. Ramsay, *Chem. Phys. Lett.* **124**, 406 (1986).
33. D. A. Ramsay, M. Barnett, and W. Huettner, *Chem. Phys. Lett.* **202**, 406 (1993).
34. Y. Ohshima and Y. Endo, *J. Mol. Spectrosc.* **172**, 225 (1995).
35. T. Suzuki, S. Saito, and E. Hirota, *J. Mol. Spectrosc.* **120**, 414 (1986).
36. J. Nakagawa, P. J. Domaille, T. C. Steimle, and D. O. Harris, *J. Mol. Spectrosc.* **70**, 374 (1978).
37. W. J. Childs and L. S. Goodman, *Phys. Rev. Lett.* **44**, 316 (1980).
38. W. J. Childs and L. S. Goodman, *Phys. Rev.* **A21**, 1216 (1980).
39. W. J. Childs, G. L. Goodman, and L. S. Goodman, *J. Mol. Spectrosc.* **86**, 365 (1981).
40. W. J. Childs, D. R. Cok, and L. S. Goodman, *J. Chem. Phys.* **78**, 3993 (1982).
41. W. E. Ernst and T. Törring, *Phys. Rev.* **A27**, 875 (1983).
42. W. E. Ernst, S. Kindt, and T. Törring, *Phys. Rev. Lett.* **51**, 979 (1983).
43. W. E. Ernst, S. Kindt, K. P. R. Nair, and T. Törring, *Phys. Rev.* **A29**, 1158 (1984).
44. W. J. Childs, D. R. Cok, G. L. Goodman, and L. S. Goodman, *J. Chem. Phys.* **75**, 501 (1981).
45. W. J. Childs, D. R. Cok, and L. S. Goodman, *J. Mol. Spectrosc.* **95**, 153 (1982).
46. W. J. Childs, L. S. Goodman, and I. Renhorn, *J. Mol. Spectrosc.* **87**, 522 (1981).
47. Y. Azuma, W. J. Childs, G. L. Goodman, and T. C. Steimle, *J. Chem. Phys.* **93**, 5533 (1990).
48. M. C. L. Gerry, A. J. Merer, U. Sassenberg, and T. C. Steimle, *J. Chem. Phys.* **86**, 4754 (1987).
49. T. C. Steimle, W. -L. Chang, and D. F. Nachman, *Chem. Phys. Lett.* **153**, 534 (1988).
50. W. J. Childs, O. Poulsen, and T. C. Steimle, *J. Chem. Phys.* **88**, 598 (1988).
51. Y. Azuma and W. J. Childs, *J. Chem. Phys.* **93**, 8415 (1990).
52. J. E. Shirley, W. L Barclay Jr., L. M. Ziurys, and T. C. Steimle, *Chem. Phys. Lett.* **183**, 363 (1991).
53. W. J. Childs and T. C. Steimle, *J. Chem. Phys.* **88**, 6168 (1988).
54. W. J. Childs, Y. Azuma, and G. L. Goodman, *J. Mol. Spectrosc.* **144**, 70 (1990).
55. T. C. Steimle, J. E. Shirley, K. Y. Jung, L. R. Russen, and C. T. Scurlock, *J. Mol. Spectrosc.* **144**, 27 (1990).
56. D. A. Fletcher, C. T. Scurlock, K. Y. Jung, and T. C. Steimle, *J. Chem. Phys.* **99**, 4288 (1993).
57. T. C. Steimle, D. A. Fletcher, K. Y. Jung, and C. T. Scurlock, J. Chem. Phys. 90, 2556 (1992).
58. Y. Matsuo, Y. Endo, E. Hirota, and T. Shimizu, *J. Chem. Phys.* **88**, 2852 (1988).
59. R. E. Smalley, L. Wharton, and D. H. Levy, *J. Chem. Phys.* **63**, 4977 (1975).
60. Y. Endo, M. Iida, and Y. Ohshima, *Chem. Phys. Lett.* **174**, 401 (1990).
61. S. A. Henck, M. A. Mason, W. -B. Yan, K. K. Lehmann, and S. L. Coy, *J. Chem. Phys.* **102**, 4772 (1995).
62. S. A. Henck, M. A. Mason, W. -B. Yan, K. K. Lehmann, and S. L. Coy, *J. Chem. Phys.* **102**, 4783 (1995).
63. R. F. Field, G. A. Capelle, and M. A. Revelli, *J. Chem. Phys.* **63**, 3228 (1975).
64. R. A. Gottscho, J. B. Koffend, and R. W. Field, *J. Chem. Phys.* **68**, 4110 (1978).
65. T. Shinazawa, T. Tokunaga, T. Ishiwata, I. Tanaka, K. Kasatani, M. Kawasaki, and H. Sato, *J. Chem. Phys.* **80**, 5909 (1984).
66. T. Shinozawa, A. Tokunaga, T. Ishiwata, and I. Tanaka, *J. Chem. Phys.* **83**, 5407 (1985).
67. T. Ishiwata, Y. Kasai, K. Obi, *Chem. Phys. Lett.*, **261**, 175 (1996).
68. T. Ishiwata, I. Fujiwara, T. Shinozawa, and I. Tanaka, *JChem. Phys.* **79**, 4779 (1983).
69. T. Ishiwata, A. Tokunaga, T. Shinozawa, and I. Tanaka, *J. Mol. Spectrosc.* **108**, 314 (1984).
70. J. P. Perrot, A. J. Bouvier, A. Bouvier, B. Femelat, and J. Chevaleyre, *J. Mol. Spectrosc.* **114**, 60 (1985).
71. T. Ishiwata, T. Kasayanagi, T. Hara, and I. Tanaka, *J. Mol. Spectrosc.* **119**, 337 (1986).
72. T. Ishiwata, and I. Tanaka, *Laser Chem.* **7**, 79 (1987).
73. A. J. Holmes, K. P. Lawley, T. Ridley, R. J. Donovan, R. R. Langridge-Smith, *J. Chem. Soc. Faraday Trans.* **87**, 15 (1991).
74. P. J. Jewsbury, K. P. Lawley, T. Ridley, and R. J. Donovan, *J. Chem. Soc. Faraday Trans.* **88**, 1599 (1992).
75. L. Li, and R. W. Field, *J. Phys. Chem.* **87**, 3020 (1983).
76. L. Li, S. F. Rice, and R. W. Field, *J. Chem. Phys.* **82**, 1178 (1985).
77. L. Li, Z. Qingshi, A. M. Lyyra, T. -J. Whang, W. C. Stwalley, R. W. Field, and M. H. Alexander, *J. Chem. Phys.* **97**, 8835 (1992), *ibid.* **98**, 8406 (1993).
78. Y. -L. Pan, L. -S. Ma, L. -E. Ding, and D. -P. Sun, *J. Mol. Spectrosc.* **162**, 178 (1993).
79. H. Wang, L. Li, A. M. Lyyra, and W. C. Stwalley, *J. Mol. Spectrosc.* **137**, 304 (1989).
80. G. Jong, H. Wang, C. -C. Tsai, W. C. Stwalley, and A. M. Lyyra, *J. Mol. Spectrosc.* **154**, 324 (1992).
81. J. T. Kim, H. Wang, J. T. Bahns, and W. C. Stwalley, *J. Chem. Phys.* **102**, 6966 (1995).
82. N. W. Carlson, A. J. Taylor, K. M. Jones, and A. L. Schawlow, *Phys. Rev. A*, **24**, 822 (1981).

83. A. J. Taylor, K. M. Jones, and A. L. Schawlow, *J. Opt. Soc. Am.* **73**, 994 (1983).
84. M. Fujii, T. Kakinuma, N. Mikami, and M. Ito, *Chem. Phys. Lett.* **127**, 397 (1986).
85. C. -C. Tsai, J. T. Bahns, T. -J. Whang, H. Wang, W. C. Stwalley, and A. M. Lyyra, *Phys. Rev. Lett.* **71**, 1152 (1993).
86. C. -C. Tsai, J. T. Bahns, and W. C. Stwalley, *Rev. Sci. Instrum.* **63**, 5576 (1992).
87. A. Goto, M. Fujii, and M. Ito, *J. Phys. Chem.* **91**, 2268 (1987).
88. M. Fujii, T. Ebata, N. Mikami, and M. Ito, *Chem. Phys. Lett.* **101**, 578 (1983).

3 Stimulated Emission Pumping (SEP) Spectroscopy

Hai-Lung Dai
Department of Chemistry, University of Pennsylvania
Philadelphia, PA 19104-6323, USA

3.1 Introduction - vibrational spectroscopy and dynamics by stimulated emission pumping

The traditional goal for the study of molecular vibrational levels has been the determination of molecular symmetry, the constituent functional groups, and eventually the equilibrium structure and the force field. This goal can be achieved primarily through examining the lower vibrational levels which are well described by the normal mode model. As the physical principles of such practices are well understood and constantly in use, the challenges for vibrational spectroscopy in the relatively low energy region are in experimental detection of vibrational levels of novel forms of molecules such as radicals, ions, molecular complexes and clusters or unstable, transient species such as reaction intermediates. For species that are transient in nature and can only be generated in small quantities, techniques with fast time response and high sensitivity are needed.

More recently, spurred by the pursuit of mode- or bond-selective chemistry since intense and coherent light sources, namely lasers, have been used to excite molecules, interest in vibrational levels have shifted toward high energy regions where vibrational excitation is expected to affect the chemical properties of molecules. The challenges here range from how to achieve non-thermal, energy specific or even bond- or mode-specific vibrational excitation to understanding the ensuing dynamics following photoexcitation. What is Intramolecular Vibrational Redistribution? Where does the energy flow within the molecule? Why? How? How fast? Can vibrational excitation be used to control unimolecular reactions? How do highly vibrationally excited molecules transfer energy during collisions? Can vibrational excitation be used to influence bimolecular reactions? To tackle these questions requires spectroscopic and excitation techniques for high vibrational levels.

There have been significant advances in vibrational spectroscopy in the last decades. The use of Fourier transform spectrometers enhances the sensitivity and resolution, and pushes the spectral range into the near-IR. Similar advantages can be achieved by a combination of IR lasers and more sensitive detection schemes such as the bolometric technique. Single-photon optical transitions allow the detection and excitation of highly anharmonic overtone levels. Laser driven Raman processes have proven to provide spatial resolution in addition to high excitation and detection efficiency for low vibrational levels. And IR multiple photon excitation can efficiently create a noncanonical distribution of highly excited and reactive molecules. These significant advances, however, will not be the subject of this chapter. Instead, the focus here is on a widely applicable multiple-resonance technique that uses an excited electronic state as a 'window' for accessing the vibrational levels on the ground or a low lying electronic state.

Most molecules have excited electronic states that are efficiently accessible by single photon excitation and would fluoresce to the ground electronic state with high quantum yields. By dispersing the fluorescence, vibrational levels of the ground state are revealed through the Franck-Condon allowed transitions. However, there has always been the consideration of sensitivity in competition with spectral resolution in dispersed fluorescence spectra recorded with a conventional dispersion instrument. Furthermore, the dynamic range of detection of weaker transitions is limited. The use of a Fourier transform spectrometer for dispersion has recently been shown to improve dramatically the fluorescence collection efficiency and therefore the sensitivity and resolution [1] for resolving the downward transitions. The other strategy, developed in the early eighties by Field, Kinsey and coworkers [2] is to use another photon to resonantly enhance a particular transition, through stimulated emission, for its detection.

The basic idea of this two-photon, double resonance process, Stimulated Emission Pumping (SEP), is illustrated in Fig. 3.1. A PUMP photon first excites the molecule of interest to an electronic excited state. A second DUMP photon then is scanned into resonance with a downward transition and stimulates emission from this excited intermediate level down to a vibrational level in the ground electronic state. The spectroscopic signal indicating the presence of the ground state vibrational level is detected by some DUMP-photon induced effect on the PUMP excitation.

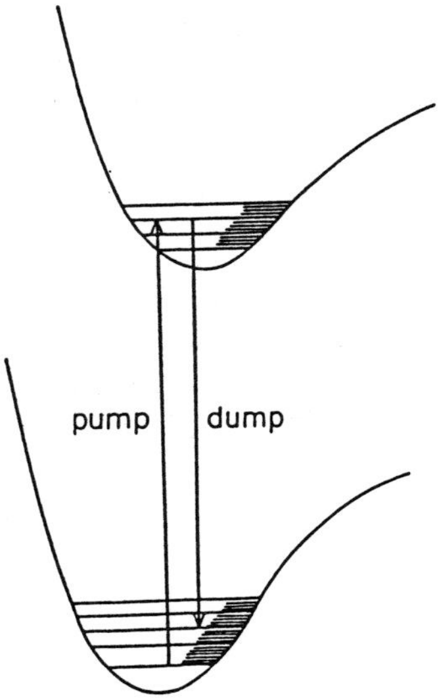

Fig. 3.1 SEP scheme.

Since the first demonstration on iodine[2], followed by p-difluorobenzene[3], formaldehyde[4], and acetylene [5], SEP has been applied as both a spectroscopic and excitation technique to a wide variety of stable molecules, radicals, atoms, ions, clusters, and weakly bound molecular complexes for the study of their intra- and inter-molecular dynamics involving vibrational excitations [6-9].

As a vibrational spectroscopic technique, sometimes referred to as Stimulated Emission Spectroscopy (SES), SEP has many unique advantages:

1. *High sensitivity.* Since electronic transitions, rather than the much weaker vibrational transitions, are used, the sensitivity is greatly enhanced. In principle, as long as laser induced fluorescence can be detected for a species, SEP can be performed. Thus, SEP has sensitivity comparable to the highly sensitive laser induced fluorescence technique.
2. *High resolution.* The spectral resolution is a convolution of the width of the PUMP and DUMP light sources. Using commercially available dye lasers with an intracavity etalon, a typical resolution of 0.05 cm-1 is reached. This resolution can be further improved by one to two orders of magnitude if single mode dye lasers are used.
3. *Extended spectral range.* The difference in the energies of the PUMP and DUMP photons determines the spectral range of probe in an SEP experiment. The use of pulsed dye lasers as PUMP-DUMP sources can result in ranges from zero to many tens of thousands of wavenumbers for vibrational spectroscopy. The higher energy vibrational levels are accessible because often the structural difference between the electronic excited and ground states brings favorable Franck-Condon factors for the DUMP transitions leading to these levels. At the other extreme, very low frequency vibrational levels, *e.g.* the inter-molecular van der Waals vibrational levels, can also be detected.
4. *Reduction of spectral complexity.* As in any double resonance techniques, the first PUMP photon may selectively excite a single quantum state. The spectrum obtained from scanning the second photon is

therefore from this selected state. For room temperature environment this selection greatly reduces rotational congestion, resulting in straightforward rotational assignments. If the sample consists of many different species, the selectivity by the PUMP photon also ensures that the SEP spectrum is from the species of interest.

5. *Transient spectroscopy.* As pulsed lasers are best used for driving the PUMP-DUMP transitions, the SEP process is completed on time scales equivalent to the laser pulse duration. Transient species with lifetimes comparable to the laser pulse length can be examined.

In addition to being a viable vibrational spectroscopic technique, SEP can also be used for preparing large populations in single, excited rovibrational levels for kinetic studies. Pulsed dye lasers in most cases can provide sufficient power to saturate both the PUMP and DUMP transitions, allowing a large percentage of population in an initial thermally populated level to be transferred to the final excited rovibrational level. The efficiency of population transfer in an incoherently driven PUMP-DUMP processes is determined by the ratio of the degeneracy of the three levels involved. Using coherent excitation processes with proper time arrangement, near complete transfer of population from the initial to the final level in small atomic/molecular systems has been demonstrated[10].

This Chapter presents a review on the studies on intra- and inter-molecular vibrational dynamics facilitated by the SEP technique within the last one and half decades. The review is unavoidably more related to the research by the author and does not aim to be complete. To begin with, a review on detection schemes developed over the years for conducting SEP experiments is given.

3.2 Experimental schemes for SEP

The first step in an SEP experiment is to lock the PUMP laser frequency onto a specific rovibronic transition. The DUMP laser is then scanned in search of resonance with the downward DUMP transitions. The SEP signal, indicating such a resonance between the DUMP laser and a DUMP transition, can be detected as a change in the PUMP laser induced effects, as a change in the DUMP laser beam characteristics, or as population transferred into the final level. The choice of experimental scheme depends on the types of lasers available and the optical properties of the species of interests.

3.2.1 Fluorescence dip

In the first experimental demonstration of SEP on a molecular system [2], the signal was detected as a decrease of the PUMP laser induced fluorescence. As this method can be used for any system that has detectable laser induced fluorescence, it is still the most versatile and widely applicable detection scheme. A detailed discussion of all the factors involved in setting up this detection scheme can be found in Ref. [11].

As the DUMP laser stimulates emission, it removes population from the PUMP laser excited intermediate level, resulting in a decrease of the fluorescence, which is illustrated in a real time experimental display in Fig. 3.2. One of the first SEP spectra recorded with the fluorescence dip scheme is shown in Fig. 3.3. The magnitude of the fluorescence loss is a direct measure of the amount of population transferred to the final level. The broadening of the line is a result of saturation. However, one major disadvantage of this scheme is that the removal of the intermediate level population can also be induced by an upward, rather than downward, transition in resonance with the DUMP laser which causes difficulties for interpreting the spectrum.

Since the signal appears as a fractional change of the PUMP excited fluorescence, it is essential that the fluorescence intensity remain stable so that even a small decrease can be detected. It is not unusual to have 10% pulse to pulse fluorescence-intensity fluctuations which constitute the major source of noise. These intensity fluctuations can be mitigated by using a reference substraction scheme. There are two ways to conduct such a reference subtraction. In one design the PUMP laser beam is split into two equal intensity

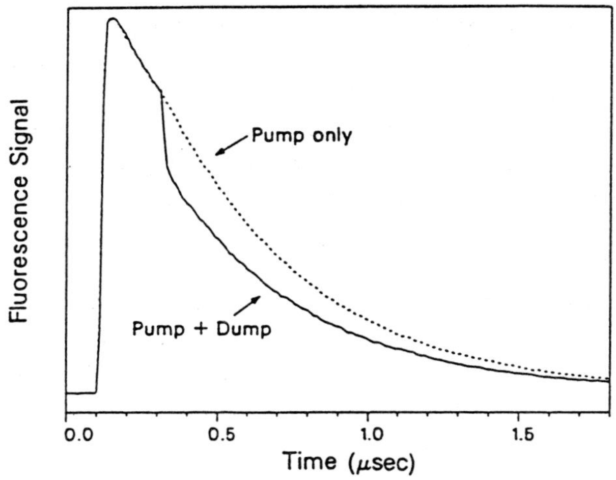

Fig. 3.2 Temporal profile of fluorescence observed in an SEP experiment of C_3 from Ref. [12].

beams for interacting with two identical samples. The fluorescence from the two samples should be identical in magnitude and fluctuations. Subtraction of one from the other yields a null signal. One of the samples is set to interact with the DUMP. As the DUMP stimulates emission and decreases the fluorescence, the subtraction gives a signal. Alternatively, two electronic gates can be used to monitor two different portions of the PUMP induced fluorescence decay signal; one gate is set after the PUMP but before the DUMP pulse and the second gate is set after the DUMP pulse. Subtraction between the fluorescence intensity integrated within the two gates respectively should give the SEP signal. The second method demands a fluorescence lifetime that is longer than the combination of the laser pulse widths and the electronic gate widths. A reference scheme involving subtraction of the laser intensity only is not adequate, as the fluorescence intensity fluctuation arises from pulse to pulse variations in both the laser intensity as well as the mode structure.

3.2.2 Ionization Dip

This scheme is essentially a variation of the fluorescence dip and is suitable for molecules that have low fluorescence quantum yields but can be ionized through a resonantly enhanced multi-PUMP photon process. For signal/noise ratio (S/N) considerations, preferably a 1+1 process is used to induce ionization

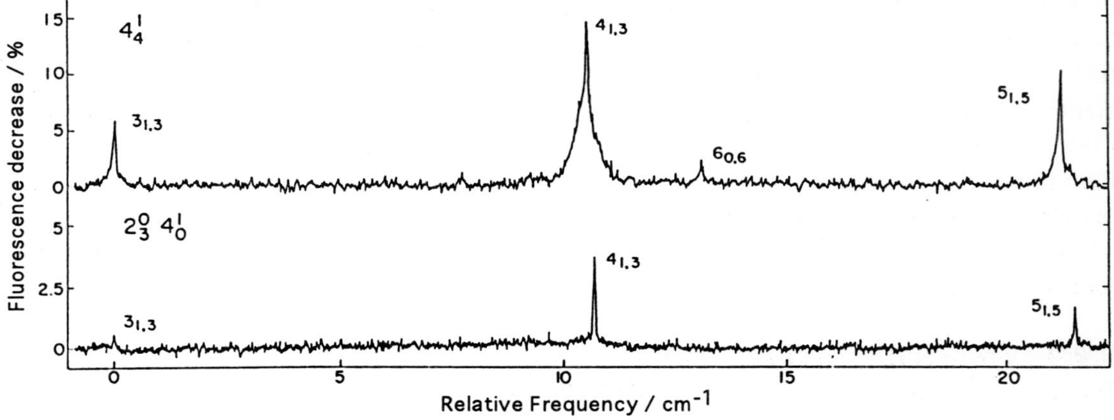

Fig. 3.3 SEP spectra of formaldehyde from Ref. [4]. The DUMP vibronic transitions and the rotational assignment are labeled.

by the PUMP. Ion intensity, instead of fluorescence, is detected. The DUMP laser is then overlapped in both time and space to compete with the PUMP induced ionization process. Whenever the DUMP is in resonance with a transition, the signal appears as a decrease in ion intensity. As detecting ions through a resonantly enhanced process can be very sensitive, this scheme has been widely used [13-17]. However, this scheme suffers from the same disadvantages as the fluorescence dip. Although it has not been demonstrated, in principle the Fourier transform technique that was applied to improve the sensitivity in ionization detection of a stimulated Raman signal [18] could be used for SEP as well.

3.2.3 DUMP Polarization Rotation

The stimulated emission photons in principle can be detected as a gain in the DUMP laser intensity. However, as the density of molecules interacting with the lasers is usually low, this gain is often extremely small. The challenge is then how does one detect the small gain in the DUMP power despite its large intensity fluctuation. This can actually be achieved by a polarization analysis of the DUMP laser beam [19]. In this detection scheme, by detecting the DUMP gain one can ensure that the SEP signal is from a downward DUMP transition.

The principles of using polarization rotation for analyzing the change in the intensity of a laser beam after passing through a medium have been illustrated in absorption spectroscopic applications in Ref. [20]. The classical description to be given here will aid the qualitative understanding of how does this scheme work in SEP. Consider the case that both the PUMP and DUMP transition dipole moments lie in the same direction in the molecular frame. A linearly polarized PUMP excites those molecules with their transition dipoles parallel to the laser polarization, i.e. the excited molecules are now aligned in space by the PUMP polarization. As the DUMP laser, which is also linearly polarized but along a different direction than the PUMP, stimulates the downward transition of the excited molecules, the photons emitted are linearly polarized along the PUMP, not the DUMP, polarization direction. The addition of the emitted photons in the same propagation direction but a different polarization direction to the DUMP laser rotates the DUMP polarization angle. Using a set of crossed polarizers with the best extinction ratio of 10^{-7}, a rotation of the polarization angle as small as 10^{-7} radian can be detected. With moderate signal averaging, this scheme may even allow the detection of a DUMP laser gain as small as 10^{-8} while the pulse to pulse fluctuation is ~10% [19,20]! With prior knowledge of the DUMP transition moment direction versus that of the PUMP, the rotation angle (the new cross angle) of the DUMP laser beam can be used to confirm if the signal arises from the DUMP transition.

SEP signal detected as the DUMP gain provides direct proof of a downward stimulated emission transition. Furthermore, knowledge of the direction of the polarization rotation can be used in assigning the transition type. Unlike the dip schemes where the signal to noise ratio is intrinsically limited by the stability of the laser intensity, the signal here can be unlimitedly high in comparison with the noise. It is important to bring up the signal magnitude with a large number of excited molecules which can be created by stronger transitions in interaction with intense laser beams. As the signal is determined by the ratio between the emitted-photons and the DUMP intensity (thus it is independent of the DUMP intensity), a cw laser with more intensity stability is preferred as the DUMP. Such a pulsed-PUMP and cw-DUMP combination has been proven highly effective [22], as the kinetics of the intermediate level population can be directly revealed in the time-resolved DUMP polarization evolution.

Instead of detecting the DUMP gain through polarization analysis, a resonator cavity can also be used for amplifying this gain and allow its detection [23].

3.2.4 Transient Grating Scattering

For molecules with strong electronic transitions and/or high third order polarizabilities the transient grating scattering technique is an effective way to achieve high S/N SEP spectra as illustrated in Chapter 4. In considering applying this detection method, one should however be cautioned with the nonlinear dependence of the signal on the sample quantity and the transition strength.

The four wave mixing technique can be applied to SEP in two different ways. 1) Following the generation of sufficient population in the intermediate excited electronic level by the PUMP laser pulse, the DUMP pulse is then used to induce a degenerate four wave mixing process through the downward transition. The DUMP laser beam is first split into three beams, two of them are used to create a transient grating resonantly enhanced by the DUMP transition and the third beam is then diffracted by this grating, generating the fourth beam to be detected as the signal. With sufficiently high molecular density and strong electronic transitions, essentially noise free SEP spectra can be acquired [24]. 2) The PUMP beam is split into two parts and recombined at the sample to create a transient grating resonantly enhanced by the PUMP transition. The DUMP laser beam is then diffracted off this grating, generating a signal beam [25]. This two color resonant four wave mixing scheme can be used even when the intermediate level lifetime is short and cannot sustain a high steady-state population, which is needed if the degenerate four wave mixing scheme is used.

3.2.5 Pump-Dump-Probe

The population accumulated in the final level - the target excited vibrational level - in the SEP process can be monitored by a third PROBE laser pulse. This third laser pulse can induce fluorescence or ionization to be detected as the SEP signal. As the time delay between the DUMP and PROBE pulses can be adjusted, the PROBE pulse also allows the time evolution of the excited vibrational population created by SEP to be monitored. Energy transfer or reaction kinetics of molecules excited in selected vibration-rotation levels can thus be characterized [26,27].

3.2.6 Other Schemes

Several other schemes have also been demonstrated for detecting the SEP signal. Following the generation of vibrationally excited molecules by SEP, vibration-rotation and vibration-translation energy transfer induces a local heating in the laser excited region. As thermalization occurs much faster than heat diffusion, the elevated temperature creates a pressure front which can be detected as acoustic waves by a microphone [28]. The caution here is that neither the electronically excited intermediate level should contain much vibrational excitation nor should it relax rapidly through internal conversion. In both situations the intermediate level dominate the acoustic signal.

3.3 Intramolecular dynamics of highly vibrationally excited molecules

The ability afforded by SEP to acquire a moderately high resolution spectrum of high vibrational levels with vibrational quantum number, rotational quantum number, or at least symmetry selectivity has stimulated great progress in our understanding of the structure and dynamics of highly vibrationally excited molecules. The acquired spectra at high energies often contain high densities of spectral lines or broadened features that are difficult or meaningless to assign in terms of the traditional normal mode or local mode quantum numbers. This challenge in understanding the complex spectra also stimulated tremendous endeavor in developing statistical theories to describe the spectra and toward understanding the correspondence between the classical mechanical models for molecular vibrational motions and the quantum mechanical spectroscopic interpretations. The following subsections highlight a few important aspects associated with the dynamical and spectroscopic behaviors of highly vibrationally excited molecules.

3.3.1 Intramolecular vibrational relaxation (IVR)

The conventionally accepted definition of IVR depicts a process in which an initially excited well-defined vibrational motion degrades into a motion which can best be described by statistical models. The degradation is propelled by the coupling of the excited zeroth-order, well-defined motion to other originally nonexcited zeroth-order motions through terms in the Hamiltonian that are not incorporated in the zeroth-order Hamiltonian used to define these well-defined motions. For vibrations associated with the same electronic state the coupling arises primarily from vibrational anharmonicity or rotational motion. The existence of multiple electronic states in the energy region of consideration introduces additional relaxation mechanisms caused by vibronic couplings.

'What is the spectroscopic manifestation of IVR?' has been one of the most debated questions in our attempt to characterize IVR. The answer to this question in fact depends on the spectroscopic method chosen by the investigator for probing the molecules. If one uses SEP as the spectroscopic probe, then the IVR of the vibrational motions which are well defined in terms of normal modes is monitored, and it may show up as increased complexity in the spectrum. This is because the vibrational levels which would show up in the SEP spectrum must have non-zero Franck-Condon overlap with the electronically excited intermediate level. As the intermediate level usually is a low vibrational level of the electronic excited state and its wavefunction is well defined in the normal mode coordinates, the Franck-Condon factors reflect the composition of the wavefunction of the SEP-probed levels in these Franck-Condon active normal modes.

In the lower vibrational energy region in the electronic ground state where normal modes can fairly well define the wavefunctions of the vibrational levels, the SEP spectrum should be very simple, i.e. each vibrational level can be approximated as a zeroth-order normal mode level and those with substantial Franck-Condon factors would appear in the SEP spectrum as single lines. As the energy increases, the zeroth order normal mode Hamiltonian becomes inadequate for defining the vibrational motions. The wavefunction of each vibrational level, a molecular eigenstate, now contains a mixture of the normal mode wavefunctions. The normal mode wavefunction of a Franck-Condon bright zeroth-order level is distributed among many vibrational levels (the molecular eigen levels) and its manifestation in the spectrum, instead of a single line, appears as a multiplicity of lines. The distribution of the spectral intensity, which reflects the distribution of the Franck-Condon bright zeroth order wavefunction among the eigenstates, can be used to deduce the coupling mechanism and strength and therefore the details of the IVR dynamics. To experimentally detect this spectral distribution requires a technique that is selective in terms of the zeroth-order vibrational quantum numbers. IVR information will be unambiguously revealed only if the spectral lines in a complex spectrum can be assigned to a known zeroth-order wavefunction. This is where SEP exhibits its power. Not only does it provide the possibility of detecting vibrational levels at high energies, being a double resonance technique SEP allows labeling of both rotational and vibrational (normal modes) quantum numbers.

The SEP spectra of acetylene with as high as 30,000 cm^{-1} of vibrational energy provide an illuminating example of the interplay between experiments and theories to describe the dynamics of the highly excited vibrational levels. The first SEP spectra of acetylene at Doppler limited resolution contained dense spectral lines that are generally unassignable [5]. However, under lower spectral resolution the lines clump together and form features some of which can be assigned with normal mode quantum numbers [29]. Figure 3.4 shows a series of acetylene spectra obtained by a combination of SEP and dispersed fluorescence techniques under different levels of resolution. At lower resolution the spectral features are assignable in terms of the Franck-Condon bright normal mode levels. However, these features contain, in each one of them, a multitude of spectral lines that are associated with individual rotation-vibration eigenstates and appear only under higher resolution.

Clearly, other than the apparent indication of strong coupling among normal modes from the complexity of the spectra, extraction of dynamical and quantitative spectroscopic information from the spectra presents a challenge that can only be met with statistical treatments. Initial statistical analysis of the nearest neighboring line energy separation in the high resolution SEP spectra of acetylene vibrational levels at high energies revealed a Wigner, rather than Poisson, distribution [30]. The Wigner distribution was proposed to arise from an inter-mode coupling scheme that is dominated by matrix elements with random magnitudes [31]. For this quantum system its corresponding dynamics in classical mechanical trajectories should be 'chaotic' [32]. Whereas the Poisson distribution comes from a noncoupled quantum system which consists of vibrational modes that have regular, well-defined energy level patterns. This quantum system also has correspondingly 'regular' well defined trajectories in classical phase space. Based on the contention that a spectrum is the Fourier transform of the time dependent evolution of a Franck-Condon active wavefunction, the acetylene SEP spectra was subjected to such statistical Fourier analyses [33] which seemed to reveal two (picosecond/sub-nanosecond) time scales in the evolution. To identify the significance of these time scales, classical trajectory calculations were performed and the specific coupling channels that lead to stepwise vibrational energy relaxation were found [34,35]. Statistical analyses on the SEP and dispersed fluorescence spectra obtained at different vibrational energies enabled the determination of the onset of the chaotic dynamics and the most important modes/motions that are responsible for IVR [36].

Many other molecules, mostly small polyatomics such as HCN[37], HCO[38,39], SO_2[40], CS_2[41], H_2CO[42,43], HFCO[44], CH_3O[45], $H_2C_2O_2$[46], *etc.*, which at high energies may have resolvable spectral features under the resolution provided by SEP, have been studied in order to understand the IVR dynamics. In addition to SEP, other spectroscopic techniques, such as dispersed fluorescence and multiple resonance absorption, have been developed or applied to these and many other molecules to gain insight into the IVR of different kinds of zeroth-order modes. On the theoretical front, statistical spectral analyses have been developed to unravel the significance of complex spectra. Reduced chi-squares analysis of spectral intensity gives a measure of the fraction of classical phase space of the vibrational motions that are strongly coupled and shows chaotic dynamics [47]. This analysis should also reveal a degree of freedom that can be interpreted as the number of regions in the spectrally sampled phase space that are isolated from each other [48]. Quantum calculations, based on realistic or model potential energy surfaces, of spectra of highly excited vibrational levels have been performed [49-52]. Classical trajectory calculations have also been performed in attempt to find correlations with the spectra [53]. Furthermore, there have been development of spectral pattern-recognition procedures and the concept of "polyad" for identifying specific resonances and coupling channels responsible for IVR [54,55].

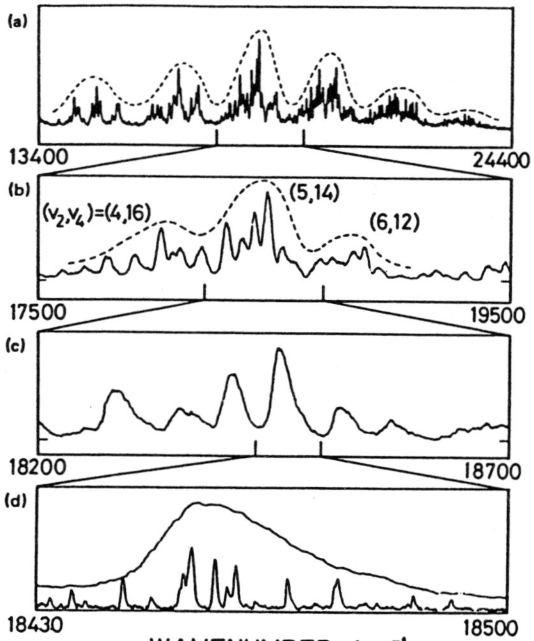

Figure 3.4 SEP and dispersed fluorescence spectra of acetylene under different resolutions. From Ref. [29].

3.3.2 Rotation induced vibrational coupling

One of the unique advantages of SEP is its capability of selecting the rotational quantum numbers for the vibrational levels that are probed. This capability allows the vibrational spectrum be studied without rotational congestion and with specified rotational quantum numbers. Vibrational level structures associated with a range of *J*, *K* values can be separately and systematically sampled. The effect of molecular rotations in inducing IVR can thus be examined. Studies on two molecules, formaldehyde and glyoxal, provide early examples of the success by SEP in this endeavor. In the case of formaldehyde [43], rotation induced vibrational mixing causes the sharing of intensity from a Franck-Condon bright vibrational mode with other vibrational modes. Extra spectroscopic lines appear at higher rotational excitation when the same set of Franck-Condon bright vibrational levels are probed. For glyoxal [46], rotation-induced (Coriolis) interaction between a pair of vibrational levels causes a systematic shift in their respective rotational level energies, affecting the rotational constants.

For a typical molecule, there are two mechanisms that are generally responsible for inducing IVR at high vibrational energies: Fermi (or anharmonic) and Coriolis (rotation induced) couplings. For the six vibrational modes of formaldehyde, of which the structure in its ground electronic state is classified as a C_{2v} point group, each vibrational level can be coupled to any other through either mechanism. For example, an A_1 symmetry level can be coupled to another A_1 level through anharmonic coupling. It may also couple to an A_2 level through rotation around the *a*-axis, to a B_1 level through the *b*-axis rotation, or to a B_2 level through the *c*-axis rotation. The anharmonic coupling conserves the vibrational symmetry and the rotational quantum numbers. On the other hand, Coriolis coupling conserves the overall rotation-vibration symmetry and *J*. The projection of *J* in the molecular frame is conserved only along the axis that is causing the vibration-rotation coupling. Rotation induced vibrational mixing can be understood as when the molecule rotates a set of Coriolis forces is generated for the atoms associated with a specific vibrational motion. As the Coriolis forces mimic the vibrational motion vectors of another vibrational mode, the two vibrational motions are mixed with each other while the molecule rotates [56].

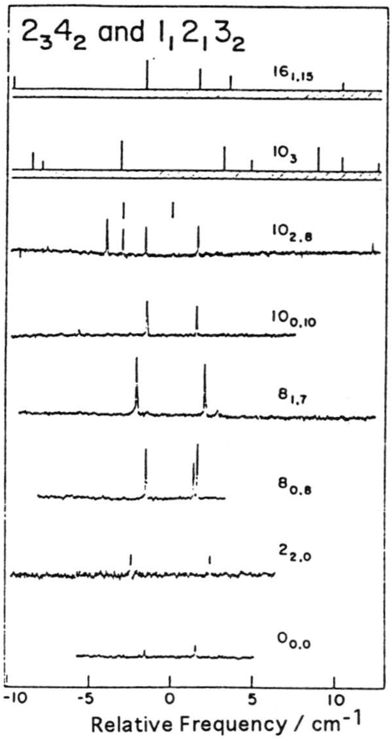

Fig. 3.5 SEP spectra of H_2CO indicating rotation induced vibrational coupling. From Ref. [43].

The matrix elements resulting from Coriolis forces are usually a fraction of the rotational constants. Thus for low vibrational levels, unless there is an accidental degeneracy between two vibrational levels, Coriolis effects are usually negligible. However, as the vibrational level density becomes larger at higher energies, the effect of Coriolis coupling on vibrational coupling becomes significant. As the Coriolis matrix elements are approximately proportional to either J or K, the effect of Coriolis coupling on IVR is particularly appreciable for higher rotational levels. One clear demonstration of this effect is the increasing complexity in the SEP vibrational spectra near the $2_3 4_2/1_1 2_1 3_2$ region shown in Fig. 3.5. The two lines in the J=0 spectrum can be unambiguously assigned to the two designated vibrational levels with detectable Franck-Condon factors. The simplicity of the spectrum at J=0 indicates how good the normal mode description is for these vibrational levels with about 7,500 cm^{-1} energy. However, as the rotational excitation increases extra spectral lines appear: the sharing of Franck-Condon factors indicates the occurrence of vibrational coupling and the breakdown of the normal mode description.

Since this first clear demonstration of the importance of molecular rotations in inducing IVR, such effects on many other molecules have been observed, some by spectroscopic techniques other than SEP [57,58]. At extremely high vibrational energies, such effects can be detected even for relatively low rotational levels [59]. It is clear from these studies that in considering IVR the role of rotations has to be included in the theoretical treatments [51,52], except for the J=0 systems.

3.3.3 Renner-Teller coupling

One type of perturbation to high vibrational levels, unique for those molecules whose electronic states arise from an orbitally degenerate electronic state with a linear configuration, is the Renner-Teller coupling. The highly reactive molecule methylene provides such an example. The two lowest singlet $\tilde{a}$ and $\tilde{b}$ states of CH_2, with bent strùctures, correlate with a degenerate, linear $^1\Delta g$ state. In the linear structure the degenerate bending vibrations produce an angular momentum along the molecular axis (K_a) that couples with the electronic angular momentum. This coupling lifts the electronic degeneracy into two non-degenerate states with bent structures. But also it causes strong coupling between the $K_a \neq 0$ levels of the low bending levels of the higher $\tilde{b}$ electronic state and the high bending levels of the lower a electronic state. There are two consequences of this coupling: The vibrational quantum numbers, defined according to the normal modes of each electronic state, of these levels are no longer good [60] and the rotational levels do not show the regular pattern expected for a near-symmetric top [60,61]. In fact since the coupling matrix elements are proportional to K_a, the A rotational constant decreases as the bending quantum number increases, instead of the increase expected of a molecule under large amplitude bending motion, for the highly excited bending levels of the lower electronic state. To confirm these effects of the Renner-Teller coupling on the highly excited bending levels, it was desirable to detect these high vibrational levels with rotational level selectivity. The degradation of vibrational quantum numbers will affect the Franck-Condon factors in the $\tilde{a}$-$\tilde{b}$ transition and should show up as extra lines in the spectra. The A rotational constant should be examined to see if it is a smooth function of bending excitation. The determination of the energies of the higher bending levels would also allow the barrier to linearity in the a state to be deduced.

As CH_2 is not a stable molecule and has to be generated by photolysis, the SEP experiments of this molecule involved three laser pulses for PHOTOLYSIS, PUMP, and DUMP. In these Flash Photolysis Stimulated Emission Pumping [62] experiments, the PHOTOLYSIS pulse first dissociated the precursor molecule ketene to produce singlet methylene. These vibration-rotationally hot CH_2 were allowed to collide with ambient gases to thermalize but the collisions were not enough to induce a crossing of CH_2 from the singlet to the triplet state before the arrival of the PUMP/PROBE pulses.

The J and K_a selected SEP spectra of methylene bending levels up to v=4 showed effects of Coriolis, Fermi, and singlet-triplet couplings [63]. In addition, unexpectedly strong ΔK_a=3 transitions were observed. These transitions arise because of the large asymmetry of the a- and b-state structures. The observation of these unexpected transitions provides a logical explanation for the huge number of observed but unassignable ã-b̃ rovibronic transitions [64]. However, the *A* constants of these vibrational levels do not show the expected decrease due to the Renner-Teller coupling. It appears that the vibronic coupling is not strong even for the v=4 level.

Even though flash photolysis-SEP with pulsed lasers is powerful for spectroscopy of excited vibrational levels of a transient species, it has apparent drawbacks in dealing with a small molecule with electronic transitions in the visible. For a molecule like CH_2, which has very large rotational constants, the DUMP laser needs to be scanned over a large frequency range to detect the high *J*, *K* levels. This increases the experimental complexity when the overlap of three laser beams is involved. But most of all, as the electronic transition in resonance with the PUMP laser is in the visible, the DUMP laser may need to be set in the near-IR region for accessing the high vibrational levels. The availability of a near-IR laser pulse for the DUMP and the experimental alignment of this laser beam may prohibit the SEP experiments. Under these circumstances, one may have to resort back to the dispersed fluorescence technique. With the development of the time-resolved Fourier transform technique [65], sufficient resolution in both time and frequency can be achieved for detecting the high vibrational levels of a transient species [1] through dispersed fluorescence. Indeed, dispersed fluorescence spectroscopy even of the small amount of methylene generated by photolysis, using the time-resolved Fourier transform emission spectroscopy, has been successful and allows the detection of the v=5 bending level at 6,400 cm^{-1} energy [66]. The *A* rotational constant clearly shows the expected decrease resulting from the Renner-Teller coupling. The magnitude of the decrease agrees well with that predicted from *ab initio* calculations including the vibronic coupling [60,61]. In fact once the *A* constant for v=5 level is measured, a comparison with all the bending level *A* constants reveals that the Renner-Teller effect can be felt for levels as low as v=3. Furthermore, the determination of the high bending level energy enables the barrier of bending motion to linearity to be set at 8,600 cm^{-1}, proving again the validity of the *ab initio* calculation [60,66].

Another excellent example of spectroscopic demonstration of the Renner-Teller effect on highly excited vibrational levels can be found on the radical NCO [67]. The vibronic coupling effect on the level energies has been carefully demonstrated in spectra obtained through a combination of the SEP and dispersed fluorescence techniques.

3.3.4 Unimolecular dissociation

One impetus that has propelled the study of highly excited vibrational levels has been the desire to understand the relationship between the form of the excitation and the induced dissociation behavior of the molecule in order to control dissociation channel and rate by manipulating the excitation of the molecule. For properly chosen polyatomic molecules, SEP spectroscopy can be used to detect vibrational levels even above the dissociation threshold. These vibrational levels, though having sufficient energy to cause dissociation, couple to the dissociation coordinate in variable strength because of the different nature of the excited vibrational wavefunctions, and therefore have different dissociation rates. The dissociation rate of a specific vibrational level can be deduced from the spectral line width. For a sufficiently small polyatomic molecule, the density of vibrational states near dissociation is usually not so high and single spectral lines can be resolved. The homogeneous portion of the linewidth, short of collisional effects, arises from the dissociation process, coupling of the vibrational level to the translational continuum along the dissociation coordinate.

As it is anticipated that highly excited vibrational levels are, in the description of the zeroth-order motions, strongly coupled to each other, the properties of these levels, including dissociation, should best be described by statistical models. Measurements on formaldehyde by Stark level crossing spectroscopy indeed showed that while there are fluctuations in the dissociation rates of isoenergetic levels at about 28,000 cm^{-1} energy, these rates can essentially be modeled by statistical distributions [59,68]. However, these studies do show a hint of correlation between the dissociation rate and the coupling strength of the level to the dissociation coordinate [68]. The SEP spectra of the high vibrational levels above dissociation limit contain the information of the Franck-Condon intensity, which is an indication of the projection of the vibrational wavefunction in a few selected normal modes, and the linewidth. An examination on the relationship between at least part of the vibrational wavefunction and the dissociation rate can thus be performed through an SEP study.

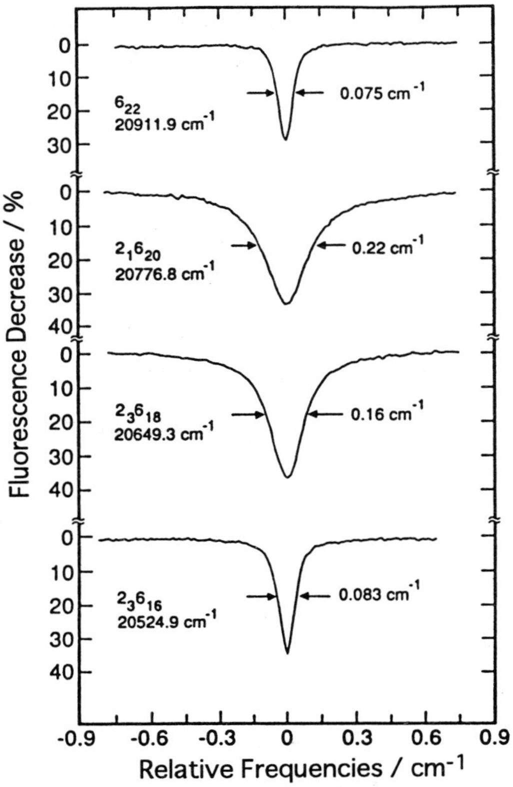

Fig. 3.6 Line width broadening of excited HFCO vibrational levels measured by SEP. From Ref. [44].

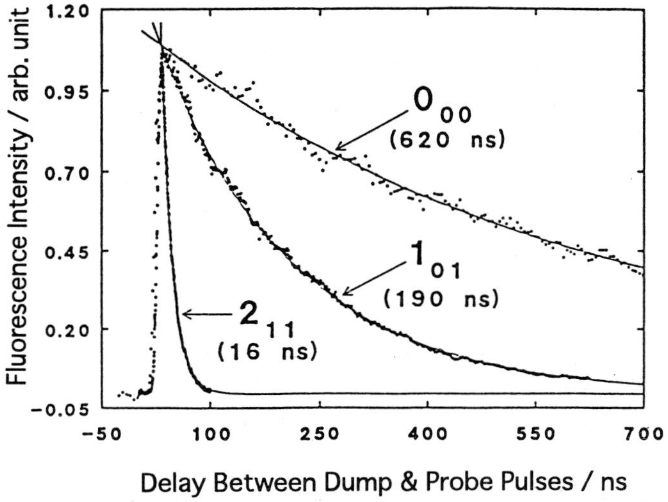

Fig. 3.7 Direct time-resolved measurement of dissociation of excited HFCO vibrational levels prepared by SEP. From Ref. [44].

The first dissociation study by SEP was performed on HFCO [44]. SEP spectra of HFCO in the range of 13,000 - 23,000 cm^{-1} have been recorded with 0.05 cm^{-1} resolution. From the Franck-Condon intensity patterns it has been deduced that IVR in this molecule near or even higher than the dissociation threshold of about 19,000 cm^{-1} for HFCO (HF + CO is non-statistical and depends on the mode of excitation, *i.e.*, levels with high quantum numbers in the CO stretch (ν_2) and out-of-plane bend (ν_6) modes are weakly coupled to the rest of the vibrational manifold and are still assignable. The dissociation rates of these and other more mixed levels were measured by two kinds of SEP-based methods. For levels which dissociate on a time scale faster than 1 nanosecond, the broadened linewidth (broader than the resolution) in the SEP spectra can be directly resolved. However, for levels which dissociate on a slower time scale and therefore have narrower linewidths, SEP was used to prepare an excited population in the selected vibrational level. A PROBE laser pulse was then introduced to monitor the decay of this population through laser induced fluorescence. Figures 3.6 and 3.7 show the line width measurements and the PUMP-DUMP-PROBE monitoring of the dissociative decay respectively.

The HFCO study has resulted in a few very interesting observations. The strong rotational dependence in the dissociation rates indicates that the Coriolis effect is essential in promoting the coupling between the ν_2-ν_6 vibrational modes and the dissociation coordinate. The strong dependence of dissociation rate on the symmetry of the vibrational levels provides information on the transition state structure. This information together with the observed vibrational mode-selectivity in IVR and dissociation enable the construction of a detailed model for understanding/predicting the reaction of vibrationally excited HFCO.

Understanding of unimolecular dissociation induced by state-specific excitation has also been achieved on HCO [38,39,69] and CH_3O [45] by using the SEP techniques. The most detailed study on HCO was conducted on a total of 73 levels spanning a 20,000 cm^{-1} range by a combination of SEP and dispersed fluorescence methods [38]. SEP, detected either as the fluorescence dip or in four wave mixing, was used to resolve the line width, particularly for the levels above the dissociation limit of HCO (H + CO at 5,000 cm^{-1}. These widths were again found to be mode-specific, showing that the levels with a combination of CH and bending vibrations have the wider width and faster dissociation rate. In the case of HCO, this information provides critical tests on the theoretically constructed potential energy surfaces. For CH_3O, levels near the CH_3O (H_2CO + H dissociation threshold of 6,900 cm^{-1} have been examined and homogeneously broadened linewidths, indicating dissociation, have been resolved [45].

3.3.5 Isomerization

Unlike dissociation, the other kind of unimolecular reaction, isomerization, does not manifest itself in the spectrum as line broadening. In fact the spectroscopic identification of a molecule undergoing isomerization reaction is still under development. Part of the reason stems from the fact that only recently, due to development of spectroscopic techniques like SEP, the levels with energies high enough for such reactions can be spectroscopically probed. To date, considerations on the issue of spectroscopic identification of isomerization can be best illustrated by the works on acetylene [70,71], HCN [37], and HCP [72].

Consider a potential energy surface that has two potential minima, each corresponding to a stable isomeric structure of the molecule and separated from each other by a discernible barrier. Below the barrier, short of strong tunneling, the vibrational levels have wavefunctions localized in the individual potential wells. Above the barrier, there will be levels which still have localized wavefunctions due to limited couplings. But there could also be levels with wavefunctions delocalized in both wells due to strong couplings. As one follows the vibrational spectra as a function of energy, concentrating on one isomer as in the SEP spectroscopy, the challenge is how to distinguish the delocalized levels, which can be used as a signature for isomerization, from the localized ones. The other indication could be a detectable increase in the level density above the barrier. However, these approaches will require that the observed levels associated with one structure can be definitively assigned to provide a good basis for detecting the existence of the delocalized levels or building an accurate knowledge of the level densities to compare with the spectra. Both are difficult to achieve at high vibrational energies. For HCN, the isomerization barrier to HNC is around 17,000 cm^{-1}. SEP spectra of HCN have been taken up to 20,000 cm^{-1} [37]. However, though some interesting phenomena that may be counted as abnormalities in the spectra have been observed, no definitive indication of isomerization in the spectra can be identified.

Unlike the HCN isomerization, where the HNC isomer correspomds to a local minimum on the potential energy surface, the HPC isomer appears as a saddle point on the HCP potential surface. The effect of this potential surface feature was probed in the SEP spectra sampling the HCP bending overtones [72]. It was found that the bending vibration exhibits regular pattern in spectra up to v_2=42. However, an abrupt change in the spectral pattern was found above v_2=36 and a sudden turning on of perturbations was observed at v_2=32. These spectral "abnormalities" were attributed to the dramatic change in the bending potential energy curve due to the presence of the isomer.

In the case of acetylene, the barrier of isomerizaton to vinylidene has been calculated to be around 14,000 cm^{-1}. The existence of the vinylidene isomer has been reported and characterized in the triplet state [73] and in the electronic ground singlet state by photoelectron detachment spectroscopy [74]. Among all the vibrational modes, the bending coordinate should be the one most closely coupled to the isomerization reaction coordinate. As the bending mode carries the strongest Franck-Condon factor in the electronic transitions, SEP should be most ideal for detecting the vibrational levels with delocalized waveunctions that characterize the isomerization. SEP spectra of acetylene from 10,000 cm^{-1} upward to about 30,000 cm^{-1} have be acquired. However, here too the spectroscopic identification of the existence of the isomer is not apparent. The high resolution SEP spectra near 14,000 cm^{-1} has already became complex and difficult to assign for the individual spectral lines. Based on the belief that the 'extra' lines that arise from the isomer should exist in all the SEP spectra taken with different intermediate acetylene levels, a cross correlation procedure was developed and applied to examine the spectral pattern associated with these common delocalized levels [70]. The result has been encouraging, but the unstable vinylidene on the electronic ground surface still awaits a more definitive confirmation through direct spectroscopic identification. Finding the signature of isomerization remains one of the most challenging issues in SEP spectroscopy.

3.4 Intermolecular interactions

SEP can be used effectively for studies related to intermolecular interactions, both as a spectroscopic technique or as a means for preparing vibrationally excited population. The spectroscopic aspect is in detecting the low frequency, intermolecular vibrational levels that exist when two molecules are bound together to form a van der Waals complex. As an excitation tool, SEP has been used to prepare a highvibrational level population the decay of which can then be monitored by a laser pulse. Energy transfer out of this level can then be characterized. There have been several successes in this endeavor. However, using SEP to excite a specific vibrational level and study the effect in bimolecular reactions remains to be illustrated. The primary obstacle here is that in SEP the DUMP laser pulse, even in saturation of the downward transition can at most transfer half of the intermediate level population into the targeted vibrational level. The other half goes into all the Franck-Condon active vibrational levels through spontaneous emission. The challenge here is then how to differentiate the effect of the selected excitation from that of the non-specific, spontaneous Franck-Condon excitation. However, this problem could be resolved as in the recent demonstration showing that by manipulating the timing of the PUMP-DUMP pulses, when both are single mode and Fourier transform limited, a near 100% transfer of excited population can be achieved [10].

3.4.1 Intermolecular (van der Waals) vibrational levels in molecular complexes

Spectroscopic characterization of a weakly bound molecular complex presents an effective way of studying intermolecular potentials and interactions. This long cherished practice has been particularly fruitful in the last two decades. A determination of the complex structure gives the location of the intermolecular potential minimum (or minima) that can be compared with theoretical calculations. On this potential surface there exist the low frequency, intermolecular, van der Waals vibrational levels, of which the determination of their energies can in principle reveal the potential energy surface away from the equilibrium position(s). As these low frequency vibrational levels have large amplitude motions, characterization of these levels may help to characterize a wide region of the intermolecular potential. Furthermore, the dynamic information associated with these levels is directly related to the intermolecular collisional processes.

Electronic spectroscopy through primarily laser induced fluorescence or resonant photo-ionization can be used for detecting van der Waals levels in electronically excited states. For studying van der Waals levels in the electronic ground state SEP is uniquely powerful. The energy difference of the PUMP-DUMP photons can be easily adjusted to match the low frequency vibrations and facilitate their detection. As SEP can be used to monitor vibrational levels over a wide energy range, van der Waals levels associated with intramolecular vibrational excitation can be systematically examined to reveal the couplings between the intramolecular and intermolecular vibrational modes. Furthermore, as the molecular complexes are usually generated in a supersonic expansion that may contain many different kinds of complexes with spectroscopic activity, the PUMP photon in SEP can label the complex of interest and simplify the assignment of the observed spectrum.

The first demonstration of SEP on a van der Waals complex was performed on glyoxal-Ar_2 [75] and later on glyoxal-Ar [76]. Figure 3.8 shows the SEP spectrum of glyoxal-Ar_2 near the intramolecular $v_5=1$ vibrational level of 550 cm^{-1}. Five van der Waals levels with energies 15.0, 20.2, 29.1, 33.7, and 39.8 cm^{-1} are clearly identifiable from the spectrum. For the glyoxal-Ar complex, where the electronic ground state complex structure is very different from the electronic excited state structure, van der Waals vibrational levels as high 80 cm^{-1} above the zero-point-level have been observed. The detection of van der Waals levels with more than half of the binding energy of the complex will allow the intermolecular potential to be probed over a large area.

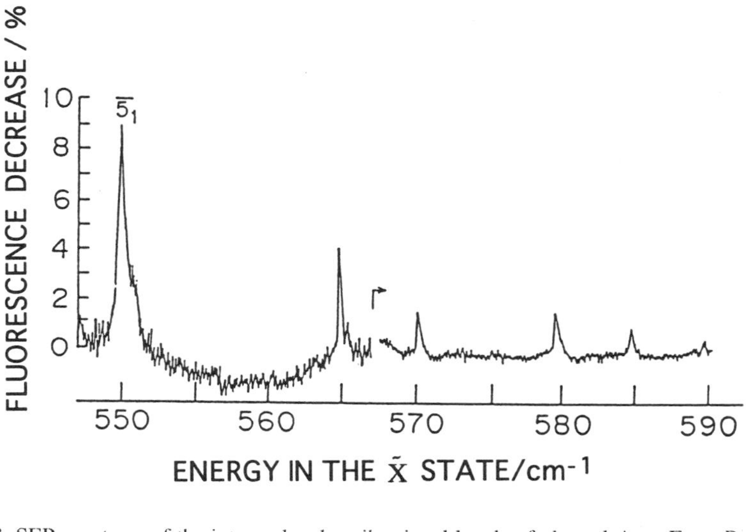

Fig. 3.8 SEP spectrum of the intermolecular vibrational levels of glyoxal-Ar_2. From Ref. [75].

The study of the OH-Ar complex presents the most successful example of using SEP to probe van der Waals levels for elucidating the potential energy surface [77]. Because of a dramatic difference in binding energies and structures of the A and X states of the complex, nearly all of the bound van der Waals levels on the X state surface, which has a ~100 cm^{-1} deep well, have been detected by SEP. Some dissociative levels with energies as high as 200 cm^{-1} were also observed. Comparison with an ab initio calculation led to assignments of the levels and refinement of an empirical potential.

SEP spectroscopy, primarily through the ion dip detection scheme, has been successful even on large molecular complexes such as Phenol-$(H_2O)_n$ (n=1-4) [78], benzene dimer [79], and benzonitrile complexes [80]. The selectivity of complex species in the supersonic expansion afforded by SEP greatly reduces spectral complexity and improves the assignability. The intermolecular level structures detected in these large clusters reveal primarily the coupling between the intramolecular vibrational modes and the van der Waals levels.

A recent study has used SEP as an excitation method to prepare complexes with intramolecular vibrational level excitation [81]. As the vibrational excitation exceeds the binding energy of the complex, vibrational predissociation occurs. The product molecules are then probed by resonant multiphoton ionization. For a large molecule, such as carbazole, with high vibrational levels even at relatively low energies, its van der Waals complex binding energy can be determined with high certainty in this way.

3.4.2 Collisional energy transfer of vibrationally excited molecules

Collisional energy transfer of vibrationally excited molecules, whether it is through vibration to vibration or vibration to translation/rotation, has been intensively studied and well characterized for lower vibrational

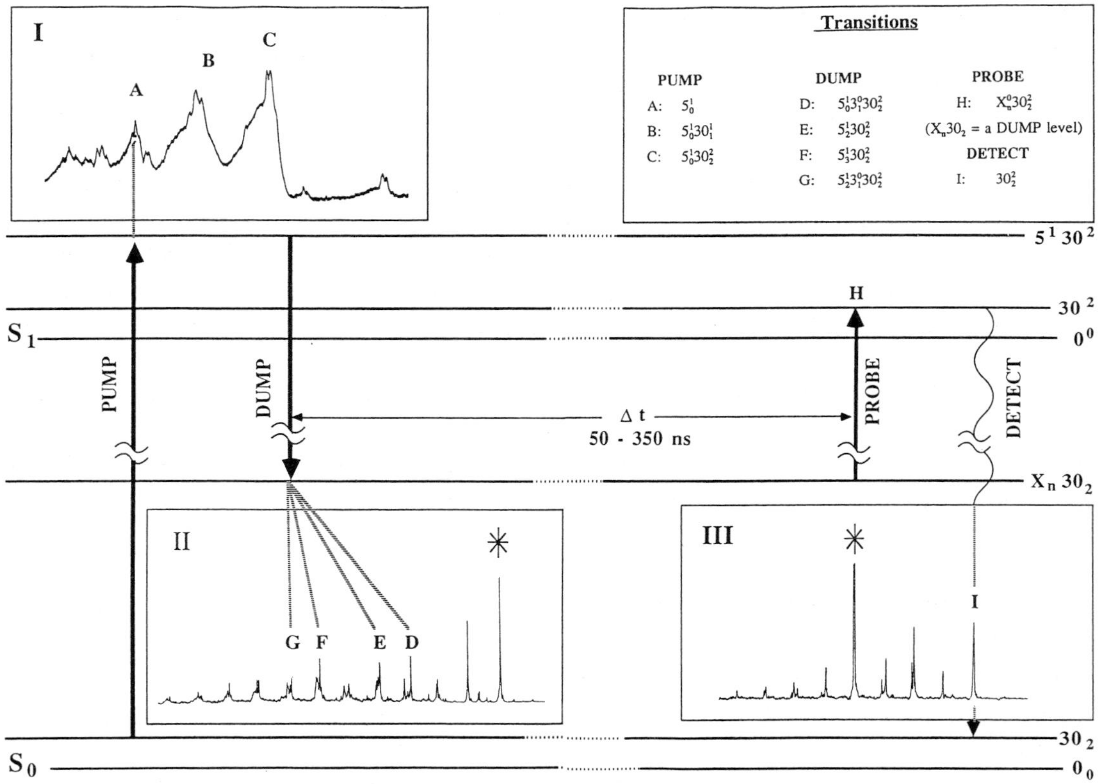

Fig. 3.9 Schematic representation of the temporal and spectroscopic characteristics in the SEP probe of energy transfer of vibrationally excited pDFB. From Ref. [82]. Panel I shows the PUMP spectrum, II the DUMP spectrum (* denotes the origin band) and III the probe spectrum (* indicates the Probe transition H). The signal is detected through ransition I which is isolated by a monochrometer.

levels where vibrational motion is essentially harmonic and well-defined. Propensity rules and the most important mechanisms have been identified. However, to understand the relationship between molecular excitation and chemical properties, particularly reactivity, it is most relevant that we understand collisional energy transfer of molecules excited to higher vibrational energies. Such studies, though desirable, have been handicapped by the lack of either the means for preparing molecules at selected high vibrational levels or detecting and characterizing the energy content of molecules during collisional processes following excitation. SEP can be used as an efficient means for vibrational excitation of selected levels. For properly chosen molecules the excitation can be as high as several tens of thousands of cm^{-1}. Even for lower vibrational levels SEP is a useful method as it can provide access to the levels unreachable by IR lasers with limited tuning ranges and power or by Raman techniques.

Despite the apparent advantage of SEP for level-selected energy transfer studies, probably due to experimental complexity, there have been only a few successful attempts in this approach. The first demonstration was on the molecule para-difluorobenzene (pDFB) [82]. Figure 3.9 shows a schematic representation of the temporal and spectroscopic characteristics of the experiments. The PUMP-DUMP pulses were tuned to resonance with two strong transitions through an S_1 vibronic level and generated a substantial population in an S_0 vibrational level with about 1300 cm^{-1} energy. These S_0 levels are difficult to access through IR transitions due to the lack of IR lasers in this region and the vibrational transition selection rules. Raman pumping is relatively inefficient for these levels as well. The population in the prepared level was then probed by a third laser pulse which induced fluorescence through another S_1 - S_0 transition. To avoid contamination from the PUMP induced fluorescence or fluorescence induced by other undesirable transitions excited by either the DUMP or PROBE, the PROBE-induced fluorescence was sent through a monochromator so only the fluorescence that clearly originates from the level of interest was monitored. The PROBE pulse was delayed with respect to the DUMP with a variable time duration. The time-dependent PROBE signal can thus give the lifetime of the selected excited level under collisional conditions. Relaxation by a variety of collisional partners and temperatures can then be measured.

A series of studies on collisional energy transfer of highly excited diatomic molecules NO and O_2 have been conducted using a similar PUMP-DUMP-PROBE approach. NO was excited to as high as v=25 at about 39,000 cm^{-1} energy [83] and O_2 as high as v=27 [84] by using SEP. The excitation of O_2 was achieved through an intermediate level that dissociates on a picosecond time scale. The PROBE pulse was used to monitor the decay of the initially excited level population as well as that of the energy-accepting level populated through collisional energy transfer in order to deduce collisional energy transfer rate constants.

The result on NO showed a dramatic increase of rate as a function of vibrational quantum number passing v=13 which cannot be understood by theories devised from experimental observations on lower vibrational levels. The O_2 energy transfer results suggest the occurrence of bimolecular reactions induced by vibrational excitation and the effect that such reactions may have in atmospheric systems.

3.5 Concluding remarks: opportunities in molecular kinetics and dynamics

Ever since the first demonstration on a molecular system, SEP has proven to be a versatile and powerful technique for spectroscopic characterization of the structure and dynamics of excited vibrational levels. The work reviewed in this chapter serves as a testament to the effectiveness of SEP spectroscopy and the resultant improved understanding of the structure and dynamics of vibrationally excited molecules achieved through its use. In intramolecular dynamics, spectroscopic characterization of isomerization processes remains a major unresolved problem and should be further explored. However, in contrast with the great strides made in spectroscopy using SEP, the apparent lack of advances in kinetic studies, particularly in bimolecular reactions, stands out.

Despite the apparent desire for demonstrating and understanding mode- or state-selective chemistry, there have been very limited activities in using SEP to populate excited species in selected states and to probe the subsequent bimolecular reactions. SEP has been used to populate metallic atoms in selected electronically excited states for the study of metal insertion reactions into hydrocarbons [85]. The strong oscillator strength associated with the PUMP transition and the relatively short radiative lifetime of the intermediate state, in the case of metal atoms, enable the SEP excitation to be efficient and clean, conditions highly desirable for studying bimolecular reactions of excited species. These conditions, however, do not necessarily exist for molecular excitations where the spontaneous emission from the intermediate level often results in populating a wide-range of Franck-Condon allowed excited vibrational levels other than the designated final level in SEP. This problem may be resolved by a variety of methods: Single-mode, Fourier-transform limited laser pulses can be used in the STIRAP scheme to achieve near total transfer of population to the final level from the intermediate level [10]. Alternatively, a combination of high degeneracy of the final level and high DUMP power may bring most of the excited population into the target level. Furthermore, SEP may be used to populate molecules in the final level with as high energy as possible so the lower levels populated through Franck-Condon pumping may be of negligible consequence in inducing reactions.

Following the recent demonstration of mode-selective excitation induced bimolecular reaction, achieved by single-photon local mode excitation in small molecules [86], there is substantial interest in examining bimolecular reaction dynamics of molecules excited to high vibration-rotation eigenstates. The latter can be effectively accessed through SEP. One of the most interesting question is that whether it is possible that by choosing a particular vibrational level with the right combination of normal mode motions one can influence the reactivity of the molecule.

There is also ample room for SEP to be an effective excitation means for energy transfer studies. Recent studies on intermolecular energy transfer of highly vibrationally excited molecules have revealed the importance of long range interactions, *e.g.* transition dipole coupling, in energy regions where vibronic coupling is strong [87]. However, these studies, have not been state specific. SEP can presumably be used to populate a specific level and a probe laser pulse can subsequently monitor the energy giving and receiving states behavior and test this qualitatively important discovery in quantitative details.

Last, but not least, is the possibility of extending SEP studies into ultrafast time scales. On this time scale, dynamics on the electronic excited state surface, the intermediate level, can be probed through the DUMP transition with a variable time delay from the PUMP. In a more complex scheme, a non-stationary wave packet can be prepared by SEP on a reactive ground state potential surface which is then probed by a PROBE pulse with a variable time delay from the PUMP-DUMP pulses. The interpretation of the observation will unavoidably be complex without prior detailed knowledge of the potential energy surfaces [88]. However, this will certainly be an effective way to reveal structure and dynamics on unstable potential energy surfaces.

Acknowledgement: This work is supported in part by the U.S. Department of Energy Chemical Sciences Division, the National Science Foundation, the University of Pennsylvania Research Foundation, the Research Corporation, the American Chemical Society Petroleum Research Fund, an Alfred P. Sloan Fellowship, the Henry and Camille Dreyfus Foundation and an Alexander von Humboldt Award to senior U.S. scientists.

References

1. G.V. Hartland, D. Qin, and H.L. Dai, *J. Chem. Phys.* **98**, 2469 (1993).
2. C. Kittrell, E. Abramson, J.L. Kinsey, S.A. McDonald, D.E. Reisner, R.W. Field, and D.H. Katayama, *J. Chem. Phys.* **75**, 2956 (1981).
3. W.D. Lawrence and A.E.W. Knight, *J. Chem. Phys.* **76**, 5637 (1982).
4. D. E. Reisner, P. H. Vaccaro, C. Kittrell, R. W. Field, J. L. Kinsey, and H. L. Dai, *J. Chem. Phys.* **77**, 573 (1982).

5. E. Abramson, R.W. Field, D. Imre, K.K. Innes, and J.L. Kinsey, *J. Chem. Phys.* **80**, 2298 (1984).
6. C.E. Hamilton, J.L. Kinsey, and R.W. Field, *Annu. Rev. Phys. Chem.* **87**, 493 (1986).
7. Special issue on *Molecular Spectroscopy and Dynamics by Stimulated Emission Pumping* (H.L. Dai ed.), *J. Opt. Soc. Am.* **B**, Vol. 7, Sept. (1990).
8. H. L. Dai, in *Advances in Multi-Photon Processes and Spectroscopy*(S. H. Lin ed.) Vol. 7, p. 169, World Scientific, River Edge, New Jersey(1991).
9. Chapters in *Molecular Dynamics and Spectroscopy by Stimulated Emission Pumping* (H.L. Dai and R.W. Field ed.) *Adv. Ser. Phys. Chem.* Vol. 4, World Scientific, River Edge, New Jersey (1995).
10. a) K. Bergman and B.W. Shore, Chapter 9 in Ref. [9]; (b) G.Z. He, A. Kuhn, S. Schiemann, and K. Bergmann, *J. Opt. Soc. Am.* **B7**, 1960 (1990).
11. C. Kitrell, Chapter 3 in Ref. [9].
12. E.A. Rohlfing and J.E.M. Goldsmith, *J. Opt. Soc. Am.* **B7**, 1915 (1990).
13. D.E. Cooper, C.M. Klimcak, and J.E. Wessel, *Phys. Rev. Lett.* **46**, 324 (1981).
14. J. Xie, G. Sha, X. Zhang, and C. Zhang, *Chem. Phys. Lett.* **124**, 99 (1986).
15. T. Ebata, M. Furukawa, T. Suzuki, and M. Ito, *J. Opt. Soc. Am.* **B7**, 1890 (1990).
16. Th. Weber, E. Riedle, and H.J. Neusser, *J. Opt. Soc. Am.* **B7**, 1875 (1990).
17. R.J. Stanley and A.W. Castleman, Jr., *J. Chem. Phys.* **92**, 5770 (1990).
18. G.V. Hartland, B.F. Henson, V.A. Venturo, R.A. Hertz, and P.M. Felker, *J. Opt. Soc. Am.* **B7**, 1950 (1990).
19. D. Frye, H. T. Liou, and H. L. Dai, *Chem. Phys. Lett.* **133**, 249 (1987).
20. W. Demtröder, *Laser Spectroscopy*, Springer-Verlag, New York (1982).
21. D. Frye, *Ph.D. Thesis*, University of Pennsylvania (1989).
22. P.H. Vacarro, R.L. Redington, J. Schmidt, J.L. Kinsey, and R.W. Field, *J. Chem. Phys.* **82**, 5755 (1985).
23. J.P. Pique, F. Stoeckel, and A. Camparque, *Appl. Opt.* **26**, 3103 (1987).
24. P.H. Vacarro, Chapter 4 in this book.
25. S. Williams, J.D. Tobiason, J.R. Dunlop, and E.A. Rohlfing, *J. Chem. Phys.* **102**, 8242 (1995).
26. S.H. Kable, W.D. Lawrence, and A.E.W. Knight, Chapter 16 in Ref. [9].
27. X. Yang, J.M. Price, J.A. Mack, C.G. Morgan, C.A. Rogaski, and A.M. Wodtke, Chapter 17 in Ref. [9].
28. D.J. Moll, G.R. Parker, and A. Kupperman, *J. Chem. Phys.* **80**, 4800 (1984).
29. K. Yamanouchi, N. Ikeda, S. Tsuchiya, D.M. Jonas, J.K. Lundberg, G.A. Adamson, and R.W. Field, *J. Chem. Phys.* **95**, 6330 (1991).
30. E. Abramson, R.W. Field, D. Imre, K.K. Innes, and J.L. Kinsey, *J. Chem. Phys.* **83**, 453 (1985).
31. S. Mukamel, J. Sue, and A. Pandey, *J. Chem. Phys.* **80**, 2298 (1984).
32. P. Pechukas, *Phys. Rev. Lett.* **51**, 943 (1983).
33. J.P. Pique, Y. Chen, R.W. Field, and J.L. Kinsey, *Phys. Rev. Lett.* **58**, 975 (1987).
34. T.A. Holme and R.D. Levine, *J. Chem. Phys.* **89**, 3379 (1988).
35. H.S. Taylor, Chapter 28 in Ref. [9].
36. D.M. Jones, S.A.B. Solina, B. RaJaram, R.J. Silbey, R.W. Field, K. Yamanouchi, and S. Tsuchiya, *J. Chem. Phys.* **97**, 2813 (1992).
37. D.M. Jonas, X. Yang, and A.M. Wodtke, *J. Chem. Phys.* **97**, 2284 (1992).
38. G.W. Adamson, Z. Zhao, and R.W. Field, *J. Mol. Spectrosc.* **160**, 11 (1993).
39. J.D. Tobiason, J.R. Dunlop, and E.A. Rohlfing, *J. Chem. Phys.* **103**, 22 (1995).
40. K. Yamanouchi, S. Takeuchi, and S. Tsuchiya, *J. Chem. Phys.* **92**, 4044 (1990).
41. P.H. Vaccaro, Chapter 1 in Ref. [9].
42. D. E. Reisner, R. W. Field, J. L. Kinsey, and H. L. Dai, *J. Chem. Phys.* **80**, 5968 (1984)
43. H.L. Dai, K.L. Korpa, J.L. Kinsey, and R.W. Field, *J. Chem. Phys.* **80**, 2298 (1984).
44. Y.S. Choi and C.B. Moore, *J. Chem. Phys.* **94**, 5414 (1991).
45. F. Temps, Chapter 10 in Ref.[9].
46. D. Frye, L. Lapierre, and H. L. Dai, *J. Chem. Phys.* **89**, 2609 (1988).
47. E.B. Stechel and E.J. Heller, *Annu. Rev. Phys. Chem.* **35**, 563 (1984).
48. R.D. Levine, *Adv. Chem. Phys.* **70**, 53 (1987).
49. C. Leforestier and R.E. Wyatt, Chapter 21 in Ref. [9].
50. M. Aoyagi, S.K. Gray, and M.J. Davis, *J. Opt. Soc. Am.* **B7**, 1859 (1990).
51. D.C. Burleigh, A.B. McCoy, and E.L. Sibert, Chapter 26 in Ref. [9].
52. T. Uzer and T. Carrington, Chapter 27 in Ref. [9].
53. J.M. Gomes Llorente, S.C. Farantos, O. Hahn, and H.S. Taylor, *J. Opt. Soc. Am.* **B7**, 1851 (1990).
54. M.J. Davis, Chapter 23 in Ref. [9].
55. M.E. Kellman, Chapter 25 in Ref. [9].
56. H.A. Jahn, *Phys. Rev.* **56**, 680 (1939).
57. D.S. Perry, *J. Chem. Phys.* **98**, 6665 (1993).
58. X. Lou and T.R. Rizzo, *J. Chem. Phys.* **93**, 8620 (1990).
59. W.F. Polik, D.R. Guyen, and C.B. Moore, *J. Chem. Phys.* **92**, 3453 (1990).
60. W.H. Green, N.C. Handy, D.J. Knowels, and S. Carter, *J. Chem. Phys.* **94**, 118 (1991).
61. G. Duxbury and Ch. Jungen, *Mol. Phys.* **63**, 981 (1988).
62. W. Xie, A. Ritter, C. Harkin, K. Kasturi, and H. L. Dai, *J. Chem. Phys.* **89**, 7033 (1988)

63. W. Xie, C. Harkin, and H.L. Dai, *J. Chem. Phys.* **93**, 4615 (1990)
64. H. Petek, D.J. Nesbitt, D.C. Darwin, and C.B. Moore, *J. Chem. Phys.* **86**, 1172 (1987).
65. G.V. Hartland and H.L. Dai, Chapter 5 in Ref. [9].
66. G.V. Hartland, D. Qin, and H.L. Dai, *J. Chem. Phys.* **102**, 6641 (1995).
67. M. Wu and T.J. Sears, Chapter 6 in Ref. [9].
68. a) H.L. Dai, in *Advances in Molecular Vibrations and Collision Dynamics* Vol. 1B, p. 305 (J.M. Bowman ed.) JAI Press, Connecticut (1991); (b) A. Ritter, *Ph.D. Thesis*, University of Pennsylvania (1994).
69. D.W. Neyer, X. Luo, I. Burak, and P.L. Houston, *J. Chem. Phys.* **102**, 1645 (1995).
70. Y. Chen, D.M. Jonas, J.L. Kinsey, and R.W. Field, *J. Chem. Phys.* **91**, 3976 (1989).
71. J.K. Lundburg, Chapter 22 in Ref. [9].
72. H. Ishikawa, Y.-T. Chen, Y. Oshima, J. Wang, and R.W. Field, *J. Chem. Phys.* **105**. 7383 (1996).
73. A. H. Laufer, *J. Chem. Phys.* **76**, 945 (1982).
74. K.M. Ervin, J. Ho, and W.C. Lineburger, *J. Chem. Phys.* **91**, 5974 (1989).
75. D. Frye, P. Arias, and H. L. Dai, *J. Chem. Phys.* **88**, 7240 (1988).
76. L. Lapierre, Ph.D. Thesis, University of Pennsylvania (1991).
77. M.I. Lester, W.H. Green, C. Chakravarty, and D.C. Clary, Chapter 18 in Ref. [9].
78. R.J. Stanley and A.W. Castleman. Chapter 19 in Ref. [9].
79. T. Ebata and M. Ito, Chapter 15 in Ref. [9].
80. M. Takayanagi and I. Hanazaki, *J. Opt. Soc. Am.* **B7**, 1899 (1990).
81. T. Droz, T. Burgi, and S. Leutwyler, *J. Chem. Phys.* **103**, 4035 (1995).
82. S.H. Kable, W.D. Lawrence, and A.E.W. Knight, Chapter 16 in Ref. [9].
83. X. Yang, E.H. Kim, and A.M. Wodtke, *J. Chem. Phys.* **96**, 5111 (1992).
84. J.A. Mack, K. Mikulecky, and A.M. Wodtke, *J. Chem. Phys.* **105**, 4105 (1996).
85. Y. Wen and J.C. Weisshaar, *J. Chem. Phys.*, in press (1997).
86. F.F. Crim, *Annu. Rev. Phys. Chem.* **44**, 397 (1993).
87. G.V. Hartland, D. Qin, and H.L. Dai, *J. Chem. Phys.* **102**, 8677 (1995).
88. L. Hunziker, P. Ludowise, W. Price, M. Morgen, M. Blackwell, and Y. Chen, Chapter 2 in Ref. [9].

4 Degenerate Four–Wave Mixing (DFWM) Spectroscopy

Patrick H. Vaccaro
Department of Chemistry, Yale University, 225 Prospect Street, New Haven, CT 06520-8107 U.S.A.

4.1 Introduction

Contemporary molecular spectroscopy is a multifaceted discipline, the diversity of which reflects the fruitful interplay that long has existed between theoretical advancements and technological innovations. In particular, the advent of tunable laser sources has led to a veritable explosion of optical techniques designed to unravel the isolated and collective behavior of molecules [1, 2]. While linear absorption spectroscopies remain unparalleled in terms of overall applicability and conceptual simplicity, their practical implementation often has been restricted by the need to extract potentially weak signals from large and fluctuating levels of baseline light. A variety of ingenious means have been conceived to alleviate this problem, however, the use of an inherently background–free technique, such as that embodied in the ubiquitous Laser–Induced Fluorescence (LIF) methodology, affords substantial advantages for both sensitivity and resolution. Indeed, the LIF scheme offers the amazing possibility of observing a single atom or molecule with complete internal quantum state specificity [3], a feat matched only by hybrid methods based upon optically–induced ionization processes (*e.g.*, Resonant–Enhanced Multi–Photon Ionization or REMPI spectroscopy). Unfortunately, any form of spectroscopy that relies upon secondary interactions (*e.g.*, fluorescence or ionization) for the detection of resonant optical transitions must be subject to certain limitations, with experiments built upon the observation of spontaneous emission being restricted to species that exhibit reasonably large fluorescence quantum yields and relatively long fluorescence lifetimes. In the case of LIF, this usually dismisses from consideration the large and important class of molecules that support unstable (*i.e.*, dissociative or predissociative) excited electronic states. Under such circumstances, a background–free, absorption–based probe, which combines high sensitivity and resolution with general species applicability, would prove to be most advantageous. Many of these desirable features can be found in various forms of nonlinear spectroscopy built upon the concepts of four–wave mixing [4, 5]. Although a considerable number of nonlinear optical techniques can fall into this rather broad classification, including conventional coherent Raman scattering processes [*e.g.*, Coherent Anti–Stokes Raman Scattering (CARS) and Coherent Stokes Raman Scattering (CSRS)], this chapter will focus on a fully resonant–enhanced variant of such schemes known as Degenerate Four–Wave Mixing (DFWM) [6-8].

Recent years have witnessed a renewed interest in degenerate four–wave mixing, a technique traditionally employed for the investigation of optical phase–conjugation processes [9, 10] and ultrafast relaxation phenomena [11]. Although exhibiting a greater–than–linear dependence upon sample number density that is the hallmark of any nonlinear form of spectroscopy [4, 5], the DFWM interaction entails a substantial degree of resonant enhancement which permits the background–free detection of molecular transitions with sensitivities that far exceed those attainable through other nonlinear schemes (*e.g.*, conventional CARS). This unique capability, coupled with a relatively straightforward experimental configuration that affords sub–Doppler spectral resolution, suggests numerous potential applications for DFWM in the field of molecular spectroscopy [6, 8]. The coherent, beam–like response of four–wave mixing also has motivated substantial efforts to exploit DFWM as a probe of trace species in luminous and/or hostile environments that are inaccessible to more conventional spectroscopic methods [7].

In 1986, building on their previous investigations of flame–entrained sodium atoms [12], Ewart, *et al.* reported the use of DFWM as a means of monitoring OH radicals produced in combustion environments [13]. Although not the first DFWM measurements performed on a molecule, this pioneering work served to stimulate much of the current interest in four–wave mixing spectroscopy, with a substantial portion of ongoing research directed towards understanding the fundamental and practical limitations imposed upon this burgeoning technique. The chemically relevant OH radical has remained the most studied of any molecular system [14-21]; however, DFWM experiments based upon resonant electronic excitation have been executed successfully on a variety of other diatomics, including both stable (*e.g.*, NO [22-28], CO [29-31], I_2 [32-37], H_2 [29, 38, 31], and O_2 [29, 31]) and transient (*e.g.*, NH [15, 39], C_2 [40], CH [41, 40], SH [42, 43], and NaH [44, 45]) species. The quantitative analysis of trace constituents in atmospheric and combustion related processes has been the primary motivation for many of these efforts, however, the potential spectroscopic applications of degenerate four–wave mixing have not gone unnoticed. For example, two–photon resonant variants of DFWM [46, 30, 47, 48, 38, 31, 49], as well as Optical–Optical Double Resonance (OODR) schemes relying upon DFWM detection [23], have permitted exploration of high–lying (one–photon forbidden) electronic states in both atomic and molecular systems. Likewise, recent extension of the DFWM methodology to diatomic vibrational spectroscopy (*e.g.*, HF [50] and HCl [51]) has demonstrated unique capabilities as well as unprecedented levels of sensitivity.

While polyatomic molecules, with their more robust and varied internal degrees of freedom, present additional difficulties (and rewards) for any spectroscopic probe, resonant DFWM measurements built upon visible/ultraviolet excitation of electronic transitions have been conducted on a number of systems including HCO [52, 53], NO_2 [54-57], H_2O [29, 31, 58], SiC_2 [59, 60], NH_3 [47, 61], C_3 [59, 60], CH_3 [62], CS_2 [63-65, 8, 66], SO_2 [67], S_2O [66], H_2CO [65, 68], glyoxal [69], benzene [70, 71], and pyrazine [65]. More importantly, non–fluorescing species (*e.g.*, azulene [72] and malonaldehyde [73, 74]), as well as weakly–bound molecular complexes (*e.g.*, glyoxal dimer [69]), have proven amenable to investigation via such techniques. The advent of tunable infrared sources based primarily on Optical Parametric Oscillator (OPO) technology [75] has made possible the direct interrogation of molecules in their ground state potential energy surfaces, with high resolution infrared DFWM spectroscopy reported for C_2H_2 [76, 77], NO_2 [51], and CH_4 [76]. Incorporation of the background–free DFWM detection scheme into the folded OODR configuration of Stimulated Emission Pumping (SEP) [78, 8] has been shown to provide a facile means for examining vibrationally highly–excited polyatomic systems under both bulk–gas [63] and free–jet conditions [64]. Two–color variants of resonant four–wave mixing have been developed [58, 59] and applied to a variety of spectroscopic problems, not the least of which is the elucidation of molecular structure and dynamics in regimes of extreme vibrational excitation [60, 53]. Although non–resonant forms of DFWM have a long tradition of use in the time–domain community [11], the recent extension of resonant–enhanced DFWM techniques into the femtosecond regime [79] should foster a new class of studies designed to unravel the nature of unimolecular and bimolecular chemical transformations.

This chapter will focus on DFWM as applied to the spectroscopic investigation of isolated molecular behavior, a situation most closely approached by species maintained under the rarefied (*i.e.*, collision–free) conditions found in low pressure bulk gases or supersonic free–jet expansions. The introduction to four–wave mixing contained in Section 4.2 is designed to provide a theoretical framework for assessing the utility and feasibility of such techniques. Building upon a weak–field perturbative treatment in which impinging monochromatic waves give rise to a third–order nonlinear polarization of the target medium [4, 80, 5], this analysis will motivate the subsequent derivation of DFWM signal expressions with their characteristic dependencies upon molecular and experimental parameters. Extensions to incorporate "real–world" complications, including Doppler and optical saturation effects, will be presented with particular attention directed towards analogies involving the concepts of volume holography and transient gratings [11]. Aside from highlighting the unique capabilities afforded by DFWM, this discussion will endeavor to demonstrate the practical limitations imposed upon the acquisition and interpretation of resonant four–wave mixing data. Specific applications of DFWM in the study of molecular structure and dynamics will be illustrated in Section 4.3 and will be followed (Section 4.4) by a brief description of related methods based upon the aforementioned OODR and/or two–color schemes. Finally, a short summary (Section 4.5) will address the future prospects of degenerate four–wave mixing as a tool for molecular spectroscopy.

4.2 Methodology and Theory

4.2.1 *Introduction*

In the case of a purely reactive (*i.e.*, non–absorbing or transparent) medium [81, 82], the phenomenon of degenerate four–wave mixing can be viewed as a form of elastic photon scattering in which the molecular sample serves as a "catalyst" designed to mediate the direction and magnitude of energy flow among various optical fields. The practical implementation of this technique entails the use of three incident beams which have identical or degenerate frequency, ω, but are otherwise distinguishable owing to their directional and/or polarization characteristics. Nonlinear interactions with the target molecules give rise to a fourth output beam, with energy conservation demanding that the frequency of this coherently "scattered" radiation be equal to ω. The direction of propagation and polarization of the emerging signal wave also follow from consideration of various conservation criteria.

As with all forms of linear and nonlinear spectroscopy [2], the utility of four–wave mixing as a probe of molecular structure stems from frequency–dependent variations in the efficiency of signal production. In particular, these changes in optical response reflect fundamental properties of the target medium (*e.g.*, transition energies and decay rates), with exceptionally strong DFWM interactions expected to accompany resonant excitation of allowed molecular transitions. Consequently, by monitoring the intensity of the emerging signal wave as a function of the incident frequency, ω, an essentially background–free, "absorption–like" spectrum can be obtained for the atom or molecule under investigation. Since the four–wave mixing process does not rely upon "secondary effects" (*e.g.*, fluorescence or ionization) to identify the location of a spectroscopic transition, such schemes should be applicable to any system that displays resonant attenuation or amplification of light.

As demonstrated most vividly by the transient grating analogies of Section 4.2.3, the spatially–distinct nature of the four–wave mixing signal beam stems from constructive and destructive interference effects that accompany the induced emission of optical radiation from a "phase–locked" *array* of target molecules. From this perspective, it is quite evident that such nonlinear schemes can never hope to attain the "single–molecule" detection limits afforded by their linear counterparts, with the best reported sensitivities for DFWM being roughly $10^8 - 10^9$ molecules / cm^3 / quantum state. However, methods based upon the coherent interaction of light with matter can provide substantial advantages whenever the intrinsic sensitivity of a linear technique is compromised by competing effects. For example, spectroscopic transitions involving excited states that undergo rapid nonradiative relaxation or target species entrained in incandescent environments that exhibit broadband optical noise can both lead to situations that undermine the utility of methods based upon the observation of fluorescence. Owing to its direct absorption–based response and remote sensing capabilities [7], DFWM spectroscopy should be applicable under such adverse conditions.

4.2.2 *General Description*

Reduced to simplest terms, the output or signal beam in a four–wave mixing interaction stems from the bulk electric polarization, $\boldsymbol{P}(t)$, induced within a target medium by virtue of its nonlinear coupling with three impinging electromagnetic waves [4, 80, 5]. In the case of DFWM, all optical fields are of identical frequency, ω, in the laboratory–fixed frame of reference. However, in order to distinguish and classify the various wave–coupling phenomena that can contribute to an observed response, it will prove convenient to label the spatial and temporal quantities associated with each beam by a distinct subscript. For experiments performed with pulsed sources of narrowband light (*e.g.*, single longitudinal mode lasers), it is most appropriate to express the time–dependent electric field vector for an individual electromagnetic wave in terms of a quasi–monochromatic approximation [4]:

$$\boldsymbol{E}_d(t) = \frac{1}{2}\left[\boldsymbol{E}_{\omega_d}(t)\mathrm{e}^{-\mathrm{i}\omega_d t} + \boldsymbol{E}_{-\omega_d}(t)\mathrm{e}^{\mathrm{i}\omega_d t}\right] \tag{4.1}$$

where the subscript "d" denotes a particular incident or generated beam of light which oscillates at the characteristic central (*i.e.*, carrier) frequency given by ω_d. The envelope functions $\boldsymbol{E}_{\omega_d}(t)$ and $\boldsymbol{E}_{-\omega_d}(t)$ contain amplitude and phase information as well as the polarization characteristics of the transverse electromagnetic wave. These quantities are assumed to be varying slowly on the timescale of an optical period (*i.e.*, $1/\omega_d$) and must satisfy the condition $\boldsymbol{E}_{-\omega_d}(t) = \boldsymbol{E}^*_{\omega_d}(t)$ in order to ensure that $\boldsymbol{E}_d(t)$ represents a classical (*i.e.*, pure real) field [83].

As shown in Section 4.2.4, the time–domain description of nonlinear optics [84, 5] provides a convenient theoretical framework for perturbative expansion of the DFWM response generated under weak–field (*i.e.*, unsaturated) conditions. From a spectroscopic point of view, however, it is often more appropriate to transcribe quantities into the frequency domain, with the Fourier Transform of $\boldsymbol{E}_d(t)$, denoted by $\tilde{\boldsymbol{E}}_d(\omega)$, characterizing the frequency spectrum spanned by an optical pulse. While $\tilde{\boldsymbol{E}}_d(\omega)$ clearly depends on the precise form of $\boldsymbol{E}_{\omega_d}(t)$, for pedagogical purposes it is usual to consider monochromatic fields defined by constant (*i.e.*, time–independent) envelope functions [4]. With the additional assumption of plane wavefronts, this leads to an electric vector having the form:

$$\begin{aligned}\boldsymbol{E}_d(t) &= \frac{1}{2}\left[\boldsymbol{E}_{\omega_d}\,\mathrm{e}^{-\mathrm{i}\omega_d t} + \boldsymbol{E}_{-\omega_d}\,\mathrm{e}^{\mathrm{i}\omega_d t}\right] \\ &= \frac{1}{2}\left[\hat{\boldsymbol{e}}_{\omega_d}\,\mathcal{E}_{\omega_d}\,\mathrm{e}^{\mathrm{i}\left(\boldsymbol{k}_{\omega_d}\cdot\boldsymbol{r}-\omega_d t\right)} + \hat{\boldsymbol{e}}_{-\omega_d}\,\mathcal{E}_{-\omega_d}\,\mathrm{e}^{\mathrm{i}\left(\boldsymbol{k}_{-\omega_d}\cdot\boldsymbol{r}+\omega_d t\right)}\right] \\ &= \frac{1}{2}\left[\hat{\boldsymbol{e}}_d\,\mathcal{E}_d\,\mathrm{e}^{\mathrm{i}\left(\boldsymbol{k}_d\cdot\boldsymbol{r}-\omega_d t\right)} + \hat{\boldsymbol{e}}^*_d\,\mathcal{E}^*_d\,\mathrm{e}^{-\mathrm{i}\left(\boldsymbol{k}_d\cdot\boldsymbol{r}-\omega_d t\right)}\right]\end{aligned} \tag{4.2}$$

where the scalar quantity $\mathcal{E}_{\omega_d}$ describes the amplitude for an electromagnetic wave propagating in the direction specified by wavevector $\boldsymbol{k}_{\omega_d}$ and $\hat{\boldsymbol{e}}_{\omega_d}$ is the polarization unit vector. In particular, $\hat{\boldsymbol{e}}_{\omega_d}$ defines the direction of polarization for the transverse wave and satisfies the normalization condition $\hat{\boldsymbol{e}}_{\omega_d}\cdot\hat{\boldsymbol{e}}^*_{\omega_d} = 1$ (*i.e.*, when expressed in a basis of Cartesian unit vectors, $\hat{\boldsymbol{e}}_{\omega_d}$ is complex for elliptically–polarized or circularly–polarized fields and pure real for linearly polarized fields [85, 86]). As in the case of Eq. (4.1), the implicit reality of classical fields [83] demands that $\boldsymbol{E}_{-\omega_d} = \boldsymbol{E}^*_{\omega_d}$ from which it follows that $\mathcal{E}_{-\omega_d} = \mathcal{E}^*_{\omega_d}$, $\hat{\boldsymbol{e}}_{-\omega_d} = \hat{\boldsymbol{e}}^*_{\omega_d}$, and $\boldsymbol{k}_{-\omega_d} = -\boldsymbol{k}_{\omega_d}$ (*i.e.*, in the absence of optical dispersion effects [83], the magnitude of $\boldsymbol{k}_{\omega_d}$ is given by ω_d/c). Owing to the simple correspondence between these quantities, it is common to drop the subscript "ω_d" in favor of the more compact notation introduced by the last equality of expression (4.2) (*viz.*, $\mathcal{E}_d \equiv \mathcal{E}_{\omega_d}$, $\hat{\boldsymbol{e}}_d \equiv \hat{\boldsymbol{e}}_{\omega_d}$, and $\boldsymbol{k}_d \equiv \boldsymbol{k}_{\omega_d}$). This terminology will be adopted for the remainder of the present chapter.

The monochromatic nature of expression (4.2) is demonstrated by Fourier Transformation into the frequency domain [4]:

$$\tilde{\boldsymbol{E}}_d(\omega) = \frac{1}{2\pi}\int_{-\infty}^{+\infty}\boldsymbol{E}_d(t)\mathrm{e}^{\mathrm{i}\omega t}\,dt = \frac{1}{2}\left[\boldsymbol{E}_{\omega_d}\,\delta(\omega-\omega_d) + \boldsymbol{E}_{-\omega_d}\,\delta(\omega+\omega_d)\right] \tag{4.3}$$

where $\delta(\omega)$ represents the conventional Dirac Delta Function [87]. The intensity of this electromagnetic wave, I_d, can be calculated from the time average of $|\boldsymbol{E}_d(t)|^2 = \boldsymbol{E}_d(t)\cdot\boldsymbol{E}^*_d(t)$ over a complete optical cycle [88]:

$$I_d = c\varepsilon_0\left\langle\left\langle|\boldsymbol{E}_d(t)|^2\right\rangle\right\rangle_t = \frac{c\varepsilon_0}{2}\left(\boldsymbol{E}_{\omega_d}\cdot\boldsymbol{E}^*_{\omega_d}\right) = \frac{c\varepsilon_0}{2}|\mathcal{E}_d|^2 \tag{4.4}$$

where ε_0 is the permittivity of free space, c is the speed of light in a vacuum, and the double set of angular brackets signifies a cycle average. Although a more precise terminology for the electric field vectors defined in Eq. (4.1) and Eq. (4.3) would be $\boldsymbol{E}_d(\boldsymbol{r},t)$ and $\tilde{\boldsymbol{E}}_d(\boldsymbol{r},\omega)$, respectively, the neglect of spatial dispersion effects in the ensuing analysis makes such a rigorous description needlessly cumbersome. Indeed, the more compact notation $\boldsymbol{E}_d$ will often be used to refer to the electromagnetic wave associated with carrier frequency ω_d. Explicit specification of the inherent time or frequency dependence of $\boldsymbol{E}_d$ will be reserved for situations that demand this additional degree of clarification.

While many variants of DFWM are possible, Fig. 4.1 schematically illustrates the so–called "phase–conjugate" experimental configuration [89, 90, 6] which has found widespread use in spectroscopic studies owing to its relative ease of implementation and sub–Doppler frequency response. In this geometry, two strong pump waves, having electric field vectors $\boldsymbol{E}_f$ and $\boldsymbol{E}_b$ (*i.e.*, subscripts denoting "forward–going" and "backward–going" pump waves, respectively), are directed through a molecular sample in a coaxial and counterpropagating fashion. A probe beam, with electric vector $\boldsymbol{E}_p$, crosses the pump fields under a small angle θ, with the region of mutual intersection forming the interaction volume for the DFWM process. These three input beams are coupled by nonlinear effects within the target medium thereby producing a signal wave, $\boldsymbol{E}_s$, which emerges in a direction that is exactly collinear and counterpropagating with respect to the incident probe radiation. A signal field generated in this manner can be shown to satisfy phase–matching or momentum conservation criteria for all angles θ with the wavevectors for the various optical fields being related by [6]:

$$\boldsymbol{k}_s = \boldsymbol{k}_f - \boldsymbol{k}_p + \boldsymbol{k}_b = -\boldsymbol{k}_p \tag{4.5}$$

where the latter equality follows from the fact that $\boldsymbol{k}_b = -\boldsymbol{k}_f$ for the phase–conjugate geometry. In a similar manner, the frequency of the emerging signal wave follows from the restrictions imposed by energy conservation [6]:

$$\omega_s = \omega_f - \omega_p + \omega_b \tag{4.6}$$

or, given the degenerate nature of the incident optical waves (*i.e.*, $\omega_d = \omega$ for $d = f, p, \text{or } b$), $\omega_s = \omega$. Furthermore, the signal beam produced by the DFWM interaction has the unique property of being a phase–conjugated or time–reversed replica of the incoming probe radiation (*i.e.*, $\boldsymbol{E}_s \propto \boldsymbol{E}_p^*$). Indeed, the counterpropagating experimental configuration of Fig. 4.1 serves as the prototypical example of a phase–conjugate mirror (or "phase–conjugator") [91] which can be exploited for the real–time correction of wavefront aberrations that accrue as a result of light propagation through inhomogenous and/or unsteady optical media [92-95, 9, 10].

As previously mentioned, the spectroscopic capabilities of DFWM stem from the enormous enhancement in the intensity of $\boldsymbol{E}_s$ (*viz.*, I_s) that is expected to accompany the resonant excitation of a molecular transition. Since the signal wave produced in the phase–conjugate geometry retraces the path traversed by the incident probe radiation, appropriate optical elements, based upon reflection (*i.e.*, a beamsplitter) or some other discriminatory mechanism (*e.g.*, polarization), must be inserted into the probe beam in order to extract the desired signal photons and direct them towards a detector. The initial placement of such collection optics is greatly facilitated by the counterpropagating nature of the experimental configuration depicted in Fig. 4.1. More specifically, the presence of $\boldsymbol{E}_s$ can be simulated by using a simple mirror to precisely retroreflect the probe light impinging upon the target medium. The "artificial signal" generated in this fashion provides a straightforward means for crudely adjusting the positions of components comprising the detection train, with more refined alignment obtained by maximizing the amplitude of observed DFWM spectral features.

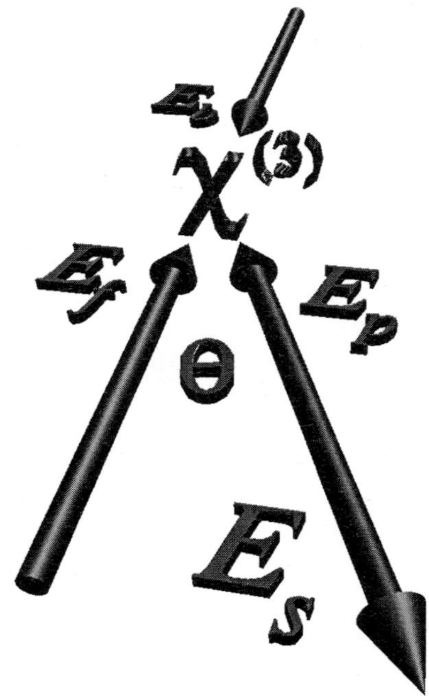

Figure 4.1 Schematic diagram of the "phase–conjugate" DFWM configuration. Two pump beams ($\boldsymbol{E}_f$ & $\boldsymbol{E}_b$) are directed through the target medium in a coaxial and counterpropagating fashion while a third incident probe beam ($\boldsymbol{E}_p$) intersects them under a small crossing angle, θ. Momentum conservation demands that the signal wave ($\boldsymbol{E}_s$) emerge in a direction that is exactly collinear with respect to the impinging probe radiation. All fields are of identical frequency with $\chi^{(3)}$ denoting the third–order susceptibility of the target medium which mediates the nonlinear wave coupling phenomenon.

For measurements conducted on rarefied media (*i.e.*, low–pressure gases or free–jet expansions), the major source of baseline noise in a properly aligned and optimized phase–conjugate DFWM configuration is the residual light scattered from optical elements used to extract signal photons from the counterpropagating probe beam [8]. While these deleterious effects can be minimized through judicious selection of high–quality, scatter–free components, they often set the lower limit of sensitivity achieved in a particular study. Fig. 4.2 illustrates two "box" variants of DFWM which eliminate the source of such incoherently scattered background light by dispensing with any need to insert collection optics into the path of an intense incident wave. Built upon the crossed–beam CARS geometry known as BOXCARS [96], both of these nonplanar schemes enable the direct observation of $\boldsymbol{E}_s$ which, owing to momentum conservation criteria, now emerges in a spatially distinct direction that is oriented along a "dark" axis of detection [22, 25, 26]. Given the numerical designation of impinging fields presented in Fig. 4.2, it is straightforward to show that the signal beam generated from the DFWM interaction region satisfies a phase–matching expression analogous to that of Eq. (4.5):

$$\boldsymbol{k}_s = \boldsymbol{k}_1 - \boldsymbol{k}_2 + \boldsymbol{k}_3 . \tag{4.7}$$

The nonlinear response produced in the box configurations is phase–conjugate in the sense that $\boldsymbol{E}_s \propto \boldsymbol{E}_2^*$; however, the emerging signal wave does not retrace the path of any incoming field (*i.e.*, a phase–conjugate mirror or phase–conjugator is not established). While the backward–box arrangement represents an obvious pyramidal distortion of the conventional (planar) counterpropagating DFWM scheme (*viz.*, $\boldsymbol{E}_b$ slightly tipped out of plane resulting in a corresponding displacement for $\boldsymbol{E}_s$ [22]), the forward–box variant follows from a more drastic mutation of the basic geometry illustrated in Fig. 4.1. Indeed, the theoretical descriptions for each of these four–wave mixing configurations are essentially identical provided that the corresponding sets of electromagnetic waves are related in an appropriate fashion (*e.g.*, $\boldsymbol{E}_f \Leftrightarrow \boldsymbol{E}_1$, $\boldsymbol{E}_p \Leftrightarrow \boldsymbol{E}_2$, and $\boldsymbol{E}_b \Leftrightarrow \boldsymbol{E}_3$).

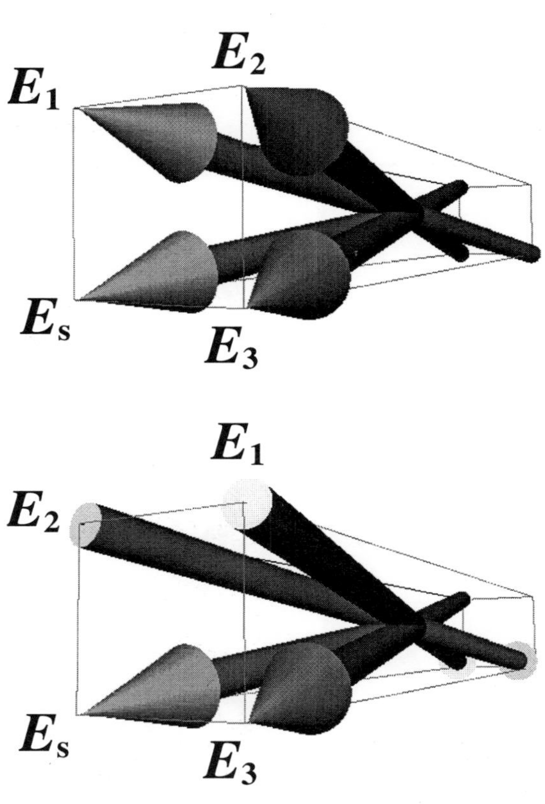

Figure 4.2 "Box" configurations for DFWM spectroscopy. Three incident waves of identical frequency (*i.e.*, designated by electric field vectors $\boldsymbol{E}_1$, $\boldsymbol{E}_2$, and $\boldsymbol{E}_3$) are directed through a molecular sample along distinct diagonals of a rectangular parallelepiped (or box). Nonlinear interactions within the target medium give rise to a fourth output or signal wave (*viz.*, $\boldsymbol{E}_s$), which emerges along the "dark" axis of detection defined by the remaining diagonal. The top panel depicts a forward–box configuration while the bottom panel illustrates a backward–box geometry, with the latter capable of being transformed into the former through a 180° rotation of $\boldsymbol{E}_1$ and $\boldsymbol{E}_2$ about the point of mutual beam intersection.

By eliminating the need to employ beamsplitters or other devices for extraction of $\boldsymbol{E}_s$, the forward–box and backward–box implementations of DFWM lead to an effective increase in the recorded signal strength (*i.e.*, potentially all signal photons are collected) and overall decrease in the level of baseline noise (*i.e.*, minimal incoherent scattering of light from optical components) [22, 25]. The constrained three–dimensional overlap of beams in such nonplanar geometries can also be exploited for spatially–resolved spectroscopic measurements. However, these advantages are obtained at the expense of experimental convenience. In particular, the box schemes require a substantially more sophisticated adjustment of the detection optics (*i.e.*, when compared to the phase–conjugate geometry) owing to the nontrivial nature of the path traversed by $\boldsymbol{E}_s$ (*viz.*, $\boldsymbol{k}_s \neq -\boldsymbol{k}_2$). Various means have been used to create "artificial signals" for alignment purposes, including the temporary insertion of colored glass filters or thin cuvettes containing dye solution into the DFWM interaction region. While absorption in these condensed–phase media can give rise to strong four–wave mixing processes with concomitant generation of signal beams having sufficient intensity to be visible to the unaided eye, great care must be taken to minimize any spatial deviations and optical distortions produced in the incident beams. These deleterious effects, when transferred into the emerging signal wave, can readily degrade the efficacy of initial alignments performed on the detection train.

4.2.3 Transient Gratings, Holography, and other Analogies

A succinct and physically satisfying interpretation for DFWM can be found in the realm of dynamic light scattering or laser–induced grating spectroscopy [11]. Long a mainstay of the time–domain community where it has been employed to describe a variety of both degenerate and non–degenerate wave–mixing phenomena [97-100], this approach attributes the creation of an emerging signal wave to two interrelated processes. First and foremost, the spatial interference of two overlapping light beams (*e.g.*, $\boldsymbol{E}_f$ and $\boldsymbol{E}_p$), in conjunction with strong matter–field interactions, gives rise to a modulation in the sample's complex index of refraction. The remaining incident beam (*e.g.*, $\boldsymbol{E}_b$) subsequently Bragg scatters from this transient diffraction grating, thereby producing the observed signal photons. As illustrated in Fig. 4.3, this concerted sequence of events mimics the formation and recreation of a volume hologram [92, 101, 102], with the "recording medium" for a gas–phase or liquid–phase experiment now consisting of a non–stationary (*i.e.*, moving) ensemble of target molecules.

While three possible scattering mechanisms can be envisioned by grouping the incident DFWM beams in various ways, it can be shown [75] that only volume holograms generated through coherent overlap of $\boldsymbol{E}_p$ with either $\boldsymbol{E}_f$ or $\boldsymbol{E}_b$ lead to four–wave mixing interactions that are one–photon resonant [*cf.*, Eq. (4.13) and accompanying discussion]. In each of these cases, the pairwise interference of optical beams leads to a *spatial* modulation in the intensity and/or polarization of the total electromagnetic field, the structure of which resembles the repetitive pattern of grooves associated with a conventional diffraction grating. The periodicity, Λ, of this effect follows from a straightforward analysis [11, 8]:

$$\Lambda = \frac{2\pi}{|\Delta \boldsymbol{k}|} = \frac{\lambda}{2\sin\left(\frac{\phi}{2}\right)} \tag{4.8}$$

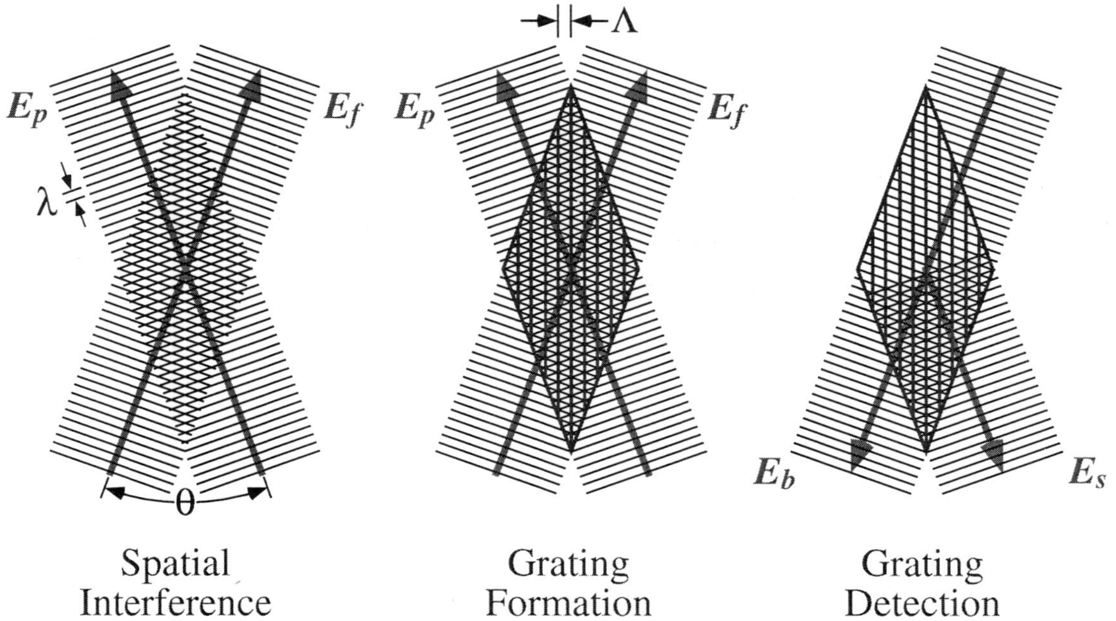

Figure 4.3 Transient grating analogy for DFWM. Upon resonant excitation of a molecular transition, the spatial interference of two electromagnetic waves ($\boldsymbol{E}_f$ and $\boldsymbol{E}_p$) having common wavelength λ leads to creation of a macroscopic modulation in the population and/or polarization distribution of the target molecules. The concomitant variations in optical properties give rise to a diffraction grating of periodicity Λ. Illumination by a third resonant electromagnetic field ($\boldsymbol{E}_b$) results in the Bragg–scattering of light into the signal path ($\boldsymbol{E}_s$). This concerted sequence of events strongly resembles the formation and recreation of a volume hologram.

where $\boldsymbol{\Delta k}$ is the difference in wavevectors for two beams of identical wavelength, λ, that cross under angle ϕ. As shown in Fig. 4.1, the counterpropagating phase–conjugate DFWM scheme has $\phi = \theta$ for the interaction of $\boldsymbol{E}_f$ with $\boldsymbol{E}_p$ and $\phi = \pi - \theta$ for the interaction of $\boldsymbol{E}_b$ with $\boldsymbol{E}_p$. Given the near–collinear geometries typically employed for experimental measurements, the $\boldsymbol{E}_f$ coupling can be associated with gratings of large Λ (*viz.*, $\Lambda \to \infty$ as $\theta \to 0$) while the analogous $\boldsymbol{E}_b$ diffraction structures entail much smaller periodicities on the order of $\lambda/2$. For monochromatic plane waves, the resulting sinusoidal interference patterns are directed parallel to a hypothetical line that bisects the crossing angle ϕ, with the crests and troughs "jutting" out of the plane formed by intersection of the incident beams (*cf.*, Fig. 4.3). Although the nonuniform transverse profiles and finite bandwidths of actual laser sources might appear to invalidate this simple description, more sophisticated treatments show that the plane wave analysis provides a reasonable first–order interpretation for the induced scattering phenomena [103, 104].

Matter–field interactions are responsible for transcribing the interference patterns, as defined in Eq. (4.8), onto the ensemble of target molecules. The resulting amplitude or phase gratings [11] of periodicity Λ coherently diffract a minute portion of the remaining incident beam into the signal field, with restrictions imposed by the first–order Bragg–scattering condition [83] reproducing the phase–matching criteria embodied in the nonlinear optical description of DFWM [*i.e.*, Eq. (4.5)] [8]. Obviously, the efficiency of the diffraction process will be a strong function of the impinging light frequency and the most pronounced effects can be expected to occur upon resonant excitation of a molecular transition. Under such circumstances, a macroscopic spatial modulation can be induced in the net population or overall polarization (*i.e.*, magnetic sublevel distribution) for molecules residing in the initial and final quantum states of the selected transition. When illuminated by resonant radiation, the concomitant variations in optical properties enable the target medium to serve as a three–dimensional array of phase–locked antennas whose emission can interfere constructively only in certain directions defined by the orientation of the material grating and the path of the impinging electromagnetic wave (*i.e.*, the Bragg–scattering or phase–matching criteria). The wavelength dependence of this scattering phenomenon is responsible for the spectroscopic utility of grating–based techniques.

4.2.4 *Nonlinear Optical Polarization*

In the absence of saturation effects, the well established susceptibility formalism of nonlinear optics [4, 80, 5] provides a systematic means for calculating the response generated by a four–wave mixing interaction. This approach relies on a perturbative expansion of the induced electric polarization vector, $\boldsymbol{P}(t)$, in terms of the total electric field, $\boldsymbol{E}(t)$, with successive members of the expansion representing higher orders of nonlinearity. Phenomena resulting from the coupling of three electromagnetic waves, such as formation of the signal beam in DFWM spectroscopy, can be described through use of the third–order polarization, $\boldsymbol{P}^{(3)}(t)$. For the general case of optical fields having finite spectral bandwidths and nonuniform transverse characteristics, the expression for the DFWM signal polarization, $\boldsymbol{P}_s^{(3)}(t)$, can become quite complicated [66]. However, the monochromatic plane wave approximation permits a succinct formulation of $\boldsymbol{P}_s^{(3)}(t)$ in a manner that highlights the key features ascribed to this quantity:

$$\boldsymbol{P}_s^{(3)}(t) = \frac{1}{2}\left[\boldsymbol{P}_{\omega_s}^{(3)}\,\mathrm{e}^{-\mathrm{i}\omega_s t} + \boldsymbol{P}_{-\omega_s}^{(3)}\,\mathrm{e}^{\mathrm{i}\omega_s t}\right] \tag{4.9}$$

with $\boldsymbol{P}_{-\omega_s}^{(3)} = \left(\boldsymbol{P}_{\omega_s}^{(3)}\right)^*$. This optically–induced polarization oscillates at the characteristic frequency of ω_s and is responsible for manifestation of the observed signal wave. The overall strength of the DFWM response is embodied in the amplitude vector $\boldsymbol{P}_{\omega_s}^{(3)}$ which can be expressed in terms of the incident electromagnetic fields (*viz.*, $\boldsymbol{E}_f$, $\boldsymbol{E}_p$, and $\boldsymbol{E}_b$) and the third–order susceptibility tensor, $\boldsymbol{\chi}^{(3)}(-\omega_s;\omega_f,-\omega_p,\omega_b)$, that mediates the wave–coupling phenomenon [75]. In particular, the inherent symmetries that characterize the DFWM interaction enable $\boldsymbol{P}_{\omega_s}^{(3)}$ to be written as [4]:

$$\begin{aligned}\boldsymbol{P}_{\omega_s}^{(3)} &= \frac{3}{4}\varepsilon_0\,\boldsymbol{\chi}^{(3)}(-\omega_s;\omega_f,-\omega_p,\omega_b)\vdots\boldsymbol{E}_{\omega_f}\,\boldsymbol{E}_{\omega_p}^*\,\boldsymbol{E}_{\omega_b} \\ &= \frac{3}{4}\varepsilon_0\left[\boldsymbol{\chi}^{(3)}(-\omega_s;\omega_f,-\omega_p,\omega_b)\vdots\hat{\boldsymbol{e}}_f\,\hat{\boldsymbol{e}}_p^*\,\hat{\boldsymbol{e}}_b\right]\mathcal{E}_f\,\mathcal{E}_p^*\,\mathcal{E}_b\,\mathrm{e}^{\mathrm{i}\left(\boldsymbol{k}_f-\boldsymbol{k}_p+\boldsymbol{k}_b\right)\cdot\boldsymbol{r}}\end{aligned} \tag{4.10}$$

where the column of three dots, $\vdots$, represents a tensor contraction and the electric field amplitude vectors ($\boldsymbol{E}_{\omega_f}$, $\boldsymbol{E}_{\omega_p}$, and $\boldsymbol{E}_{\omega_b}$) have been defined previously through expression (4.2). The algebraic signs preceding the frequency arguments of $\boldsymbol{\chi}^{(3)}(-\omega_s;\omega_f,-\omega_p,\omega_b)$ reflect the imposition of momentum and energy conservation criteria, with the requisite matter–field interaction for each impinging wave being associated with a photon annihilation operator (for $+\omega_d$) or a photon creation operator (for $-\omega_d$) [5]. Individual Cartesian components of the induced polarization are denoted by $\left(\boldsymbol{P}^{(3)}_{\omega_s}\right)_j$, with $j = x, y$, or z, and can be obtained from expansion of Eq. (4.10):

$$\left(\boldsymbol{P}^{(3)}_{\omega_s}\right)_j = \hat{e}_j^* \cdot \boldsymbol{P}^{(3)}_{\omega_s} = \frac{3}{4}\varepsilon_0 \sum_{k,l,m} \chi^{(3)}_{jklm}(-\omega_s;\omega_f,-\omega_p,\omega_b)\left(\boldsymbol{E}_{\omega_f}\right)_k \left(\boldsymbol{E}^*_{\omega_p}\right)_l \left(\boldsymbol{E}_{\omega_b}\right)_m \tag{4.11}$$

where the projection of $\boldsymbol{E}_{\omega_q}$ along the k–axis ($k = x, y$, or z as does l and m) is represented by $\hat{e}_k^* \cdot \boldsymbol{E}_{\omega_q} = \left(\boldsymbol{E}_{\omega_q}\right)_k$ and, formally, $\chi^{(3)}_{jklm}(-\omega_s;\omega_f,-\omega_p,\omega_b) = \hat{e}_j^* \cdot \boldsymbol{\chi}^{(3)}(-\omega_s;\omega_f,-\omega_p,\omega_b) \vdots \hat{e}_k \, \hat{e}_l^* \, \hat{e}_m$. In particular, the quantity $\hat{e}_s^* \cdot \boldsymbol{P}^{(3)}_{\omega_s}$ [or $\hat{e}_s^* \cdot \boldsymbol{P}^{(3)}_s(t)$] yields the component of the induced polarization which gives rise to a signal field having transverse polarization characteristics specified by unit vector $\hat{e}_s$. As expected, $\boldsymbol{P}^{(3)}_{\omega_s}$ is found to be proportional to the complex conjugate (or phase–conjugate) of the amplitude vector for the incident probe radiation, with the spatial $\mathrm{e}^{\mathrm{i}(\boldsymbol{k}_f-\boldsymbol{k}_p+\boldsymbol{k}_b)\cdot\boldsymbol{r}} = \mathrm{e}^{\mathrm{i}\boldsymbol{k}_s\cdot\boldsymbol{r}}$ term responsible for the phase–matching condition embodied in expression (4.5) [8].

An explicit relationship between the nonlinear polarization formalism for DFWM and the transient grating description presented in Section 4.2.3 follows from the symmetry restrictions imposed upon $\boldsymbol{\chi}^{(3)}\left(-\omega_s;\omega_f,-\omega_p,\omega_b\right)$. For an isotropic medium, the 81 possible elements of a general fourth–rank susceptibility tensor collapse to yield 21 nonzero terms, only three of which are linearly independent [105, 4]. The identical frequencies exploited for the DFWM interaction place additional constraints on $\boldsymbol{\chi}^{(3)}\left(-\omega_s;\omega_f,-\omega_p,\omega_b\right)$, thereby permitting all Cartesian components of this quantity to be expressed by means of two unique tensor elements designated as $\chi^{(3)}_{\eta\eta\varsigma\varsigma}$ and $\chi^{(3)}_{\eta\varsigma\eta\varsigma}$ [82, 80]:

$$\chi^{(3)}_{jklm}\left(-\omega_s;\omega_f,-\omega_p,\omega_b\right) = \left[\delta_{jk}\,\delta_{lm} + \delta_{jm}\,\delta_{lk}\right]\chi^{(3)}_{\eta\eta\varsigma\varsigma} + \delta_{jl}\,\delta_{km}\,\chi^{(3)}_{\eta\varsigma\eta\varsigma} \tag{4.12}$$

where the Kronecker Delta symbol has the usual definition (*i.e.*, $\delta_{jk} = 1$ if $j = k$ and zero otherwise [87]) with the subscripts η and ς equal to x, y, or z subject to the condition $\eta \neq \varsigma$. Incorporation of this symmetry restriction into expression (4.10) or (4.11) enables the amplitude vector for the DFWM signal polarization to be reformulated as a sum of three distinct terms [82, 75]:

$$\boldsymbol{P}^{(3)}_{\omega_s} = \mathcal{A}\left(\boldsymbol{E}_{\omega_b} \cdot \boldsymbol{E}^*_{\omega_p}\right)\boldsymbol{E}_{\omega_f} + \mathcal{B}\left(\boldsymbol{E}_{\omega_f} \cdot \boldsymbol{E}^*_{\omega_p}\right)\boldsymbol{E}_{\omega_b} + \mathcal{C}\left(\boldsymbol{E}_{\omega_f} \cdot \boldsymbol{E}_{\omega_b}\right)\boldsymbol{E}^*_{\omega_p} \tag{4.13}$$

where the coefficients $\mathcal{A}$, $\mathcal{B}$, and $\mathcal{C}$ depend on the linearly independent components of the susceptibility tensor introduced by expression (4.12). This form for the induced nonlinear polarization readily lends itself to interpretation of the overall DFWM process as diffraction from optically–induced volume holograms. In particular, the creation of a transient molecular grating by interference of two incident beams is described by the scalar product of the corresponding electric field amplitude vectors. Similarly, Bragg scattering of photons from the third impinging wave into the signal field is implied by multiplication with the remaining $\boldsymbol{E}_{\omega_d}$ factor.

The first two terms of expression (4.13) stem from scalar coupling of the probe beam with one of the incident pump waves, thereby leading to the formation of spatial interference patterns having periodicities described by Λ [*cf.* Eq. (4.8)]. The coherent light scattering processes resulting from such interactions are resonantly–enhanced by one–photon molecular transitions [82] and are responsible for signals observed in the majority of DFWM experiments. In contrast, the last term of Eq. (4.13) corresponds to a two–photon resonant interaction which gives rise to a spatially invariant transient grating that oscillates in time at twice the input frequency (*viz.*, 2ω) [82]. This phenomenon has been exploited for high resolution two–photon spectroscopy [46, 30, 29, 47, 31, 49]. Expression (4.13) also provides the basis of simple "conservation rules" [106] which enable the polarization properties for an emerging DFWM signal beam to be predicted from the transverse characteristics of the impinging electromagnetic waves (*e.g.*, if $\boldsymbol{E}_{\omega_f} \| \boldsymbol{E}^*_{\omega_p} \perp \boldsymbol{E}_{\omega_b}$, then $\boldsymbol{P}^{(3)}_{\omega_s} \propto \boldsymbol{E}_{\omega_b}$ and, therefore, $\boldsymbol{E}_{\omega_s} \| \boldsymbol{E}_{\omega_b}$). Several authors have discussed the limitations imposed upon this rudimentary scalar theory for the induced nonlinear polarization [107, 82], with more sophisticated treatments required to account fully for the polarization–specific effects associated with magnetic sublevel degeneracy.

The intensity of the signal beam emerging from a DFWM interaction region, I_s, can be derived from the solution of a coupled set of differential equations describing the evolution of incident and generated electromagnetic waves within the target medium [4]. For such calculations, the induced polarization, $\boldsymbol{P}^{(3)}_s(t)$, provides the source term leading to creation of the emerging signal field. In the "weak response" limit, where the pump and probe beams propagate through an optically–thin sample of length ℓ without experiencing significant attenuation or amplification, an effective decoupling of the differential equation system can be obtained with the resulting expression for I_s reducing to [10, 8]:

$$I_s = \frac{\omega_s^2 \ell^2}{8c\varepsilon_0} \left| \hat{e}_s^* \cdot \boldsymbol{P}^{(3)}_{\omega_s} \right|^2 = \left(\frac{3\omega_s \ell}{4c^2 \varepsilon_0} \right)^2 \left| \hat{e}_s^* \cdot \boldsymbol{\chi}^{(3)}\left(-\omega_s;\omega_f,-\omega_p,\omega_b\right) \vdots \hat{e}_f \hat{e}_p^* \hat{e}_b \right|^2 I_f I_p I_b = R I_p \tag{4.14}$$

where I_q ($q = f, p, b$, or s) represents the time averaged intensity of $\boldsymbol{E}_q(t)$ [*cf.*, Eq. (4.4)] and R denotes the phase–conjugate reflectivity. For the counterpropagating experimental configuration depicted in Fig. 4.1, the final equality of this expression reinforces the description of DFWM in terms of a phase–conjugate mirror capable of reflecting a time–reversed replica of the impinging probe radiation [91].

Eq. (4.14) shows that the unsaturated DFWM signal strength scales as the square of the interaction length, ℓ, and as the product of intensities for all three incident beams, $I_f I_p I_b$. Since the degenerate pump and probe waves typically are derived from a common source of tunable light, the weak–field magnitude of I_s can be expected to exhibit a cubic dependence upon applied laser power, with small variations in the incident intensity translating into large fluctuations of observed signal amplitude. The quadratic dependence on ℓ also suggests that a significant enhancement in detection sensitivity can be realized through use of experimental configurations that maximize effective sample length.

The magnitude of the DFWM signal predicted by Eq. (4.14) also depends on the third–order susceptibility tensor, a complex, frequency–dependent quantity that embodies the spectroscopic response of the molecular sample. In the absence of optical saturation effects, this fourth–rank tensor is independent of the applied field amplitudes. However, as demonstrated by expressions (4.11) and (4.12), the polarization characteristics of the incident electromagnetic waves, in conjunction with the inherent symmetry properties of the target medium, will determine which elements of $\boldsymbol{\chi}^{(3)}\left(-\omega_s;\omega_f,-\omega_p,\omega_b\right)$ are required for the description of a particular four–wave mixing process [4].

The functional form of $\boldsymbol{\chi}^{(3)}\left(-\omega_s;\omega_f,-\omega_p,\omega_b\right)$, with its explicit dependence on the characteristic transition energies and decay rates of the target medium, can be obtained from the third–order perturbative expansion used to define the induced signal polarization, $\boldsymbol{P}^{(3)}_s(t)$. While tedious in nature, this calculation is accomplished most efficiently by means of a density operator formalism [5], with various diagrammatic techniques [108-110] serving both to organize and to interpret each molecule–field coupling. Assuming that the molecular sample can be treated as an isolated ensemble of independent and distinguishable particles that interact with electromagnetic radiation under the dipole approximation, $\boldsymbol{P}^{(3)}_s(t)$ can be expressed as the quantum mechanical expectation value of the dipole moment operator, $\boldsymbol{\mu}$ [111, 75]:

$$\boldsymbol{P}^{(3)}_s(t) = N\,\mathrm{Tr}\left[\rho^{(3)}(t)\,\boldsymbol{\mu}\right] \tag{4.15}$$

where N is the total number density of target species, $\rho^{(3)}(t)$ denotes the third–order term arising from a perturbative expansion of the density operator [111, 84, 5], and the trace operation [85], represented by the symbol $\mathrm{Tr}[\cdots]$, is taken over the zero–order eigenstates of the molecular ensemble. When represented in this energy eigenbasis, the diagonal and off–diagonal matrix elements of $\rho^{(3)}(t)$ describe molecular populations and optical coherences, respectively, that are created and/or modified through successive matter–field interactions [85].

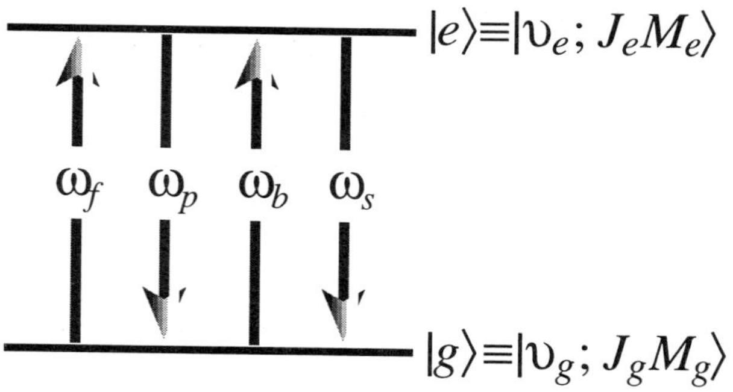

Figure 4.4 Energy level diagram for resonant DFWM spectroscopy. An isolated rovibronic transition having rest angular frequency ω_{eg} and transition dipole moment matrix element $\boldsymbol{\mu}_{eg} = \langle e|\boldsymbol{\mu}|g\rangle$ connects ground and excited states designated by kets $|g\rangle$ and $|e\rangle$, respectively. The angular momentum and magnetic quantum numbers for these energy eigenstates are designated by J and M with υ denoting all other rovibronic quantities needed to fully specify the system. The degeneracy of magnetic sublevels implies that each distinct energy be associated with a manifold of $2J+1$ states: $\{|\upsilon; JM\rangle; M = -J, -J+1, \cdots, J\}$. The three incident beams of a resonant DFWM process, labeled as ω_f, ω_p, and ω_b, are coupled by the strong nonlinear effects accompanying the $|e\rangle \leftrightarrow |g\rangle$ transition, thereby resulting in formation of the emerging signal wave, ω_s.

Fig. 4.4 presents a schematic energy level diagram for a near–resonant DFWM interaction, where coupling of the incident and generated electromagnetic waves is dominated by the nonlinear effects associated with an isolated spectroscopic transition. The depicted four–wave mixing scheme is one–photon resonant, with the pertinent optical transitions connecting ground and excited states designated by $|g\rangle$ and $|e\rangle$, respectively. Since the states of this hypothetical system are presumed to form the eigenbasis of the zero–order (unperturbed) molecular Hamiltonian, H_0, they can be formulated as $|\upsilon; JM\rangle$ where M (the magnetic quantum number) signifies the projection of total angular momentum J on the space–fixed axis of quantization and υ denotes all other (rovibronic) quantum numbers. The unperturbed energies and population decay rates for eigenstates are given by E_α and $\Gamma_{\alpha\alpha}$ where the subscript α can assume values of "g" or "e" with primes serving to distinguish individual magnetic sublevels in each $(2J_\alpha+1)$–fold degenerate manifold (*e.g.*, $E_g = E_{g'}$ where E_g and $E_{g'}$ are the energies for states $|g\rangle \equiv |\upsilon_g; J_g M_g\rangle$ and $|g'\rangle \equiv |\upsilon_g; J_g M'_g\rangle$, respectively). In particular, the longitudinal relaxation time (T_{1_α}) [112] or effective lifetime (τ_α) for state $|\alpha\rangle$ is equal to the inverse of $\Gamma_{\alpha\alpha}$, $T_{1_\alpha} = \tau_\alpha = \Gamma_{\alpha\alpha}^{-1}$, and the rest angular frequency for the $|\beta\rangle \leftrightarrow |\alpha\rangle$ resonance is defined by $\omega_{\beta\alpha} = (E_\beta - E_\alpha)/\hbar$. It must be stressed that the transition arrows contained in Fig. 4.4 do not imply a specific temporal ordering for the impinging optical fields. Indeed, proper description of the DFWM response in a frequency domain study requires that all possible time–orderings of matter–field interactions (*i.e.*, that are consistent with momentum and energy conservation criteria) be taken explicitly into account [113, 110].

The present analysis assumes that only the ground energy state of an isolated molecular transition is populated prior to the DFWM interaction, with the distribution of probability amplitude among degenerate magnetic sublevels specified by the matrix elements of the zero–order density operator: $\rho_{gg'}^{(0)} = \langle \upsilon_g; J_g M_g | \rho^{(0)} | \upsilon_g; J_g M'_g \rangle$. In order to describe the thermalized (*i.e.*, incoherent and isotropic) ensembles investigated in the majority of DFWM spectroscopic measurements, $\rho^{(0)}$ must have the form [85, 5]:

$$\rho^{(0)} = \frac{e^{-H_0/k_B T}}{Z} \tag{4.16}$$

where k_B is the Boltzmann constant, T is the absolute temperature of the target medium, and Z is the canonical partition function [114] defined by the condition $\mathrm{Tr}[\rho^{(0)}] = 1$. When multiplied by total number density N, the general matrix element for $\rho^{(0)}$ yields:

$$N\langle\beta|\rho^{(0)}|\alpha\rangle = N\frac{e^{-E_\alpha/k_B T}}{Z}\delta_{\beta\alpha} = N_\alpha\,\delta_{\beta\alpha} \tag{4.17}$$

where N_α is the number density of species residing in state $|\alpha\rangle$ with the condition $E_e \gg E_g$ ensuring that N_e is negligible compared to N_g. Optical coherences created between various molecular eigenstates by resonant matter–field interactions are characterized by the off–diagonal components of matrices corresponding to higher order terms in the density operator expansion [111, 75]: $\rho_{\alpha\beta}^{(n)}$ with $n > 0$. Under quite general conditions, the temporal dissipation of such coherences can be related to depopulation rates for the pertinent states [112]:

$$\Gamma_{\beta\alpha} = \frac{1}{2}\left(\Gamma_{\beta\beta} + \Gamma_{\alpha\alpha}\right) + \Gamma_{\beta\alpha}^{\Phi} = \frac{1}{T_{2_{\beta\alpha}}} \tag{4.18}$$

where $T_{2_{\beta\alpha}}$ is the transverse relaxation time for the transition between $|\beta\rangle$ and $|\alpha\rangle$. The quantity $\Gamma_{\beta\alpha}^{\Phi}$ represents the pure dipole dephasing rate which stems from processes (*e.g.*, elastic collisions) that destroy molecular coherence without disrupting the corresponding molecular populations [115]. In the "weak collision" limit [112, 2], the damping parameter $\Gamma_{\beta\alpha}$ can be identified as the Half–Width at Half–Maximum (HWHM) height for the homogeneously–broadened natural linewidth of the $|\beta\rangle \leftrightarrow |\alpha\rangle$ resonance.

As demonstrated by expressions (4.10) and (4.15), calculation of the DFWM signal polarization, $\boldsymbol{P}_s^{(3)}(t)$, and associated susceptibility tensor, $\boldsymbol{\chi}^{(3)}\left(-\omega_s;\omega_f,-\omega_p,\omega_b\right)$, requires knowledge of the third–order density operator $\rho^{(3)}(t)$. The explicit form for $\rho^{(3)}(t)$ follows from perturbative solution of the quantum mechanical Liouville equation which describes temporal evolution of the system under the influence of both applied electromagnetic fields and intrinsic relaxation processes [85, 84, 5]. The requisite perturbation analysis is illustrated schematically by the double–sided Feynman diagrams of Fig. 4.5 [110, 5, 116]. Each of the depicted panels represents a specific sequence of matter–field interactions that contribute to the overall four–wave mixing response, with the only significant difference among the various diagrams being the relative time–ordering of impinging waves. While this figure highlights $\rho^{(3)}(t)$ terms corresponding to diffraction of $\boldsymbol{E}_b$ from the transient gratings created through spatial interference of $\boldsymbol{E}_f$ and $\boldsymbol{E}_p$, an analogous set of perturbation diagrams, obtained by interchanging the roles of the forward–going and backward–going pump beams (*i.e.*, interchanging subscripts "f" and "b"), must also be considered in order to fully describe the one–photon resonant DFWM spectrum for an isolated molecular transition.

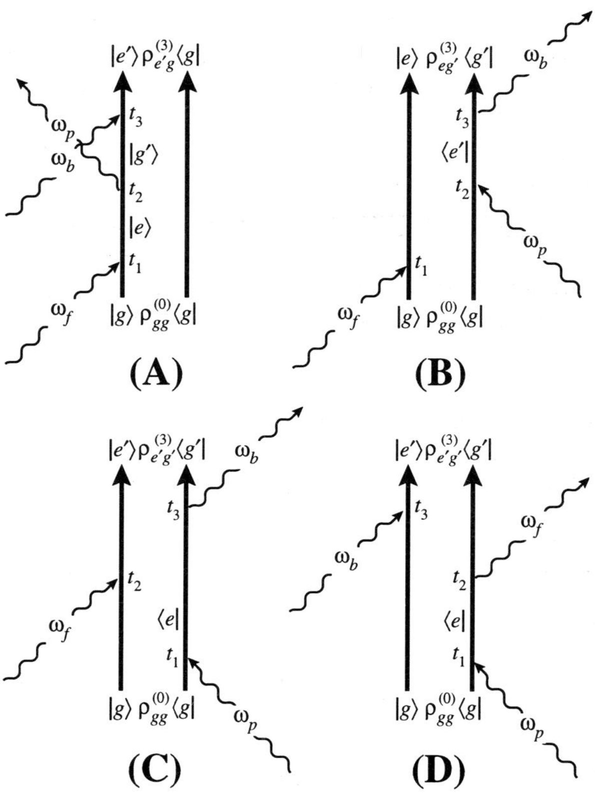

Figure 4.5 Double–Sided Feynman diagrams for DFWM spectroscopy. The density operator for an unsaturated DFWM process is expanded as a perturbative power series in the applied electromagnetic field with the ground and excited states of an isolated spectroscopic transition, $|e\rangle \leftrightarrow |g\rangle$, providing the basis for a matrix representation. Each diagram portrays a distinct time–ordering of resonant matter–field interactions (wavy lines) that leads to creation of an induced signal polarization through correlated evolution of bras and kets (*NB.*, time increases from bottom to top in each diagram). While this set of perturbation graphs suggests Bragg–diffraction of $\boldsymbol{E}_b$ from volume holograms formed by interference of $\boldsymbol{E}_f$ and $\boldsymbol{E}_p$, an analogous figure describing transient gratings originating from spatial overlap of $\boldsymbol{E}_b$ with $\boldsymbol{E}_p$ can be obtained by interchanging the forward-going and backward–going pump fields ($\omega_f \leftrightarrow \omega_b$).

The double–sided Feynman diagrams of Fig. 4.5 provide a pictorial representation for the transformation of $\rho^{(0)}$ into $\rho^{(3)}(t)$. More specifically, the density operator is denoted by two vertical arrows which symbolize the correlated temporal evolution of the dual vector spaces [87, 5] corresponding to the ket (on the left) and the bra (on the right) of a matrix representation. Time is defined to increase from bottom to top and interactions with electromagnetic radiation are designated by wavy arrows which point either towards or away from the diagram center so as to indicate absorption or emission processes, respectively. The Hermitian Conjugate relationship between bras and kets [87] requires that incoming arrows on the left and outgoing arrows on the right be associated with the photon annihilation operator while oppositely directed arrows imply action of the photon creation operator [108, 5, 117]. The molecular quantum states involved in an optically–induced transition are indicated above and below each interaction arrow with the nth matter–field coupling defined to take place at time t_n where $-\infty \le t_1 \le t_2 \le t_3 \le t$.

The eight matter–field interaction sequences embodied in Fig. 4.5 (four depicted diagrams plus four others obtained by interchange of "f" and "b") are the only combinations of one–photon resonant processes capable of generating a DFWM signal wave that exhibits the spatial and temporal characteristics defined by Eq. (4.5) and Eq. (4.6) [116]. Their interpretation relies on the fact that each coupling with an applied electromagnetic field gives rise to a successively higher order member in the perturbative expansion of the density operator. For example, the matrix representations for the $\rho^{(n)}(t)$ terms corresponding to diagrams (A) and (C) of Fig. 4.5 lead to the following temporal progressions:

$$\begin{aligned}
&\text{(A):}\quad \rho_{gg}^{(0)} \xrightarrow[t_1]{\hbar\omega_f} \rho_{eg}^{(1)}(t_1 \le t \le t_2) \xrightarrow[t_2]{\hbar\omega_p} \rho_{g'g}^{(2)}(t_2 \le t \le t_3) \xrightarrow[t_3]{\hbar\omega_b} \rho_{e'g}^{(3)}(t \ge t_3) \\
&\text{(C):}\quad \rho_{gg}^{(0)} \xrightarrow[t_1]{\hbar\omega_p} \rho_{ge}^{(1)}(t_1 \le t \le t_2) \xrightarrow[t_2]{\hbar\omega_f} \rho_{e'e}^{(2)}(t_2 \le t \le t_3) \xrightarrow[t_3]{\hbar\omega_b} \rho_{e'g'}^{(3)}(t \ge t_3)
\end{aligned} \tag{4.19}$$

where the time interval for existence of each $\rho^{(n)}(t)$ element has been specified explicitly. Starting with an isotropic and incoherent target medium defined by $\rho_{gg}^{(0)}$ [*cf.*, Eq. (4.16) and Eq. (4.17)], resonant interaction with either $\boldsymbol{E}_f(t)$ or $\boldsymbol{E}_p(t)$ at time t_1 [*i.e.*, for diagrams (A) or (C), respectively] leads to formation of a coherence between the ground and excited energy states described by the first–order density matrix elements $\rho_{eg}^{(1)}(t)$ or $\rho_{ge}^{(1)}(t)$. Subsequent coupling with either $\boldsymbol{E}_p(t)$ or $\boldsymbol{E}_f(t)$ at time t_2 [*i.e.*, for diagrams (A) or (C), respectively] transforms this optical coherence into ground or excited state probability amplitudes [*viz.*, $\rho^{(2)}(t)$ components such as $\rho_{g'g}^{(2)}(t)$ and $\rho_{e'e}^{(2)}(t)$ representing magnetic sublevel populations and/or coherences] which encode the interference structure derived from overlapping electromagnetic waves. The final interaction with $\boldsymbol{E}_b(t)$ at time t_3 transforms this effective modulation of optical properties into a new ground state/excited state coherence, with the resulting third–order density operator $\rho^{(3)}(t)$ displaying the temporal and spatial characteristics required for production of the signal field. Similar descriptions can be obtained for diagrams (B) and (D), with interchange of the forward–going and backward–going pump fields causing the initial grating formation to involve $\boldsymbol{E}_b$ rather than $\boldsymbol{E}_f$.

The Feynman diagrams of Fig. 4.5 provide a convenient mechanism for transcribing time–ordered sequences of matter–field interactions into perturbative expansions of the corresponding density operator [108, 109, 111, 84, 5]. For example, inspection of diagram (C) leads to the following expression for $\rho^{(3)}(t)$:

$$\begin{aligned}
\rho^{(3)}(t) &= \left(\frac{-i}{\hbar}\right)^3 \int_{-\infty}^{t} dt_3 \int_{-\infty}^{t_3} dt_2 \int_{-\infty}^{t_2} dt_1 U_0(t,t_2) W_f(t_2) U_0(t_2,-\infty)\rho(-\infty)U_0^\dagger(t_1,-\infty)W_p(t_1)U_0^\dagger(t_3,t_1)W_b(t_3)U_0^\dagger(t,t_3) \\
&= \left(\frac{-i}{\hbar}\right)^3 \int_{-\infty}^{t} dt_3 \int_{-\infty}^{t_3} dt_2 \int_{-\infty}^{t_2} dt_1 U_0(t,t_2) W_f(t_2) U_0(t_2,t_1)\rho^{(0)} W_p(t_1)U_0^\dagger(t_3,t_1)W_b(t_3)U_0^\dagger(t,t_3)
\end{aligned} \tag{4.20}$$

where the integrals are performed subject to the constraint $t_1 \le t_2 \le t_3 \le t$ with $\rho(-\infty)$ signifying the density operator in the distant past. Since diagram (C) involves transitions taking place both in the ket and the bra spaces, the perturbation Hamiltonian for interaction of electromagnetic wave $\boldsymbol{E}_d(t)$ at time t_n, denoted by $W_d(t_n)$, appears both to the left and to the right of $\rho(-\infty)$. In contrast, the analogous $\rho^{(3)}(t)$ expression for diagram (A) will have all three $W_d(t_n)$ factors to the left of $\rho(-\infty)$. Field–free evolution of the system during the intervals between interactions is governed by the zero–order time translation operator $U_0(t_n,t_{n-1})$ which describes a unitary time transformation [87] of $\rho(t)$: $\rho(t_n) = U_0(t_n,t_{n-1})\rho(t_{n-1})U_0^\dagger(t_n,t_{n-1})$. In particular, $U_0(t_n,t_{n-1})$ satisfies a combination rule [87] of the form $U_0(t_n,t_{n-1}) = U_0(t_n,\tau)U_0(\tau,t_{n-1})$ $(t_n \ge \tau \ge t_{n-1})$ which has been exploited for derivation of the second equality in Eq. (4.20):

$$U_0(t_2,-\infty)\rho(-\infty)U_0^\dagger(t_1,-\infty) = U_0(t_2,t_1)\left[U_0(t_1,-\infty)\rho(-\infty)U_0^\dagger(t_1,-\infty)\right] = U_0(t_2,t_1)\rho(t_1) \tag{4.21}$$

where $\rho(t_1)=\rho^{(0)}$ signifies the density operator at the time of the first matter–field interaction (*viz.*, t_1). The removal of all temporal dependence in the notation $\rho^{(0)}$ reinforces the static nature of the equilibrium density operator used to describe a thermalized ensemble of target molecules [*cf.*, Eq. (4.16)].

Further specification of the third–order density operator corresponding to diagram (C) of Fig. 4.5 requires that Eq. (4.20) be represented in terms of an appropriate basis which can most conveniently be constructed from the energy eigenstates of the zero–order (unperturbed) Hamiltonian. This task is facilitated by judicious placement of unit operators, *1*, which can be resolved into a summation over normalized projectors of the eigenbasis [87]: $1=\sum_\alpha |\alpha\rangle\langle\alpha|$ where the label "α" can assume all values consistent with state labels for the two degenerate manifolds in Fig. 4.4 (*i.e.*, "α" can equal "g" and "e" as well as all primed values of "g" and "e" required for the identification of individual magnetic sublevels). Taking into account that the matrices for $W_d(t_n)$ and $U_0(t_n,t_{n-1})$ are, respectively, off–diagonal and diagonal in the $\{|g\rangle,|e\rangle\}$ basis (*i.e.*, $\langle g|W_d(t_n)|g\rangle=\langle e|W_d(t_n)|e\rangle=0$ and $\langle g|U_0(t_n,t_{n-1})|e\rangle=\langle e|U_0(t_n,t_{n-1})|g\rangle=0$ independent of magnetic quantum number) [5], the desired expression for $\rho^{(3)}(t)$ becomes:

$$\begin{aligned}
\rho^{(3)}(t) &= \sum_{g'e'} |e'\rangle\rho^{(3)}_{e'g'}(t)\langle g'| \\
&= \left(\frac{-i}{\hbar}\right)^3 \sum_{\substack{gg'\\ee'}} \int_{-\infty}^{t} dt_3 \int_{-\infty}^{t_3} dt_2 \int_{-\infty}^{t_2} dt_1 |e'\rangle\langle e'|U_0(t,t_2)|e'\rangle\langle e'|W_f(t_2)|g\rangle\langle g|U_0(t_2,t_1)|g\rangle\langle g|\rho^{(0)}|g\rangle \\
&\times \quad \langle g|W_p(t_1)|e\rangle\langle e|U_0^\dagger(t_3,t_1)|e\rangle\langle e|W_b(t_3)|g'\rangle\langle g'|U_0^\dagger(t,t_3)|g'\rangle\langle g'| \\
&= \left(\frac{-i}{\hbar}\right)^3 \sum_{\substack{gg'\\ee'}} |e'\rangle\rho^{(0)}_{gg}\langle g'|e^{-i\Omega_{e'g'}t} \int_{-\infty}^{t} dt_3 \langle e|W_b(t_3)|g'\rangle e^{-i(\Omega_{e'e}-\Omega_{e'g'})t_3} \\
&\times \quad \int_{-\infty}^{t_3} dt_2 \langle e'|W_f(t_2)|g\rangle e^{-i(\Omega_{ge}-\Omega_{e'e})t_2} \int_{-\infty}^{t_2} dt_1 \langle g|W_p(t_1)|e\rangle e^{i\Omega_{ge}t_1}
\end{aligned} \tag{4.22}$$

where the equilibrated nature of the initial density operator, as defined in expressions (4.16) and (4.17), has been used to eliminate the off–diagonal elements of $\rho^{(0)}$ as well as contributions from the unpopulated excited state (*viz.*, $\rho^{(0)}_{ee}=0$). The final equality of Eq. (4.22) follows from the effect of $U_0(t_n,t_{n-1})$ on the energy eigenbasis [87, 5]:

$$U_0(t_n,t_{n-1})|\beta\rangle\langle\alpha|U_0^\dagger(t_n,t_{n-1}) = |\beta\rangle\langle\alpha|e^{-i(\omega_{\beta\alpha}-i\Gamma_{\beta\alpha})(t_n-t_{n-1})} = |\beta\rangle\langle\alpha|e^{-i\Omega_{\beta\alpha}(t_n-t_{n-1})} \tag{4.23}$$

where the complex factor $\Omega_{\beta\alpha}=\omega_{\beta\alpha}-i\Gamma_{\beta\alpha}$ incorporates both the angular frequency of a spectroscopic transition, $\omega_{\beta\alpha}$, and the associated phenomenological damping rate, $\Gamma_{\beta\alpha}$. For matter–field interactions occurring under the electric dipole approximation [118], the perturbation operators in Eq. (4.22) can be formulated as:

$$\begin{aligned}
W_f(t) &= -\boldsymbol{\mu}\cdot\boldsymbol{E}_f(t) = -\boldsymbol{\mu}\cdot\boldsymbol{E}_{\omega_f}e^{-i\omega_f t} \\
W_p(t) &= -\boldsymbol{\mu}\cdot\boldsymbol{E}_p(t) = -\boldsymbol{\mu}\cdot\boldsymbol{E}_{-\omega_p}e^{+i\omega_p t} \\
W_b(t) &= -\boldsymbol{\mu}\cdot\boldsymbol{E}_b(t) = -\boldsymbol{\mu}\cdot\boldsymbol{E}_{\omega_b}e^{-i\omega_b t}
\end{aligned} \tag{4.24}$$

where $\boldsymbol{\mu}$ denotes the dipole transition operator for the $|e\rangle\leftrightarrow|g\rangle$ resonance. These expressions for $W_d(t)$, with their inherent disparity between interactions driven by the pump and probe waves, reflect both momentum and energy conservation criteria as applied to the overall DFWM process. Furthermore, when considered from the viewpoint of quantized electromagnetic fields [86], the amplitude vectors $\boldsymbol{E}_{\omega_d}$ and $\boldsymbol{E}_{-\omega_d}$ of monochromatic field $\boldsymbol{E}_d(t)$ [*cf.*, Eq. (4.1) and Eq. (4.2)] are found to be proportional to photon annihilation and photon creation operators, respectively. Consequently, interactions involving $\boldsymbol{E}_{\omega_d}$ ($\boldsymbol{E}_{-\omega_d}$) must be associated with absorption (emission) events in the ket space and emission (absorption) events in the dual bra space [5, 117].

Substitution of the monochromatic field perturbation operators defined by Eq. (4.24) into expression (4.22) enables the three integrals over time to be evaluated in a straightforward manner. The third–order density operator for the sequence of interactions described by diagram (C) of Fig. 4.5 is thus given by:

$$\rho^{(3)}(t)=\frac{1}{\hbar^3}\sum_{\substack{gg'\\ee'}}|e'\rangle\rho_{gg}^{(0)}\langle g'|\frac{\langle e'|\boldsymbol{\mu}\cdot\boldsymbol{E}_{\omega_f}|g\rangle\langle g|\boldsymbol{\mu}\cdot\boldsymbol{E}_{-\omega_p}|e\rangle\langle e|\boldsymbol{\mu}\cdot\boldsymbol{E}_{\omega_b}|g'\rangle}{\left(\Omega_{ge}+\omega_p\right)\left(\Omega_{e'e}+\omega_p-\omega_f\right)\left(\Omega_{e'g'}-\omega_b+\omega_p-\omega_f\right)}\,\mathrm{e}^{-\mathrm{i}\left(\omega_f-\omega_p+\omega_b\right)t}. \tag{4.25}$$

The corresponding induced signal polarization follows from Eq. (4.15):

$$\hat{\boldsymbol{e}}_s^*\cdot\boldsymbol{P}_s^{(3)}(t)=\frac{N}{\hbar^3}\sum_{\substack{gg'\\ee'}}\rho_{gg}^{(0)}\frac{\langle g'|\boldsymbol{\mu}\cdot\hat{\boldsymbol{e}}_s^*|e'\rangle\langle e'|\boldsymbol{\mu}\cdot\hat{\boldsymbol{e}}_f|g\rangle\langle g|\boldsymbol{\mu}\cdot\hat{\boldsymbol{e}}_p^*|e\rangle\langle e|\boldsymbol{\mu}\cdot\hat{\boldsymbol{e}}_b|g'\rangle}{\left(\Omega_{ge}+\omega_p\right)\left(\Omega_{e'e}+\omega_p-\omega_f\right)\left(\Omega_{e'g'}-\omega_b+\omega_p-\omega_f\right)}\mathcal{E}_f\mathcal{E}_p^*\mathcal{E}_b\,\mathrm{e}^{\mathrm{i}(\boldsymbol{k}_s\cdot\boldsymbol{r}-\omega_s t)} \tag{4.26}$$

where the vector field amplitudes $\boldsymbol{E}_{\pm\omega_d}$ have been expressed in terms of their scalar amplitudes ($\mathcal{E}_d$ or $\mathcal{E}_d^*$), polarization unit vectors ($\hat{\boldsymbol{e}}_d$ or $\hat{\boldsymbol{e}}_d^*$), and spatial dispersion factors ($\mathrm{e}^{\mathrm{i}\boldsymbol{k}_d\cdot\boldsymbol{r}}$ or $\mathrm{e}^{-\mathrm{i}\boldsymbol{k}_d\cdot\boldsymbol{r}}$). In particular, the resulting product of complex exponentials has been combined to yield a single term containing the frequency, ω_s, and wavevector, $\boldsymbol{k}_s$, of the signal beam as defined in Eq. (4.5) and Eq. (4.6). While expression (4.26) and the steps leading to it provide a specific example for the calculation of $\boldsymbol{P}_s^{(3)}(t)$ in the case of diagram (C), analogous derivations can be presented for the other interaction sequences of Fig. 4.5.

The DFWM susceptibility tensor for an isolated one–photon resonant spectroscopic transition can be obtained by comparing the calculated signal polarization [as demonstrated above for diagram (C)] with the general definition of Eq. (4.10). In particular, $\boldsymbol{\chi}^{(3)}\left(-\omega_s;\omega_f,-\omega_p,\omega_b\right)$ is found to be a summation of eight distinct terms corresponding to the four diagrams of Fig. 4.5 and the analogous set of matter–field interactions obtained by interchanging the forward–going and backward–going pump waves [5, 119]:

$$\hat{\boldsymbol{e}}_s^*\cdot\boldsymbol{\chi}^{(3)}\left(-\omega_s;\omega_f,-\omega_p,\omega_b\right)\vdots\hat{\boldsymbol{e}}_f\hat{\boldsymbol{e}}_p^*\hat{\boldsymbol{e}}_b=\chi_{\mathrm{A}}+\chi_{\mathrm{B}}+\chi_{\mathrm{C}}+\chi_{\mathrm{D}}+\chi_{\mathrm{A}}'+\chi_{\mathrm{B}}'+\chi_{\mathrm{C}}'+\chi_{\mathrm{D}}' \tag{4.27}$$

where:

$$\begin{aligned}
\chi_A &\propto \frac{N}{\varepsilon_0\hbar^3}\sum_{\substack{gg'\\ee'}}\rho_{gg}^{(0)}\frac{\langle g|\boldsymbol{\mu}\cdot\hat{\boldsymbol{e}}_s^*|e'\rangle\langle e'|\boldsymbol{\mu}\cdot\hat{\boldsymbol{e}}_b|g'\rangle\langle g'|\boldsymbol{\mu}\cdot\hat{\boldsymbol{e}}_p^*|e\rangle\langle e|\boldsymbol{\mu}\cdot\hat{\boldsymbol{e}}_f|g\rangle}{\left(\Omega_{eg}-\omega_f\right)\left(\Omega_{g'g}-\omega_f+\omega_p\right)\left(\Omega_{e'g}-\omega_f+\omega_p-\omega_b\right)}\\
\chi_B &\propto \frac{N}{\varepsilon_0\hbar^3}\sum_{\substack{gg'\\ee'}}\rho_{gg}^{(0)}\frac{\langle g|\boldsymbol{\mu}\cdot\hat{\boldsymbol{e}}_p^*|e'\rangle\langle e'|\boldsymbol{\mu}\cdot\hat{\boldsymbol{e}}_b|g'\rangle\langle g'|\boldsymbol{\mu}\cdot\hat{\boldsymbol{e}}_s^*|e\rangle\langle e|\boldsymbol{\mu}\cdot\hat{\boldsymbol{e}}_f|g\rangle}{\left(\Omega_{eg}-\omega_f\right)\left(\Omega_{ee'}-\omega_f+\omega_p\right)\left(\Omega_{eg'}-\omega_f+\omega_p-\omega_b\right)}\\
\chi_C &\propto \frac{N}{\varepsilon_0\hbar^3}\sum_{\substack{gg'\\ee'}}\rho_{gg}^{(0)}\frac{\langle g|\boldsymbol{\mu}\cdot\hat{\boldsymbol{e}}_p^*|e\rangle\langle e|\boldsymbol{\mu}\cdot\hat{\boldsymbol{e}}_b|g'\rangle\langle g'|\boldsymbol{\mu}\cdot\hat{\boldsymbol{e}}_s^*|e'\rangle\langle e'|\boldsymbol{\mu}\cdot\hat{\boldsymbol{e}}_f|g\rangle}{\left(\Omega_{ge}+\omega_p\right)\left(\Omega_{e'e}-\omega_f+\omega_p\right)\left(\Omega_{e'g'}-\omega_f+\omega_p-\omega_b\right)}\\
\chi_D &\propto \frac{N}{\varepsilon_0\hbar^3}\sum_{\substack{gg'\\ee'}}\rho_{gg}^{(0)}\frac{\langle g|\boldsymbol{\mu}\cdot\hat{\boldsymbol{e}}_p^*|e\rangle\langle e|\boldsymbol{\mu}\cdot\hat{\boldsymbol{e}}_f|g'\rangle\langle g'|\boldsymbol{\mu}\cdot\hat{\boldsymbol{e}}_s^*|e'\rangle\langle e'|\boldsymbol{\mu}\cdot\hat{\boldsymbol{e}}_b|g\rangle}{\left(\Omega_{ge}+\omega_p\right)\left(\Omega_{gg'}-\omega_f+\omega_p\right)\left(\Omega_{e'g'}-\omega_f+\omega_p-\omega_b\right)}
\end{aligned} \tag{4.28}$$

and the primed quantities in Eq. (4.27) are obtained by interchanging the subscripts "f" and "b" in the corresponding unprimed susceptibility terms.

The third–order susceptibility tensor is found to be proportional to the quantity $N\rho_{gg}^{(0)}=N_g$ [*cf.*, Eq. (4.17)] and to the product of four one–photon dipole moment matrix elements. As shown in expression (4.10), the intensity of the DFWM signal, I_s, is related to the square modulus of $\hat{\boldsymbol{e}}_s^*\cdot\boldsymbol{\chi}^{(3)}\left(-\omega_s;\omega_f,-\omega_p,\omega_b\right)\vdots\hat{\boldsymbol{e}}_f\hat{\boldsymbol{e}}_p^*\hat{\boldsymbol{e}}_b$. Consequently, under the weak–field conditions appropriate for the perturbation analysis, I_s can be expected to exhibit a $N_g^2|\mu_{eg}|^8$ dependence on sample number density (N_g) and electronic transition moment (μ_{eg}). This behavior is quite distinct from the $N_g|\mu_{eg}|^2$ response encountered in most one–photon techniques and leads to unsaturated DFWM spectra that display radically different intensity patterns from those obtained through application of conventional (linear) spectroscopic probes. In addition, the quadratic scaling of I_s with target concentration implies a reduced sensitivity for the detection of trace species in rarefied environments.

The spectral response of the four–wave mixing interaction is embodied in the products of three frequency–dependent factors that appear in the various denominators of Eq. (4.28). Given the degenerate nature of the incident waves (*viz.*, $\omega_f = \omega_p = \omega_b = \omega$), each of these terms has the general form of a complex Half–Lorentzian: $(\Omega_{\beta\alpha} - \omega)^{-1}$ [120]. Since $I_s \propto |\hat{e}_s^* \cdot \chi^{(3)}(-\omega_s;\omega_f,-\omega_p,\omega_b):\hat{e}_f \hat{e}_p^* \hat{e}_b|^2$, the DFWM process will give rise to a Lorentzian–cubed lineshape centered at $\omega = \omega_{eg}$ and characterized by an effective width that scales in proportion to the rates for population and coherence dissipation. However, this simple description is complicated by the fact that complete specification of $\chi^{(3)}(-\omega_s;\omega_f,-\omega_p,\omega_b)$ for an isolated spectroscopic resonance entails the coherent summation of eight complex entities with phase differences among these quantities manifesting themselves as a modulation in the total signal pattern. Such interference effects are especially pronounced in the case of overlapping rovibronic transitions, where cross terms incurred by multilevel contributions to the susceptibility tensor can result in severe distortion of measured spectral profiles [25, 53].

When compared to other forms of nonlinear spectroscopy, DFWM has the distinct advantage of being fully resonant–enhanced [113]. In the context of Eq. (4.27) and Eq. (4.28), this means that all three denominators factors in each term of the susceptibility tensor display resonance–like behavior (*viz.*, become small in magnitude) as the common frequency of the impinging optical fields is tuned to the vicinity of a rovibronic line. The enormous increase in the magnitude of $\chi^{(3)}(-\omega_s;\omega_f,-\omega_p,\omega_b)$ that occurs upon resonant excitation of a molecular transition is responsible for the trace species sensitivity of DFWM (*i.e.*, despite the above mentioned N_g^2 scaling) and also provides for an essentially background–free response that exhibits no evidence of off–resonance interference effects which often plague other nonlinear schemes (*e.g.*, CARS [121, 122]).

By exploiting spherical tensor algebraic methods to recouple the product of four scalar operators (*viz.*, terms of the form $\hat{e}_d \cdot \boldsymbol{\mu}$) appearing in the numerator of each susceptibility element [Eq. (4.28)], Williams, Rahn, and Zare [116] have developed a compact and elegant formulation of near–resonant DFWM spectroscopy. This analysis assumes that all relaxation processes are independent of magnetic quantum number ($\Gamma_{\beta\alpha} \approx \Gamma_{\beta'\alpha} \approx \Gamma_{\beta\alpha'} \approx \Gamma_{\beta'\alpha'}$, *etc.*) and that the initial population of target species is isotropic (*i.e.*, all magnetic sublevels in the initial state are populated equally with no explicit phase relationships existing among them). Under such conditions, the DFWM signal strength is found to be proportional to the product of a concentration related factor, a one–photon molecular transition factor encoding the body–fixed properties of the selected spectroscopic resonance (*e.g.*, angular momentum quantum numbers, rotational branch assignments, and other vibronic properties), and a geometric factor containing the laboratory–fixed polarization characteristics for the impinging optical fields. The final expression for I_s entails multiplication of these quantities by $I_f I_p I_b$ and by appropriate lineshape functions derived from the denominator terms of $\chi^{(3)}(-\omega_s;\omega_f,-\omega_p,\omega_b)$. As found in conventional applications of the Wigner–Eckart Theorem [123], the separation of molecular and experimental parameters (*i.e.*, of dynamical and geometrical information) afforded by this treatment provides a convenient means for calculating the response of a DFWM interaction and for quantitatively comparing DFWM data sets obtained through use of different experimental configurations (*e.g.*, different polarization arrangements). A similar theoretical approach has been reported for the two–color variants of four–wave mixing discussed in Section 4.4.2 [60].

4.2.5 *Doppler Effects*

The thermal motion of molecules in a gaseous or liquid environment introduces additional complications into the description of four–wave mixing interactions [124, 107], with the spectral, temporal, and angular behavior of a gas–phase or liquid–phase measurement departing significantly from predictions based upon a stationary medium. A physical understanding for such phenomena follows from the grating analogy of DFWM spectroscopy (*cf.*, Section 4.2.3) which attributes the observed signal photons to the diffraction of light from optically–induced volume holograms. From this perspective, the one–photon resonant response for a motionless ensemble of molecules contains contributions from two *essentially equivalent* scattering amplitudes, the origins of which can be traced to periodic modulations in optical properties created through pairwise interference of the probe wave with each of the pump beams [*cf.*, first two terms of Eq. (4.13)]. The situation changes dramatically when motional effects are taken into account. Provided that secondary processes arising from the bulk transport of heat and mass can be neglected [11], the random translation of target species will lead to a temporal dissipation or "washout" of macroscopic spatial structures. This manifests itself as an overall reduction in diffraction efficiency and provides a mechanism for distinguishing the roles played by individual scattering channels. In particular, the effective lifetime for each grating, τ, will scale in proportion to the time required for a molecule to traverse the corresponding fringe spacing, Λ:

$$\tau = \Lambda / v_{\mathrm{mp}} \tag{4.29}$$

where v_{mp} denotes the most probable molecular speed. Since Λ is a function of the angular separation between the beams responsible for grating formation [*cf.*, Eq. (4.8)], the four–wave mixing response of a nonstationary medium can be expected to exhibit a pronounced dependence upon experimental geometry. In the case of the phase–conjugate DFWM scheme, where all incident and generated electromagnetic waves propagate in a nearly collinear fashion, typical bulk–gas velocities will lead to rapid washout of the small periodicity structure ascribed to the interference between $\boldsymbol{E}_p$ and $\boldsymbol{E}_b$. Since the timescale for such dissipation processes is comparable to the nanosecond duration of laser pulses employed for the majority of spectroscopic studies, the observed one–photon resonant signal will be generated primarily by Bragg–scattering from the large Λ grating induced through coupling of the forward–going pump and probe beams. Indeed, the effective reflectivity of the phase–conjugator (*i.e.*, magnitude of I_s) should decrease as the pump–probe crossing angle, θ, increases, attaining a minimum value at $\theta = \pi/2$ where the fringe spacings for the two spatial gratings are equivalent ($\Lambda = \lambda/\sqrt{2}$) [124, 8]. As discussed below, the overall efficiency of the diffraction process also will be influenced by the velocity–dependent shifts in molecular transition frequency that are responsible for inhomogeneous Doppler broadening.

The validity of Eq. (4.29) is contingent upon the mean free path for molecular motion exceeding the periodicity of the induced grating structures, with the characteristic rates for transverse and longitudinal relaxation of the target medium being sufficiently slow so as not to introduce further constraints. These conditions are usually well satisfied by gas–phase molecules maintained under low pressure static bulb or rarefied free–jet expansion conditions. However, for environments dominated by collisions (*e.g.*, high pressure gases and condensed phase), phenomena that critically depend upon the bulk thermodynamic properties of the sample, including the diffusive transport of heat and mass, can lead to the formation of secondary transient gratings (*e.g.*, thermal phase gratings) which exhibit their own distinct behavior for diffractive scattering and temporal dissipation [11]. These processes have been investigated extensively in liquids and solids where they are responsible for a major portion of the signals observed without recourse to time–resolved detection. Recently, Danehy, Paul, and Farrow [26, 125] have reported detailed experimental and theoretical studies designed to isolate the relative contributions made to a resonant gas–phase DFWM response by diffraction originating either from population gratings (*i.e.*, the spatial modulation of quantum state occupancy) or from thermal gratings (*i.e.*, the spatial modulation of overall density). Measurements conducted on room temperature samples of NO entrained in various buffer gases revealed that thermally–induced scattering mechanisms, attributable to modulation of the bulk refractive index, where dominant when total pressures exceeded a value ranging from 0.5 atm to 1.0 atm depending upon the efficiency of collisional quenching. Consequently, thermal grating diffraction can be expected to play a significant role for DFWM performed in high density regimes where collision–induced transfer of energy among internal and translational degrees of freedom leads to rapid degradation (*i.e.*, thermalization) of an initially selective optical excitation. In keeping with the spectroscopic orientation of this chapter, the ensuing discussion will focus on nonstationary ensembles of "isolated" molecules that are not subject to such secondary scattering events.

In order to participate in a four–wave mixing process, the individual molecules of an inhomogeneously–broadened medium must be able to interact simultaneously with all of the incident and generated electromagnetic fields. For a gas–phase ensemble, the frequency shifts associated with the first–order Doppler effect [1] will place restrictions on the velocity vector, $\boldsymbol{v}$, of species that can contribute successfully to the generation of a resonant nonlinear signal. While the four beams of DFWM spectroscopy are degenerate in the stationary, laboratory–fixed frame of reference (*i.e.*, all have identical frequency ω), the random rectilinear motion of individual molecules gives rise to velocity–dependent offsets in frequency that can be correlated to the direction of propagation for the corresponding light wave. Consequently, a particle moving with velocity $\boldsymbol{v}$ will experience $\boldsymbol{E}_f$ and $\boldsymbol{E}_p$ fields having effective optical frequencies of:

$$\begin{aligned} \omega_f &= \omega - \boldsymbol{k}_f \cdot \boldsymbol{v} \\ \omega_p &= \omega - \boldsymbol{k}_p \cdot \boldsymbol{v} \end{aligned} \tag{4.30}$$

where ω_f and ω_p now represent the frequencies of the forward–going pump and probe waves in the moving, molecule–fixed frame of reference. Similar expressions can be obtained for the frequency shifts associated with $\boldsymbol{E}_b$ and $\boldsymbol{E}_s$:

$$\begin{aligned} \omega_b &= \omega - \boldsymbol{k}_b \cdot \boldsymbol{v} = \omega + \boldsymbol{k}_f \cdot \boldsymbol{v} \\ \omega_s &= \omega - \boldsymbol{k}_s \cdot \boldsymbol{v} = \omega + \boldsymbol{k}_p \cdot \boldsymbol{v} \end{aligned} \tag{4.31}$$

where the latter equalities are specific to the counterpropagating, phase–conjugate experimental configuration where $\boldsymbol{k}_b = -\boldsymbol{k}_f$ and $\boldsymbol{k}_s = -\boldsymbol{k}_p$.

All of the beams employed in DFWM spectroscopy have wavevectors of identical magnitude ($|\boldsymbol{k}| = k = 2\pi/\lambda = \omega/c$), however, their spatial directions are usually distinct. For fully–resonant interactions to occur, the molecule–fixed frequencies of each optical wave [*cf.*, Eq. (4.30) and Eq. (4.31)] must coincide with the rest frequency of a molecular transition, ω_{eg}. For the hypothetical limit of extreme inhomogenous broadening, where monochromatic fields impinge upon a medium whose Doppler width ($\propto k v_{\mathrm{mp}}$) far exceeds its natural or homogeneous linewidth ($\propto \Gamma_{eg}$) [6], the counterpropagating phase–conjugate geometry can satisfy this criterion rigorously only for target species that are either stationary (*i.e.*, $\boldsymbol{v} = 0$) or moving precisely along the unique axis aligned perpendicular to all wavevectors (*i.e.*, $\boldsymbol{k} \cdot \boldsymbol{v} = 0$). More realistically, the resonance condition requires that the subset of molecules engaged in the DFWM process have small velocity components along the nearly collinear path of wave propagation, thereby implying a bias for motion orthogonal to the plane defined by the crossed pump and probe beams (*cf.*, top panel of Fig. 4.15). This preferential sensitivity for individual velocity groups gives rise to a sub–Doppler response [126, 124, 127, 6] which manifests itself as a spectral feature located precisely at ω_{eg} with a weak–field linewidth determined either by the effects of homogeneous broadening [124] or by the bandwidth of the excitation source [128, 66]. Although similar considerations apply to the backward–box DFWM scheme, the copropagating nature of the forward–box configuration enables all target species to contribute in the generation of a nonlinear signal. In this case, the offset between incident and resonance frequencies (*i.e.*, $\omega - \omega_{eg}$) specifies which molecules of a Doppler–broadened medium participate in the four–wave mixing interaction, with all velocity groups selected sequentially as ω tunes across the rovibronic transition.

In the weak–field regime, the DFWM signal polarization induced by monochromatic excitation of a nonstationary medium follows from the averaging of Eq. (4.14) over target motion [113, 6, 75]. For gas–phase molecules having a normalized distribution of velocities denoted by $f(\boldsymbol{v})$, the intensity and spectral profile of the nonlinear response evoked from the inhomogeneously–broadened ensemble is now given by:

$$I_s = \left(\frac{3\omega\ell}{4c^2\varepsilon_0}\right)^2 \left|\int_{\boldsymbol{v}} f(\boldsymbol{v})\, \hat{\boldsymbol{e}}_s^* \cdot \boldsymbol{\chi}^{(3)}\left(-\omega_s;\omega_f,-\omega_p,\omega_b\right) \vdots\, \hat{\boldsymbol{e}}_f\, \hat{\boldsymbol{e}}_p^*\, \hat{\boldsymbol{e}}_b \,\mathrm{d}\boldsymbol{v}\right|^2 I_f I_p I_b \tag{4.32}$$

where the elements of the third–order susceptibility tensor, as defined in Section 4.2.4, must be reformulated in terms of the effective frequencies experienced by the moving molecules [*cf.*, expressions (4.30) and (4.31)]. Even for a relatively simple Maxwellian distribution of velocities, exact solution of the integrals embodied in Eq. (4.32) is a formidable task. As discussed below, analytical forms are available for the limiting cases obtained when either Doppler or homogenous broadening dominates [6], with the explicit assumption of collinearity (*i.e.*, $\theta = 0$ in Fig. 4.1) enabling the calculated DFWM response to be cast in terms of plasma dispersion functions [129] or complex error functions [25]. For situations where these approximations are not valid or for systems characterized by more complicated, non–Maxwellian velocity distributions (*e.g.*, the nascent products of a chemical transformation), the only recourse for accurate evaluation of expression (4.32) may reside in the methods of numerical integration [119].

Abrams, *et al.* [6] have applied the perturbation treatment of nonlinear optics to the description of phase–conjugate DFWM spectroscopy in a nonstationary medium. Assuming monochromatic fields impinging upon an ensemble of ideal two–level systems (*i.e.*, no magnetic sublevel degeneracy) characterized by a Maxwellian distribution of velocities (*i.e.*, $v_{\text{mp}} = \sqrt{2k_{\text{B}}T/m}$ for particles of mass m and temperature T), the four–wave mixing signal intensity, I_s, is found to be:

$$I_s \propto (\Delta N_0)^2 \left|\mu_{eg}\right|^8 \left|r_{fp} + r_{bp}\right|^2 I_f I_b I_p \tag{4.33}$$

where ΔN_0 denotes the initial (unperturbed) population difference between the ground and excited states of an isolated molecular resonance having transition dipole moment matrix element μ_{eg}. The complex quantities r_{fp} and r_{bp} are reflectivity amplitudes which describe the dynamics of Bragg scattering from the volume holograms created through spatial overlap of the probe wave with the forward–going and backward–going pump beams, respectively. In particular, the square moduli of r_{fp} and r_{bp} are proportional to the distinct phase–conjugate reflectivities exhibited by each of the transient gratings involved in a one–photon resonant DFWM process. However, the possibility of constructive and destructive interference between these diffraction processes is suggested by the fact that, in general, $\left|r_{fp} + r_{bp}\right|^2 \neq \left|r_{fp}\right|^2 + \left|r_{bp}\right|^2$.

For a presumed collinear interaction (*i.e.*, $\theta = 0$) with a two–level medium having equal population decay rates for the ground and excited states (*i.e.*, $\Gamma_{gg} = \Gamma_{ee}$), simple analytical expressions can be derived for the complex reflectivity amplitudes r_{fp} and r_{bp} in the limit of extreme Doppler broadening ($kv_{\text{mp}} \gg \Gamma_{eg}, \Gamma_{gg}, \Gamma_{ee}$) and small detuning from resonance ($kv_{\text{mp}} \gg \Delta\omega$) [6]:

$$r_{fp} \cong \frac{2\sqrt{\pi}}{kv_{\text{mp}}} \frac{1/\Gamma_{gg}}{\Gamma_{eg} + \mathrm{i}\Delta\omega} \qquad r_{bp} \cong \frac{2\sqrt{\pi}}{\left(kv_{\text{mp}}\right)^3} \tag{4.34}$$

where $\Delta\omega = \omega - \omega_{eg}$. Under these asymptotic conditions, the portion of the DFWM signal resulting from $\left|r_{fp}\right|^2$ takes the form of a sub–Doppler Lorentzian lineshape centered at ω_{eg} and exhibiting a width (HWHM) equal to the natural linewidth: $\Delta\omega_{\text{HWHM}} = \Gamma_{eg}$. In contrast, $\left|r_{bp}\right|^2$ gives rise to a frequency–independent response commensurate with the large inhomogenous broadening of the extreme Doppler limit. At line center ($\Delta\omega = 0$), the relative amplitudes of contributions made to I_s by each of the transient gratings can be expressed as:

$$\left|\frac{r_{bp}}{r_{fp}}\right| \approx \frac{\Gamma_{gg}\Gamma_{eg}}{\left(kv_{\text{mp}}\right)^2}. \tag{4.35}$$

This quantity is typically much smaller than unity, a result in keeping with the temporal dissipation effects produced by motion of the target molecules. In particular, rapid washout of the small periodicity structure induced by interference of the backward–going pump and probe waves allows scattering processes associated with r_{fp} to dominate the overall DFWM response, thereby leading to sub–Doppler spectral profiles.

As previously mentioned, a pronounced dependence of DFWM signal strength on pump–probe crossing angle, θ, is expected to accompany implementation of the counterpropagating, phase–conjugate geometry in a nonstationary medium. Provided that optical pumping and saturation effects can be neglected, Wandzura [130] has suggested that finite Doppler width systems exhibit an angular response that scales as $I_s(\theta) \propto (\sin\theta)^{-4}$ where the apparent singularity at $\theta = 0$ disappears when the full $I_s(\theta)$ expression is considered. Ducloy and Bloch [124], employing a perturbative analysis similar to that embodied in Eq. (4.32), have confirmed this precipitous decline in signal with increasing θ and have derived an analytical form for the ratio of DFWM intensities obtained at orthogonal ($\theta = \pi/2$) and collinear ($\theta = 0$) experimental configurations:

$$\frac{I_s(\theta = \pi/2)}{I_s(\theta = 0)} = 4\theta_\circ^4 \tag{4.36}$$

where $\theta_\circ = \Gamma_{eg}/kv_{\mathrm{mp}}$ denotes the characteristic angle at which the mean free path of molecular motion ($D \approx v_{\mathrm{mp}}T_{2_{eg}} = v_{\mathrm{mp}}/\Gamma_{eg}$) approximately equals the periodicity [$\Lambda = \lambda/2\sin(\theta/2) \approx \lambda/\theta$] of the grating formed by interference of $\boldsymbol{E}_f$ and $\boldsymbol{E}_p$. Evaluation of this monochromatic field expression for typical values of experimental parameters reveals a $10^6 - 10^7$ decrease in signal strength as the geometry changes from collinear to orthogonal [124, 8]. Since the derivation of Eq. (4.36) relies upon an infinite plane wave approximation in which the length, ℓ, of the four–wave mixing interaction region remains constant, the limited transverse extent of actual laser sources will lead to an even more drastic weakening of $I_s(\theta)$ with increasing value of θ (*i.e.*, owing to the concomitant reduction in ℓ). While effects arising from finite optical bandwidths and short molecular lifetimes (*i.e.*, large homogeneous broadening) will mitigate this pronounced angular dependence [66], maximum DFWM response in a Doppler–broadened medium demands use of the smallest pump–probe crossing angle consistent with convenient extraction of the emerging signal wave and efficient discrimination against incoherently scattered background light.

The frequency response of phase–conjugate DFWM spectroscopy in a nonstationary medium can also be a strong function of the pump–probe crossing angle. A perturbative weak–field treatment of the collinear geometry ($\theta = 0$), such as that embodied in Eq. (4.33), yields a sub–Doppler Lorentzian lineshape of natural width (HWHM) Γ_{eg}. As θ increases, this spectral profile broadens and evolves until the Gaussian–like shape obtained in the orthogonal beam configuration ($\theta = \pi/2$) exhibits a linewidth comparable to the Doppler–limited value (*viz.*, $\Delta\omega_{\mathrm{HWHM}} \approx kv_{\mathrm{mp}}$) [8]. Ducloy and Bloch [124] have predicted a residual inhomogeneous broadening proportional to $\sin\theta$, with the angular dependence of the spectral width exhibiting a tripartite form:

$$\frac{\Delta\omega_{\mathrm{HWHM}}}{\Gamma_{eg}} = \begin{cases} 1 + \dfrac{3}{2}\left(\dfrac{\theta}{\theta_\circ}\right)^2 & \theta \ll \theta_\circ \\ 1.19\dfrac{\theta}{\theta_\circ} & \theta_\circ \ll \theta \ll 1 \\ \dfrac{1.17}{\theta_\circ} & \theta = \dfrac{\pi}{2} \end{cases} \tag{4.37}$$

where the characteristic angle $\theta_\circ$ has been defined in conjunction with Eq. (4.36). In particular, $\theta \ll \theta_\circ$ corresponds to a quasi–collinear sub–Doppler configuration while $\theta \gg \theta_\circ$ implies a large angular separation between the pump and probe beams. The spectral response of four–wave mixing measurements performed in inhomogeneously–broadened media is modified significantly by the presence of strong matter–field interactions (*e.g.*, optical saturation and optical pumping), with power–dependent broadening and bifurcation of resonant lineshapes reported [131-134].

4.2.6 *Optical Saturation Effects*

Much of the theoretical work performed on DFWM has been based on the nonlinear susceptibility formalism of Section 4.2.4, a treatment which suffices for unsaturated, weak–field situations where optically–induced Stark shifts and population transfers can justifiably be neglected. Aside from computational simplicity, this approach affords several conceptual advantages including the ability to unravel and classify various processes contributing to the wave–mixing phenomenon. However, the strong matter–field interactions that accompany resonant excitation of molecular transitions can rapidly degrade convergence of the perturbation expansions employed for calculation of the induced signal polarization, $\boldsymbol{P}_s^{(3)}(t)$ [82, 4]. Under such conditions, it is more appropriate to make use of a non–perturbative scheme that simultaneously considers all orders of nonlinearity and explicitly incorporates the dynamic coupling mechanisms responsible for optical saturation effects. The predictions obtained from this more general level of theory should asymptotically approach those derived from the conventional susceptibility analysis of four–wave mixing in the limit of low incident laser intensities.

A complete theoretical description of four–wave mixing in the strong–field limit, including explicit incorporation of effects arising from level degeneracies (*i.e.*, magnetic sublevels and their differential coupling through various light polarizations) and Doppler shifts (*i.e.*, non–stationary ensembles of target molecules), is far from trivial. One viable approach to this problem, made possible by the advent of massive computational resources, is direct numerical solution of the Liouville equation which governs temporal and spatial evolution of the molecular density operator [135-137]. While this "brute force" method promises to provide the best predictions for comparison with experimental data, greater insight and understanding of underlying physical processes can often be gleaned from analytical treatments based upon somewhat rudimentary assumptions. In particular, Abrams and Lind [138, 139, 6] have developed a relatively simple saturable absorber model for interpreting the near–resonant response of DFWM spectroscopy. Although derived for the specific case of a stationary, homogeneously–broadened medium comprised of hypothetical two–level atoms, a variety of experimental measurements [22, 140, 41, 137, 51, 56, 76] have demonstrated the feasibility and utility of extending this formalism to Doppler–broadened ensembles of molecules with their explicit magnetic sublevel degeneracies. Ewart and co–workers [141-143] have proposed a quite analogous theory for describing DFWM interactions induced by broad bandwidth sources of optical radiation.

As originally formulated by Abrams and Lind [138, 139, 6], the saturable absorber model focuses upon a system of motionless two–level atoms, with the common spectroscopic transition between their non–degenerate ground (*i.e.*, $|g\rangle$) and excited (*i.e.*, $|e\rangle$) states having resonant frequency ω_{eg} and dipole moment matrix element μ_{eg} (*cf.*, Fig. 4.4). The corresponding rates for dissipation of population and coherence are denoted by Γ_0 and Γ_{eg}, respectively, where the latter quantity can be defined in a manner analogous to that of expression (4.18) while the appropriate longitudinal relaxation parameter (for an homogeneously–broadened medium [144, 112]) is given by the inverse of the average lifetimes for the ground and excited states: $\Gamma_0^{-1} = \left(\Gamma_{gg}^{-1} + \Gamma_{ee}^{-1}\right)/2$. The incident and generated electromagnetic waves are presumed to be monochromatic and of identical frequency ω. The saturable response induced by pump fields of arbitrarily large intensity is described through full (non–perturbative) solution of the density operator equations. Subsequent application of first–order perturbation theory enables the effects of the presumed weak probe and signal fields to be taken into account. In the limit of small overall absorption or gain, with the two pump beams having a common intensity, $I = I_f = I_b$, and not subject to attenuation or amplification as a result of their propagation through the target medium, a simple analytical form can be obtained for the magnitude of the signal wave, I_s, emerging from a DFWM interaction region of length ℓ [140, 8]:

$$I_s = R I_p = \frac{(\alpha_\circ \ell)^2}{1+\delta^2} \frac{(2I/I_{\mathrm{sat}})^2}{(1+4I/I_{\mathrm{sat}})^3} I_p \tag{4.38}$$

where R signifies the phase–conjugate reflectivity experienced by the incoming probe radiation of intensity I_p. The DFWM signal strength increases quadratically with the field absorption coefficient at line center, $\alpha_\circ$,

$$\alpha_\circ = \frac{\omega_{eg} |\mu_{eg}|^2}{2\hbar c \varepsilon_0 \Gamma_{eg}} \Delta N_0 \tag{4.39}$$

where $\Delta N_0 = N_g - N_e$ denotes the initial population difference (per unit volume) that exists between the ground and excited states of the ensemble. In particular, a positive value of ΔN_0 corresponds to the equilibrium situation of net absorption ($N_g > N_e$) while negative ΔN_0 implies an inverted or gain medium ($N_g < N_e$). The frequency response of the four–wave mixing interaction is embodied in the normalized detuning parameter, δ, defined as the difference between incident and resonance frequencies, $\Delta\omega = \omega - \omega_{eg}$, scaled by the homogenous decay rate (or natural linewidth) of the optical transition, Γ_{eg}:

$$\delta = \Delta\omega/\Gamma_{eg} = \left(\omega - \omega_{eg}\right)/\Gamma_{eg}. \tag{4.40}$$

This dimensionless quantity also appears in the defining expression for the optical saturation intensity, I_{sat}:

$$I_{\mathrm{sat}} = I_{\mathrm{sat}}^{\circ}\left(1+\delta^2\right) \tag{4.41}$$

where the saturation intensity at line center, I_{sat}°, depends upon the characteristic properties of the $|e\rangle \leftrightarrow |g\rangle$ spectroscopic transition:

$$I_{\mathrm{sat}}^{\circ} = \frac{\hbar^2 c\varepsilon_0 \Gamma_{eg}\Gamma_0}{2\left|\mu_{eg}\right|^2} = \frac{\hbar\omega_{eg}}{4}\left(\frac{\Delta N_0\,\Gamma_0}{\alpha_\circ}\right). \tag{4.42}$$

In particular, I_{sat}° provides a measure of the field strength for which the optically–induced oscillation of population between ground and excited states of a rovibronic transition (*i.e.*, the Rabi frequency) becomes comparable to the rates of relaxation/dephasing processes [145, 146, 112]. As illustrated below, the functional form of a resonant DFWM response is correlated strongly with the ratio of incident pump and line–center saturation intensities, $I/I_{\mathrm{sat}}^{\circ}$.

Eq. (4.38) predicts the four–wave mixing signal strength to scale as the square of the field absorbance at line center, $I_s \propto (\alpha_\circ \ell)^2$, thereby leading to the expected quadratic dependence on sample number density and interaction path length. The saturable absorber model also provides an analytical expression for the DFWM spectral profile, with Brown, *et al.* [16] demonstrating that reformulation of I_s in terms of a power–broadening rate parameter, Γ_{PB},

$$\Gamma_{PB} = \Gamma_{eg}\sqrt{1+4I/I_{\mathrm{sat}}^{\circ}}, \tag{4.43}$$

leads to a physically–intuitive form for the lineshape of an isolated resonance:

$$I_s = RI_p = (2\alpha_\circ \ell)^2 \left(\frac{I}{I_{\mathrm{sat}}^{\circ}}\right)^2 \left[\frac{\Gamma_{eg}^2}{(\Delta\omega)^2+\Gamma_{PB}^2}\right]^3 I_p. \tag{4.44}$$

In keeping with the weak–field predictions of the nonlinear susceptibility analysis, Eq. (4.44) suggests a Lorentzian–cubed spectral profile. However, the saturable absorber lineshape broadens in proportion to the common intensity of the incident pump waves with the effective width (HWHM), $\Delta\omega_{\mathrm{HWHM}}$, for an isolated resonance given by:

$$\Delta\omega_{\mathrm{HWHM}} = \Gamma_{PB}\sqrt{2^{1/3}-1} = \Gamma_{eg}\sqrt{\left(2^{1/3}-1\right)\left(1+4I/I_{\mathrm{sat}}^{\circ}\right)}. \tag{4.45}$$

For weak incident pump field (*i.e.*, $I \ll I_{\mathrm{sat}}^{\circ}$), this expression predicts $\Delta\omega_{\mathrm{HWHM}} \approx \Gamma_{eg}\sqrt{2^{1/3}-1}$, which corresponds to roughly half of the $|e\rangle \leftrightarrow |g\rangle$ natural linewidth. In contrast, the DFWM lineshape for $I \gg I_{\mathrm{sat}}^{\circ}$ is characterized by $\Delta\omega_{\mathrm{HWHM}} \approx \Gamma_{eg}\sqrt{I/I_{\mathrm{sat}}^{\circ}}$, thereby implying a near constant (*i.e.*, frequency independent) response in the "infinite intensity" regime.

Brown, Rahn, and co–workers [16, 19, 134] have reported a detailed investigation of saturation processes in molecular DFWM spectra recorded via the counterpropagating, phase–conjugate experimental configuration. This work entailed high resolution measurements of OH radicals under conditions where Doppler and homogeneous broadening were of comparable magnitude, with the intensity of the probe beam held fixed at a non–saturating level (*i.e.*, $I_p \leq 0.2I_{\mathrm{sat}}^{\circ}$) while the pump intensities were varied systematically over a wide range. For equivalent pump fields ($I_f = I_b = I$), the Abrams and Lind theory was found to be consistent with spectral bandwidths observed for $I_{\mathrm{sat}}^{\circ} \leq I \leq 8I_{\mathrm{sat}}^{\circ}$. At lower intensities, the measured lineshape had a Lorentzian–like form which, although sharper than predictions based upon the infinite–Doppler limit [*cf.*, Eq. (4.33) and Eq. (4.34)], was well described by a perturbative two–level analysis that incorporated effects arising from finite molecular velocities. Deviation from the saturable absorber model in this weak–field regime can be attributed to the presence of residual Doppler broadening which is neglected in the stationary Abrams and Lind analysis.

In the case of high pump powers (*i.e.*, $I > 8I_{\mathrm{sat}}^{\circ}$), Brown, *et al.* [134] reported DFWM spectral profiles that departed significantly from the Abrams and Lind prediction of a Lorentzian–cubed frequency dependence.

Aside from flat–topped lineshapes encountered for equivalent pump beams (*i.e.*, $I_f = I_b >> I^\circ_{sat}$), a pronounced directional anisotropy was noted, with saturation produced by $\boldsymbol{E}_f$ leading to a completely different behavior than that induced by $\boldsymbol{E}_b$. In particular, as reported in previous studies [131-133], saturation by the backward–going pump wave resulted in bifurcated spectral features exhibiting a distinct decrease or dip in signal production at line center. Such phenomena have been addressed by theoretical analyses performed in the Doppler–broadened limit through both non–perturbative density operator [147-149, 136] and dressed–atom [150, 132, 133] methods. These efforts suggest that the observed saturation response is intrinsic to the inhomogeneous–broadening of gas–phase media, with the differential matter–field interactions experienced by individual velocity groups responsible for the manifestation of spectral splittings. Since the motionless target molecules of the saturable absorber model provide no mechanism for distinguishing the roles played by the counterpropagating pump beams, such velocity–dependent effects cannot be predicted from the Abrams and Lind treatment.

In order to provide a viable tool for spectroscopic measurements, the dependence of DFWM signal strength upon experimental parameters (*e.g.*, incident light intensities) and molecular properties (*e.g.*, transition moments and decay rates) must be understood at a level of detail commensurate with the quantitative analysis of observed spectral features. Some insight into the behavior of four–wave mixing processes follows from consideration of the saturable absorber model in the limits of low and high pump intensity (*i.e.*, measured relative to I°_{sat}). The appropriate asymptotic forms for the DFWM response at line–center, I°_s, can be obtained from Eq. (4.38) by substituting $\delta = 0$ and performing a series expansion of the intensity–dependent denominator [8]:

$$I \ll I^\circ_{sat}; \qquad I^\circ_s = (\alpha_\circ \ell)^2 \left(\frac{2I}{I^\circ_{sat}} \right)^2 I_p \propto (\Delta N_0 \ell)^2 \frac{|\mu_{eg}|^8}{\Gamma_0^2 \Gamma_{eg}^4} I^2 I_p \tag{4.46}$$

$$I \gg I^\circ_{sat}; \qquad I^\circ_s = \left(\frac{\alpha_\circ \ell}{4} \right)^2 \frac{I^\circ_{sat}}{I} I_p \propto (\Delta N_0 \ell)^2 \frac{|\mu_{eg}|^2 \Gamma_0}{\Gamma_{eg}} \frac{I_p}{I} \tag{4.47}$$

where the latter proportionalities stem from the definitions for $\alpha_\circ$ and I°_{sat}. Since most DFWM experiments derive the pump and probe beams from a common source of optical radiation, it is useful to consider the situation obtained when $I_p \propto I$. Under this condition, the saturable absorber response for $I \ll I^\circ_{sat}$ scales as the cube of the incident intensity, a result in keeping with predictions derived from the nonlinear susceptibility formalism. In contrast, the high–power limit for I°_s displays a behavior that is essentially independent of both I and I_p. This suggests a possible advantage for working in the saturated regime since inevitable amplitude fluctuations on the incoming laser beams will not be transferred appreciably into the detected DFWM signal. Furthermore, expression (4.38) shows that the highest phase–conjugate reflectivity, R, is attained when $I = I_{sat}/2 = I^\circ_{sat}(1+\delta^2)/2$, thereby yielding:

$$I = \frac{I^\circ_{sat}}{2}; \qquad I^\circ_s = \frac{(\alpha_\circ \ell)^2}{27} I_p \propto (\Delta N_0 \ell)^2 \frac{|\mu_{eg}|^2}{\Gamma_{eg}^2} I_p. \tag{4.48}$$

Detailed analysis of the exact Abrams and Lind theory [6] confirms this maximization of DFWM response at near–saturation intensities and also suggests the possibility of achieving enhanced signal production and spectral resolution for inverted gain media (*i.e.*, ΔN_0 or $\alpha_\circ$ negative) [8].

The asymptotic expressions for I°_s display significant differences in their dependence upon transition dipole moment matrix element, μ_{eg}. While the low intensity regime is characterized by the same $|\mu_{eg}|^8$ scaling predicted by the perturbative susceptibility analysis, the four–wave mixing response at high intensities is found to be proportional to $|\mu_{eg}|^2$. Farrow, *et al.* [140] have reported a systematic investigation of resonant gas–phase DFWM interactions designed to assess the role of transition dipole magnitude. Experiments were performed on individual rovibronic lines in the NO $A\,^2\Sigma^+ - X\,^2\Pi$ (0,0) band, with the effects of finite laser bandwidth taken into account by integrating the saturable absorber model over all possible frequency detuning [*i.e.*, integration of

Eq. (4.38) over $-\infty \le \delta \le +\infty$]. The resulting spectrally–integrated form for the DFWM signal strength, I_s^{int}, is given by:

$$I_s^{\text{int}} = \frac{3\pi}{2}(\alpha_\circ \ell)^2 \, \Gamma_{eg} \frac{\left(I/I_{\text{sat}}^\circ\right)^2}{\left(1+4I/I_{\text{sat}}^\circ\right)^{5/2}} I_p \tag{4.49}$$

where Γ_{eg} represents the natural (homogeneous) linewidth of the optical transition. A straightforward analysis yields the corresponding asymptotic expressions for I_s^{int} in the limits of low and high pump intensity:

$$I \ll I_{\text{sat}}^\circ; \qquad I_s^{\text{int}} = \frac{3\pi}{2}(\alpha_\circ \ell)^2 \, \Gamma_{eg} \left(\frac{I}{I_{\text{sat}}^\circ}\right)^2 I_p \propto (\Delta N_0 \ell)^2 \frac{|\mu_{eg}|^8}{\Gamma_0^2 \Gamma_{eg}^3} I^2 I_p \tag{4.50}$$

$$I \gg I_{\text{sat}}^\circ; \qquad I_s^{\text{int}} = \frac{3\pi}{64}(\alpha_\circ \ell)^2 \, \Gamma_{eg} \sqrt{\frac{I_{\text{sat}}^\circ}{I}} \, I_p \propto (\Delta N_0 \ell)^2 |\mu_{eg}|^3 \sqrt{\frac{\Gamma_0}{\Gamma_{eg}}} \frac{I_p}{\sqrt{I}}. \tag{4.51}$$

While the spectrally–integrated and line–center expressions for I_s are nearly identical when $I \ll I_{\text{sat}}^\circ$, they differ substantially in the strong–field limit with I_s^{int} scaling as $|\mu_{eg}|^3$ rather than the $|\mu_{eg}|^2$ proportionality suggested by I_s°. Farrow and co–workers [140] found their experimental results to be in reasonable accord with the saturable absorber response predicted by I_s^{int}. In particular, measured signal strengths were well described by a power law of the form $I_s \propto (\mu_{eg})^m$ where, owing to saturation effects, the empirically determined value of the exponent m initially exhibited a rapid decrease with increasing incident intensity followed by a more gradual fall off that asymptotically approached a high–power $m = 3.6$ plateau. This behavior has important ramifications for practical DFWM spectroscopy which typically involves the observation of rovibronic features having widely different transition moments and, consequently, subject to varying degrees of optical saturation. More specifically, power–induced changes in the $|\mu_{eg}|$ dependence of I_s, combined with an intrinsic quadratic scaling on population number density, can lead to recorded spectral patterns that bear little resemblance to those obtained through more conventional linear techniques. Recently, Germann and Rakestraw [51] have exploited this differential response to implement a multiplex approach for the *in situ* determination of transition moment magnitudes and absolute concentrations.

The saturable absorber model predicts the dependence of DFWM signal strength upon population and coherence dissipation rates to change dramatically as a function of incident laser intensity, with the line–center values of I_s scaling as $\Gamma_0^{-2}\Gamma_{eg}^{-4}$ and $\Gamma_0\Gamma_{eg}^{-1}$ in the low–power and high–power limits, respectively. Provided that pure dephasing processes can be neglected ($\Gamma_{eg}^{\Phi} \approx 0$) and relaxation of the excited state population occurs substantially faster than that of the corresponding ground state ($\Gamma_{ee} \gg \Gamma_{gg}$), then $\Gamma_0 \approx \Gamma_{gg} = \tau_g^{-1}$ and $\Gamma_{eg} \approx \Gamma_{ee} = \tau_e^{-1}$ where τ_α denotes the effective lifetime for state $|\alpha\rangle$. Under such conditions, the Abrams and Lind theory suggests that I_s° (or I_s^{int}) should be proportional to $(\Gamma_{ee})^{-m} = (\tau_e)^m$ where $m = 4$ ($m = 3$) for $I \ll I_{\text{sat}}^\circ$ and $m = 1$ ($m = \frac{1}{2}$) for $I \gg I_{\text{sat}}^\circ$. Consequently, the promise of absorption–based DFWM techniques to enable facile interrogation of "dark" species possessing strongly predissociated excited states (small values of τ_e) will be offset by the decrease in signal production expected to accompany the nonradiative relaxation channels responsible for suppression of observable fluorescence [66]. While these deleterious effects can be mitigated by operating in a strong–field regime, the commensurate increase in power required to achieve saturation (*i.e.*, $I_{\text{sat}}^\circ \propto \Gamma_0\Gamma_{eg}$) might prohibit successful implementation of this approach.

For collision–dominated environments, where the rates of population and coherence dissipation are directly proportional to the pressure, p, of foreign gas molecules, the pressure dependence of the line–center DFWM signal amplitude is expected to vary from p^{-6} to p^0 as incident intensities increase from completely non–saturating to strongly–saturating levels. Danehy, *et al.* [137] have examined the influence of collisional quenching on resonant DFWM interactions, confirming previous experimental assertions that saturated four–wave mixing is relatively insensitive to quenching rates. In the weak–field limit ($I \ll I_{\text{sat}}^\circ$), measurements of I_s as a function of p were found to be in excellent agreement with predictions derived from the Abrams and Lind treatment. However, quantitative description of DFWM data sets obtained under conditions of extreme optical saturation necessitated numerical solution of the time–dependent density operator equations with explicit incorporation of effects arising from target motion and transient laser pulse propagation.

While the primary inadequacies of the saturable absorber model for gas–phase DFWM spectroscopy stem from its complete neglect of effects arising from Doppler broadening and magnetic sublevel degeneracy, the above discussion has highlighted the utility of exploiting this simple analytical approach for the quantitative interpretation of resonant four–wave mixing processes. Various extensions of the Abrams and Lind analysis have been reported, including treatments designed to account for unequal pump fields [151, 152], pump absorption/depletion [153, 154], overlapping transitions [153], and pump–probe frequency detuning [155, 156, 6]. Several attempts have been made to cobble the advantages of perturbation schemes, with their explicit incorporation of velocity–dependent frequency shifts and polarization–specific magnetic sublevel interactions, onto the saturable absorber model. Recently, Williams, Zare, and Rahn [116, 157] have employed both theory and experiment to unravel the influence of optical polarization on molecular DFWM spectra acquired over a wide range of saturation conditions. Measurements performed on the CH $A\,^2\Delta - X\,^2\Pi$ (0,0) band [41, 157] revealed that the line–center saturation intensity, I_{sat}°, for an individual rovibronic line was relatively independent of polarization geometry. In addition, the substantial polarization–induced differences in signal strength that exist between rotational branches were not affected strongly by changes in the incident laser power. Consequently, weak–field geometric factors [116] were found to provide an adequate description of spatial characteristics for DFWM experiments involving incident beam intensities as high as twice I_{sat}°, with appropriate scaling of the transition moment dependence (*e.g.*, as suggested by the saturable absorber model) leading to estimated errors of only 10–30% in the calculated response of a saturated four–wave mixing interaction.

4.3 Applications

4.3.1 *Experimental Configurations*

The experimental configuration illustrated in Fig. 4.6 is typical of those employed for phase–conjugate DFWM studies of molecules possessing one–photon resonant ultraviolet transitions. In this sub–Doppler geometry, the incident and generated beams form a plane with their region of mutual intersection defining the sample interaction volume. Very similar approaches can be implemented for the investigation of infrared transitions (*i.e.*, substitution of infrared sources/components) and two–photon resonances (*i.e.*, focusing of incident beams). While the backward–box DFWM scheme can be obtained by tipping the backward–going pump and signal waves out–of–plane so as to form a pyramidal phase–matching geometry (*cf.*, Fig. 4.2), the forward–box configuration would necessitate a somewhat more drastic rearrangement of the depicted experimental apparatus.

For studies performed in the visible and ultraviolet regions of the spectrum, the single color of tunable light required by the DFWM detection scheme typically is obtained by using one of the harmonics from a pulsed Nd:YAG laser to pump a high resolution dye laser. An injection seeder on the Nd:YAG oscillator and an intracavity etalon in the dye laser greatly facilitate in the production of high quality tunable radiation approaching ideal "single frequency" (*i.e.*, Fourier Transform–limited, single longitudinal mode) performance. Aside from maximizing the spectroscopic benefits derived through implementation of sub–Doppler (laser bandwidth–limited) DFWM geometries, such light sources exhibit substantial coherence lengths that enable phase–matched nonlinear interactions to be established over long sample distances [158], thereby leading to enhanced signal generation. If necessary, the dye laser output is frequency doubled by means of a servo–locked harmonic crystal and the resulting ultraviolet light is isolated via a set of four Brewster angle prisms oriented to cancel any wavelength–dependent spatial displacement in the emerging beam. As shown in Fig. 4.6, small portions of the rejected fundamental radiation are directed through a monitor etalon and an I_2 (or Te_2) absorption cell. While the former serves both to assess the spectral quality of the dye laser output and to provide relative frequency markers, the latter, when used in conjunction with published atlases [159], furnishes a convenient absolute frequency calibration.

The frequency–doubled radiation initially passes through a variable attenuator which permits the emerging pulse energy to be varied in a continuous and reproducible fashion. Aside from being essential for the saturation studies that must precede any quantitative attempt at DFWM signal interpretation, this device can be incorporated into a feedback loop designed to stabilize laser power over the entire tuning range spanned by a particular spectrum [22, 140, 7]. The ultraviolet output subsequently is directed through a Keplerian telescope where it is spatially filtered and recollimated to a diameter appropriate for the ongoing experiment (*i.e.*, typically $0.2-0.4$cm diameter). The high quality of wavefronts in the resulting beam greatly enhances the efficacy of the four–wave mixing process and increases overall detection sensitivity by minimizing background noise that arises from incoherently scattered laser light.

A set of achromatic beamsplitters is used to separate the tunable ultraviolet radiation into three equivalent beams destined to become the input waves for implementation of the phase–conjugate DFWM geometry. In addition, a small portion of the frequency–doubled light is diverted towards a calibrated photodiode which serves to monitor the energy of individual laser pulses. For the depicted configuration, the forward–going pump ($\boldsymbol{E}_f$) and probe ($\boldsymbol{E}_p$) waves have vertical polarization (*viz.*, linear polarization orthogonal to the plane formed by the optical table surface) while the backward–going pump beam ($\boldsymbol{E}_b$) has its direction of linear polarization rotated to the horizontal plane (*i.e.*, $\hat{e}_b \perp \hat{e}_f \parallel \hat{e}_p$) by passage through a double (half–wave) Fresnel Rhomb. Conservation criteria imposed upon the DFWM interaction demand that the resulting signal wave ($\boldsymbol{E}_s$) emerge with a corresponding horizontal polarization (*i.e.*, $\hat{e}_s \parallel \hat{e}_b$), thereby facilitating its discrimination and extraction from the coaxial probe radiation. Often referred to as the "cross–population" geometry [6], this scheme permits the collection of virtually all generated signal photons and provides for the efficient rejection of scattered background light. While these desirable characteristics are offset by the reduction in signal strength expected to accompany the use of cross–polarized fields in the four–wave mixing interaction [116], the signal–to–noise advantages afforded by polarization–based detection usually outweigh any disadvantages [60]. This holds true even when considering the intrinsically low–background, non–planar variants of DFWM illustrated in Fig. 4.2.

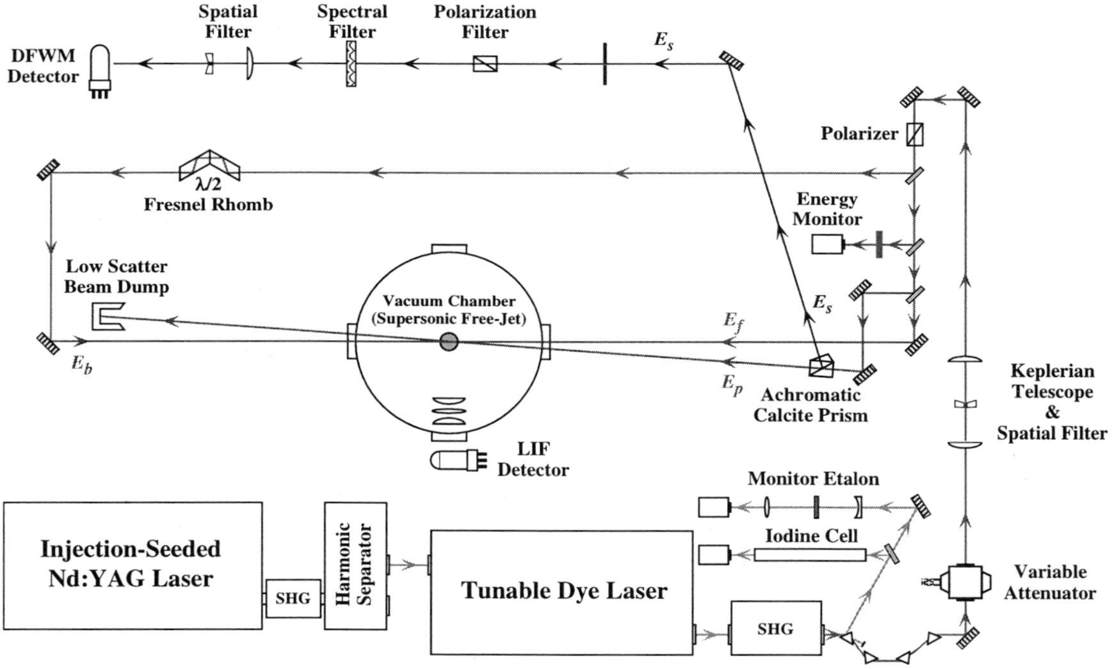

Figure 4.6 Experimental configuration for phase–conjugate DFWM spectroscopy. Three linearly polarized beams of identical frequency form a plane with their region of mutual intersection located in a supersonic free–jet expansion of target molecules. The forward–going pump ($\boldsymbol{E}_f$) and probe ($\boldsymbol{E}_p$) waves are vertically polarized and cross under a small angle while the backward–going pump wave ($\boldsymbol{E}_b$) is horizontally polarized and aligned to precisely retrace the path of $\boldsymbol{E}_f$. The four–wave mixing signal field ($\boldsymbol{E}_s$), which emerges as a coherent beam that is coaxial and counterpropagating with respect to the incident probe radiation, is also horizontally polarized and can be isolated through use of polarization–selective components (*e.g.*, an achromatic calcite prism).

The counterpropagating pump waves (*i.e.*, $\boldsymbol{E}_f$ and $\boldsymbol{E}_b$) are directed along the central axis of the sample vessel with appropriate delay lines used to ensure their temporal overlap within the DFWM interaction region. The probe beam intersects the pump fields under a small angle (*i.e.*, typically $\leq 1^\circ$ with larger angles serving to reduce background noise at the expense of signal amplitude) and ultimately impinges upon an efficient "black–body" absorber designed to minimize backscattered light. An achromatic prism polarizer is inserted into the optical train of the probe radiation. While the vertically polarized probe light is transmitted through this component essentially unscathed, the emerging signal wave, which retraces the path of the probe beam, is deflected totally owing to the horizontal orientation of its linear polarization vector. In this manner, the desired $\boldsymbol{E}_s$ photons can readily be discriminated from the substantially more intense probe radiation. For situations that are not amenable to polarization–based detection schemes, a simple beamsplitter can be employed for extraction of the counterpropagating signal field. Since this approach leads to a substantial decrease in overall collection efficiency, non–planar experimental geometries, with their isolation of the signal wave along a distinct spatial direction, can prove to be advantageous under such conditions [22, 25, 26].

After passing through polarization, spectral, and spatial filters designed to reject stray light, the DFWM signal beam impinges upon the photocathode of a simple (uncooled) photomultiplier tube. The resulting photocurrent is preamplified and directed to a gated integrator system that enables the signal intensity to be measured as a function of laser wavelength. Absolute and relative frequency calibrations, as well as the incident pulse energy, are recorded simultaneously. Pulse–to–pulse fluctuations in the DFWM signal amplitude can be minimized through use of an active normalization procedure in which the response produced by each shot of the laser system is divided by an appropriate function of the corresponding pulse energy (*e.g.*, divided by power cubed in the absence of optical saturation effects). Even under the best of circumstances, the nonlinear nature of the four–wave mixing interaction demands use of appreciable signal averaging for the collection of high quality spectral data (*e.g.*, averaging of response produced by ≥ 20 pulses for each increment of dye laser frequency).

The experimental configuration depicted in Fig. 4.6 is typical of that encountered in a large number of reported DFWM studies, however, several modifications can be made to reduce the overall complexity of this apparatus. In particular, the separate optical train for manipulation of the backward–going pump beam can be eliminated by allowing the forward–going pump light that emerges from the sample vessel to propagate through a single (quarter–wave) Fresnel Rhomb before impinging upon a mirror that precisely retroreflects it back along its initial path [63, 64]. In this manner, collinear and counterpropagating pump fields of orthogonal polarization (*i.e.*, $\hat{\boldsymbol{e}}_b \perp \hat{\boldsymbol{e}}_f$) can readily be established with a minimum number of optical components. Unfortunately, this scheme leads to a temporal mismatch of the two pump waves, with the delay between their arrival times at the DFWM interaction region adversely affecting studies of short–lived or transient species.

Although the target medium probed via DFWM techniques could, in principle, exist in any state (*i.e.*, solid, liquid, or gas), Fig. 4.6 depicts the frequently encountered situation of a molecular sample entrained in a free–jet expansion. An impressive body of work has documented the utility of exploiting supersonic beams for the spectroscopic interrogation of molecules under essentially collision–free conditions [160]. Aside from providing a viable means for the preparation and retention of weakly–bound molecular complexes [161], the substantial cooling of internal and translational degrees of freedom afforded by such environments leads to enormous simplification of the spectral features exhibited by massive polyatomic species [162, 163]. However, the high stream velocities of molecular beam sources require that certain precautions be taken when incorporating them into four–wave mixing studies. While the velocity discrimination afforded by sub–Doppler DFWM schemes is partially responsible for this prudence, additional constraints are imposed by the directional specificity of dissipation processes acting upon optically–induced transient gratings.

As demonstrated by the two–color, time–resolved measurements of Butenhoff and Rohlfing [55, 164], the relative directions of propagation for molecular and optical beams can profoundly influence both the magnitude and the form of signals observed in a four–wave mixing process. This effect stems primarily from the vastly different rates of transient grating washout (*cf.*, Section 4.2.5) expected when the molecular stream velocity, $\boldsymbol{v}_\circ$, is pointing either along or across the corresponding pattern of fringes, with the latter condition ($\boldsymbol{v}_\circ$ across fringes) leading to a substantially more rapid decay of optically–induced periodic structures. Maximum signal production in the phase–conjugate DFWM scheme should be realized through implementation of the

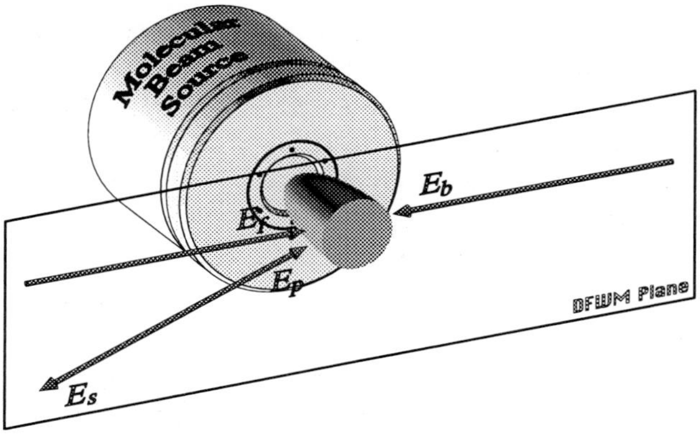

Figure 4.7 Implementation of phase–conjugate DFWM spectroscopy in a supersonic free–jet expansion. Maximum signal production demands that the fringes of optically–induced transient gratings be directed along, rather than perpendicular to, the high velocity axis of molecular flow.

experimental arrangement depicted in Fig. 4.7. In particular, this configuration has the high velocity axis of supersonic molecular flow oriented parallel to the crests and troughs of the "grooves" comprising the one–photon resonant gratings that are formed by pairwise interference of $\boldsymbol{E}_p$ with either $\boldsymbol{E}_f$ or $\boldsymbol{E}_b$. Consequently, volume holograms created within the target medium will be "long–lived" with the only translational contributions to their dissipation arising from the relatively slow transverse motion of molecules entrained in the free–jet expansion. In contrast, orthogonal geometries which have $\boldsymbol{v}_\circ$ crossing the fringe spacing will lead to degradation of diffraction efficiency on a timescale given by $\Lambda/|\boldsymbol{v}_\circ|$ [expression (4.29)]. Typical values of grating periodicity and stream velocity suggest that an appreciable reduction in signal strength can take place over the nanosecond duration of laser pulses employed for the majority of DFWM studies. The spatial characteristics of a supersonic molecular beam can also affect spectroscopic response, with the velocity–collimated output from a slit–shaped nozzle orifice found to greatly enhance the signal magnitude generated in linear sub–Doppler schemes [165-172]. Recent experiments conducted on the $\mathrm{V}\,{}^1\mathrm{B}_2({}^1\Delta_u) - \tilde{\mathrm{X}}\,{}^1\Sigma_g^+$ 10V band [173] of CS_2 molecules entrained in helium carrier gas have demonstrated over two orders of magnitude increase in DFWM signal strength upon replacing a pulsed pinhole expansion with a pulsed slit source of moderate dimensions ($\sim 1\,\mathrm{cm}$ slit length) [174].

4.3.2 *Spectroscopy*

The unique capabilities that four–wave mixing techniques bring to bear for studies of molecular spectroscopy are demonstrated most succinctly by Fig. 4.8 which presents LIF and DFWM spectra recorded simultaneously for a low pressure (*i.e.*, $\sim 0.02\,\mathrm{Torr}$) bulk–gas sample of NO [7, 22]. The acquisition of these data sets entailed use of the counterpropagating phase–conjugate experimental configuration with the requisite tunable ultraviolet radiation derived from a single–frequency source based upon pulsed–amplification of a continuous–wave ring dye laser ($\sim 0.004\,\mathrm{cm}^{-1}$ bandwidth). The depicted features correspond to individual rovibronic transitions in the $\mathrm{O}_{12}(\mathrm{N})$ bandhead region of the NO $\mathrm{A}\,{}^2\Sigma^+ - \mathrm{X}\,{}^2\Pi$ (0,0) system.

The signal–to–noise ratios displayed by the two spectral traces of Fig. 4.8 are quite comparable, highlighting the background–free response of the absorption–based four–wave mixing detection scheme. The recorded LIF features exhibit linewidths of roughly $0.1\,\mathrm{cm}^{-1}$ which are commensurate with the Doppler broadening expected for a room temperature ensemble of NO molecules. In contrast, the sub–Doppler resolution afforded by the nearly collinear (*i.e.*, $\theta \approx 2.5^\circ$), phase–conjugate implementation of DFWM results in laser–limited spectral profiles situated at the rest

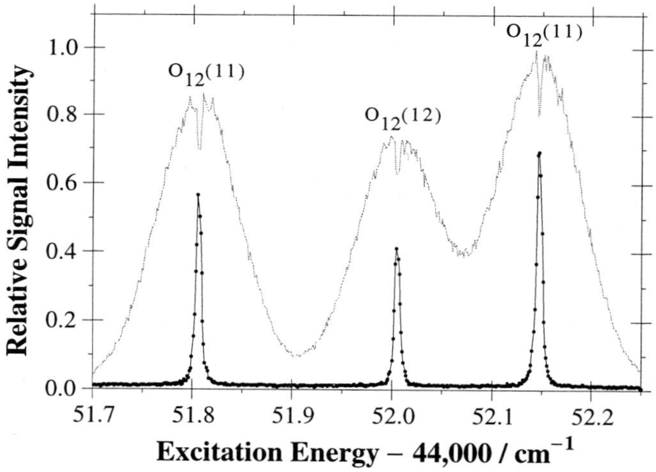

Figure 4.8 Example of high resolution gas–phase DFWM spectroscopy. Rovibronic features in the $\mathrm{O}_{12}(\mathrm{N})$ bandhead region of the NO $\mathrm{A}\,{}^2\Sigma^+ - \mathrm{X}\,{}^2\Pi$ (0,0) system are examined under low pressure (*i.e.*, $\sim 0.02\,\mathrm{Torr}$) bulk–gas conditions through simultaneous use of the DFWM (solid symbols and connecting line) and LIF (grayed line) techniques. The sub–Doppler response afforded by the counterpropagating phase–conjugate four–wave mixing geometry is clearly evident. (courtesy of D. J. Rakestraw and R. L. Farrow)

frequencies for each rovibronic transition. Lamb dips [2], arising from nonlinear saturation processes induced by the counterpropagating pump fields, also are observed on the tops of individual fluorescence peaks.

By minimizing the manifestation of inhomogenous broadening effects, the phase–conjugate DFWM scheme provides an effective resolution enhancement which can be beneficial for the analysis of dense spectral regions containing numerous closely spaced or overlapping transitions. As discussed below, such capabilities also can be exploited for the direct measurement of translational anisotropies in the nascent products of a photochemical reaction. However, this sub–Doppler response is accompanied by a decrease in overall sensitivity since only a small subset of the target species have velocity vectors that enable them to contribute successfully to the signal generation process. For example, under the conditions employed for acquisition of the data presented in Fig. 4.8, roughly 3% of the room temperature NO molecules residing in a given ground state rovibronic level can participate in the four–wave mixing interaction [7]. Provided that a reduction in spectral resolution is acceptable, improved detection limits can be achieved either through the use of excitation sources that have bandwidths more comparable to the target Doppler width or by the implementation of experimental geometries that afford less velocity discrimination. In particular, the latter situation can best be realized through the forward–box DFWM configuration depicted in the top panel of Fig. 4.2.

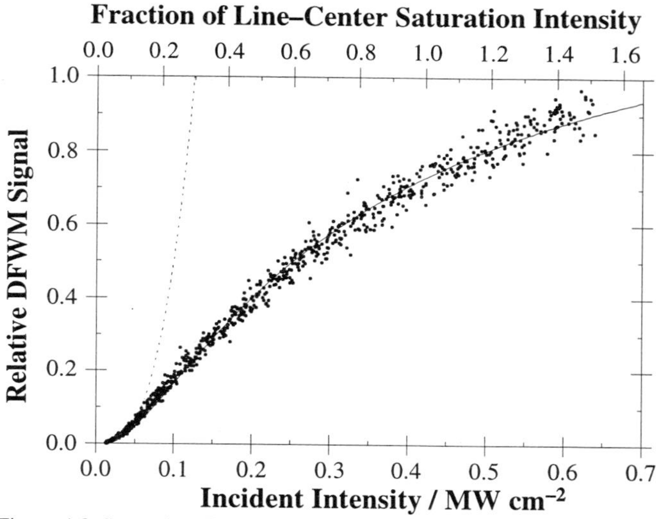

Figure 4.9 Saturation Behavior of Resonant DFWM Spectroscopy. A single–mode excitation source tuned to the $R_1(20)$ line of the NO $A\,^2\Sigma^+ - X\,^2\Pi$ (0,0) vibronic band was used to record the power dependence of the DFWM signal emerging from a bulk–gas sample consisting of 100 mTorr NO in 9.3 Torr of N_2. A least–squares regression based upon the Abrams and Lind saturable absorber model yields the solid curve with an effective line–center saturation intensity of $I^{\circ}_{sat} = 0.422\,\mathrm{MW/cm^2}$. The dashed line, obtained from the weak–field limit of the Abrams and Lind formalism (*i.e.*, $I \ll I^{\circ}_{sat}$), is designed to suggest the behavior expected in the absence of optical saturation effects. (courtesy of R. L. Farrow and D. J. Rakestraw)

The unusual saturation behavior inherent to DFWM spectroscopy is highlighted in Fig. 4.9 which depicts the strength of the four–wave mixing signal recorded for a bulk–gas NO sample as a function of impinging light intensity [22]. Each symbol represents the response generated by an individual pulse of a Fourier–Transform limited, single mode excitation source ($\sim 0.004\,\mathrm{cm^{-1}}$ bandwidth) which had been tuned into resonance with the $R_1(20)$ line of the NO $A\,^2\Sigma^+ - X\,^2\Pi$ (0,0) system at $\sim 44379.9\,\mathrm{cm^{-1}}$. This work entailed use of a target medium composed of 100 mTorr NO in 9.3 Torr of N_2 buffer gas, with the latter designed to enhance collision–induced dephasing processes and therefore increase the effective magnitude of the line–center saturation intensity, I°_{sat} [*cf.*, Eq. (4.42)].

The power dependence exhibited in Fig. 4.9 clearly deviates from the cubic scaling of DFWM signal strength on incident light intensity, I, as would otherwise be suggested by perturbative (weak–field) treatments. The solid curve is the result of a nonlinear least–squares analysis built upon the resonant behavior predicted from the Abrams and Lind saturable absorber model [*viz.*, Eq. (4.38) with $\delta = 0$]. This procedure yields a line–center saturation intensity of $I^{\circ}_{sat} = 0.422\,\mathrm{MW/cm^2}$ which is in good accord with the value calculated from Eq. (4.42) by making reasonable estimates for the effective NO transition dipole moment and the collision–dominated molecular relaxation rates. In contrast, the dashed line is obtained by substituting the optimized regression parameters into the asymptotic form of the saturable absorber model which holds for $I \ll I^{\circ}_{sat}$ [*cf.*, Eq. (4.46)]. The cubic power dependence derived in this manner is found to coincide with the experimentally measured response only over a very limited range of incident intensities, thereby suggesting that strong–field effects can play an important role in DFWM processes for all but the lowest powers commensurate with the acquisition of usable spectroscopic data.

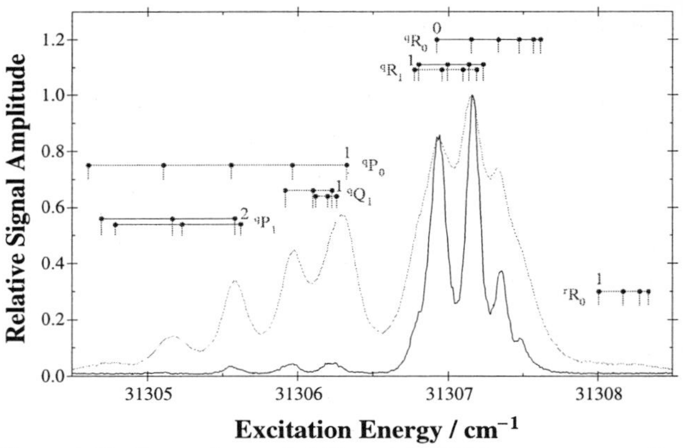

Figure 4.10 Comparison of LIF and DFWM spectroscopy for transient species in a supersonic free–jet expansion. The solid and dotted curves represent DFWM and LIF spectra, respectively, recorded for the 2_0^4 vibronic band of the S_2O $\tilde{C}\,^1A' - \tilde{X}\,^1A'$ system. Both data sets were acquired simultaneously under free–jet expansion conditions corresponding to effective rotational temperatures of ~1K with $\leq 10\,\mu J$/pulse in each of the incident four–wave mixing beams (~0.35cm diameter). The inherent width of spectroscopic transitions reflects the predissociation lifetime of ~60ps ascribed to the $v_2 = 4$ vibronic level of the electronically–excited $\tilde{C}\,^1A'$ state. Branch assignments follow from a nonlinear least–squares regression based upon model asymmetric rotor Hamiltonians and guided by the resolved structure displayed in lower–lying members of the 2_0^{υ} progression.

The ability of four–wave mixing schemes to interrogate transient species is demonstrated by Fig. 4.10 which presents LIF and DFWM spectra recorded simultaneously, under supersonic free–jet expansion conditions, for the highly reactive disulfur monoxide (S_2O) molecule [66]. Following a long history of mistaken identity and muddled spectral interpretation, the equilibrium ground state configuration for this molecule has definitively been established as being unsymmetrical and bent (*i.e.*, S—S—O where $\angle SSO \approx 118°$) [175-179], with several higher–lying isomeric forms theoretically predicted to exist on the same $\tilde{X}\,^1A'$ potential energy surface [180-182]. For acquisition of these data, a continuous stream of S_2O entrained in helium carrier gas (*e.g.*, <2% S_2O in ~5atm total stagnation pressure) was prepared *in situ* by means of a heterogeneous reaction [183, 184] and rapidly flowed past the nozzle orifice of a pulsed free–jet source based upon the current–loop–actuated design [185]. Detailed investigation of the massive S_2O species (*i.e.*, molecular weight comparable to that of benzene) demands the reduction in spectral congestion afforded by a supersonic molecular beam environment, with least–squares analysis of spectra obtained in the DFWM interaction region (*viz.*, located ~1–2cm downstream from the 0.05cm diameter circular nozzle orifice) suggesting effective rotational temperatures on the order of 1 K [186]. The depicted features correspond to overlapping rovibronic transitions in the 2_0^4 band of the S_2O $\tilde{C}\,^1A' - \tilde{X}\,^1A'$ electronic system, where v_2 denotes the S—S stretching mode. The large oscillator strength of this fully allowed $\pi^* \leftarrow \pi$ electron promotion ($f \approx 0.026$ [186]) enables fluorescence to be observed despite rapid predissociation of the pertinent excited state levels. The 2_0^4 band is primarily type–*a* in character ($\Delta K_a = 0$ transitions predominate) with additional weak type–*b* features ($\Delta K_a = \pm 1$ transitions) arising from an axis switching effect induced by a ~3° rotation of inertial axes upon electronic excitation [186].

The S_2O results of Fig. 4.10 were obtained through use of a phase–conjugate experimental configuration with polarization–based detection providing for the efficient rejection of residual scattered light (*cf.*, Fig. 4.6). For such strongly predissociated levels, the recorded DFWM and LIF traces exhibit comparable signal–to–noise ratios despite substantial differences in the relative intensities and widths of individual spectral features. However, rovibronic transitions terminating on bound excited state levels (*e.g.*, S_2O spectra obtained for the 2_0^{υ} bands where $0 \leq \upsilon \leq 3$) could be detected with much greater sensitivity and convenience through linear fluorescence methods, a fact attributed primarily to the low density of target species entrained in the molecular beam environment [66]. As discussed in conjunction with Fig. 4.8, the sub–Doppler response inherent to the counterpropagating phase–conjugate geometry can also severely restrict the number of molecules capable of contributing to the four–wave mixing interaction. In particular, the frequency characteristics of the high resolution laser utilized for the S_2O measurements, combined with the Doppler shifts expected to accompany the *transverse* distribution of molecules in an axially–symmetric free–jet expansion, suggest that effective sample interaction lengths were considerably shorter than $\ell = 0.1$cm. Given the ℓ^2 dependence of DFWM signal strength [Eq. (4.14) and Eq. (4.38)], a significant increase in sensitivity should be realized through use of velocity–collimated beam sources built upon rectangular or slit–shaped nozzle apertures [174].

The experimental results of Fig. 4.10 were acquired under "low intensity" conditions with incident laser energies attenuated to $< 10\,\mu J$/pulse. Individual rovibronic transitions display homogeneously–broadened lineshapes which can be attributed to rapid predissociation of the pertinent S_2O $\tilde{C}\,^1A'$ levels. Spectral simulations, incorporating the finite bandwidth of the laser excitation source, suggest a rotation–independent excited state lifetime of $63 \pm 10\,\text{ps}$ for the 2_0^4 band [186]. The highly resolved appearance of the DFWM trace stems primarily from the Lorentzian–cubed frequency response expected in an unsaturated regime. More importantly, the LIF and DFWM data sets differ significantly in the relative intensities of their corresponding R–branch and P–branch features, with the later being unusually weak in the recorded four–wave mixing spectrum. A partial explanation for these discrepancies can be found in the weak–field $\left(N_g\right)^2 \left|\mu_{eg}\right|^8$ dependence of DFWM signal strength upon population number density (N_g) and transition dipole moment (μ_{eg}). This scaling, in conjunction with the additional spectral discrimination afforded by the cross–polarization detection scheme, can lead to DFWM intensity patterns that depart considerably from those obtained via linear fluorescence techniques. However, detailed analyses reveal that quantitative interpretation of the four–wave mixing spectrum demands explicit consideration of the coherence properties for the incident light fields and the unusually rapid dissipation of transient gratings created within the predissociative $\tilde{C}\,^1A'$ state [66].

As demonstrated by the dual spectral traces contained both in Fig. 4.8 and Fig. 4.10, DFWM provides a viable spectroscopic tool for situations where more conventional linear methods, including those build upon fluorescence detection, are applicable. Under such circumstances, the advantages and disadvantages of each technique must be weighed in order to establish the most expedient and useful approach for a particular investigation. In contrast, Fig. 4.11 presents one example where the absorption–based four–wave mixing methodology succeeds while traditional LIF schemes do not. Depicted is a rovibronically–resolved spectrum of the $\tilde{A}\,^1B_1 - \tilde{X}\,^1A_1$ $\left(\pi^* \leftarrow n\right)$ electronic transition for malonaldehyde, the most stable isomer of 3–hydroxy–2–propenal, which contains an intramolecular hydrogen bond that adjoins hydroxylic (proton donor) and ketonic (proton acceptor) oxygen atoms. Owing to its small size and relative simplicity, this polyatomic molecule has served as the prototypical system for numerous theoretical [187-195] and experimental [196-205] studies designed to elucidate the interrelated phenomena of hydrogen bonding and proton transfer. Extensive microwave and infrared measurements have shown conclusively that the isolated gas–phase species exists almost completely as the chelated enol tautomer, with the $\tilde{X}\,^1A_1$ ground electronic state supporting a symmetric double–minimum potential well along the $O-H\cdots O$ coordinate that separates two stable isomers of C_s symmetry:

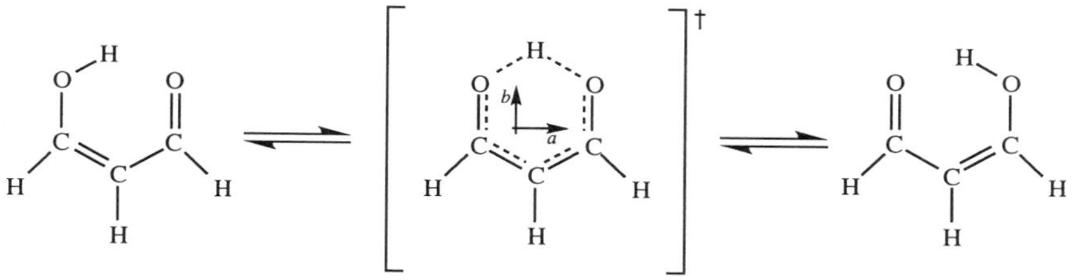

(4.52)

Although hindered by a finite barrier having an estimated height of $2000\,\text{cm}^{-1}$ [202], rapid interconversion between the equivalent C_s structures of the ground electronic state can occur via quantum mechanical proton tunneling, thereby producing a characteristic doubling of observed spectroscopic transitions (*e.g.*, the tunneling–induced splitting for the vibrationless $\tilde{X}\,^1A_1$ level is $\sim 21.585\,\text{cm}^{-1}$ [202]). Rather than depending solely on the motion of the "shuttling" proton, the magnitude of such spectral splittings contains information regarding the concomitant displacement of all other nuclei in the molecular framework [193] as well as the redistribution of charge density among the various atoms [206, 207].

The only previously reported gas–phase investigation of the malonaldehyde $\tilde{A}\,^1B_1 - \tilde{X}\,^1A_1$ system is based upon conventional linear absorption spectroscopy [208-210] which, owing to the small oscillator strength of the $\pi^* \leftarrow n$ electronic excitation and practical limitations imposed upon sample number density, necessitated the use of long pathlengths $(50 - 100\,\text{m})$ and moderate spectral resolution $(\sim 5\,\text{cm}^{-1})$. Nevertheless, these pioneering efforts resulted in several important discoveries including the $^1B_1$ symmetry of the excited state and the perpendicular orientation of the $\tilde{A} - \tilde{X}$ transition dipole moment with respect to the molecular plane (*i.e.*, giving rise to electronically–allowed type–*c* transitions). In contrast, initial attempts to observe undispersed

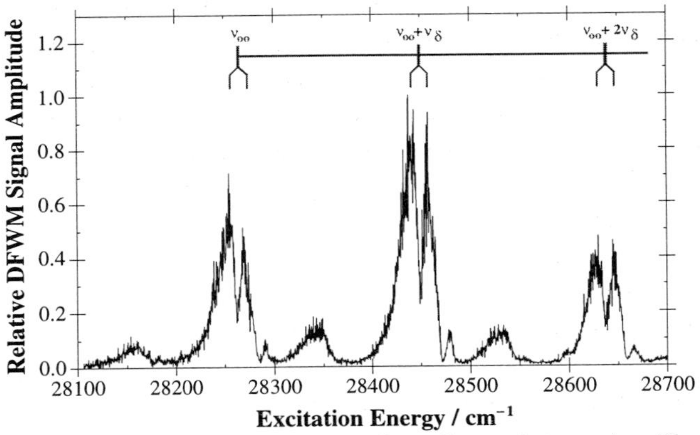

Figure 4.11 DFWM spectroscopy of nonfluorescing species. The backwards–box DFWM configuration was used to record a portion of the malonaldehyde $\tilde{A}\,^1B_1 - \tilde{X}\,^1A_1\left(\pi^* \leftarrow n\right)$ system under bulk–gas conditions with an excitation source of $\leq 0.2\,\text{cm}^{-1}$ bandwidth. Owing to the weak nature of the pertinent electronic transition, incident energies approaching 1 mJ/pulse (~ 0.4 cm beam diameter) were found still to be below the threshold required for manifestation of optical saturation effects. The illustrated vibronic progression has been attributed to a $\nu_\delta \approx 190\,\text{cm}^{-1}$ in–plane skeletal vibration of the $\tilde{A}\,^1B_1$ state.

fluorescence following resonant excitation of $\tilde{A}\,^1B_1$ malonaldehyde have not been successful [74]. The spectrum presented in Fig. 4.11 was obtained under ambient, bulk–gas conditions (<500 mTorr sample pressure and <5 cm interaction length) by means of a backward–box DFWM geometry that incorporated cross–polarization detection in order to further discriminate against incoherently scattered background light. The tunable laser source employed for this work yields an effective resolution of $0.1-0.2\,\text{cm}^{-1}$, with higher resolution measurements demonstrating that much of the noise–like structure observed on prominent spectral features stems from rovibronic congestion. Indeed, taking into account the large number of internal quantum states populated in a room temperature ensemble of malonaldehyde molecules, the detection limits derived from Fig. 4.11 are quite comparable to those achieved in other DFWM studies ($<10^{11}$ molecules / cm^3 / quantum state). A cubic dependence of the four–wave mixing signal amplitude upon incident optical power confirmed that these data were acquired in a "saturation–free" regime.

The gross features of the malonaldehyde DFWM spectrum are in reasonable accord with those obtained in the long pathlength linear absorption measurements [210]. For example, the vibronic progression highlighted in Fig. 4.11 has been attributed to an in–plane skeletal bending mode of the $\tilde{A}\,^1B_1$ state having a frequency of $\nu_\delta \approx 190\,\text{cm}^{-1}$. Previous work [210] also identified a minute $7\,\text{cm}^{-1}$ splitting on each member of this progression, an observation equated with a decrease in tunneling splitting or increase in proton transfer barrier height following $\pi^* \leftarrow n$ electronic excitation. In contrast, the four–wave mixing data, obtained at substantially higher spectral resolution, display a pronounced $20\,\text{cm}^{-1}$ bifurcation of these vibronic structures. While additional work remains to be performed, at this juncture it appears that the DFWM results can be explained, in part, through the rotational band contours expected for vibronically–induced transitions. The implications of this conjecture, as well as those derived from recent *ab initio* predictions of small or non–existent proton transfer barriers in the $\tilde{A}\,^1B_1\left(\pi^* n\right)$ state [211], are the subject of ongoing theoretical and experimental investigations [74].

The malonaldehyde results presented above illustrate the ability of absorption–based DFWM schemes to bring the advantages of laser spectroscopic techniques to bear upon molecular systems that are not amenable to the ubiquitous LIF methodology. In a similar manner, Dunlop and Rohlfing [72] have exploited two–color variants of four–wave mixing to probe the $S_1\left(\tilde{A}\,^1B_1\right) - S_0\left(\tilde{X}\,^1A_1\right)$ electronic transition of azulene under supersonic free–jet conditions. While nonradiative internal conversion of the S_1 state on a sub–picosecond timescale prohibits efficient detection of spontaneous emission, this work demonstrated that coherent scattering processes, originating from spatial depletion gratings created among ground state molecules upon resonant $S_1 - S_0$ excitation with crossed optical beams, can provide a facile means for interrogating such non–fluorescing species. The same approach enabled Butenhoff and Rohlfing [55] to investigate the rovibronic structure of NO_2 above its predissociation threshold. The absorption–like spectra recorded in this manner were found to be in excellent agreement with action spectra obtained by monitoring the yield of nascent $O\left(^3P\right)$ photofragments as a function of photolysis wavelength.

While the majority of previous and ongoing DFWM studies have entailed the use of one–photon resonant electronic transitions residing in the visible or ultraviolet regions of the electromagnetic spectrum, experiments based upon the infrared transitions that occur between molecular vibrational levels have also been reported.

Initial efforts employed fixed–frequency or line–tunable infrared lasers (*e.g.*, CO_2 lasers) whose fortuitous overlap with polyatomic rotation–vibration resonances (*e.g.*, the ν_3 fundamental band of SF_6) provided a facile means for investigating the fundamental nature of four–wave mixing processes [106, 212, 153, 152, 213]. More recently, Rakestraw and co–workers [50, 214, 51, 76] have pioneered measurements with scanning sources of infrared radiation, including multiple–mode difference–frequency generators and single–mode optical parametric oscillators/amplifiers. These studies have demonstrated unequivocally the feasibility of exploiting infrared DFWM spectroscopy for the detection and identification of trace species.

Since long spontaneous emission lifetimes usually make fluorescence–based schemes impractical for the study of pure rovibrational transitions, absorption techniques have evolved to a position of prominence in the field of vibrational spectroscopy. While the general species applicability and experimental simplicity of such methods provide distinct advantages, their sensitivity suffers from the differential nature of the linear absorption process wherein potentially weak signals must be discriminated from substantially larger baseline levels of superfluous light. With its high sensitivity and background–free response, infrared DFWM promises to provide a powerful tool for the investigation of polyatomic vibrations both in fundamental studies of molecular structure and in the elucidation of energy disposal among the nascent products of collisional or reactive encounters. The temporal and spatial specificities afforded by various four–wave mixing geometries also provide new capabilities that are not available with the pathlength–integrated techniques of traditional absorption spectroscopy.

As a tool for investigation of molecular structure and dynamics, infrared DFWM offers multiple advantages [76], not the least of which is the ability to interrogate species having excited electronic states that are either inaccessible or unamenable to optical probes (*i.e.*, ultraviolet/visible LIF or DFWM). When compared to rovibronic resonances, the large dipole moments and small Doppler widths of rotation–vibration transitions promise to yield substantial improvements in nonlinear signal strength, especially for experimental configurations based upon sub–Doppler schemes (*cf.*, Fig. 4.8 and accompanying discussion). In addition, Eq. (4.8) shows that the long wavelength of infrared radiation provides for increased fringe spacing of the transient gratings that give rise to the four–wave mixing response, thereby minimizing deleterious effects ascribed to motional dissipation.

Vander Wal, *et al.* [50] have exploited the counterpropagating phase–conjugate DFWM scheme to interrogate HF molecules by means of their $(1-0)$ and $(3-0)$ vibrational transitions located in the vicinity of $4000\,cm^{-1}$ and $11500\,cm^{-1}$, respectively. The strong resonant enhancement of the fundamental band [*i.e.*, $|\mu_{eg}| \approx 0.082\,D$ for the HF $(1-0)$ transition] led to a reported detection sensitivity of $< 10^{10}$ molecules/cm^3, with well–resolved rovibrational features easily recorded at total pressures of $< 0.1\,mTorr$. In the case of the substantially weaker overtone band [*i.e.*, $|\mu_{eg}| \approx 0.0016\,D$ for the HF $(3-0)$ transition], however, measurable DFWM signals could be obtained only through use of focused incident beams and sample pressures in excess of $10\,Torr$. Time–resolved and polarization–based analyses revealed that the overtone response stems primarily from thermal gratings produced subsequent to the collision–induced degradation of internal molecular energy into translational degrees of freedom. In a similar manner, secondary scattering processes arising from both resonant (*e.g.*, thermal density gradients) [215, 216] and

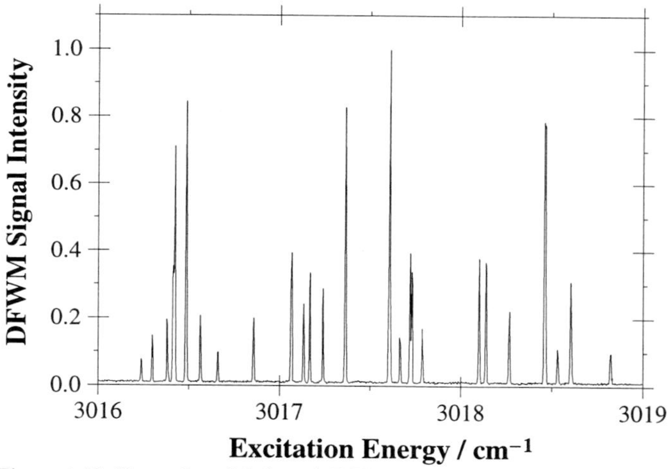

Figure 4.12 Example of Infrared DFWM spectroscopy. The phase-conjugate DFWM configuration was implemented with a near Fourier–Transform limited infrared excitation source in order to record a portion of the methane infrared spectrum under 10 mTorr bulk–gas conditions. The observed transitions correspond to the Q(7) through Q(1) multiplets of the ν_3 C—H stretching fundamental band. The use of incident beams having energies of $\sim 5\,\mu J$/pulse was found to produce a small degree of saturation broadening (courtesy of D. J. Rakestraw).

nonresonant (*e.g.*, electrostriction) [217, 216] interactions have been shown to play significant roles in the high pressure overtone signals derived through application of two–color four–wave mixing schemes in Vibrationally–Mediated Photodissociation (VMP) experiments.

Fig. 4.12, adapted from the work of Germann, *et al.* [76], presents an example of infrared DFWM spectroscopy performed on methane, a molecule that has proven difficult to interrogate through conventional visible/ultraviolet techniques. The depicted spectral trace corresponds to the Q – branch region of the fundamental ν_3 C – H stretching band and was obtained at a total bulk–gas pressure of 10 mTorr. The exceptional signal–to–noise ratio of this data set highlights the zero–background nature and strong resonant enhancement of the four–wave mixing detection scheme. Despite slight saturation broadening induced by incident pump and probe energies of ~ 5 μJ/pulse (*i.e.*, 0.4 – 0.6 cm diameter unfocused beams), the sub–Doppler response inherent to the phase–conjugate experimental configuration, in conjunction with the use of a single mode light source (*i.e.*, < 0.007 cm^{-1} measured scanning bandwidth), results in observed rovibrational features that are better resolved than those obtained through traditional high resolution (Doppler–limited) linear absorption techniques. While methane spectra measured in the optically saturated limit (*i.e.*, $I \geq I^{\circ}_{sat}$) were found to be in good accord with predictions derived from the stationary absorber model of Abrams and Lind (*cf.*, Section 4.2.6), intensity related discrepancies of unknown origin were noted under low power conditions. Data of comparable quality have been recorded for the 3000 cm^{-1} vibrational transitions of acetylene [214], with recent reports by Tang and Reid [77] demonstrating the ability to probe such bands under the rarefied conditions afforded by a supersonic free–jet expansion.

4.3.3 *Photodissociation Dynamics*

With its implicit goal of understanding the simplest of chemical transformations, the field of molecular photodissociation dynamics has matured to a level of sophistication where theory and experiment are capable of providing sufficiently refined information so as to mutually illuminate one another [218, 219]. This enviable situation can be attributed to a variety of factors, not the least of which is the myriad of optical probes that have been developed and employed successfully for the elucidation of photofragmentation phenomena. While direct spectroscopic interrogation of the parent molecule through either linear absorption or resonance Raman techniques [220] can furnish substantial insight regarding initial stages of the bond rupturing event, complementary data can be gleaned from investigation of the nascent product species formed upon completion of the photolysis reaction. In the latter case, measurements of both scalar properties (*e.g.*, energy partitioning and state distributions) and vector correlations [221-224] (*e.g.*, translational and rotational anisotropies) have been shown to yield valuable mechanistic details. Ongoing developments in the creation and manipulation of ultrashort laser pulses (*i.e.*, pulse duration comparable to the period of molecular vibration) have opened entirely new vistas in the realm of chemical dynamics, with the real–time investigation of photo–induced bond cleavage now being transformed from a theoretical abstraction to an experimental observable [225]. Since four–wave mixing signals stem from an essentially instantaneous absorption–based interaction, the DFWM methodology should be applicable over a wide range of photodissociation timescales and, as demonstrated in the recent work of Zewail, *et al.* [79], should be capable of yielding femtosecond temporal resolution. However, in keeping with the spectroscopic nature of this chapter, the ensuing discussion will focus on the asymptotic (long time) aspects of photofragmentation processes as revealed through detailed analysis of the newly formed photoproducts.

The most prevalent optical means for examining the products of molecular photodissociation are built upon the exquisitely sensitive techniques of LIF. As in the case of pure spectroscopy, the absorption–based methods of four–wave mixing provide a viable alternative for such studies while simultaneously offering the possibility to interrogate species that are not amenable to direct fluorescence detection. The high sensitivity and large dynamic range of the background–free DFWM scheme should permit facile determination of relative internal quantum state populations for the emerging photofragments, with the "instantaneous" nature of the nonlinear interaction moderating the severity of experimental constraints required to ensure nascent or collision–free conditions. In addition, the sub–Doppler response of the counterpropagating phase–conjugate geometry can be exploited as a state–specific probe of the translational anisotropy which accompanies the photolysis of precursor molecules with linearly polarized light [42, 119]. The spatial distribution of photoproduct rotational angular

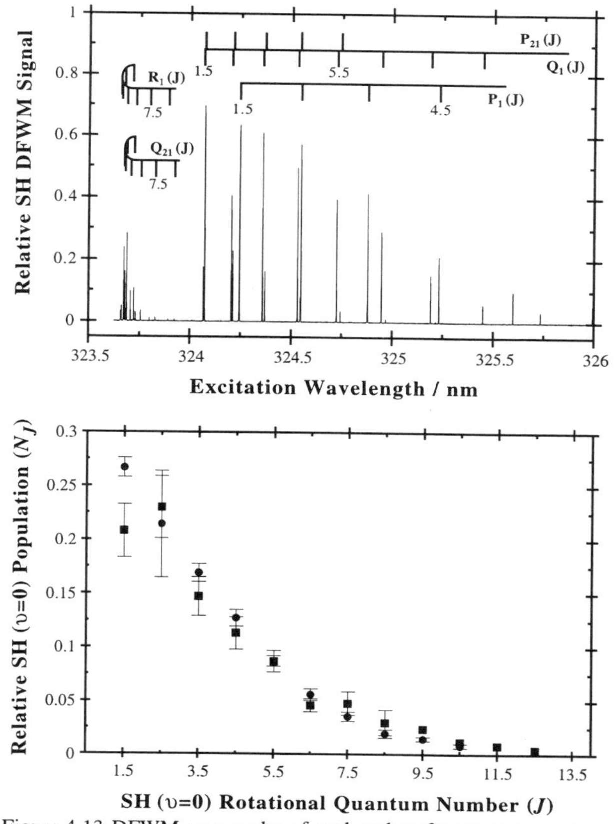

Figure 4.13 DFWM as a probe of scalar photofragment properties. The top panel shows a portion of the four–wave mixing spectrum obtained by using the SH $A\,^2\Sigma^+ - X\,^2\Pi$ (0,0) band to probe nascent mercapto radicals generated in the 266 nm photodissociation of a bulk–gas H_2S sample. Each peak corresponds to the formation of product molecules in a specific rotational level of the vibrationless ground electronic state. The bottom panel compares $SH(\upsilon = 0)$ rotational distributions measured under two distinct sets of experimental conditions: (●) for 266 nm photolysis under ambient, bulk–gas conditions via DFWM [43] and (■) for 193 nm photolysis under jet–cooled conditions via LIF [228].

momentum (*i.e.*, alignment and/or orientation), as well as the speed of the departing fragments, can also be extracted from such DFWM measurements [21, 119].

The feasibility of exploiting four–wave mixing techniques for the interrogation of nascent photoproducts is demonstrated by the top panel of Fig. 4.13 which presents a portion of the SH $A\,^2\Sigma^+ - X\,^2\Pi$ (0,0) spectrum recorded subsequent to the 266 nm photolysis of bulk–gas hydrogen sulfide (H_2S) [42, 43]. The small photodissociation cross section of H_2S at this wavelength ($\sigma_{266} \approx 6\times10^{-22}\ \mathrm{cm}^2$ [226]), coupled with the sub–nanosecond predissociation lifetimes reported for the $A\,^2\Sigma^+$ state of the mercapto radical [227], encumbers the implementation and interpretation of LIF spectroscopy. The depicted results were obtained by means of a counterpropagating phase–conjugate DFWM geometry built upon the "cross–population" scheme [*i.e.*, $(\hat{e}_s \| \hat{e}_b) \perp (\hat{e}_f \| \hat{e}_p)$], with the unfocused incident beams of ~0.4 cm diameter attenuated to energies of ≤ 200 nJ/pulse so as to minimize saturation effects. While relatively high sample pressures were required for these studies (*viz.*, ~ 400 mTorr), a series of measurements conducted with varying time delays between the optical pulses responsible for precursor photolysis and product detection confirmed the collision–free nature of the experiment. Taking into account the low H_2S photofragmentation yield at 266 nm [226], unsaturated data sets such as that displayed in Fig. 4.13 translate into estimated sensitivities of $< 10^{11}$ molecules / cm^3 / quantum state with significantly better detection limits attained under strong–field (saturated) conditions. Active shot–by–shot subtraction of baseline noise and normalization of the DFWM signal amplitude (*i.e.*, by the cube of incident power) were deemed essential for reliable extraction of quantitative information from the SH spectra.

Given the HS – H bond dissociation energy of 3.91 ± 0.2 eV [229], the 266 nm photolysis of a room temperature H_2S sample will result in roughly 0.8 eV (*i.e.*, ~ 77 kJ/mole) of excess energy to be partitioned among accessible product channels. This quantity of energy is insufficient to create electronically–excited mercapto radicals but can lead to moderate rovibrational excitation of the SH $X\,^2\Pi$ photoproducts including the possible occupation of vibrational states up to $\upsilon_{prod} = 2$ and of vibrationless rotational levels as high as $J_{prod} = 25.5$. While the $(N_g)^2 |\mu_{eg}|^8$ scaling of the unsaturated DFWM signal strength precludes extraction of quantitative internal state distributions from a cursory inspection of Fig. 4.13, spectral features are observed only for nascent SH molecules residing in vibrationless levels of the $X\,^2\Pi_{3/2}$ and $X\,^2\Pi_{1/2}$ spin–orbit doublet, with rotational states having quantum numbers beyond $J_{prod} = 10.5$ found to contain negligible population. This suggests that the major fraction of available energy (*i.e.*, >95%) appears as relative translational motion of the departing fragments, a result in keeping with those obtained in previous time–of–flight [230-232] and LIF [233, 234, 228] measurements. In particular, successful execution of the latter work required the use of intense fluorescence excitation pulses (*i.e.*, leading to optical saturation effects) and photolysis wavelengths in closer proximity to the 196 nm maximum of the lowest–lying H_2S absorption band (*i.e.*, increased photofragmentation yield).

Reduction of the DFWM data to relative populations for SH internal quantum states was based upon a perturbative treatment in which the integrated areas of individual spectral peaks were normalized by the square modulus of the induced third–order signal polarization, $\boldsymbol{P}_s^{(3)}(t)$ [*cf.*, Eq. (4.9) and Eq. (4.10)]. In brief, calculation of $\boldsymbol{P}_s^{(3)}(t)$ entailed direct evaluation of the nonlinear susceptibility, $\boldsymbol{\chi}^{(3)}\left(-\omega_s;\omega_f,-\omega_p,\omega_b\right)$, for each SH rovibronic transition (*i.e.*, including effects associated with incident light polarizations, excited state predissociation, and finite bandwidth of the excitation source), followed by numerical integration of the resulting expression over model distributions of photofragment velocities. Detailed analysis of the DFWM spectra revealed a non–Boltzmann rotational distribution of nascent products created exclusively within the vibrationless levels of the $X\,^2\Pi$ manifold, with a non–statistical preference for the lower–lying spin–orbit component but no significant population disparity between the members of a lambda–doublet [43].

The SH $(\upsilon = 0)$ rotational distribution measured (via DFWM) for the 266 nm photodissociation of H_2S under ambient bulk–gas conditions ($\sim 300\,\mathrm{K}$) [43] is presented in the lower panel of Fig. 4.13 along with analogous results obtained (via LIF) from 193 nm experiments performed on jet–cooled samples ($\sim 10\,\mathrm{K}$) [228]. Despite substantial differences in the total available energy and the initial degree of H_2S rovibrational excitation, these data sets exhibit rotational distributions for the vibrationless photofragments that are near identical in shape and extent. This observation provides a textbook example of the rotational "Franck–Condon" principle which explains the invariant "Gaussian–like" nature of such distributions as a manifestation of the (Gaussian) vibrational wavefunction for ground state H_2S bending motion being projected onto the individual eigenfunctions of the free SH rotor [235, 218]. In particular, such an interpretation implies the absence of anisotropic forces in the (dissociative) electronically–excited potential energy surface(s) of H_2S, with negligible influence of parent rotation on the overall dynamics of photodissociation.

Even in situations where the methods of LIF spectroscopy are applicable, the unique capabilities of four–wave mixing can provide advantages for the extraction of photoproduct internal state distributions. This has been demonstrated most clearly by the studies of Butenhoff and Rohlfing [55] which exploited two–color laser–induced grating techniques (Section 4.4.2) to examine the NO fragments formed subsequent to near–threshold photodissociation of jet–cooled NO_2. While a minute NO contamination in the molecular beam source significantly complicated interpretation of recorded LIF data, the doubly–resonant nature of the transient grating scheme provided an effective means for discriminating against ambient NO, thereby permitting the direct acquisition of state–resolved spectra for the nascent photoproducts.

Fig. 4.14 compares high resolution LIF and DFWM spectra recorded simultaneously, under weak–field conditions, for the nascent hydroxyl radicals emerging from the 266 nm photolysis of bulk–gas HOOH ($\sim 70\,\mathrm{mTorr}$ pressure) [21]. As in the case of H_2S photodissociation, a major portion of the available energy ($\sim 250\,\mathrm{kJ/mole}$) appears as relative translational motion of the departing fragments, thereby giving rise to OH velocities that readily exceed $3500\,\mathrm{m/s}$. Thus, while the LIF trace contains a broad, *Doppler–limited* spectral feature ($> 0.5\,\mathrm{cm}^{-1}$ HWHM) that masks underlying rovibronic structure, the DFWM data exhibit *sub–Doppler* linewidths which are commensurate with the estimated bandwidth of the ultraviolet excitation source. The two weaker peaks in this region can be assigned to the $Q_1(1.5)$ and $P_{21}(1.5)$ lines of the OH $A\,^2\Sigma^+ - X\,^2\Pi$ (0,0) band. Equally spaced between these transitions is an intense crossover resonance which stems from the counterpropagating nature of the pump beams employed in a phase–conjugate DFWM geometry. Such artifacts, well known in the field of sub–Doppler laser spectroscopy [1, 2], occur anytime two adjacent lines share a common upper or lower quantum state [*i.e.*, the $Q_1(1.5)$ and $P_{21}(1.5)$ transitions originate from the same $J_{prod} = 1.5$ level of the vibrationless OH $X\,^2\Pi_{3/2}$ state] and have mutually overlapping inhomogeneously–broadened lineshapes. The velocity dependence of the Doppler effect enables both of these transitions to participate in a combined four–wave mixing interaction when the frequency of the incident radiation is tuned to the exact mean of the corresponding (rest) resonance frequencies.

As discussed by Friedmann–Hill, *et al.* [25], the prominence of the crossover feature appearing in Fig. 4.14 can be attributed to the polarization–based scheme employed for detection of the four–wave mixing process. Such artifacts must be taken into account for the quantitative interpretation and simulation of DFWM

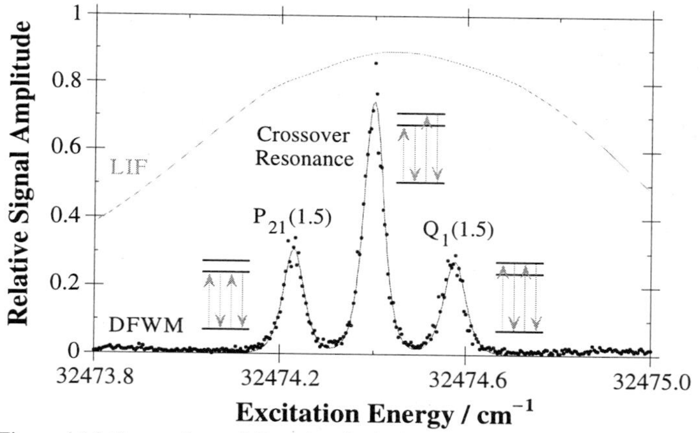

Figure 4.14 Comparison of Doppler–limited LIF and sub–Doppler DFWM spectroscopy as probes of photodissociation dynamics. The nascent hydroxyl radicals formed upon 266 nm photolysis of bulk–gas HOOH were examined by recording spectra for the $Q_1(1.5)$ and $P_{21}(1.5)$ lines of the OH $A\,^2\Sigma^+ - X\,^2\Pi$ (0,0) system. Since these transitions share a common lower rovibronic state, their relative intensities provide a measure for the quadrupolar alignment tensor, $A_0^{(2)}(1.5)$, which describes the directional correlation between the rotational angular momentum of the photofragment and the transition dipole moment of the parent molecule. The large peak labeled as a crossover resonance is a characteristic feature of sub–Doppler laser spectroscopy which arises from the overlapping inhomogeneous profiles of the adjacent $Q_1(1.5)$ and $P_{21}(1.5)$ lines. Insets highlight the spectroscopic origins of resonances observed in the DFWM data set.

data sets, especially in cases where insufficient spectral resolution precludes the isolation and identification of individual rovibronic transitions. When monitored as a function of the J_{prod} – dependent frequency separation between adjacent lines of main and satellite branches [*e.g.*, $Q_1(J)$ and $P_{21}(J)$ in the OH $A - X$ system], the presence or absence of crossover resonances also can provide a crude estimate for the maximum speed of the target species. Valuable information regarding photofragment translational anisotropy can also be gleaned from the dependence of crossover signal strength on the direction of linear polarization for the photolysis source [119].

In addition to scalar properties, such as quantum state distributions, a detailed understanding of molecular photodissociation dynamics demands knowledge of any relationships that might exist among various vector entities including the internal angular momentum ($\boldsymbol{J}_{prod}$) and the relative velocity ($\boldsymbol{v}_{prod}$) of the nascent products [218, 219]. Both of these quantities can be correlated to the body–fixed transition dipole moment of the parent molecule ($\boldsymbol{\mu}_{par}$) which, in turn, is coupled to the laboratory–fixed frame of reference through its interaction with the electric field of the photolysis beam ($\boldsymbol{E}_{\text{diss}}$) [236, 123]. The presence of directional correlations among $\boldsymbol{\mu}_{par}$, $\boldsymbol{J}_{prod}$, and $\boldsymbol{v}_{prod}$ can provide a quantitative signature for the anisotropic forces that mediate a chemical transformation [237, 222-224, 238]. Four–wave mixing spectroscopy, with its inherent polarization specificity and velocity discrimination, can be exploited for state–specific measurement of the spatial anisotropies that result from these correlations.

Reduced to simplest terms, the ability of any frequency domain scheme to probe the spatial distribution of rotational angular momentum vectors stems from the inherent polarization properties of resonant matter–field interactions [123, 239]. For a given spectroscopic transition, symmetry constraints demand that the transition dipole moment of the target species (*e.g.*, a photoproduct), denoted as $\boldsymbol{\mu}_{prod}$, have a specific orientation in the molecular framework. Since body–fixed projections of the rotational angular momentum vector, $\boldsymbol{J}_{prod}$, are restricted to certain quantized values, a directional correlation will exist between $\boldsymbol{\mu}_{prod}$ and $\boldsymbol{J}_{prod}$. An anisotropic spatial distribution of rotational angular momentum will therefore manifest itself as a corresponding anisotropy in $\boldsymbol{\mu}_{prod}$. The $\boldsymbol{\mu}_{prod} \cdot \boldsymbol{E}$ scaling of the dipole interaction, where $\boldsymbol{E}$ represents the transverse electric vector for an interrogating (laboratory–fixed) electromagnetic wave, leads to a pronounced dependence of the photon coupling efficiency (*i.e.*, the amount of absorption or emission) upon the spatial arrangement of transition moments which, in turn, reflects the anisotropy of $\boldsymbol{J}_{prod}$. Consequently, a systematic investigation of polarization effects in resonant absorption and emission processes can provide detailed information on any directional specificity imparted to the rotational motion of a nascent reaction product [240, 241, 239]. This data, in conjunction with angular momentum conservation criteria, can yield valuable insights into the mechanism and timescale of a chemical transformation [238, 218]. When compared to more conventional (linear) absorption and fluorescence methods, DFWM spectroscopy provides special advantages for the analysis of rotational anisotropies, with the near simultaneous interaction of four optical fields (each having its own unique polarization characteristics) enhancing the sensitivity, resolution, and scope of such measurements.

In the case of photodissociation, where the initial orientation of the transition dipole moment in the parent molecule is tethered to the laboratory–fixed frame of reference by the $\left|\boldsymbol{\mu}_{par} \cdot \boldsymbol{E}_{diss}\right|^2$ dependence of the photoabsorption probability [236, 123, 239], determination of the (laboratory–fixed) directional anisotropy for photofragment rotational angular momentum is tantamount to measuring the $\boldsymbol{\mu}_{par} - \boldsymbol{J}_{prod}$ correlation. For experiments performed with linearly polarized sources of photolysis light, this relationship is commonly referred to as the molecular alignment [85, 123] and conventionally parameterized by the quadrupolar tensor component $A_0^{(2)}\left(J_{prod}\right)$. The classical definition for $A_0^{(2)}\left(J_{prod}\right)$ can be formulated as [239]:

$$A_0^{(2)}\left(J_{prod}\right) = 2\left\langle\left\langle P_2\left(\hat{\boldsymbol{J}}_{prod} \cdot \hat{\boldsymbol{E}}_{diss}\right)\right\rangle\right\rangle_{\Omega_{J_{prod}}} = 2\left\langle\left\langle P_2\left(\cos\left(\theta_{\hat{\boldsymbol{J}}_{prod},\hat{\boldsymbol{E}}_{diss}}\right)\right)\right\rangle\right\rangle_{\Omega_{J_{prod}}} \tag{4.53}$$

where $\hat{\boldsymbol{J}}_{prod}$ and $\hat{\boldsymbol{E}}_{diss}$ are unit vectors pointing along the directions of product rotational angular momentum and photolysis source polarization, respectively, with $\theta_{\hat{\boldsymbol{J}}_{prod},\hat{\boldsymbol{E}}_{diss}}$ signifying their included angle. In particular, $\hat{\boldsymbol{E}}_{diss}$ defines a laboratory–fixed axis of cylindrical symmetry about which the alignment is measured. The double angular brackets in Eq. (4.53) signify that the second order Legendre function, $P_2\left(\hat{\boldsymbol{J}}_{prod} \cdot \hat{\boldsymbol{E}}_{diss}\right)$, is to be averaged over the angular distribution, $f\left(\boldsymbol{\Omega}_{J_{prod}}\right)$, of $\boldsymbol{J}_{prod}$. From a quantum mechanical perspective, the alignment tensor reflects the differential occupation of degenerate magnetic sublevels in a particular rovibronic state (*i.e.*, having rotational quantum number J_{prod}) with levels having the same absolute value of magnetic quantum number constrained to be populated equally [85, 123, 239]. In particular, limiting case positive values of $A_0^{(2)}\left(J_{prod}\right)$ imply a bias for $\hat{\boldsymbol{J}}_{prod}$ to lie along $\hat{\boldsymbol{E}}_{diss}$ (*viz.*, $\hat{\boldsymbol{J}}_{prod} \| \hat{\boldsymbol{E}}_{diss}$ or $-\hat{\boldsymbol{J}}_{prod} \| \hat{\boldsymbol{E}}_{diss}$) while negative values indicate a preference for $\hat{\boldsymbol{J}}_{prod} \perp \hat{\boldsymbol{E}}_{diss}$. Since the largest photodissociation yield in an ensemble of randomly oriented species is obtained for the subset of parent molecules having $\boldsymbol{\mu}_{par}$ parallel or anti–parallel to $\boldsymbol{E}_{diss}$ (recall the $\left|\boldsymbol{\mu}_{par} \cdot \boldsymbol{E}_{diss}\right|^2$ photoabsorption probability), these asymptotic high $- J_{prod}$ limits for $A_0^{(2)}\left(J_{prod}\right)$ can be transcribed readily into the corresponding relationships between $\boldsymbol{\mu}_{par}$ and $\boldsymbol{J}_{prod}$ [239].

While there are several conceivable methods for implementing DFWM detection of $\boldsymbol{\mu}_{par} - \boldsymbol{J}_{prod}$ correlations, the simplest approach builds upon the same spectroscopic properties exploited in analogous LIF studies [240, 241]. For the special case of nascent photofragments created through linearly polarized photolysis and probed in a phase–conjugate/cross–population geometry [*i.e.*, $\left(\hat{e}_f \| \hat{e}_p\right) \perp \left(\hat{e}_b \| \hat{e}_s\right)$ with all fields linearly polarized], the relative four–wave mixing signal strengths for transitions sharing a common initial rovibronic state (*i.e.*, having the same overall rovibrational population) will provide a measure for the spatial anisotropy ascribed to molecular alignment [21]. This conclusion can be verified through direct numerical evaluation of the susceptibility tensor [119] and reflects the fact that the zero–order density operator no longer satisfies the equilibrium conditions imposed by Eq. (4.16) and Eq. (4.17) owing to the non–uniform distribution of population among degenerate magnetic sublevels of the target medium. Analysis of the data presented in Fig. 4.14 for the $Q_1(1.5)$ and $P_{21}(1.5)$ lines of the OH $A\,^2\Sigma^+ - X\,^2\Pi$ (0,0) band yields a quadrupolar tensor component of $A_0^{(2)}(1.5) = -0.06 \pm 0.02$, with other low–lying rotational levels of the vibrationless photoproducts displaying equally small values of this quantity [21]. Such results are in good accord with those derived from previous 266 nm photolysis work [242, 243] based upon the deconvolution of LIF Doppler profiles [244] and suggest that degenerate magnetic sublevels for individual rotational states of the OH$(\upsilon = 0)$ fragments are populated in a near–uniform fashion [*i.e.*, as measured by $A_0^{(2)}\left(J_{prod}\right)$]. This lack of a clear preference for the direction of OH internal angular momentum with respect to the parent HOOH transition dipole moment has been ascribed to the opposing forces of bending and torsional motion during the photodissociation process [245-247].

Any parent molecule will display a well–defined angular relationship between the transition dipole moment giving rise to a bound–free spectroscopic transition, $\boldsymbol{\mu}_{par}$, and the axis of the chemical bond destined to undergo photo–induced cleavage [123]. Since fragment recoil usually takes place along the rupturing bond, a directional correlation can thus be established between $\boldsymbol{\mu}_{par}$ and $\boldsymbol{v}_{prod}$ [222, 223]. As in the case of molecular alignment, the characteristic $\left|\boldsymbol{\mu}_{par} \cdot \boldsymbol{E}_{diss}\right|^2$ dependence of photoabsorption probability will transcribe this $\boldsymbol{\mu}_{par} - \boldsymbol{v}_{prod}$ correlation from the molecular framework into the laboratory–fixed frame of reference, thereby resulting in an anisotropic spatial distribution of the nascent photofragments. For linearly polarized photolysis, the center–of–mass angular distribution of product molecules, $N(\theta_{\hat{v}_{prod},\hat{E}_{diss}})$, can be described in terms of the well known anisotropy parameter, β [236, 123, 239]:

$$N\left(\theta_{\hat{v}_{prod},\hat{E}_{diss}}\right)=\frac{1}{4\pi}\left[1+\beta P_2\left(\hat{v}_{prod}\cdot\hat{E}_{diss}\right)\right]=\frac{1}{4\pi}\left[1+\beta P_2\left(\cos\left(\theta_{\hat{v}_{prod},\hat{E}_{diss}}\right)\right)\right] \tag{4.54}$$

where $\theta_{\hat{v}_{prod},\hat{E}_{diss}}$ denotes the angle between unit vectors $\hat{\boldsymbol{v}}_{prod}$ and $\hat{\boldsymbol{E}}_{diss}$. The quantity β can take on values ranging from -1 to $+2$, with the former giving rise to a cylindrically–symmetric $\sin^2(\theta_{\hat{v}_{prod},\hat{E}_{diss}})$ distribution of $\hat{\boldsymbol{v}}_{prod}$ about $\hat{\boldsymbol{E}}_{diss}$ (*i.e.*, $\boldsymbol{\mu}_{par}$ orthogonal to the rupturing bond implying a perpendicular transition) while the latter leads to a $\cos^2(\theta_{\hat{v}_{prod},\hat{E}_{diss}})$ dependence (*i.e.*, $\boldsymbol{\mu}_{par}$ along the rupturing bond implying a parallel transition). Measurement of the recoil velocity anisotropy (or translational anisotropy) thus provides information on the symmetry of the transition dipole moment in the parent molecule from which a symmetry–based identification of the pertinent excited electronic state can be made [218]. Since manifestation of the $\boldsymbol{\mu}_{par}-\boldsymbol{J}_{prod}$ and $\boldsymbol{\mu}_{par}-\boldsymbol{v}_{prod}$ correlations depends upon the alignment of $\boldsymbol{\mu}_{par}$ in the laboratory–fixed frame for their manifestation (*viz.*, alignment induced by coupling with $\boldsymbol{E}_{diss}$), they are influenced strongly by internal motion of the dissociating molecule, with the most pronounced directional properties observed when the timescale of bond fission is much shorter than that of the parent rotational period [222-224]. In the case of photofragment translational anisotropy, departure of measured angular distributions from the form suggested by the known (or hypothesized) orientation of $\boldsymbol{\mu}_{par}$ in the molecular framework can yield an estimate of the effective lifetime for the dissociative potential energy surface.

While the Doppler–free capabilities of DFWM are of great importance for high resolution spectroscopy, they can also be exploited for the extraction of directional information [248, 249]. As discussed in Section 4.2.5, the individual molecules participating in a four–wave mixing process must be able to interact simultaneously with all incident and generated electromagnetic waves. This fact, in conjunction with the spectral shifts inherent to the Doppler effect, constrains detection to only a small subset of the nonstationary species comprising a gas–phase ensemble of target molecules. The top panel of Fig. 4.15 shows that, to a first approximation, only those molecules moving orthogonal to the plane defined by the impinging DFWM waves (*i.e.*, only molecules with velocity vectors residing primarily in the "Zero–Velocity" plane) are amenable to direct interrogation. This preferential sensitivity for species traveling in

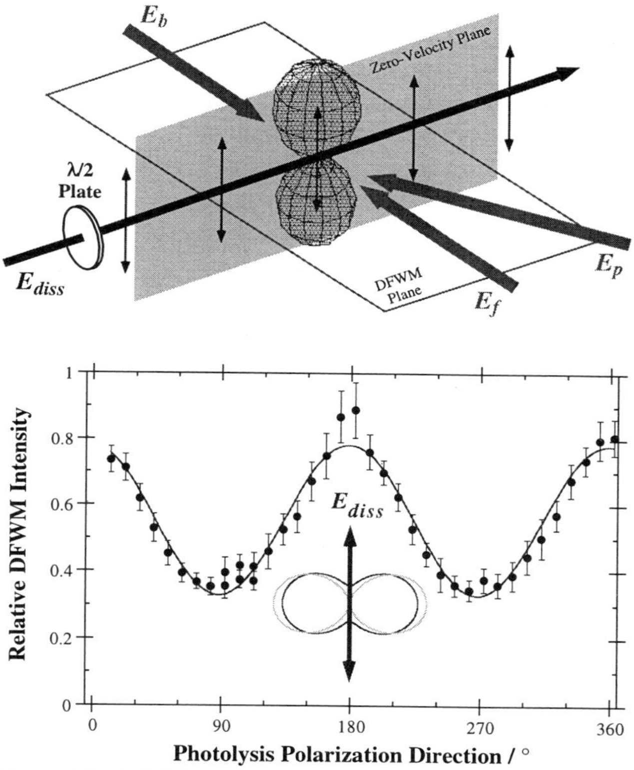

Figure 4.15 DFWM spectroscopy as a probe of photofragment translational anisotropy. The top panel highlights the scheme used for measuring the spatial distribution of recoil velocities in nascent photoproducts. The preferential sensitivity for specific velocity groups inherent to the phase–conjugate DFWM interaction leads to predominant signal generation from molecules moving in the "zero–velocity" plane. As shown in the lower panel, rotation of the linearly polarized photolysis source (zero photolysis polarization direction defined for $\boldsymbol{E}_{diss}$ in the DFWM plane) leads to a corresponding rotation in the cylindrically–symmetric angular distribution of photofragments, which manifests itself as a modulation in the observed nonlinear response. The depicted data were obtained by probing the $Q_1(6.5)$ rovibronic transition in the $A\,^2\Sigma^+-X\,^2\Pi$ (0,0) band of the OH radicals produced upon 266 nm photolysis of HOOH. Analysis reveals an anisotropy parameter of $\beta=-0.65\pm0.15$ (dark curve of inset) which departs from the $\beta=-1$ limit of a perpendicular transition (light curve of inset) owing to a finite lifetime of the electronically–excited parent molecule.

certain spatial directions can be exploited as a quantum state specific probe of the translational anisotropy which accompanies the photofragmentation of precursor molecules via linearly polarized light [42]. A more detailed analysis of the phase–conjugate DFWM configuration reveals that both the magnitude and orientation of the photoproduct recoil velocity are encoded in the dependence of the four–wave mixing response upon various experimental quantities, including the direction of linear photolysis polarization and the pump–probe crossing angle [119].

The lower panel of Fig. 4.15 provides an example of the results obtained by exploiting four–wave mixing spectroscopy to probe the spatial distribution of recoil velocities for the hydroxyl radicals emerging from the 266 nm photolysis of HOOH [119]. For this measurement, the common wavelength of the beams employed in a phase–conjugate DFWM configuration was tuned to coincide with the rest frequency of the $Q_1(6.5)$ transition in the OH $A\,^2\Sigma^+ - X\,^2\Pi$ (0,0) band. Rotation of the electric vector for the linearly polarized photolysis source, $\boldsymbol{E}_{diss}$, about the line of intersection connecting the "DFWM" and "Zero–Velocity" planes (*cf.*, upper panel of Fig. 4.15) produces a corresponding rotation in the cylindrically–symmetric angular distribution of the photofragments which, in turn, manifests itself as an angle dependent variation in the intensity of the observed DFWM response. Quantitative analysis of such data is based upon a computational model which incorporates translational motion of the OH target molecules into a perturbative treatment of the induced third–order nonlinear signal polarization. Least squares analysis suggests that the unrelaxed photoproducts are characterized by an effective anisotropy parameter of $\beta = -0.65 \pm 0.15$ and a velocity distribution that peaks sharply in the vicinity of 3.85 km/s. These results, with their prediction of a near perpendicular electronic transition in HOOH that imparts substantial translational energy to the departing OH fragments, are in good accord with conclusions derived from the more conventional deconvolution of Doppler–broadened spectral profiles as recorded through LIF methods [242, 243]. The deviation of β from the limiting value of -1 (*i.e.*, $\boldsymbol{\mu}_{par}$ orthogonal to rupturing O – O bond), in conjunction with reasonable estimates for parent rotational motion, place an upper limit of roughly 60 fs on the lifetime of the initially excited (dissociative) state [242].

4.4 Two–Color Schemes and Related Techniques

4.4.1 *Hybrid Double–Resonance Spectroscopy: SEP–DFWM & SEP–OODR*

Owing to its relative ease of implementation and minimal requirements for optical access to the target medium, DFWM detection can readily be incorporated into a variety of multiply–resonant techniques designed to probe molecular states that are inaccessible through single–photon transitions. As illustrated in Fig. 4.16, this permits the development of hybrid double–resonance schemes that combine the spectroscopic selectivity afforded by the presence of two resonance conditions with the background–free response inherent to the four–wave mixing interaction. Folded and unfolded variants of double–resonance are possible, with the former providing a means to investigate vibrationally–excited polyatomic species in the manner of Stimulated Emission Pumping (*viz.*, SEP–DFWM spectroscopy) [63, 64, 8] while the latter builds upon more traditional Optical–Optical Double Resonance methods so as to make possible the interrogation of higher–lying potential energy surfaces (*viz.*, OODR–DFWM spectroscopy) [23]. Two–photon DFWM spectroscopy, capable of accessing highly–excited electronic states through more convenient (although, usually, less selective and sensitive) single–color methods, has also been demonstrated on both atomic [46, 48, 38, 49] and molecular [30, 29, 47, 31] systems.

Both of the energy level diagrams depicted in Fig. 4.16 employ an intense source of PUMP light having frequency ω_{PUMP} in order to transfer a significant fraction of molecules from an initially populated level of the ground state, $|g\rangle$, to a preselected rotation–vibration level of an intermediate state, $|i\rangle$. Before this transient molecular population can decay through either radiative or nonradiative channels, a nearly instantaneous DFWM interaction is used to record an "absorption–like" spectrum terminating on target levels of the excited state under investigation, $|e\rangle$. While $|g\rangle$, $|i\rangle$, and $|e\rangle$ could represent rovibronic levels in three distinct electronic manifolds, conventional SEP spectroscopy requires that $|e\rangle$ be a vibrationally–excited eigenstate of the same ground state potential surface that supports $|g\rangle$. Infrared variants of the OODR–DFWM and SEP–DFWM techniques also can be envisioned where transitions take place exclusively between the vibrational levels of a single electronic state.

The intermediate state $|i\rangle$ of any double resonance scheme serves as a window to mediate the visibility or intensity of individual spectroscopic features, with the selection rules imposed upon resonant matter–field interactions responsible for alleviating much of the spectral complexity normally ascribed to the optical transitions of polyatomic species. In addition, the polarized nature of the optical pumping process induced by the PUMP radiation will create a nonuniform distribution of population among the degenerate magnetic sublevels of $|i\rangle$. Consequently, the four–wave mixing process initiates from an anisotropic molecular ensemble (*e.g.*, net alignment and/or orientation) which can be characterized by a zero–order density matrix having inequivalent diagonal elements and, perhaps, non–zero off–diagonal terms (*i.e.*, coherences). As well documented in the case of polarization spectroscopy [250, 251], the unique intensity patterns arising from such magnetic sublevel anisotropies can be exploited for the assignment of rotational branch structure and angular momentum quantum numbers in regions of extreme vibronic congestion.

Both folded and unfolded variants of DFWM–detected double resonance have been demonstrated experimentally, with the SEP–DFWM scheme found to provide a facile means for interrogating vibrationally highly–excited polyatomic molecules [63, 64, 8]. As discussed in Chapter 3 of this monograph, conventional SEP studies rely upon the depletion of spontaneous emission from intermediate state $|i\rangle$ to signify resonant interaction with a "downward–going" $|i\rangle \rightarrow |e\rangle$ transition and thus identify the location of vibrationally–excited target level $|e\rangle$. This "fluorescence–dip" approach abrogates the zero–background advantage of LIF spectroscopy and leads to a situation analogous to that encountered in traditional linear absorption measurements: potentially weak signals must be discriminated and extracted from a large baseline level of superfluous light. In contrast, the background–free response inherent to the absorption–based SEP–DFWM methodology enables the sensitive detection of $|i\rangle \rightarrow |e\rangle$ resonances with additional improvements in resolution following from implementation of the Doppler–free, phase–conjugate configuration. Since SEP creates a population inversion between the intermediate and final states (*i.e.*, $N_i > N_e$ with $|e\rangle$ usually unoccupied under ambient conditions), the overall efficiency of the four–wave mixing process is expected to be enhanced further by resonant amplification of the signal photons as they propagate through the transitory gain medium [8].

4.4.2 *TC–LIGS Techniques*

Section 4.2.3 has described the response of DFWM in terms of diffractive scattering from optically–induced volume holograms (*cf.*, Fig. 4.3). The distinct preparation and detection steps involved in this process suggest several intriguing possibilities for implementing doubly–resonant forms of spectroscopy. While such transient

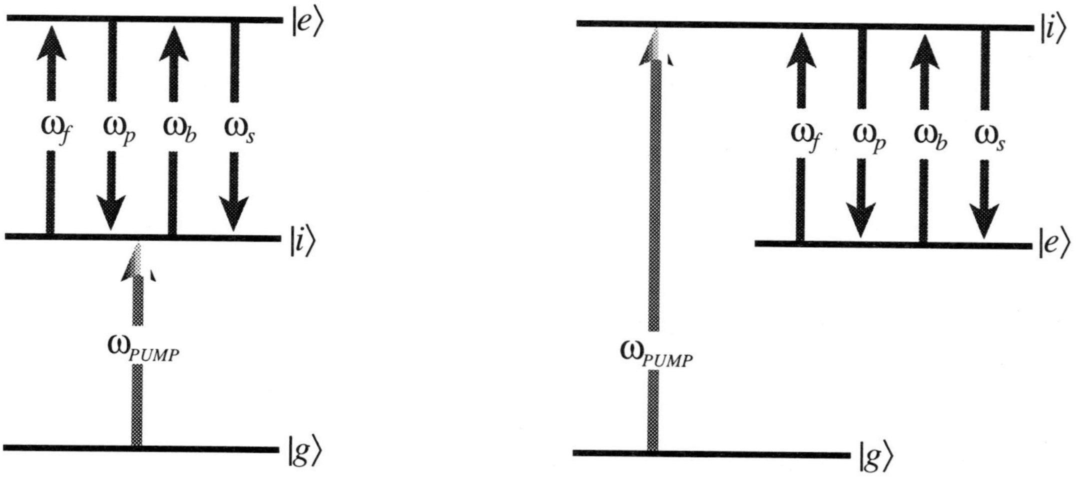

Figure 4.16 Double–Resonance schemes incorporating DFWM detection. An intense PUMP wave of frequency ω_{PUMP} transfers molecular population from ground state $|g\rangle$ to intermediate state $|i\rangle$ with the target state $|e\rangle$ probed through absorption–based DFWM spectroscopy. The left panel depicts unfolded OODR (*i.e.*, OODR–DFWM spectroscopy) while the right panel illustrates the inverted SEP configuration (*i.e.*, SEP–DFWM spectroscopy).

grating techniques are well established in the time–domain community [11, 100], their use for frequency–resolved measurements of molecular structure and dynamics has emerged only recently following the demonstration of Two–Color Laser–Induced Grating Spectroscopy (TC–LIGS) by Hayden, Chandler, and co–workers [58, 215, 61]. These pioneering efforts highlighted the ability to probe vibrationally–excited diatomic and polyatomic molecules through two–color variants of resonant four–wave mixing built upon the concepts of conventional Stimulated Emission Pumping [58] and Vibrationally–Mediated Photodissociation [215, 61]. Subsequent refinements introduced by Rohlfing, *et al.* [59, 55, 164, 72, 53] have led to numerous applications of the TC–LIGS methodology, including experiments designed to interrogate non–fluorescing species [55, 72] and to elucidate primary photochemical events [55, 164].

Although not rigorously classifiable as a degenerate form of four–wave mixing (*i.e.*, two distinct optical frequencies are involved), TC–LIGS nevertheless shares many of the unique advantages afforded by DFWM spectroscopy owing to the fully resonant–enhanced nature of the requisite matter–field interactions (*i.e.*, all optical waves are tuned into resonance with molecular transitions). In particular, this leads to an "absorption–based" double–resonance scheme having sufficient sensitivity for direct optical detection of trace species in rarefied environments (*viz.*, low pressure gases and free–jet expansions). The techniques of DFWM and CARS have often been referred to as the nonlinear analogs of resonance fluorescence and spontaneous Raman processes, respectively. From this perspective, folded variants of TC–LIGS might be considered to represent the nonlinear equivalent of resonance Raman scattering, with direct electronic excitation of the target medium providing spectroscopic access to the entire manifold of ground state rovibrational levels.

Reduced to most fundamental terms, all transient grating techniques are based upon the spatial interference of two time–coincident optical pulses of identical frequency [11]. For notational convenience, the frequencies and wavevectors for these grating–forming beams will be denoted as $(\omega_1, \boldsymbol{k}_1)$ and $(\omega_2, \boldsymbol{k}_2)$, respectively, where $\omega_1 = \omega_2$ and the included angle between $\boldsymbol{k}_1$ and $\boldsymbol{k}_2$ determines the fringe periodicity as per Eq. (4.8). The resulting modulation in intensity or polarization of the total electromagnetic field is transcribed onto the spatial distribution of molecular population and/or orientation, with the most pronounced effects observed upon resonant excitation of an allowed rovibronic transition. The volume hologram created in this manner can be probed sensitively through diffraction of a third incident beam [*i.e.*, corresponding to $(\omega_3, \boldsymbol{k}_3)$] which gives rise to a signal wave [*i.e.*, corresponding to $(\omega_s, \boldsymbol{k}_s)$] that emerges in a direction $\hat{\boldsymbol{k}}_s$ specified by the Bragg scattering condition [83]. While the degenerate nature of the probe and signal beams is obvious (*i.e.*, $\omega_s = \omega_1 - \omega_2 + \omega_3 = \omega_3$), the first–order Bragg criterion for a thick material grating (*viz.*, grating dimensions substantially larger than the periodicity) is found to reproduce exactly the nonlinear phase–matching constraint embodied in expression (4.7).

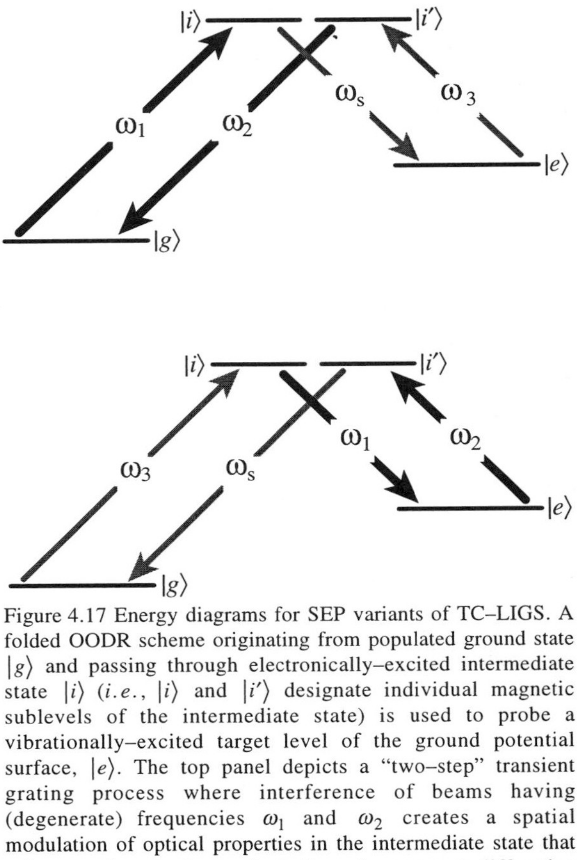

Figure 4.17 Energy diagrams for SEP variants of TC–LIGS. A folded OODR scheme originating from populated ground state $|g\rangle$ and passing through electronically–excited intermediate state $|i\rangle$ (*i.e.*, $|i\rangle$ and $|i'\rangle$ designate individual magnetic sublevels of the intermediate state) is used to probe a vibrationally–excited target level of the ground potential surface, $|e\rangle$. The top panel depicts a "two–step" transient grating process where interference of beams having (degenerate) frequencies ω_1 and ω_2 creates a spatial modulation of optical properties in the intermediate state that subsequently can be probed through resonant diffractive scattering of a third incident pulse (ω_3) to give a signal wave of identical frequency $\omega_s = \omega_3$. In contrast, the bottom panel illustrates a concerted four–wave mixing interaction that requires the temporal overlap of all impinging fields.

Fig. 4.17, adapted from the work of Williams, *et al.* [60], presents energy level diagrams for two possible implementations of TC–LIGS designed to probe the rotation–vibration structure of highly–excited polyatomic molecules. While these schemes build upon the inverted geometry of conventional SEP, extension to other (unfolded) forms of OODR should be obvious. In each case, the initially populated ground state $|g\rangle$ is connected to the vibrationally–excited target level $|e\rangle$ by a pair of resonant transitions involving electronically–excited intermediate state, $|i\rangle$ (*NB*, the symbols $|i\rangle$ and $|i'\rangle$ in Fig. 4.17 designate individual magnetic sublevels in a preselected rovibronic level of the intermediate state). However, the depicted TC–LIGS configurations do exhibit important differences both in the frequency of the emerging signal photon (*i.e.*, $\omega_s = \omega_{ie}$ in the top panel while $\omega_s = \omega_{ig}$ in the bottom panel) and in the sequence of matter–field interactions governing the induced nonlinear response.

A detailed theoretical analysis of TC–LIGS has been reported by Williams and co–workers [60], with accompanying laboratory measurements highlighting the advantages and disadvantages of various experimental arrangements. In particular, these authors have made a clear distinction between the SEP schemes of Fig. 4.17 and have suggested that the spectroscopic utility of TC–LIGS can be maximized by selecting the signal field to be resonant with the stronger of the two pertinent transitions (*i.e.*, $\omega_s = \omega_{ie}$ as in the top panel or $\omega_s = \omega_{ig}$ as in the bottom panel). Both frequency– and time–domain studies were employed to show that the upper energy level diagram of Fig. 4.17 corresponds primarily to the formation and subsequent detection of a spatial modulation in the concentration and/or orientation of molecules within the electronically–excited intermediate state. The temporal evolution of this diffraction phenomenon is governed by the two–photon decay rate $\Gamma_{ii'}$ which, in the absence of collisions, can be equated to the reciprocal of the lifetime for state $|i\rangle$ (*viz.*, $\Gamma_{ii'} \approx \Gamma_{ii} = 1/\tau_i$). On the other hand, the lower portion of Fig. 4.17 depicts the coherent Raman excitation of ground state levels $|g\rangle$ and $|e\rangle$ which can only occur when all incident fields are time–coincident. Dissipation of this two–photon coherence takes place on a timescale commensurate with the relaxation rate Γ_{eg}. As demonstrated by Williams, *et al.* [60], signals derived from the aforementioned excited–state transient grating scheme (*i.e.*, upper energy level diagram) also have contributions from such concerted Raman processes which manifest themselves as a pronounced "coherence spike" when all optical pulses overlap in time.

Compared to the techniques discussed in Section 4.4.1, TC–LIGS has the distinct advantage of generating a true doubly–resonant response. From the perspective of transient gratings, this implies that efficient diffraction of the probe beam (ω_3) into the signal field (ω_s) ensues only upon resonant interaction with molecules whose spatial concentration and/or orientation has been modulated by the grating–forming pulses (ω_1 and ω_2). In contrast, both SEP and SEP–DFWM produce strong signals whenever spectroscopic transitions involving populated eigenstates are encountered, regardless of whether or not the pertinent molecules have been labeled through prior interaction with the PUMP radiation (ω_{PUMP}; *cf.*, Fig. 4.16). As first demonstrated in the work of Buntine, *et al.* [58], and commented upon in subsequent studies [55, 36], this unique feature of TC–LIGS can prove advantageous for the discrimination and extraction of double resonance signals in regions of extreme spectral congestion. For ω_1 and ω_2 tuned into coincidence with a preselected $|i\rangle \leftrightarrow |g\rangle$ transition, careful inspection of the energy level diagram depicted in the top portion of Fig. 4.17 shows that scanning of the probe frequency can result in two clearly distinguishable scattering processes [58]: (1) resonant diffraction of ω_3 from population gratings induced in the intermediate state $|i\rangle$ and (2) resonant diffraction of ω_3 from depletion gratings created in the ground state $|g\rangle$. While the former mechanism corresponds to the "conventional" SEP (or, by extension, unfolded OODR) scheme [59, 60, 53], the latter phenomenon has been exploited for the interrogation of non–fluorescing excited states in polyatomic species [72].

Aside from obvious spectroscopic advantages, the doubly–resonant nature of TC–LIGS provides unusual capabilities for the elucidation of primary photochemical events. As demonstrated by the work of Butenhoff and Rohlfing [55, 164], a combination of frequency– and time–resolved transient grating techniques can be exploited for extraction of scalar properties and vector correlations from nascent photoproducts. Fig. 4.18 presents an example of translational anisotropy measurements made possible by applying the TC–LIGS methodology to the photodissociation of jet–cooled NO_2 molecules [164]. Owing to the near–threshold photolysis wavelengths employed for this study, only the $NO(X\,^2\Pi)+O(^3P)$ product channel is energetically accessible, with the depicted results corresponding to vibrationless NO fragments formed with $125.9\,\text{cm}^{-1}$ (*i.e.*, $< 0.016\,\text{eV}$) of available excess energy.

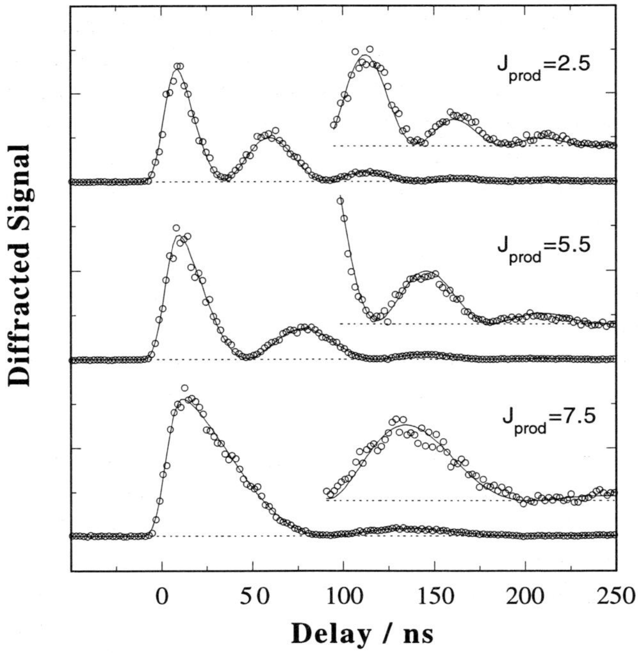

Figure 4.18 TC–LIGS as a probe of photofragment translational anisotropy. A free–jet expansion of NO_2 molecules is photolyzed at $25254.4\,cm^{-1}$ ($125.9\,cm^{-1}$ excess energy) by means of crossed ultraviolet beams, thereby resulting in the creation of a photofragment transient grating characterized by a fringe spacing of $\Lambda = 16.6\,\mu m$. The nascent NO products formed in the $J_{prod} = 2.5$, 5.5, and 7.5 rotational levels of the vibrationless $X\,^2\Pi_{1/2}$ state are probed through resonant diffraction processes involving the overlapping $R_{11}(J)$ and $R_{21}(J)$ lines. All optical beams are linearly polarized in a direction orthogonal to that of the grating fringes. The time dependence of the observed four–wave mixing signals (circular symbols) reflects the translational anisotropy of the departing fragments with the solid lines representing the results of least–squares analyses based on a model that incorporates the recoil velocity (speed and anisotropy) convoluted with distributions for parent molecule velocity and laser pulse duration. Each decay curve is normalized to unit area and the insets (vertically offset for clarity) display a $10\times$ magnification of the long–time behavior. (courtesy of E. A. Rohlfing)

The data in Fig. 4.18 were obtained by a variant of TC–LIGS in which a transient grating created in the parent NO_2 species is transcribed by photodissociation onto the daughter NO fragments where it can be probed through resonant diffractive scattering [164]. In brief, two time–coincident photolysis beams having the same frequency and direction of linear polarization are crossed in the target medium so as to produce a spatial modulation in the concentration of electronically–excited NO_2 molecules. The dissociative nature of the excited potential surface leads to abrupt cleavage of the O – NO bond, thereby resulting in the formation of $NO(X\,^2\Pi)$ and $O(^3P)$ products which (initially) display the same spatial characteristics imposed upon the parent molecules. Therefore, as discussed in connection with Eq. (4.8), the fringe spacing, Λ, of the photofragment grating generated in this fashion will be determined by the wavelength of the two photolysis pulses, λ_{diss}, and their included angle, ϕ_{diss}, with the experimental conditions employed for the acquisition of Fig. 4.18 (*viz.*, $\tilde{\nu}_{diss} = 1/\lambda_{diss} = 25254.4\,cm^{-1}$ and $\phi_{diss} = 1.36°$) yielding $\Lambda = 16.6\,\mu m$. The vibrationless, ground state NO molecules emerging from this near threshold photodissociation process can be interrogated by the phase–matched and resonant–enhanced diffraction of a third incident wave tuned into coincidence with individual rovibronic features of the NO $A\,^2\Sigma^+ - X\,^2\Pi$ (0,0) band at $\sim 226\,nm$. It is particularly interesting to note that this implementation of TC–LIGS further distinguishes the preparation and detection steps for the pertinent volume hologram by making use of different molecular species (*viz.*, parent NO_2 molecules for grating formation and daughter NO molecules for grating detection).

For a constant time delay between the incident photolysis and probe pulses, the experimental scheme outlined in the previous paragraph suggests two distinct possibilities for frequency–resolved measurements. With the wavelength of the photolysis source fixed at an above–threshold value, scanning of the probe frequency will generate a spectrum of the nascent NO photofragments from which the distribution of population among internal quantum states can be deduced. Alternatively, by tuning the probe frequency into resonance with a specific NO rovibronic feature and varying the photolysis wavelength, an "action spectrum" corresponding to the relative yield of state–selected photoproducts can be obtained. Such information can be used to accurately identify the dissociation limit and to explore the nature of continuum structures ascribed to the formation of scattering–like resonances in the unbound NO_2 complex [55].

The time–resolved data of Fig. 4.18 were obtained by recording the TC–LIGS intensity as a function of the time delay, Δt, between photolysis and probe pulses, with the wavelength of the photolysis source fixed at an above threshold value while the frequency of the probe beam was tuned into resonance with transitions originated from the indicated rotational levels of the vibrationless NO $X\,^2\Pi_{1/2}$ product. Following creation of the photofragment grating, on a timescale commensurate with the duration of impinging laser pulses, a dissipation of diffraction efficiency for the probe radiation is evident. This phenomenon can be attributed to "washout" of the optically–induced volume hologram as the target molecules move from regions of high concentration (grating crests) to those of low concentration (grating nulls). For bulk–gas samples characterized by isotropic Maxwellian speed distributions and not subject to collisional relaxation, Rose, *et al.* [98, 252-254] have shown that the observed signal intensity should exhibit a Gaussian scaling with respect to the ratio of time delay to fringe spacing, $\Delta t/\Lambda$. However, the anisotropic and strongly peaked distribution of nascent recoil velocities encountered in photodissociation processes [*cf.*, Eq. (4.54) and accompanying discussion] can lead to otherwise unexpected time–dependent features. In particular, the diffraction efficiency can rebound as "collective" translational displacement of the target medium results in a reformation of the initial volume hologram. Such is the case in Fig. 4.18, where the oscillatory behavior of the TC–LIGS response reflects periodic rephasing of the photofragment grating as a consequence of the inherent specificity imposed upon the direction and magnitude of NO velocities.

Detailed interpretation of the time–domain signals in Fig. 4.18 requires appropriate modeling both for the preparation and subsequent dissipation of the optically–induced transient grating. In particular, the TC–LIGS decay curves encode the magnitude and spread of NO recoil speeds as well as the angular distribution of photoproduct velocity vectors, $N(\theta_{\hat{\boldsymbol{v}}_{prod},\hat{\boldsymbol{E}}_{diss}})$. As demonstrated in Eq. (4.54), this quantity is expected to be cylindrically–symmetric about the direction of linear polarization for the photolysis source, $\boldsymbol{E}_{diss}$, and can be described in terms of the anisotropy parameter, β, which gives information on the transition moment orientation and excited state lifetime of the parent molecule [236, 218]. The experimental traces of Fig. 4.18 were all obtained with $\boldsymbol{E}_{diss}$ parallel to the probe beam polarization and orthogonal to the grating fringes. Analogous measurements performed with $\boldsymbol{E}_{diss}$ directed along the grating fringes were found to give time–resolved profiles that displayed negligible recurrence of diffraction efficiency [164]. These observations suggest positive values of β commensurate with the $\cos^2(\theta_{\hat{\boldsymbol{v}}_{prod},\hat{\boldsymbol{E}}_{diss}})$ translation anisotropy of a parallel photodissociation process. More refined analyses, incorporating convolutions over the initial motion of parent species and the finite duration of incident optical pulses, yield quantitative photofragment velocity distributions (*i.e.*, speed and anisotropy) in good accord with expectations for the near–threshold photolysis of NO_2. When compared to conventional studies based upon the deconvolution of LIF Doppler profiles, it is important to recognize that the TC–LIGS scheme permits the extraction of $\boldsymbol{\mu}_{par} - \boldsymbol{v}_{prod}$ vector correlations despite the fact that the exceptionally low translational excitation of the NO fragments (*e.g.*, only $7.3\,\mathrm{cm}^{-1}$ of translational energy is partitioned into the vibrationless NO product in $J_{prod} = 7.5$) leads to barely resolvable Doppler broadening. Consequently, the photofragment transient grating technique provides information complementary to that obtained through more traditional fluorescence methods.

To some extent, the unique advantages of TC–LIGS are compromised by restrictions associated with the conservation of linear momentum. More specifically, a given arrangement of incident and generated electromagnetic waves can satisfy phase–matching conditions exactly only for a single set of grating–forming and grating–probing frequencies (*i.e.*, only a single selection of $\omega_1 = \omega_2$ and $\omega_3 = \omega_s$) with any other combinations leading to a pronounced reduction in overall signal amplitude. For a wavevector mismatch of $\boldsymbol{\Delta k} = (\boldsymbol{k}_1 - \boldsymbol{k}_2 + \boldsymbol{k}_3) - \boldsymbol{k}_s$, a straightforward analysis reveals that the TC–LIGS response should scale in proportion to $\mathrm{sinc}^2(\Delta k\,\ell/2)$ where $\mathrm{sinc}(x) = \sin(x)/x$, $\Delta k = |\boldsymbol{\Delta k}|$, and ℓ is the sample interaction length [75, 4, 80]. This situation is to be contrasted with that obtained for the phase–conjugate DFWM scheme of Fig. 4.1 which satisfies momentum conservation criteria precisely (*i.e.*, $\Delta k = 0$) for all choices of the common (degenerate) wavelength and the pump–probe crossing angle [8].

The restrictions imposed upon TC–LIGS by momentum conservation can be mitigated through judicious selection of experimental configuration, with the widest frequency range for "approximate" phase–matching obtained when the incident beams of varying frequency copropagate [255]. Butenhoff and Rohlfing [55] have addressed this problem under the reasonable assumption that the signal wave emerges in a direction which tends to minimize the value of Δk. For a planar geometry in which the grating–forming beams remain fixed at $\omega_1 = \omega_2$ while the frequency of the nearly copropagating probe wave ω_3 is scanned (*cf.*, top panel of

Fig. 4.17), the effective phase–match bandwidth $\Delta\omega$, defined as the <u>F</u>ull–<u>W</u>idth at <u>H</u>alf–<u>M</u>aximum (FWHM) height for the central lobe of the $\text{sinc}^2(\Delta k\,\ell/2)$ function, has the form:

$$\Delta\omega \propto \frac{(\omega_3/\omega_1)^2}{\ell\sin^2(\theta_{21}/2)} \tag{4.55}$$

where θ_{21} is the angle between the two grating–forming beams. The predicted increase in phase–matching bandwidth with decreasing sample interaction length ℓ is consistent with relaxation of the Bragg scattering condition expected to accompany contraction of a thick material grating [11]. The validity of this expression, as well as that of an analogous formula derived for the case in which the probe wave remains fixed in frequency while the grating–forming beams are scanned, has been demonstrated experimentally by Butenhoff and Rohlfing [55]. By incorporating additional imaging optics into the TC–LIGS detection path so as to minimize the angular spread of the emerging signal beam, values of $\Delta\omega$ in excess of $1000\,\text{cm}^{-1}$ have been reported for SEP [36, 60, 53] and OODR [23, 37, 27] measurements performed through use of the transient grating methodology.

4.5 Conclusions and Future Directions

As demonstrated by the numerous contributions to this monograph, forms of spectroscopy based upon the nonlinear interaction of light with matter have played pivotal roles in a wide variety of research efforts designed to elucidate the structure and dynamics of molecules. Once considered to be esoteric curiosities confined to the realm of optical physics, these techniques have evolved to become eminently practical and exceptionally useful weapons in the arsenal of molecular science. To a large extent, this transformation has been driven by the incessant advances in optical/electronic technology, with the commercial availability of tunable, high resolution laser sources responsible for the successful integration of such nonlinear schemes into the repertoire of the modern–day spectroscopist.

This chapter has been focused on the conceptual and practical foundations of a rather mature nonlinear optical scheme, degenerate four–wave mixing, which has emerged as a powerful and versatile tool for the elucidation of molecular structure. The ultimate utility of any spectroscopic probe depends upon a myriad of factors, with the advantages and shortcomings ascribed to each technique serving to determine its viability for individual research problems. Detection sensitivity, dynamic range, signal interpretation, species applicability, and experimental complexity, as well as a host of other, less obvious properties, must be considered when evaluating the relative merits of a particular method. The material presented above has attempted to assess each of these characteristics in reference to DFWM spectroscopy. Indeed, much of the ongoing interest in this technique stems from the unique capabilities that it affords in situations where more conventional linear and nonlinear probes tend to falter. Furthermore, as demonstrated most succinctly by newly–developed nonlinear variants of Stimulated Emission Pumping (*i.e.*, SEP–DFWM [63, 64, 8] and TC–LIGS [58, 59, 53]), the concepts of four wave–mixing, as well as their characteristic advantages, can readily be incorporated into multiply–resonant optical schemes.

While early interest in the DFWM process centered about its unusual phase–conjugation properties [93, 256, 9, 10], recent spectroscopic uses for this nonlinear scheme have built upon the absorption–based nature of the four–wave mixing interaction, which combines high detection sensitivity with a background–free, sub–Doppler spectral response. In particular, by eliminating any reliance upon secondary processes (*e.g.*, spontaneous emission or multiphoton ionization) to locate resonant transitions, DFWM should be applicable to a variety of molecular systems that have escaped prior investigation. For example, DFWM promises to enable the direct interrogation of species that are inaccessible to traditional LIF methods owing to small fluorescence quantum yields and/or short fluorescence lifetimes exhibited by their excited electronic states [72-74, 62]. However, the rapid decrease in DFWM signal strength expected to accompany increasing depopulation/dephasing rates implies that this capability of probing "dark" molecules will be tempered by the same nonradiative relaxation processes (*e.g.*, predissociation, internal conversion, or intersystem crossing) responsible for the suppression of observable spontaneous emission. On the other hand, the inherent difficulties associated with fluorescence measurement in the infrared portion of the electromagnetic spectrum suggest that infrared DFWM spectroscopy could play an important role both for the investigation of polyatomic vibrational dynamics and for the elucidation of chemical reactivity (*viz.*, as revealed in the internal state distributions and vector correlations of nascent product molecules).

When compared to the ubiquitous LIF methodology (as implemented through visible or ultraviolet transitions), absorption–based forms of molecular spectroscopy usually exhibit substantially reduced sensitivity. This fact stems primarily from the differential nature of the absorption process which necessitates the identification and extraction of potentially weak signals superimposed on a much larger baseline of superfluous light. In contrast, the enormous resonant enhancement inherent to the DFWM interaction provides a viable means for the detection of molecular transitions with essentially zero background, thereby combining the versatility and general applicability of absorption probes with a sensitivity and dynamic range approaching that of idealized, background–free fluorescence techniques. However, the strong matter–field interactions responsible for this augmented sensitivity lead to pronounced dependencies upon both transition moments and relaxation rates, with optical saturation effects frequently playing a significant role in the acquisition and interpretation of DFWM spectra.

For many applications, especially those involving the interrogation of molecules entrained in hostile and/or luminous environments, DFWM has the distinct advantage of generating a response that emerges in the form of a coherent and well–collimated beam of light. Aside from providing a viable mechanism for the rejection of background interference, this remote sensing capability enables spectroscopic measurements to be performed in a manner that requires minimal optical access to the target region. In contrast, the majority of linear techniques (*e.g.*, LIF and spontaneous Raman scattering) result in signals that radiate in an essentially isotropic fashion, thereby greatly diminishing the efficiency with which they can be collected and discriminated. The exceptional spatial and temporal selectivity accompanying the four–wave mixing interaction can also be exploited for the investigation of transient species and their distribution in a non–uniform medium. Recent work has combined the high spectral resolution of DFWM with the image forming capabilities inherent to its phase–conjugate nature so as to produce two–dimensional, quantum state–specific maps of the atoms and molecules formed in combustion processes [257, 258, 54, 259]. This unique capability, in conjunction with burgeoning multiplex schemes that enable the "instantaneous" acquisition of an entire DFWM spectrum (*i.e.*, during a single broadband laser pulse) [141-143], promises to revolutionize our ability to investigate turbulent systems (*e.g.*, flames) where spatial and temporal inhomogeneities preclude the use of conventional signal averaging procedures.

While the applications of resonant four–wave mixing in the fields of molecular spectroscopy and chemical reaction dynamics are just beginning to emerge, it is evident that the unique abilities afforded by such schemes will make them the method of choice for a variety of fundamental and practical measurements. Although not capable of the "single–particle" detection limits achieved by LIF and ionization techniques, DFWM offers the possibility of directly interrogating *any* absorbing species (*i.e.*, any atom/molecule having accessible optical or infrared transitions) through use of relatively straightforward experimental configurations that combine high spectral, temporal, and spatial resolution with the remote sensing properties inherent to a coherent response. These desirable characteristics, with their obvious potential for exploring the spectroscopy of both isolated and interacting molecules, should foster the continued development of four–wave mixing probes for the elucidation of molecular structure and dynamics.

Acknowledgments

The author wishes to thank T. A. W. Wasserman, A. A. Arias, Q. Zhang, T. Müller, S. A. Kandel, D. Hsu, B. Grzybowski, P. Staats, Dr. B. R. Johnson (Rice Quantum Institute), and Dr. P. Dupré (CNRS, Grenoble) for their assistance in the collection and interpretation of results presented in this chapter. He is especially indebted to Drs. D. J. Rakestraw, R. L. Farrow, and E. A. Rohlfing of Sandia National Laboratory for providing the material contained in figures 4.8, 4.9, 4.12, and 4.18. This work was performed under the auspices of grants provided by the United States National Science Foundation, Experimental Physical Chemistry Program (CHE – 9158107, CHE – 9207835, and CHE – 9523286) with additional resources provided by the NATO International Scientific Exchange Programme (CRG 940214; in collaboration with Dr. H.–G. Rubahn, MPI, Göttingen). Acknowledgment is made to the Camille and Henry Dreyfus Foundation for a Camille Dreyfus Teacher–Scholar Award and to The David and Lucile Packard Foundation for support through a Packard Fellowship for Science and Engineering.

References

1. W. Demtröder, *Laser Spectroscopy: Basic Concepts and Instumentation* (Springer–Verlag, Berlin, 1996).
2. M. D. Levenson and S. Kano, *Introduction to Nonlinear Laser Spectroscopy* (Academic Press, Boston, 1988).
3. J. L. Skinner and W. E. Moerner, *J. Phys. Chem.* **100**, 13251 (1996).
4. P. N. Butcher and D. Cotter, *The Elements of Nonlinear Optics* (Cambridge University Press, Cambridge, 1990).
5. S. Mukamel, *Principles of Nonlinear Optical Spectroscopy* (Oxford University Press, New York, 1995).
6. R. L. Abrams, J. F. Lam, R. C. Lind, D. G. Steel, and P. F. Liao, in *Optical Phase Conjugation* R. A. Fisher, Ed. (Academic Press, San Diego, 1983) p. 211.
7. R. L. Farrow and D. J. Rakestraw, *Science* **257**, 1894 (1992).
8. P. H. Vaccaro, in *Molecular Dynamics and Spectroscopy by Stimulated Emission Pumping* H. L. Dai and R. W. Field, Eds. (World Scientific, Singapore, 1995) p. 1.
9. T. R. O'Meara, D. M. Pepper, and J. O. White, in *Optical Phase Conjugation* R. A. Fisher, Ed. (Academic Press, San Diego, 1983) p. 537.
10. R. W. Boyd and G. Grynberg, in *Contemporary Nonlinear Optics* G. P. Agrawal and R. W. Boyd, Eds. (Academic Press, Boston, 1992) p. 85.
11. H. J. Eichler, P. Günter, and D. W. Pohl, *Laser–Induced Dynamic Gratings* (Springer–Verlag, Berlin, 1986).
12. P. Ewart and S. V. O'Leary, *J. Phys. B* **15**, 3669 (1982).
13. P. Ewart and S. V. O'Leary, *Opt. Lett.* **11**, 279 (1986).
14. T. Dreier and D. J. Rakestraw, *Opt. Lett.* **15**, 72 (1990).
15. T. Dreier and D. J. Rakestraw, *Appl. Phys. B* **B50**, 479 (1990).
16. M. S. Brown, L. A. Rahn, and T. Dreier, *Opt. Lett.* **17**, 76 (1992).
17. H. Bervas, B. Attal–Trétout, L. Labrunie, and S. Le Boiteux, *Nuovo Cim.* **14**, 1043 (1992).
18. D. A. Feikema, E. Domingues, and M. J. Cotereau, *Appl. Phys. B.* **55**, 424 (1992).
19. L. A. Rahn and M. S. Brown, *Opt. Lett.* **19**, 1249 (1994).
20. A. Dreizler, R. Tadday, A. A. Suvernev, M. Himmelhaus, T. Dreier, and P. Foggi, *Chem. Phys. Lett.* **240**, 315 (1995).
21. T. A. W. Wasserman, A. A. Arias, S. A. Kandel, D. Hsu, and P. H. Vaccaro, *SPIE* **2548**, 220 (1995).
22. R. L. Vander Wal, R. L. Farrow, and D. J. Rakestraw, *High–Resolution Investigation of Degenerate Four–Wave Mixing in the $\gamma(0,0)$ Band of Nitric Oxide*, Twenty–Fourth (International) Symposium on Combustion (Combustion Institute, 1992) p. 1653.
23. E. F. McCormack, S. T. Pratt, P. M. Dehmer, and J. L. Dehmer, *Chem. Phys. Lett.* **211**, 147 (1993).
24. E. F. McCormack, S. T. Pratt, P. M. Dehmer, and J. L. Dehmer, *Chem. Phys. Lett.* **227**, 656 (1994).
25. E. J. Friedman–Hill, L. A. Rahn, and R. L. Farrow, *J. Chem. Phys.* **100**, 4065 (1994).
26. P. M. Danehy, P. H. Paul, and R. L. Farrow, *J. Opt. Soc. Am. B* **12**, 1564 (1995).
27. E. F. McCormack, P. M. Dehmer, J. L. Dehmer, and S. T. Pratt, *J. Chem. Phys.* **102**, 4740 (1995).
28. E. Konz, V. Fabelinsky, G. Marowsky, and H. G. Rubahn, *Chem. Phys. Lett.* **247**, 522 (1995).
29. G. Meijer and D. W. Chandler, *Chem. Phys. Lett.* **192**, 1 (1992).
30. N. Georgiev, U. Westblom, and M. Aldén, *Opt. Comm.* **94**, 99 (1992).
31. M. Versluis, G. Meijer, and D. W. Chandler, *Appl. Opt.* **33**, 3289 (1994).
32. I. Aben, W. Ubachs, P. Levelt, G. van der Zwan, and W. Hogervorst, *Phys. Rev. A* **44**, 5881 (1991).
33. J. Gumbel and W. Kiefer, *Chem. Phys. Lett.* **189**, 231 (1992).
34. J. Gumbel and W. Kiefer, *J. Opt. Soc. Am. B* **9**, 2206 (1992).
35. N. Böhm and W. Kiefer, *Appl. Spectrosc.* **47**, 246 (1993).
36. M. D. Wheeler, I. R. Lambert, and M. N. R. Ashfold, *Chem. Phys. Lett.* **211**, 381 (1993).
37. M. D. Wheeler, I. R. Lambert, and M. N. R. Ashfold, *Chem. Phys. Lett.* **229**, 285 (1994).
38. J. A. Gray and R. Trebino, *Chem. Phys. Lett.* **216**, 519 (1993).
39. D. J. Rakestraw, T. Dreier, and L. R. Thorne, *Detection of NH Radicals in Flames Using Degenerate Four–Wave Mixing*, Twenty–Third (International) Symposium on Combustion (Combustion Institute, 1991) p. 1901.
40. D. S. Green, T. G. Owano, S. Williams, D. G. Goodwin, R. N. Zare, and C. H. Kruger, *Science* **259**, 1726 (1993).
41. S. Williams, D. S. Green, S. Sethuraman, and R. N. Zare, *J. Am. Chem. Soc.* **114**, 9122 (1992).
42. T. A. W. Wasserman, A. A. Arias, T. Müller, and P. H. Vaccaro, *Hydrogen Sulfide Photodissociation Dynamics at the Red Edge of the First Absorption Band*, D. M. Neumark, Ed., 1995 Conference on the Dynamics of Molecular Collisions, Asilomar, CA (1995) p. PB91.
43. T. A. W. Wasserman, A. A. Arias, T. Müller, and P. H. Vaccaro, *Chem. Phys. Lett.* **262**, 329 (1996).
44. M. Motzkus, G. Pichler, M. Dillmann, K. L. Kompa, and P. Hering, *Appl. Phys. B* **57**, 261 (1993).
45. A. Klamminger, M. Motzkus, S. Lochbrunner, G. Pichler, K. L. Kompa, and P. Hering, *Appl. Phys. B* **61**, 311 (1995).
46. D. C. Haueisen, *Opt. Comm.* **28**, 183 (1979).

47. N. Georgiev and M. Aldén, *Appl. Phys. B* **56**, 281 (1993).
48. J. A. Gray, J. E. M. Goldsmith, and R. Trebino, *Opt. Lett.* **18**, 444 (1993).
49. E. Konz, G. Maroswky, and H. G. Rubahn, *Opt. Comm.* **134**, 75 (1997).
50. R. L. Vander Wal, B. E. Holmes, J. B. Jeffries, P. M. Danehy, R. L. Farrow, and D. J. Rakestraw, *Chem. Phys. Lett.* **191**, 251 (1992).
51. G. J. Germann and D. J. Rakestraw, *Science* **264**, 1750 (1994).
52. G. Hall, A. G. Suits, and B. J. Whitaker, *Chem. Phys. Lett.* **203**, 277 (1993).
53. J. D. Tobiason, J. R. Dunlop, and E. A. Rohlfing, *J. Chem. Phys.* **103**, 1448 (1995).
54. B. A. Mann, S. V. O'Leary, A. G. Astill, and D. A. Greenhalgh, *Appl. Phys. B* **54**, 271 (1992).
55. T. J. Butenhoff and E. A. Rohlfing, *J. Chem. Phys.* **98**, 5460 (1993).
56. A. P. Smith, G. Hall, B. J. Whitaker, A. G. Astill, D. W. Neyer, and P. A. Delve, *Appl. Phys. B* **60**, 11 (1995).
57. B. A. Mann, R. F. White, and R. J. S. Morrison, *Appl. Opt.* **35**, 475 (1996).
58. M. A. Buntine, D. W. Chandler, and C. C. Hayden, *J. Chem. Phys.* **97**, 707 (1992).
59. T. J. Butenhoff and E. A. Rohlfing, *J. Chem. Phys.* **97**, 1595 (1992).
60. S. Williams, J. D. Tobiason, J. R. Dunlop, and E. A. Rohlfing, *J. Chem. Phys.* **102**, 8342 (1995).
61. M. N. R. Ashfold, D. W. Chandler, C. C. Hayden, R. I. McKay, and A. J. R. Heck, *Chem. Phys.* **201**, 237 (1995).
62. V. Sick, M. N. Bui–Pham, and R. L. Farrow, *Opt. Lett.* **20**, 2036 (1995).
63. Q. Zhang, S. A. Kandel, T. A. W. Wasserman, and P. H. Vaccaro, *J. Chem. Phys.* **96**, 1640 (1992).
64. S. A. Kandel, T. A. W. Wasserman, Q. Zhang, H. Wang, A. A. Arias, and P. H. Vaccaro, *SPIE* **1858**, 126 (1993).
65. A. Leone and P. H. Vaccaro, *Evaluation of Four–Wave Mixing Spectroscopy as a Probe of Trace Molecular Species*, Internal Report for: Lockheed Missiles and Space Company, Inc. (Palo Alto, CA, 1993).
66. Q. Zhang, Ph. D., Yale University (1996).
67. A. Okazaki, T. Ebata, and N. Mikami, *Chem. Phys. Lett.* **241**, 275 (1994).
68. T. Müller, A. A. Arias, and P. H. Vaccaro, *(work in progress).*
69. P. DeRose, H. L. Dai, and P. Y. Cheng, *Chem. Phys. Lett.* **220**, 207 (1994).
70. T. Ebata, A. Okazaki, Y. Inokuchi, and N. Mikami, *J. Mol. Struct.* **352/353**, 533 (1995).
71. R. M. Helm, R. Neuhauser, and H. J. Neusser, *Chem. Phys. Lett.* **249**, 365 (1996).
72. J. R. Dunlop and E. A. Rohlfing, *J. Chem. Phys.* **100**, 856 (1993).
73. A. A. Arias, T. A. W. Wasserman, and P. H. Vaccaro, *Nonlinear Optical Spectroscopy of Malonaldehyde: An Investigation of Proton–Transfer Dynamics in the First Excited Singlet State*, D. M. Neumark, Ed., 1995 Conference on the Dynamics of Molecular Collisions, Asilomar, CA (1995) p. PA113.
74. A. A. Arias and P. H. Vaccaro, *(work in progress).*
75. Y. R. Shen, *The Principles of Nonlinear Optics* (John Wiley and Sons, New York, 1984).
76. G. J. Germann, R. L. Farrow, and D. J. Rakestraw, *J. Opt. Soc. Am. B* **12**, 25 (1995).
77. Y. Tang and S. A. Reid, *Chem. Phys. Lett.* **248**, 476 (1996).
78. C. E. Hamilton, J. L. Kinsey, and R. W. Field, *Ann. Rev. Phys. Chem.* **37**, 493 (1986).
79. M. Motzkus, S. Pedersen, and A. H. Zewail, *J. Phys. Chem.* **100**, 5620 (1996).
80. R. W. Boyd, *Nonlinear Optics* (Academic Press, Boston, 1992).
81. D. C. Hanna, M. A. Yuratich, and D. Cotter, *Nonlinear Optics of Free Atoms and Molecules* (Springer–Verlag, Berlin, 1979).
82. J. F. Reintjes, *Nonlinear Optical Parametric Processes in Liquids and Gases* (Academic Press, New York, 1984).
83. J. D. Jackson, *Classical Electrodynamics* (John Wiley and Sons, New York, 1975).
84. S. Mukamel and R. F. Loring, *J. Opt. Soc. Am. B* **3**, 595 (1986).
85. K. Blum, *Density Matrix Theory and Applications* (Plenum Press, New York, 1996).
86. R. Loudon, *The Quantum Theory of Light* (Oxford University Press, Oxford, 1990).
87. J. J. Sakurai, *Modern Quantum Mechanics* (Addison–Wesley Publishing, Reading, Massachusetts, 1994).
88. A. Yariv, *Optical Electronics* (Holt, Rinehart, and Winston, New York, 1985).
89. A. Yariv and D. M. Pepper, *Opt. Lett.* **1**, 16 (1977).
90. R. W. Hellwarth, *J. Opt. Soc. Am.* **67**, 1 (1977).
91. A. Yariv and R. A. Fisher, in *Optical Phase Conjugation* R. A. Fisher, Ed. (Academic Press, San Diego, 1983) p. 1.
92. A. Yariv, *IEEE J. Quantum Electron.* **14**, 650 (1978).
93. D. M. Pepper, *Opt. Eng.* **21**, 156 (1982).
94. T. R. O'Meara and A. Yariv, *Opt. Eng.* **21**, 237 (1982).
95. J. O. White and A. Yariv, *Opt. Eng.* **21**, 224 (1982).
96. A. C. Eckbreth, *Appl. Phys.* **32**, 421 (1978).
97. J. R. Salcedo, A. E. Siegman, D. D. Dlott, and M. D. Fayer, *Phys. Rev. Lett.* **41**, 131 (1978).
98. T. S. Rose, W. L. Wilson, G. Wäckerle, and M. D. Fayer, *J. Chem. Phys.* **86**, 5370 (1987).
99. Y. Cui, M. Zhao, G. S. He, and P. N. Parsad, *J. Phys. Chem.* **95**, 6842 (1991).
100. J. T. Fourkas and M. D. Fayer, *Acc. Chem. Res.* **25**, 227 (1992).
101. R. J. Collier, C. B. Burkhardt, and L. H. Lin, *Optical Holography* (Academic Press, New York, 1971).
102. L. Solyman and D. J. Cooke, *Volume Holography and Volume Gratings* (Academic Press, London, 1981).
103. D. B. Brayton, *Appl. Opt.* **13**, 2346 (1974).
104. A. E. Siegman, *J. Opt. Soc. Am.* **67**, 545 (1977).
105. R. W. Hellwarth, *Prog. Quant. Electr.* **5**, 1 (1977).
106. D. G. Steel, R. C. Lind, J. F. Lam, and C. R. Giuliano, *Appl. Phys. Lett.* **35**, 376 (1979).
107. M. Ducloy and D. Bloch, *Phys. Rev. A* **30**, 3107 (1984).
108. T. K. Yee and T. K. Gustafson, *Phys. Rev. A* **18**, 1597 (1978).

109. J. P. Uyemura, *IEEE J. Quantum Electron.* **13**, 472 (1980).
110. Y. Prior, *IEEE J. Quantum Electron.* **QE–20**, 37 (1984).
111. P. Ye and Y. R. Shen, *Phys. Rev. A* **25**, 2183 (1982).
112. S. Stenholm, *Foundations of Laser Spectroscopy* (John Wiley and Sons, New York, 1984).
113. J. L. Oudar and Y. R. Shen, *Phys. Rev. A* **22**, 1141 (1980).
114. D. A. McQuarrie, *Statistical Mechanics* (Harper and Row, New York, 1976).
115. P. R. Berman, *Phys. Rev. A* **5**, 927 (1972).
116. S. Williams, R. N. Zare, and L. A. Rahn, *J. Chem. Phys.* **101**, 1072 (1994).
117. Consider a number state for the quantized radiation field, $|n_k\rangle$, which specifies the number of photons occupying the electromagnetic cavity mode of wavevector $\boldsymbol{k}$. The corresponding creation ($a_k^\dagger$) and annihilation (a_k) operators describe interactions leading to the formation and destruction of one quantum of energy $\hbar\omega_k$ ($\omega_k = c|\boldsymbol{k}|$) according to: $a_k^\dagger|n_k\rangle = |a_k^\dagger n_k\rangle = \sqrt{n_k+1}\,|n_k+1\rangle$ and $a_k|n_k\rangle = |a_k n_k\rangle = \sqrt{n_k}\,|n_k-1\rangle$. Thus, when acting on the ket $|n_k\rangle$, $a_k^\dagger$ and a_k can be associated with the processes of photon emission and photon absorption, respectively. These roles are reversed, however, when considering the action of creation and annihilation operators on the dual bra $\langle n_k|$ since Hermitian Conjugation [with $(a_k^\dagger)^\dagger = a_k$] yields: $\langle n_k|a_k = \langle a_k^\dagger n_k| = \sqrt{n_k+1}\,\langle n_k+1|$ and $\langle n_k|a_k^\dagger = \langle a_k n_k| = \sqrt{n_k}\,\langle n_k-1|$.
118. M. Weissbluth, *Atoms and Molecules* (Academic Press, San Diego, 1978).
119. T. A. W. Wasserman, B. R. Johnson, and P. H. Vaccaro, *(work in progress).*
120. This description is not rigorously correct for a stationary target medium since the middle denominator factor in each susceptibility term of Eq. (4.28) has the form $(\Omega_{\beta\alpha} - \omega_f + \omega_p)^{-1} = \Omega_{\beta\alpha}^{-1}$. However, when motional effects are taken into account, ω_f is no longer generally equal to ω_p and the complex Half–Lorentzian form is obtained.
121. S. A. J. Druet, B. Attal, T. K. Gustafson, and J. P. Taran, *Phys. Rev. A* **18**, 1529 (1978).
122. S. A. J. Druet and J. P. Taran, *Prog. Quant. Electr.* **7**, 1 (1981).
123. R. N. Zare, *Angular Momentum: Understanding Spatial Aspects in Chemistry and Physics* (John Wiley and Sons, New York, 1988).
124. M. Ducloy and D. Bloch, *J. Physique* **42**, 711 (1981).
125. P. H. Paul, R. L. Farrow, and P. M. Danehy, *J. Opt. Soc. Am. B* **12**, 384 (1995).
126. A. K. Popov and V. M. Shalaev, *Appl. Phys.* **21**, 93 (1980).
127. J. F. Lam, *Opt. Eng.* **21**, 219 (1982).
128. Q. Zhang, P. Dupré, T. Müller, and P. H. Vaccaro, *Linear and Nonlinear Optical Probes of S_2O Dynamics*, D. M. Neumark, Ed., 1995 Conference on the Dynamics of Molecular Collisions, Asilomar, CA (1995) p. PB107.
129. B. D. Fried, *Plasma Dispersion Function: The Hilbert Transform of the Gaussian* (Academic Press, New York, 1961).
130. S. M. Wandzura, *Opt. Lett.* **4**, 208 (1979).
131. D. Bloch, R. K. Raj, K. S. Peng, and M. Ducloy, *Phys. Rev. Lett.* **49**, 719 (1982).
132. P. Verkerk, M. Pinard, and G. Grynberg, *Phys. Rev. A* **34**, 4008 (1986).
133. M. Pinard, P. Verkerk, and G. Grynberg, *Phys. Rev. A* **35**, 4679 (1987).
134. M. S. Brown, L. A. Rahn, and R. P. Lucht, *Appl. Opt.* **34**, 3274 (1995).
135. R. P. Lucht, R. Trebino, and L. A. Rahn, *Phys. Rev. A* **45**, 8209 (1992).
136. R. P. Lucht, R. L. Farrow, and D. J. Rakestraw, *J. Opt. Soc. Am. B* **10**, 1508 (1993).
137. P. M. Danehy, E. J. Friedman–Hill, R. P. Lucht, and R. L. Farrow, *Appl. Phys. B.* **57**, 243 (1993).
138. R. L. Abrams and R. C. Lind, *Opt. Lett.* **2**, 94 (1978).
139. R. L. Abrams and R. C. Lind, *Opt. Lett.* **3**, 205 (1978).
140. R. L. Farrow, T. Dreier, and D. J. Rakestraw, *J. Opt. Soc. Am. B* **9**, 1770 (1992).
141. G. Alber, J. Cooper, and P. Ewart, *Phys. Rev. A* **31**, 2344 (1985).
142. J. Cooper, A. Charlton, D. R. Meacher, P. Ewart, and G. Alber, *Phys. Rev. A* **40**, 5705 (1989).
143. M. Kaczmarek, D. R. Meacher, and P. Ewart, *J. Mod. Opt.* **37**, 1561 (1990).
144. M. Sargent III, M. O. Scully, and W. E. Lamb Jr., *Laser Physics* (Addison–Wesley Publishing, London, 1974).
145. P. G. Pappas, M. M. Burns, D. D. Hinshelwood, M. S. Feld, and D. E. Murnick, *Phys. Rev. A* **21**, 1955 (1980).
146. M. S. Feld, M. M. Burns, T. U. Kühl, P. G. Pappas, and D. E. Murnick, *Opt. Lett.* **5**, 79 (1980).
147. M. Ducloy and D. Bloch, *J. Opt. Soc. Am.* **73**, 635 (1983).
148. M. Ducloy and D. Bloch, *J. Opt. Soc. Am.* **73**, 1844 (1983).
149. M. Pinard, B. Kleinmann, G. Grynberg, D. Bloch, and M. Ducloy, *J. Physique* **46**, 149 (1985).
150. G. Grynberg, M. Pinard, and P. Verkerk, *J. Physique* **47**, 617 (1986).
151. G. P. Agrawal, A. Van Lerberghe, P. Aubourg, and J. L. Boulnois, *Opt. Lett.* **7**, 540 (1982).
152. G. J. Dunning and D. G. Steel, *IEEE J. Quantum Electron.* **18**, 3 (1982).
153. D. G. Steel, R. C. Lind, and J. F. Lam, *Phys. Rev. A* **23**, 2513 (1981).
154. W. P. Brown, *J. Opt. Soc. Am.* **73**, 629 (1983).
155. T.–Y. Fu and M. Sargent, *Opt. Lett.* **4**, 366 (1979).
156. J. Nilsen and A. Yariv, *J. Opt. Soc. Am.* **71**, 180 (1981).
157. S. Williams, R. N. Zare, and L. A. Rahn, *J. Chem. Phys.* **101**, 1093 (1994).
158. L. Mandel and E. Wolf, *Optical Coherence and Quantum Optics* (Cambridge University Press, Cambridge, 1995).
159. S. Gerstenkorn and P. Luc, *Rev. Phys. Appl.* **14**, 791 (1979).
160. D. M. Lubman, C. T. Rettner, and R. N. Zare, *J. Phys. Chem.* **86**, 1129 (1982).
161. Z. Bacic and R. E. Miller, *J. Phys. Chem.* **100**, 12945 (1996).
162. R. E. Smalley, L. Wharton, and D. H. Levy, *Acc. Chem. Res.* **10**, 139 (1977).
163. A. R. Skinner and D. W. Chandler, *Am. J. Phys.* **48**, 8 (1980).
164. T. J. Butenhoff and E. A. Rohlfing, *J. Chem. Phys.* **98**, 5469 (1993).
165. C. M. Lovejoy, M. D. Schuder, and D. J. Nesbitt, *J. Chem. Phys.* **85**, 4890 (1986).

166. C. M. Lovejoy and D. J. Nesbitt, *J. Chem. Phys.* **86**, 3151 (1987).
167. C. M. Lovejoy and D. J. Nesbitt, *Rev. Sci. Instum.* **58**, 807 (1987).
168. S. W. Sharpe, R. Sheeks, C. Wittig, and R. A. Beaudet, *Chem. Phys. Lett.* **151**, 267 (1988).
169. S. W. Sharpe, Y. P. Zeng, C. Wittig, and R. A. Beaudet, *J. Chem. Phys.* **92**, 943 (1990).
170. S. W. Sharpe, D. Reifschneider, C. Wittig, and R. A. Beaudet, *J. Chem. Phys.* **94**, 233 (1991).
171. M. A. Suhm, J. T. Farrell Jr., A. McIlroy, and D. J. Nesbitt, *J. Chem. Phys.* **97**, 5341 (1992).
172. M. D. Schuder and D. J. Nesbitt, *J. Chem. Phys.* **100**, 7250 (1994).
173. C. Jungen, D. N. Malm, and A. J. Merer, *Can. J. Phys.* **51**, 1471 (1973).
174. T. Müller and P. H. Vaccaro, *Chem. Phys. Lett.* **266**, 575 (1997).
175. D. J. Meschi and R. J. Myers, *J. Mol. Spectrosc.* **3**, 405 (1959).
176. R. L. Cook, *J. Mol. Spectrosc.* **46**, 276 (1973).
177. E. Tiemann, J. Hoeft, F. J. Lovas, and D. R. Johnson, *J. Chem. Phys.* **60**, 5000 (1974).
178. J. Lindenmayer, *J. Mol. Spectrosc.* **116**, 315 (1986).
179. J. Lindenmayer, H. D. Rudolph, and H. Jones, *J. Mol. Spectosc.* **119**, 56 (1986).
180. J. N. Murrell, W. Craven, M. Vincent, and Z. H. Zhu, *Mol. Phys.* **56**, 839 (1985).
181. R. O. Jones, *Chem. Phys. Lett.* **125**, 221 (1986).
182. T. Fueno and R. J. Buenker, *Theor. Chim. Acta* **73**, 123 (1988).
183. P. W. Schenk and R. Steudel, in *Inorganic Sulphur Chemistry* G. Nickless, Ed. (Elsevier Publishing, Amsterdam, 1968) p. 10.
184. A. R. Vasudeva–Muthy, N. Kutty, and D. K. Sharma, *Int. J. Sulfur Chem. B.* **6**, 161 (1971).
185. W. R. Gentry, in *Atomic and Molecular Beam Methods: Volume I* G. Scoles, Ed. (Oxford University Press, New York, 1988), vol. 1, p. 54.
186. Q. Zhang, P. Dupré, B. Grzybowski, and P. H. Vaccaro, *J. Chem. Phys.* **103**, 67 (1995).
187. E. M. Fluder and J. R. de la Vega, *J. Am. Chem. Soc.* **100**, 5265 (1978).
188. J. R. de la Vega, *Acc. Chem. Res.* **15**, 185 (1982).
189. J. Bicerano, H. F. Schaefer III, and W. H. Miller, *J. Am. Chem. Soc.* **105**, 2550 (1983).
190. T. Carrington and W. H. Miller, *J. Chem. Phys.* **81**, 3942 (1984).
191. T. Carrington and W. H. Miller, *J. Chem. Phys.* **84**, 4364 (1986).
192. J. S. Hutchinson, *J. Phys. Chem.* **91**, 4495 (1987).
193. N. Shida, P. F. Barbara, and J. E. Almlöf, *J. Chem. Phys.* **91**, 4061 (1989).
194. T. D. Sewell and D. L. Thompson, *Chem. Phys. Lett.* **193**, 347 (1992).
195. Y. Guo, T. D. Sewell, and D. L. Thompson, *Chem. Phys. Lett.* **224**, 470 (1994).
196. W. F. Rowe, R. W. Duerst, and E. B. Wilson, *J. Am. Chem. Soc.* **98**, 4021 (1976).
197. S. L. Baughcum, R. W. Duerst, W. F. Rowe, Z. Smith, and E. B. Wilson, *J. Am. Chem. Soc.* **103**, 6296 (1981).
198. Z. Smith, E. B. Wilson, and R. W. Duerst, *Spectrochim. Acta.* **39A**, 1117 (1983).
199. S. L. Baughcum, Z. Smith, E. B. Wilson, and R. W. Duerst, *J. Am. Chem. Soc.* **106**, 2260 (1984).
200. P. Turner, S. L. Baughcum, S. L. Coy, and Z. Smith, *J. Am. Chem. Soc.* **106**, 2265 (1984).
201. D. W. Firth, P. F. Barbara, and H. P. Trommsdorff, *Chem. Phys.* **136**, 349 (1989).
202. D. W. Firth, K. Beyer, M. A. Dvorak, S. W. Reeve, A. Grushow, and K. R. Leopold, *J. Chem. Phys.* **94**, 1812 (1991).
203. T. Chiavassa, P. Roubin, L. Pizzala, P. Verlaque, A. Allouche, and F. Marinelli, *J. Phys. Chem.* **96**, 10659 (1992).
204. T. Chiavassa, P. Verlaque, L. Pizzala, A. Allouche, and P. Roubin, *J. Phys. Chem.* **97**, 5917 (1993).
205. C. J. Seliskar and R. E. Hoffman, *J. Mol. Spectrosc.* **96**, 146 (1982).
206. P. F. Barbara, P. K. Walsh, and L. E. Brus, *J. Phys. Chem.* **93**, 29 (1989).
207. H. Ozeki, M. Takahashi, K. Okuyama, and K. Kimura, *J. Chem. Phys.* **99**, 56 (1993).
208. C. J. Seliskar and R. E. Hoffman, *Chem. Phys. Lett.* **43**, 481 (1976).
209. C. J. Seliskar and R. E. Hoffman, *J. Am. Chem. Soc.* **99**, 7072 (1977).
210. C. J. Seliskar and R. E. Hoffman, *J. Mol. Spectrosc.* **88**, 30 (1981).
211. K. Luth and S. Scheiner, *J. Phys. Chem.* **98**, 3582 (1994).
212. D. G. Steel and J. F. Lam, *Phys. Rev. Lett.* **43**, 1588 (1979).
213. A. Dreizler, T. Dreier, and J. Wolfrum, *Chem. Phys. Lett.* **233**, 525 (1995).
214. G. J. Germann, A. McIlroy, T. Drier, R. L. Farrow, and D. J. Rakestraw, *Ber. Bunsenges. Phys. Chem.* **97**, 1630 (1993).
215. M. A. Buntine, D. W. Chandler, and C. C. Hayden, *J. Chem. Phys.* **102**, 2718 (1995).
216. J. A. Booze, D. E. Govoni, and F. F. Crim, *J. Chem. Phys.* **103**, 10484 (1995).
217. D. E. Govoni, J. A. Booze, A. Sinha, and F. F. Crim, *Chem. Phys. Lett.* **216**, 525 (1993).
218. R. Schinke, *Photodissociation Dynamics: Spectroscopy and Fragmentation of Small Polyatomic Molecules* (Cambridge University Press, Cambridge, 1993).
219. L. J. Butler and D. M. Neumark, *J. Phys. Chem.* **100**, 12801 (1996).
220. B. R. Johnson, C. Kittrell, P. B. Kelly, and J. L. Kinsey, *J. Phys. Chem.* **100**, 7743 (1996).
221. G. E. Hall, N. Sivakumar, R. Ogorzalek, G. Chawla, H. P. Haerri, P. L. Houston, I. Burak, and J. W. Hepburn, *Faraday Discuss. Chem. Soc.* **82**, 13 (1986).
222. P. Houston, *J. Phys. Chem.* **91**, 5388 (1987).
223. G. E. Hall and P. L. Houston, *Ann. Rev. Phys. Chem.* **40**, 375 (1989).
224. P. Houston, *Acc. Chem. Res.* **22**, 309 (1989).
225. A. H. Zewail, *J. Phys. Chem.* **100**, 12701 (1996).
226. J. Zoval, D. Imre, P. Ashjian, and V. A. Apkarian, *Chem. Phys. Lett.* **197**, 549 (1992).
227. W. Ubachs and J. J. ter Meulen, *J. Chem. Phys.* **92**, 2121 (1990).
228. B. R. Weiner, H. B. Levene, J. J. Valentini, and A. P. Baronavski, *J. Chem. Phys.* **90**, 1403 (1989).

229. H. Okabe, *Photochemistry of Small Molecules* (John Wiley & Sons, New York, 1978).
230. G. N. A. Van Veen, K. A. Mohamed, T. Baller, and A. E. De Vries, *Chem. Phys.* **74**, 261 (1983).
231. X. Xie, L. Schneider, H. Wallmeier, R. Boettner, K. H. Welge, and M. N. R. Ashfold, *J. Chem. Phys.* **92**, 1608 (1990).
232. Z. Xu, B. Koplitz, and C. Wittig, *J. Chem. Phys.* **87**, 1062 (1987).
233. W. G. Hawkins and P. L. Houston, *J. Chem. Phys.* **73**, 297 (1980).
234. W. G. Hawkins and P. L. Houston, *J. Chem. Phys.* **76**, 729 (1982).
235. R. Schinke, *Ann. Rev. Phys. Chem.* **39**, 39 (1988).
236. R. N. Zare, *Mol. Photochem.* **4**, 1 (1972).
237. J. P. Simons, *J. Phys. Chem.* **88**, 1287 (1984).
238. J. P. Simons, *J. Phys. Chem.* **91**, 5378 (1987).
239. A. J. Orr–Ewing and R. N. Zare, *Ann. Rev. Phys. Chem.* **45**, 315 (1994).
240. C. H. Greene and R. N. Zare, *Ann. Rev. Phys. Chem.* **33**, 119 (1982).
241. C. H. Greene and R. N. Zare, *J. Chem. Phys.* **78**, 6741 (1983).
242. S. Klee, K. H. Gericke, and F. J. Comes, *J. Chem. Phys.* **85**, 40 (1986).
243. K. H. Gericke, S. Klee, and F. J. Comes, *J. Chem. Phys.* **85**, 4463 (1986).
244. R. N. Dixon, *J. Chem. Phys.* **85**, 1866 (1986).
245. M. P. Docker, A. Hodgson, and J. P. Simons, *Faraday Discuss. Chem. Soc.* **82**, 25 (1986).
246. S. Klee, K. H. Gericke, and F. J. Comes, *Ber. Bunsenges. Phys. Chem.* **92**, 429 (1988).
247. F. J. Comes, K. H. Gericke, A. U. Grunewald, and S. Klee, *Ber. Bunsenges. Phys. Chem.* **92**, 273 (1988).
248. R. B. Williams, P. Ewart, and A. Dreizler, *Opt. Lett.* **19**, 1486 (1994).
249. P. M. Danehy and R. L. Farrow, *Appl. Phys. B* **62**, 407 (1996).
250. R. E. Teets, F. V. Kowalski, W. T. Hill, N. Carlson, and T. W. Hänsch, *SPIE* **113**, 88 (1977).
251. P. H. Vaccaro, Ph.D., Massachusetts Institute of Technology (1986).
252. T. S. Rose, W. L. Wilson, G. Wäckerle, and M. D. Fayer, *J. Phys. Chem.* **91**, 1704 (1987).
253. T. S. Rose, V. J. Newell, J. S. Meth, and M. D. Fayer, *Chem. Phys. Lett.* **145**, 475 (1988).
254. T. R. Brewer, J. T. Fourkas, and M. D. Fayer, *Chem. Phys. Lett.* **203**, 344 (1993).
255. R. Trebino and A. E. Siegman, *Opt. Comm.* **56**, 297 (1985).
256. D. M. Pepper and A. Yariv, in *Optical Phase Conjugation* R. A. Fisher, Ed. (Academic Press, San Diego, 1983) p. 23.
257. P. Ewart, P. Snowdon, and I. Magnusson, *Opt. Lett.* **14**, 563 (1989).
258. D. J. Rakestraw, R. L. Farrow, and T. Dreier, *Opt. Lett.* **15**, 709 (1990).
259. K. Nyholm, R. Fritzon, and M. Aldén, *Appl. Phys. B* **59**, 37 (1994).

5 Multiphoton Ionization (MPI) and Resonance Enhanced Multiphoton Ionization (REMPI) Spectroscopy

Michael N.R. Ashfold
School of Chemistry, University of Bristol, Bristol BS8 1TS, U.K.

5.1 Introduction

Multiphoton ionization (MPI), as the name implies, involves the ionization of a molecule brought about by the absorption of more than one photon, *i.e.* in the case of non-dissociative ionization:

$$M + nh\nu \rightarrow M^{+} + e^{-}, \tag{5.1}$$

where M and M^+ are, respectively, the parent neutral and ion of interest, and n is the number of photons (of frequency) required for the ionization process. This Chapter opens with a survey of some of the many variants of the 'multiphoton ionization spectroscopy' experiment, and the kinds of information that may be gleaned from each. Almost all such experiments make use of the concept of *resonance enhanced* multiphoton ionization (REMPI), and almost all are devoted to gas phase problems. Some of the many spectroscopic opportunities afforded by the technique are illustrated in Fig. 5.1.

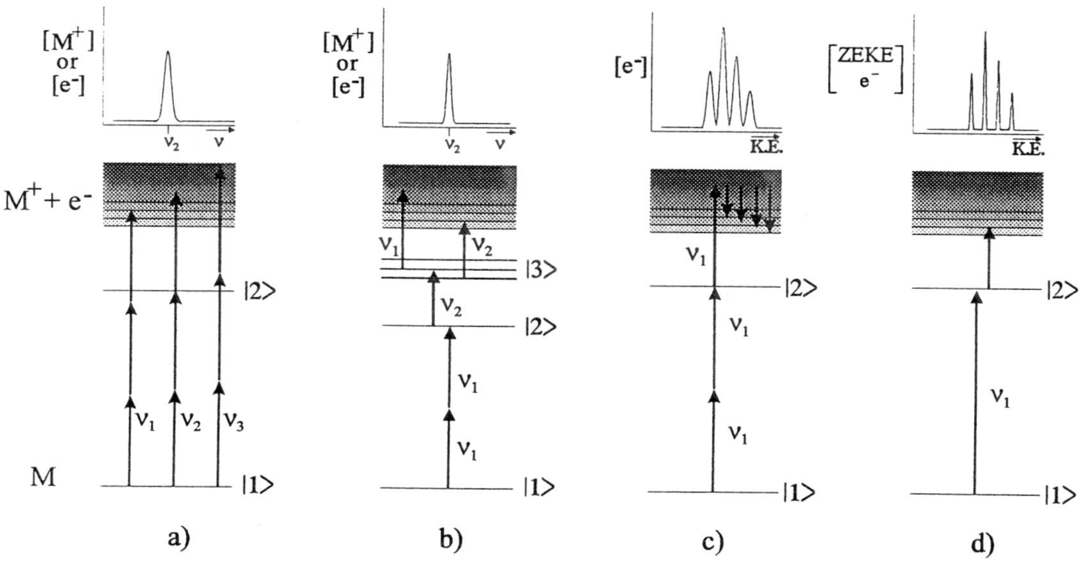

Fig. 5.1 Four possible 'multiphoton' excitation schemes above each of which are shown the type of spectrum that might result; a) conventional one colour MPI resonance enhanced at the two photon energy; b) a double resonance variant of REMPI; ν_1 is fixed so as to excite the |2> ← |1> two photon transition and ν_2 is scanned to probe the spectroscopy of the |3> ← |2> transition; c) one colour REMPI-PES; ν_1 is fixed so as to excite the |2> ← |1> two photon transition and the kinetic energies of the resulting photoelectrons are analysed; d) two colour ZEKE-PES; in this illustration ν_1 is fixed so as to excite |2> ← |1> via a one photon transition and ν_2 is scanned to reveal the various ionization thresholds.

The first panel - Fig. 5.1(a) - outlines one of the simplest (and most practiced) types of REMPI process. The energetics in this particular example are such that the molecule M needs to absorb a minimum of three photons in order for it to be excited from its ground state |1> to an energy above its first ionization limit. The probability of such a three photon excitation can be considerably enhanced if there is a real excited state of M resonant at the energy of one or two absorbed photons. The particular case shown in Fig. 5.1(a) assumes the existence of an excited state |2> resonant at the energy $2h\nu_2$, *i.e.* this is an example of a so called 2+1 REMPI process. Resonance is a necessary but not sufficient condition for state |2> to cause an enhanced MPI probability following excitation at frequency ν_2. We also require that the |2> ← |1> two photon excitation has a non-zero transition probability and that the state |2> be sufficiently long lived that the subsequent one photon ionization step has reasonable probability. If these conditions are met, then the spectrum obtained by measuring the ion yield (or the yield of the accompanying photoelectrons) as a function of excitation wavelength will provide a signature of the |2> ← |1> two photon transition of the neutral M; analysis can provide structural (and, in some cases, dynamical) information about the excited state |2>.

Just as in more traditional one photon spectroscopies, symmetry is crucial in determining the magnitude of multiphoton transition probabilities [1-5]. A multiphoton transition between two states |1> and |2> is 'allowed' if the transition moment $<2|T_q^k(\hat{O})|1>$ is non-zero; *i.e.* if the direct product of the irreducible representations for the wavefunctions of states |1> and |2> and that of $T_q^k(\hat{O})$ - the qth component of the spherical tensor of rank k representing the multiphoton transition operator Ô - contains the totally symmetric representation. Symmetry considerations ensure that only spherical tensors of either odd or even rank will contribute to any one colour multiphoton excitation. Thus, for example, whereas one photon electric dipole transitions must be carried by components of rank one, only components of rank k=0 and/or 2 can contribute to two photon transitions brought about using photons of identical frequency and polarization. The k=0 component (a scalar) can only contribute to a two photon transition connecting states of the same symmetry. Identification of k=0 components in two photon excitation spectra is generally rather straightforward since they are forbidden (and thus 'disappear') when the spectrum is recorded using circularly polarized light. Sensitivity to the polarization state of the exciting radiation is one important feature distinguishing one photon and multiphoton transitions. As Tables 5.1 and 5.2 show, for all but the least symmetric molecules, at least some of the $k \neq 1$ components will span representations different from (or additional to) those of the dipole moment operator. As a result, multiphoton excitations can provide a means of populating excited states via transitions 'forbidden' in traditional one photon absorption spectroscopy. Illustrative examples are considered in later sections of this Chapter.

Less restrictive selection rules are just one of several benefits that can result from the use of multiphoton excitation methods. Another is experimental convenience: A multiphoton excitation brought about using visible or near ultraviolet (UV) photons often provides a ready means of populating an excited state lying at energies which, in one photon absorption, would fall in the technically much more demanding vacuum ultraviolet (VUV) spectral region. Multiphoton excitation cross-sections are small; representative values for two and three photon absorption cross-sections are 10^{-50} cm^4 s^{-1} and 10^{-84} cm^6 s^{-2} respectively [2], *c.f.* a typical cross-section of *ca.* 10^{-17} cm^2 for an electric dipole allowed one photon excitation between two bound states of a molecule. High incident light intensities are thus necessary in order to compensate for these small multiphoton absorption cross-sections. For most spectroscopic applications sufficient intensities can usually be achieved by focussing the output of conventional (nanosecond pulse duration) tunable dye lasers. This points to two other potential benefits of using multiphoton excitation methods. Firstly, the interaction is concentrated in a localized volume - the focal volume - thus rendering the technique ideally matched for use with supersonic molecular beams; numerous hitherto impenetrable molecular spectra have been successfully interpreted after application of multiphoton excitation methods to jet-cooled samples of the molecule of interest. Secondly, given that high light intensities are necessary in

TABLE 5.1. Allowed changes in some of the more important quantum numbers and symmetry descriptions for atoms and molecules undergoing one-colour multiphoton transitions involving one (k=1), two (k=0 and 2) and three (k=1 and 3) photons.

quantum number / property of interest	rank of transition tensor k			
	0	1	2	3
(a) *atoms*				
orbital angular momentum, ℓ of electron being excited	$\Delta\ell=0$	$\Delta\ell=\pm1$	$\Delta\ell=0,\pm2$ (but s$\nleftrightarrow$s)	$\Delta\ell=\pm1,\pm3$ (but s $\nleftrightarrow$ p)
(b) *linear molecules [case (a)/(b)]*:				
axial projection of electronic orbital angular momentum, Λ	$\Delta\Lambda=0$	$\Delta\Lambda=0,\pm1$	$\Delta\Lambda=0,\pm1,\pm2$	$\Delta\Lambda=0,\cdots,\pm3$
linear molecules [case (c)]:				
axial projection of total electronic angular momentum, Ω	$\Delta\Omega=0$	$\Delta\Omega=0,\pm1$	$\Delta\Omega=0,\pm1,\pm2$	$\Delta\Omega=0,\cdots,\pm3$
(c) *centrosymmetric molecules*:				
inversion symmetry, u/g	u$\leftrightarrow$u g$\leftrightarrow$g	u$\leftrightarrow$g	u$\leftrightarrow$u g$\leftrightarrow$g	u$\leftrightarrow$g
(d) *atoms and molecules*:				
total angular momentum, J	$\Delta J=0$	$\Delta J=0,\pm1$ (but $J=0\nleftrightarrow J=0$)	$\Delta J=0,\pm1,\pm2$ (but $J=0\nleftrightarrow J=0,1$)	$\Delta J=0,\cdots,\pm3$ (but $J=0\nleftrightarrow J=0,1,2$; and $J=1\nleftrightarrow J=1$)
total parity, +/-	$+\leftrightarrow+$ $-\leftrightarrow-$	$+\leftrightarrow-$	$+\leftrightarrow+$ $-\leftrightarrow-$	$+\leftrightarrow-$
electron spin, S	$\Delta S=0$	$\Delta S=0$	$\Delta S=0$	$\Delta S=0$

TABLE 5.2. Representations of the spherical tensor components $T_q^k(\hat{O})$ of the one-colour, n-photon (n=1-3) transition operator.

number of photons, n	k	q	$D_{\infty h}$ [a]	D_{6h}	D_{3h} [b]
1	1	0	Σ_u^+	A_{2u}	A_2''
	1	±1	Π_u	E_{1u}	E'
2	0	0	Σ_g^+	A_{1g}	A_1'
	2	0	Σ_g^+	A_{1g}	A_1'
	2	±1	Π_g	E_{1g}	E''
	2	±2	Δ_g	E_{2g}	E'
3	1	0	Σ_u^+	A_{2u}	A_2''
	1	±1	Π_u	E_{1u}	E'
	3	0	Σ_u^+	A_{2u}	A_2''
	3	±1	Π_u	E_{1u}	E'
	3	±2	Δ_u	E_{2u}	E''
	3	±3	Φ_u	$B_{1u} + B_{2u}$	$A_1' + A_2'$

[a]Assuming Hund's case (a) or (b) coupling. Ignore u/g labels for non-centrosymmetric linear molecules.
[b]A_1' and A_2'' reduce to A_1, A_2' becomes A_2, and E' and E'' both transform as E in C_{3v} molecules.

order to drive the improbable |2> ← |1> (multiphoton) excitation step, it is almost inevitable that some of the molecules excited to state |2> will absorb one or more further photons and ionize. This can be a huge benefit since the resulting particles are charged and can be collected with far higher efficiency than could, for example, laser induced fluorescence (LIF) from the excited state |2>. This benefit manifests itself not just in high *sensitivity*, but also offers an additional dimension when it comes to species *selectivity* allowing, for example, mass analysis of the resulting M^+ ions or kinetic energy analysis of the accompanying photoelectrons.

As we have already seen both in Chapter 2 and in the latter part of Chapter 4, double resonance techniques can also provide a powerful means of spectral simplification. Figure 5.1(b) illustrates just one example of a two colour doubly resonance enhanced MPI process. In this particular case, one level (or set of levels) associated with excited state |2> is populated via a coherent two photon excitation; that portion of the |3> ← |2> excitation spectrum that originates from the populated level(s) is then observed through the wavelength dependent resonance enhancement it provides to what becomes an overall four photon ionization (*i.e.* a two colour 2+1+1 doubly resonance enhanced MPI process in this case). Precisely this excitation scheme has been used, for example, to characterise highly excited *ungerade* Rydberg states of H_2 following an initial two photon excitation to the much studied double minimum E, F $^1\Sigma_g^+$ state [6-8], and to identify Franck-Condon disfavoured vibronic levels of the $\tilde{B}$ and $\tilde{C}'$ states of ammonia following two photon excitation via its first excited ($\tilde{A}^1A_2''$) electronic state [9,10]. Numerous variants of this scheme can be envisaged. Many have been demonstrated, and many of these are illustrated by the examples cited in Chapter 2. Here we highlight just a couple of examples, which complement the studies mentioned above. Two colour 1+1+1 double resonance enhanced MPI, in which the first (VUV) photon populates selected rovibronic levels of the $B^1\Sigma_u^+$ excited state, has been used to characterise series of *gerade* Rydberg states of H_2 [11]. Other notable double resonance MPI experiments involving the ammonia molecule include (i) further investigations of the pattern of vibronic levels associated with the $\tilde{B}$ and $\tilde{C}'$ Rydberg states using a 1+2+1 double resonance enhanced MPI scheme in which the first photon (in the infrared) was used to 'tag' one (or a few) selected rovibronic levels of the electronic ground state [12] and (ii) studies of the predissociation broadened linewidths of individual rovibronic levels of the $\tilde{A}^1A_2''$ state observed via stimulated emission pumping (SEP) from quantum state selected levels of the $\tilde{C}'$ electronic state (*i.e.* a folded variant of the double resonance excitation scheme depicted in Fig. 5.1(b)) [13,14].

Up to this point we have paid scant attention to the photoelectrons formed in the REMPI process, and the information they may carry. Clearly, it is possible to measure their kinetic energies, as in conventional photoelectron spectroscopy (PES), and thereby derive information about the energy levels of the partner cation (Fig. 5.1(c)). One distinction should be noted: in all cases involving a bound excited state |2>, the REMPI–PES experiment [15-19] will yield the photoelectron spectrum of the resonant intermediate state, not of the ground state as in the more traditional one photon (*e.g.* He I) photoelectron spectroscopy. This offers several advantages, including much improved energy resolution and, in favourable cases, the possibility of forming cations in well defined, user selected, internal quantum states.

The reasons for these many benefits are fairly obvious. REMPI spectroscopy tends to discriminate against short lived, dissociative excited states and in favour of the more long lived resonant intermediate states - *i.e.* those excited states for which further photon absorption (leading to ionization) is a significant mode of population loss. Most of the excited states in this latter category tend to be Rydberg states. Molecular Rydberg states are conveniently (though often rather approximately) pictured in terms of a positive ion core – consisting of the nuclei and all the valence electrons – plus a non-bonding Rydberg electron in a large, spatially diffuse orbital. The Rydberg states can be arranged in series, each of which is characterised by a particular value of quantum defect, and each of which converges to the ionization limit appropriate to its ion core. For any one Rydberg state, its term value, $\tilde{\nu}$, and its quantum defect, δ, will be related via the following version of the Rydberg formula [20]

$$\tilde{\nu} = E_i - R/(n - \delta)^2, \quad (5.2)$$

where E_i is the ionization energy for forming the appropriate ion core and R is the suitably mass corrected value of the Rydberg constant. Obviously, as written, $\tilde{\nu}$, E_i, and R must share common units, *e.g.* cm^{-1}. The magnitude of δ provides an indication of the extent to which the wavefunction of the Rydberg electron penetrates into the core region; typical values for molecules composed entirely of first row atoms (*i.e.* for molecules whose lowest energy Rydberg orbitals will have principal quantum number n=3) are $\delta(ns)$ = 1.0-1.5, $\delta(np)$ = 0.4-0.8 whilst for nd and all higher ℓ functions δ is normally *ca.* 0. Such qualitative ideas have proved to be very useful in interpreting the patterns of excited states observed in many families of polyatomic molecules [20], though modifications due to configuration interaction (*i.e.* mixing between zero-order states sharing a common symmetry species but arising from different electronic configurations) can complicate such simple expectations.

Given this simple picture, we should expect the core configuration of any unperturbed Rydberg state to be the same as that of the ionic state that lies at the convergence limit of the series to which it belongs. The Rydberg states and the ion would therefore be expected to have very similar geometries and thus, given the Franck-Condon principle, we might further anticipate that the final ionizing step in a REMPI process via such a Rydberg state would be diagonal - *i.e.* involve a v=0 transition - leading to selective formation of ions with the same vibrational quantum number(s) as in the resonance enhancing intermediate state. Since the photoionization is brought about using some known integer number of photons, the photoelectrons accompanying such state specific ion formation will necessarily have a narrow range of kinetic energies. In favourable cases the spectrum of these photoelectron kinetic energies can be resolved (generally by time-of-flight methods) to the extent that individual rovibrational states of the ion are revealed. This explains the continuing appeal of REMPI–PES as a means of determining accurate ionization thresholds and of investigating photoionization dynamics in simple molecular systems [15-19,21].

The last few years have witnessed the emergence of another form of photoelectron spectroscopy which, typically, offers a further one (or more) order of magnitude improvement in energy resolution (*i.e.* cm^{-1} resolution). This technique is now generally referred to as zero kinetic energy (ZEKE) photoelectron spectroscopy [22,23]. Chapter 7 contains numerous examples which serve to illustrate the many attributes of this technique so here, just for completeness, we limit discussion to a few sentences outlining the essentials of the method. In conventional photoelectron spectroscopy, and in REMPI-PES, one learns about the quantum states of the ion by measuring the photoelectron kinetic energies as accurately as possible. ZEKE-PES also reveals the energy eigenstates of the cation, but is based on a different philosophy. The principle of the method is illustrated in Fig. 5.1(d). In the particular double resonant variant shown, one laser is tuned so as to populate a (known) excited state |2> of the species of interest. A second laser pulse is then used to excite this population to the energetic threshold for forming one of the allowed quantum states of the ion. Any energetic electrons (*e.g.* those formed via an autoionization process) will quickly recoil from the interaction region. The ZEKE (threshold) electrons can be detected by application of a suitably delayed pulsed extraction field. A ZEKE-PES spectrum is obtained by measuring the multiphoton (or, if tunable VUV radiation is available, the one photon) excitation spectrum for forming photoelectrons with zero kinetic energy; precise energy eigenvalues are obtained because the spectral resolution is determined, ultimately, by the bandwidth of the exciting laser. It is now recognised that such a description, whilst appealing, actually oversimplifies the physics. The 'ZEKE' electrons detected in an experiment as described actually derive from pulsed field ionization of very high Rydberg states belonging to series converging to the threshold of interest [22,23]. As a result, the ionization thresholds determined via this type of experiment will all be subject to a small, systematic shift to low energy. However, as discussed more fully in Chapter 7, the magnitude of this shift scales with the applied extraction voltage so the true thresholds can be recovered by recording such spectra using a number of different pulsed extraction voltages and extrapolating the observed line frequencies to zero applied field.

5.2 Experimental Techniques

The 'basic' MPI experiment is simplicity itself. Figure 5.2 shows a typical experimental arrangement. Tunable dye laser light is focussed into a cell equipped with two electrodes containing a low pressure (typically a few Torr) of the molecular gas of interest. The electrodes are biased and an MPI spectrum recorded simply by measuring the total ion, or the total photoelectron, yield as a function of excitation wavelength. As we have already commented, any structure observed reflects the resonance enhancements provided by the various rovibrational levels of the resonant intermediate electronic state(s) of the neutral, and may be analyzed to provide spectroscopic (and thus structural) information about the excited neutral molecule. Of course, this 'basic' form of the experiment has several serious limitations and much of the experimental effort in the last decade has gone into improving both the selectivity and sensitivity of the technique.

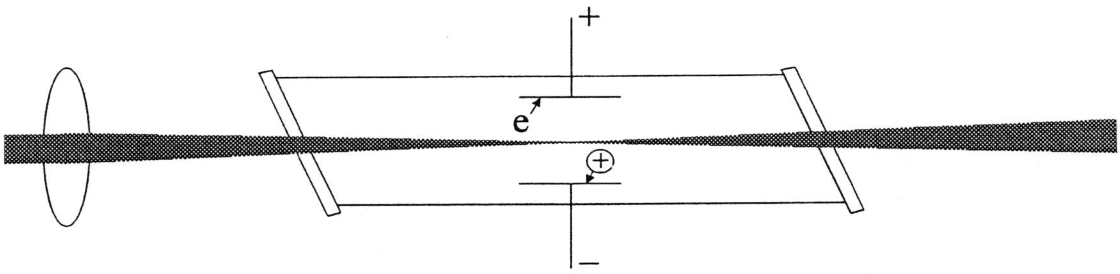

Fig. 5.2 Essential features of the basic MPI experiment. A laser beam is focussed into the gaseous sample of interest between two biased electrodes, and the resulting ion (or electron) yield measured.

One obvious deficiency of the basic experiment as described is the fact that all ions will be measured, irrespective of their masses, or that all photoelectrons will be counted, irrespective of their kinetic energies. Thus, for example, an MPI spectrum recorded in such a static cell arrangement might well include superimposed features due not just to REMPI of the parent molecule of interest, but also to REMPI of neutral fragments arising from unintentional photodissociation of the parent molecule. Such potential ambiguities can usually be resolved if one is able to mass resolve the resulting ions, and most present day REMPI experiments include just such a facility. Ion mass resolution has been achieved by performing the REMPI process in the source region of a suitably modified quadrupole mass spectrometer. More commonly, however, the required mass separation is achieved by time-of-flight (TOF) methods using either a linear TOF mass spectrometer (usually involving two acceleration stages configured so as to satisfy the Wiley-McLaren focussing conditions [24]), or a reflectron TOF mass spectrometer [25]. A variety of fast charged particle detectors (*e.g.* a channel electron multiplier or microchannel plates) can be used, together with suitable time gated signal processing electronics, to monitor the REMPI spectrum associated with formation of any single, user selected, ion mass. In this way it will usually be possible to distinguish spectral features associated with the parent from those arising from REMPI of neutral photofragments, or to distinguish different isotopomeric variants of the same parent. Note that all of these variants of mass resolved REMPI spectroscopy must, of necessity, be performed under collision free conditions. Thus in all such experiments the precursor of interest is introduced into the source region of the mass spectrometer in the form of a molecular beam. The rotational 'cooling' that accompanies any supersonic expansion can offer a further experimental advantage; namely, spectral simplification. This advantage can be particularly helpful in the case of REMPI spectroscopy since, as indicated in Table 5.1 and illustrated in more detail in Section 5.3, multiphoton excitation spectra tend to be rather complex, exhibiting up to five (in the case of a two photon resonant MPI spectrum) or seven (for a 3+1 REMPI spectrum) rotational branches per vibronic transition.

Often it proves advantageous to be able to measure the electron kinetic energies also. Several justifications for such measurements (*e.g.* accurate determination of ionization potentials, insight into the vibronic character of the resonance enhancing intermediate state of the neutral, or simply a more fundamental interest in the photoionization dynamics itself) were outlined in the Introduction, and will be explored further in the next Section. Additionally, such a measurement would normally suffice to distinguish whether a given daughter ion, B^+ for example, observed in the TOF spectrum of the ions resulting from REMPI of an AB parent, arises from photodissociation of the neutral parent followed by one (or more) photon ionization of the fragment, *i.e.*

$$AB + nh\nu \rightarrow A + B \qquad (5.3a)$$

$$B + nh\nu \rightarrow B^+, \qquad (5.3b)$$

from MPI followed by photodissociation of the resulting parent ion, *i.e.*

$$AB + nh\nu \rightarrow AB^+ \qquad (5.4a)$$

$$AB^+ + h\nu \rightarrow A + B^+, \qquad (5.4b)$$

or as a result of direct dissociative ionization of the parent, *i.e.*

$$AB + nh\nu \rightarrow (AB^+)^* \rightarrow A + B^+. \qquad (5.5)$$

Given the pulsed nature of the REMPI process, the electron kinetic energies are almost always measured by TOF methods, either using a conventional (mu metal shielded) TOF spectrometer [21] or a so called magnetic bottle photoelectron spectrometer [26,27] which offers the advantage of much higher collection efficiency with comparatively little loss of ultimate kinetic energy resolution.

5.3 Illustrative Examples

In the remainder of this Chapter we survey a few of the many applications of REMPI spectroscopy involving, firstly, diatomic molecules, and then progressively larger species – small polyatomics, then larger polyatomics, clusters and complexes – highlighting, in each case, the rich variety of structural and dynamical information that can be derived from such studies. For convenience, most of the small molecule examples are drawn from recent work involving the Bristol group.

5.3.1 Diatomic Molecules

Even a quick glance at the selection rules for electric dipole transitions summarised in Tables 5.1 and 5.2 should suffice to show that traditional one photon absorption and/or emission spectroscopies are unlikely to reveal the whole manifold of excited electronic states of a molecule. Almost inevitably, the application of *multi*photon excitation methods will reveal new spectroscopic information about any chosen target molecule. Nowhere is this more obvious than in the case of homonuclear diatomic molecules like H_2, N_2, O_2, and the halogens. These all have *gerade* electronic ground states; as a consequence, traditional one photon absorption spectroscopy has tended to discriminate in favour of their *ungerade* excited states. Two photon spectroscopy, in contrast, will excite $g \leftrightarrow g$ (and $u \leftrightarrow u$) transitions; a wealth of new spectroscopic data on the *gerade* excited states of each of the above homonuclear diatomics has been derived in just this way. The REMPI technique is finding increasing application outside the spectroscopic community. For example, MPI of H_2 and its various isotopomers, resonance enhanced at the two photon energy by levels of the E,F$^1\Sigma_g^+$ double minimum state, is being used to detect the H_2 products arising in photolyses (*e.g.* the vacuum ultraviolet photolysis of CH_4 [28]) and in simple bimolecular reactions, *e.g.* the H+H_2 [29,30] and H+HI [31] reactions. Similarly, two photon resonant MPI studies of $O_2(^1\Sigma_g)$ photofragments resulting from, for example, the near UV photolysis of O_3 have provided much new insight into the detailed dissociation dynamics of this very topical molecule [32,33].

We use recent studies of the NH [34-41] and SH [42-44] radicals to illustrate several other facets of multiphoton excitation methods in general, and the REMPI technique in particular. In both radicals the highest occupied molecular orbitals in the ground state are a pair of degenerate non-bonding orbitals (the barely perturbed $2p_x$ and $2p_y$ orbitals of atomic N, or the $3p_x$ and $3p_y$ orbitals in the case of S). Electric dipole allowed transitions in *atoms* generally satisfy the Laporte (parity) selection rule, $\Delta \ell = \pm 1$. Extending the atomic analogy, we might have expected that Rydberg states of NH and SH resulting from $ns \leftarrow \pi$ and $nd \leftarrow \pi$ electronic promotions (with, in the case of NH, $n \geq 3$) would have been identified by traditional one photon absorption spectroscopy. In fact, all of the available experimental data concerning Rydberg states of the NH radical has come from REMPI studies [34-41]. Table 1 reminds us that atomic *two* photon excitations should satisfy the selection rules $\Delta \ell = 0$ or ± 2. Consistent with this, most of the newly identified Rydberg states of NH (both singlets and triplets) have been assigned in terms of $3p \leftarrow 1\pi$ excitations [37-39]. Figure 5.3 shows the origin bands of three such transitions of NH – all originating from the metastable a$^1\Delta$ excited state, and all resonance enhanced at the two photon energy by, respectively, the f$^1\Pi$, g$^1\Delta$ and h$^1\Sigma^+$ Rydberg states. As anticipated in Table 1, O and S branches (transitions involving $\Delta J = -2$ and $+2$, respectively) are clearly evident in these two photon REMPI spectra; the relative intensities of the various branches and, of course, the J number associated with the first line in the various rotational branches, provide clear indications as to the various excited state symmetries. Careful inspection of the spectrum shown in fig. 5.3a reveals another noteworthy feature: all transitions involving levels with rotational quantum number J'=7 or 8 appear with anomalously low intensity. This has been rationalised [38] in terms of a very localised predissociation of the f $^1\Pi(v=0)$ level of NH.

We can understand why predissociation of the resonance enhancing intermediate state should so attenuate a REMPI spectrum as follows. As a first approximation, we generally assume that the relative intensities, $I_{J'J''}$, of the lines appearing within a given vibronic band in an m+1 REMPI spectrum (recorded using constant incident laser intensity) are given by an expression of the form:

$$I_{J'J''} \propto N_{J''} S_{J'J''}, \tag{5.6}$$

where $N_{J''}$ is the population in the particular absorbing level and $S_{J'J''}$ is the appropriate m-photon $J' \leftarrow J''$ rotational line strength. Implicit in this approach is the assumption that the rate of the final one photon ionization step is independent of the particular resonance enhancing rovibrational level. This is often justifiable since (a) the ionization step often involves a free $\leftarrow$ bound transition (the cross-section for which may not be markedly wavelength dependent) and (b) this transition will very likely be saturated given the high photon fluxes prevailing in the focal volume of a typical REMPI experiment. However, constancy of

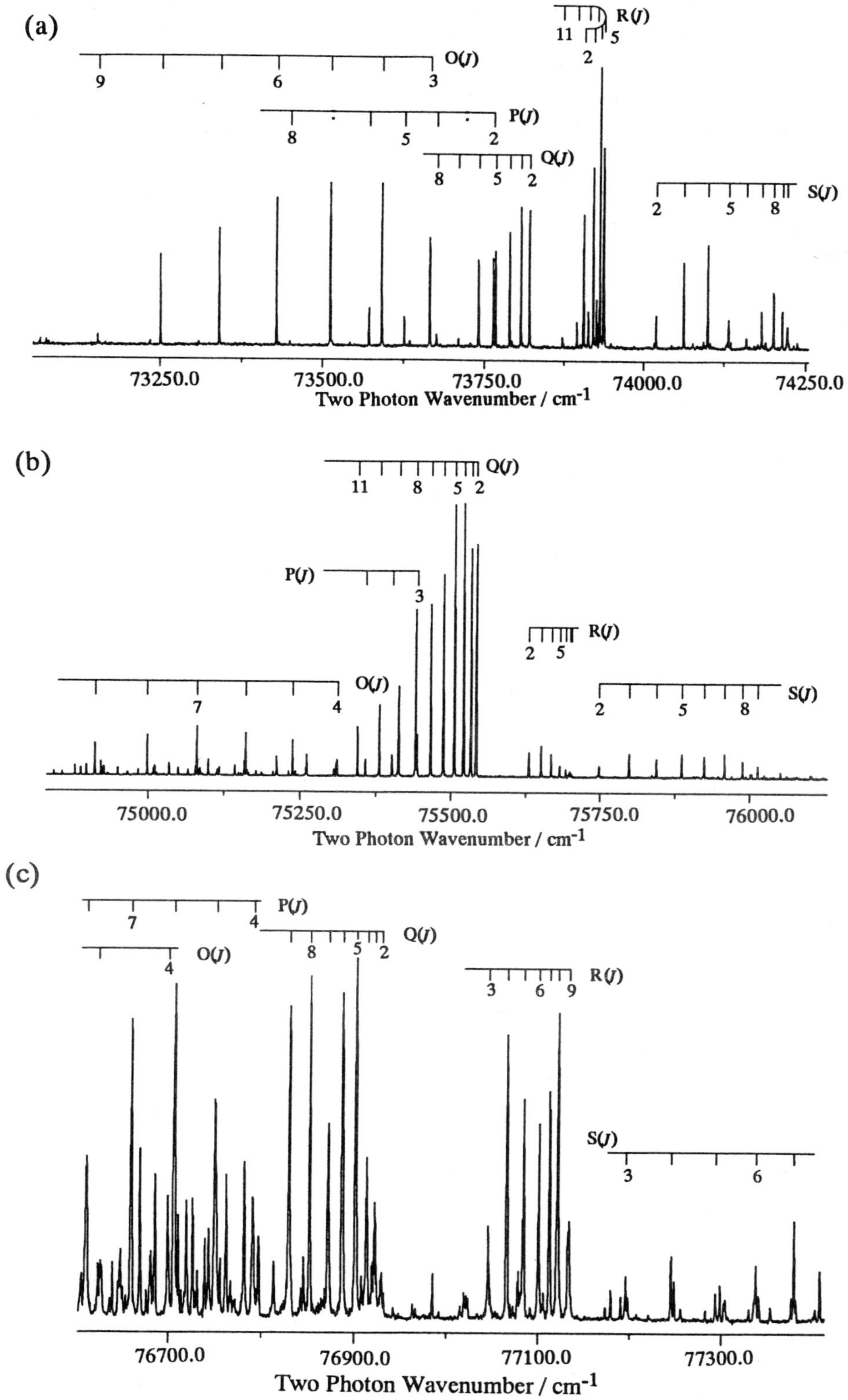

Fig. 5.3 Two-photon resonant MPI spectra of the origin bands of (a) the $f^1\Pi \leftarrow a^1\Delta$ ($3p\sigma \leftarrow 1\pi$), (b) $g^1\Delta \leftarrow a^1\Delta$ ($3p\pi \leftarrow 1\pi$) and (c) $h^1\Sigma^+ \leftarrow a^1\Delta$ ($3p\pi \leftarrow 1\pi$) transitions of the NH radical, each excited using linearly polarized laser radiation. Each spectrum was recorded by monitoring the m/z=15 ion mass channel as a function of the laser wavelength. Individual line assignments are indicated via the combs superimposed above each spectrum.

ionization *rate* does not necessarily imply a constant ionization *efficiency*. The efficiency of the ionization step must depend on the relative magnitudes of the ionization rate and the rates for all other population loss neutral molecule predissociates at a rate comparable to, or greater than, the ionization rate, this competition must lead to a reduced ionization probability and a relative diminution of the eventual ion yield; multiphoton excitations proceeding via such predissociated levels will thus appear with reduced relative intensity in the REMPI spectrum. In extreme cases the lines may show lifetime broadening as well. Careful analysis of such discrepancies between observed REMPI transition intensities and those predicted through use of the appropriate multiphoton rotational linestrengths have provided much insight into the detailed predissociation dynamics of a range of hydrides, *e.g.* H_2O [45-47], H_2S [47,48], CH_3 [49] and NH_3 (see below).

We now focus attention on the photoelectrons that accompany the REMPI process, and the additional information that can be gleaned by careful measurements of their kinetic energies. Such measurements provide one, certainly the most direct, means of actually confirming the number of photons involved in the overall ionization process. This is a not insignificant point. Recall Fig. 5.1, which depicted an MPI process resonance enhanced at the two photon energy, in which three photons provided sufficient energy to exceed the lowest ionization threshold. Such a system will generally be found referred to in the literature as a 2+1 REMPI process, even in instances where there is no kinetic energy analysis to confirm the '+1' part

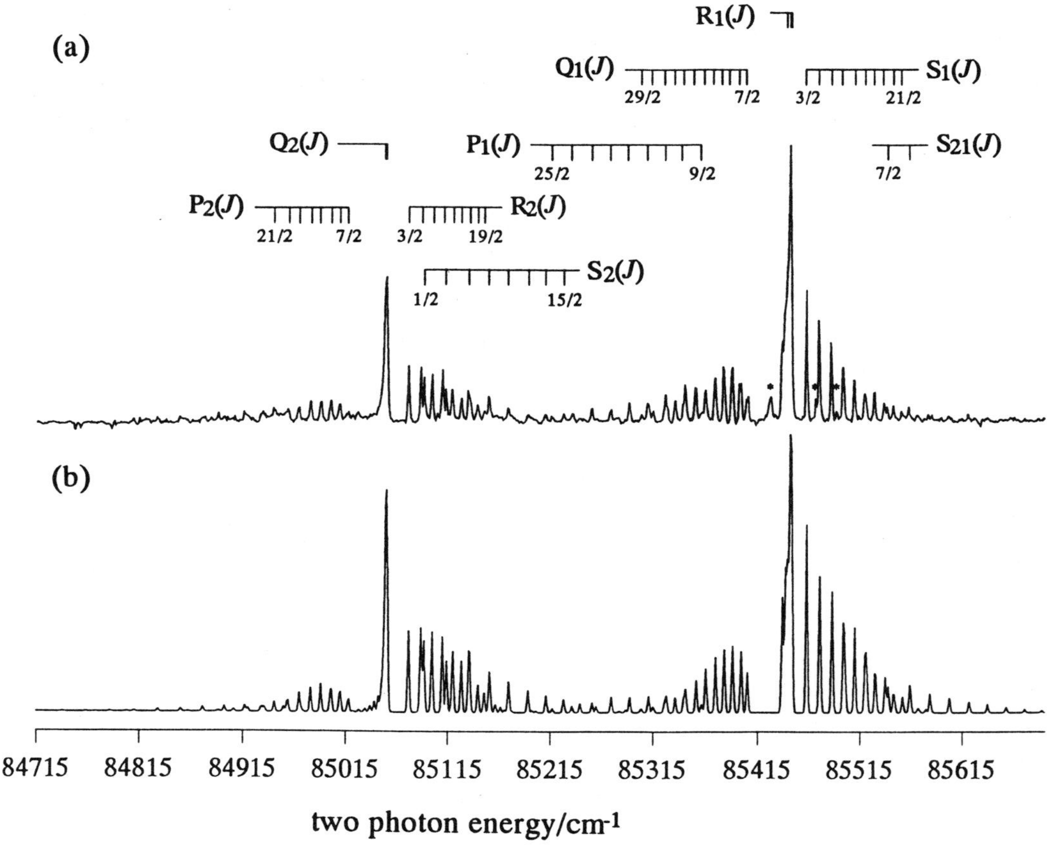

Fig. 5.4 Two photon excitation spectrum of the $[a^1\Delta]5p,{}^2\Phi \leftarrow X^2\Pi$ origin transition of the SD radical. Individual line assignments are indicated by the combs above the spectrum. (a) Experimental spectrum obtained by monitoring just that fraction of the total photoelectron yield that has a kinetic energy (4.1–4.3 eV) appropriate for a 2+1 REMPI process terminating on the zero-point level of the $SD^+(a^1\Delta)$ ion. The weak resonances marked with asterisks are due to SH impurities. (b) Simulation of this band employing spectroscopic constants listed in [43].

of the description. How then, ought we to view the REMPI spectrum of the SD($X^2\Pi$) radical shown in Fig. 5.4(a)? The accompanying simulation (Fig. 5.4(b)) leaves us in no doubt that the resonance enhancement is provided by a state of $^2\Phi$ symmetry lying at the energy of two absorbed photons, but simple energetic considerations tell us that this state lies above the first ionization limit of SD (*ca.* 84045 cm^{-1})! Does the observed REMPI signal derive from autoionisation at the two photon energy, or is a further photon absorption required? In this particular case, REMPI-PES measurements reveal that a substantial part of the ionisation signal still comes as a result of a 2+1 REMPI process [43] and that the $^2\Phi$ state is a member of a Rydberg series converging to the second ($a^1\Delta$) ionisation limit of the SD radical; quantum defect considerations then suggest that this state results from the electronic configuration [$a^1\Delta$]$5p^1$.

As hinted in the Introduction, accurate measurements of the photoelectron kinetic energies, especially in cases where the resolution is sufficient to reveal rovibrational levels of the ion, can yield much else besides. Firstly, of course, such measurements offer the opportunity of deducing accurate ionisation potentials: *e.g.* 108804 ± 5 cm^{-1} in the case of NH, measured relative to the lowest rovibronic level of the ground ($X^3\Sigma^-$) state [37]. Secondly, they can provide c lues as to the vibronic nature of the resonance enhancing level. This aspect is nicely illustrated by the two REMPI–PE spectra of the NH radical shown in Fig. 5.5. Figure 5.5 (a) shows the rather simple photoelectron spectrum obtained when the MPI is resonance enhanced, at the two photon energy, by the $v' = 1$ level of the $g^1\Delta$ state of NH. The kinetic energy of the major peak is consistent with the partner NH^+ ions being formed in the $v^+=1$ level of their ($X^2\Pi$) ground state – precisely the kind of v=0 propensity in the final ionization step that we might expect on the basis of Franck-Condon considerations were we to assume that the $g^1\Delta$ state is a Rydberg state built on the ground state

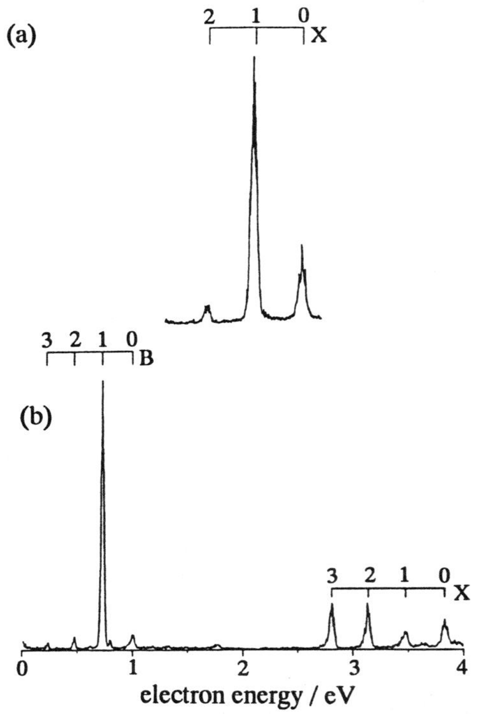

Fig. 5.5 Photoelectron kinetic energy spectra observed following 2+1 REMPI via (a) the v'=1 level of the $g^1\Delta$ state and (b) the v'=1 level of the $j^1\Delta$ state of the NH radical. Combs indicating the vibronic levels of the accompanying NH^+ ion are drawn above each spectrum.

ion core. Similar observations in the case of 2+1 REMPI via levels of the f $^1\Pi$ and h$^1\Sigma^+$ states of NH indicate that these, too, should be viewed as 'regular' Rydberg states sharing the same core configuration as the eventual ion. Contrast this with the considerably more complicated REMPI–PE spectrum shown in Fig. 5.5(b), obtained following 2+1 REMPI via the v'=1 level of the higher energy j$^1\Delta$ state of the NH radical. As Fig. 5.5(b) shows, ionisation results in formation both of ground state ions (in a spread of v^+ vibrational levels) and electronically excited (B $^2\Delta$) state ions, primarily in their v^+=1 level. This apparently non-Franck-Condon behaviour upon ionisation has been rationalised [39] by assuming that the electronic wavefunction for the resonant intermediate j$^1\Delta$ state contains major contributions both from the configuration ...$3^2 1^1[^2\Pi]3d^1$ (*i.e.* a Rydberg configuration built on the ground state ion core) and the core excited configuration ...$3^1 1^2[^2\Delta]3s^1$. This example highlights one of the particular virtues of the REMPI-PES method – the way in which it can provide insight into the electronic character of the resonance enhancing state of the neutral.

Finally it is worth pointing out that REMPI-PES, especially when applied to simple diatomics and light hydride molecules, offers a (still rather rare) opportunity to induce photoionization from a single, user selectable, rovibrational level (albeit of the intermediate Rydberg state rather than the ground state). Thus it offers a route to investigating the dynamics of the photoionization process itself, in hitherto unprecedented detail. As in any other photoinduced transition, the requirements that angular momentum, parity *etc.* must be conserved necessarily impose constraints on the allowed changes in rotational quantum number upon ionization, but the availability of *rotationally resolved* photoelectron spectra has served to inspire and test much of the necessary angular momentum theory. For example, if we assume Hund's case (a) coupling, the relevant selection rules [50-52] are encapsulated in the following expression

$$\Delta J + \Delta S + \Delta p + \Delta q + \ell = \text{even}\,, \tag{5.7}$$

where $\Delta J = J^+ - J'$ represents the difference between the quantum number for the total angular momentum in the ion and in the state selected excited neutral respectively, $\Delta S = S^+ - S'$ is the difference in the corresponding quantum numbers for the total spin, $\Delta p = p^+ - p'$ (where p is the parity index of the *e*/*f* levels), $\Delta q = q^+ - q'$ (where $q = -1$ for Σ^- states, otherwise $q = 0$) and represents a partial wave component of the ejected photoelectron. In the Hund's case (b) limit Eq. (5.7) reduces to

$$\Delta N + \Delta p + \Delta q + \ell = \text{odd}, \tag{5.8}$$

where $\Delta N = N^+ - N'$, and N^+ and N' are the quantum numbers for the total angular momentum (excluding electron spin) in the ionic and intermediate states respectively.

We end this section by considering, as a specific example, the one photon ionization of the f$^1\Pi$ state of NH that is itself prepared via a two photon excitation from the low lying a$^1\Delta$ state. As we saw earlier, the f state has been attributed to the electron promotion 3p←1π. The simple atomic picture would suggest that photoionization of a Rydberg p-electron (ℓ=1) would result in s- (ℓ=0) and d- (ℓ=2) photoelectron partial waves. Equation (5.8) would then lead us to expect a marked propensity for ΔN = odd transitions in the photoionization process. In fact, the experimental observation [37,53] is just the opposite: the rotationally resolved REMPI-PES is dominated by the ΔN=0 peak, with ΔN=±2 features more prominent than those associated with ΔN=±1 changes! One can envisage a number of factors that could conspire to contradict expectations based solely on the foregoing angular momentum considerations. Firstly, of course, NH is not an atom. The molecular potential is not spherically symmetric; consequently, ℓ is not a rigorously good quantum number. Core anisotropy can induce 'ℓ-mixing' (so called) in the electronic continuum, leading to formation of photoelectron partial waves other than those expected on the basis of simple atomic selection rules. Similarly, our description of the initial Rydberg state is somewhat simplistic. *Ab initio* calculations [53] confirm that its electronic wavefunction does indeed have dominant 3p character near the equilibrium configuration, but that it acquires increasing 3s character as the N–H bond extends. However, these same calculations reveal that the factor most responsible for the observed but unexpected propensity for N = even transitions in the near ultraviolet photoionization of the NH($f^1\Pi$) state are Cooper minima in the various components of the ionization channel involving the ℓ=2 wave [53]. Cooper minima (*i.e.* local minima in the photoionization matrix element) occur at those energies for which there is maximum destructive interference between the radial parts of the wavefunctions for the initial (Rydberg) and final (continuum) electronic states, and appear to be a fairly ubiquitous feature in those small molecules whose photoionization has thus far been investigated by REMPI-PES, having been advanced to account for the, at first sight, unexpected rotational branching ratios observed in the photoionization of not just the $f^1\Pi$ state of NH, but also in the photoionization of the neighbouring $g^1\Delta$ and $h^1\Sigma^+$ Rydberg states [53] and a number of other diatomics, *e.g.* the $D^2\Sigma^-$ state of OH [54], another $^2\Phi$ Rydberg state of the SH radical (attributed to the configuration [$a^1\Delta$]3d) [44], and the $C^2\Pi$ and $D^2\Sigma^+$ states of NO [55,56]. Measurement of the angular distributions of the resulting photoelectrons could provide support for such interpretations, but no such experimental determinations have yet been reported for NH or, indeed, for any transient species. Theory [53] predicts that the photoelectron angular distributions (PADs) arising in the 2+1 REMPI of NH via the $f^1\Pi$ state will depend sensitively not just upon v^+, the vibrational state of the ion, but also on N^+. Measurements of the PADs accompanying formation of vibrationally state selected diatomic ions (of, for example, H_2, O_2, NO and CO) are by now relatively commonplace [18,57-59], but experimental measurements of PADs associated with individual v^+, N^+ states of the ion remain rare [60].

5.3.2 Small Polyatomic Molecules

In this section we focus on triatomic and tetratomic species, especially examples which illustrate ideas additional to those summarized above. Wavelength resolved REMPI spectra have been reported for a large number of such systems - generally internally cooled via supersonic expansion. The analysis of such spectra has yielded a wealth of new data concerning the spectroscopy, structure and, in many cases, the dynamics of excited electronic states of stable molecules - *e.g.* H_2O [45,46], H_2S [48,61], NO_2 [62], CO_2 [63], OCS [64,65], CS_2 [66,67], C_2H_2 [68-73] and NH_3 (see below) - and of radical species like HCO [74,75], CH_2 [76,77] and CH_3 [13,78,79].

Despite many conventional absorption studies [80], our understanding of the vertical electronic spectrum of the ammonia molecule remained rather limited prior to the multiphoton era. The reasons for this are not hard to discern. Ground state ammonia is *pyramidal* (belonging to the C_{3v} point group), with a doubly occupied $3a_1$ nitrogen lone pair orbital as its highest occupied molecular orbital. All excited states identified in the optical spectrum arise as a result of electron promotion from this $3a_1$ orbital to Rydberg orbitals with $n\geq3$; all have *planar* (D_{3h}) equilibrium geometries. Thus each electronic transition appears in the form of a long vibronic progression associated with excitation of the out-of-plane bending mode, ν_2. The resulting spectral congestion, further compounded by the fact that all of the known excited states of NH_3 are to some extent predissociated, proved a major hindrance to definitive assignment.

Multiphoton studies of the ammonia molecule were first reported in the late 1970's [81,82]; these early works provided what is still the most complete 'global' description of the electronic spectrum of ammonia. The multiphoton studies clearly revealed one hitherto unobserved progression of sharply structured bands - associated with three photon excitation to the (so called) $\tilde{C}'$ state of ammonia [81,82]. The Bristol group has further investigated both the spectroscopy and the decay dynamics of these $\tilde{C}'(v_2')$ levels in NH_3 and ND_3 using both 3+1 [83] and 2+1 MPI techniques - the latter at sub-Doppler resolution [84]. Linewidth measurements show predissociation of the $\tilde{C}'$ state to be more developed in NH_3 than in ND_3, and that its rate scales with increase in the v_2' quantum number, whilst careful comparisons between the observed, rotationally resolved multiphoton excitation spectra and those predicted using the appropriate multiphoton rotational linestrengths indicate that, within a given ν_2' level, predissociation is enhanced by parent rotation about the in-plane (x,y) axes. All of the spectroscopic analyses, and concurrent REMPI-PES experiments [85-87] - which revealed a strong propensity for the final one photon ionization step to be Franck-Condon diagonal in all vibrational modes - indicated that the $\tilde{C}'^1A_1'$-$\tilde{X}^1A_1$ system is associated with electron promotion to the $3pa_2''$ Rydberg orbital. Much more recently, 2+1 REMPI spectroscopy has enabled a similar detailed characterization [88] of the analogous $^1A_1'$ excited state resulting from the $4pa_2''\leftarrow 3a_1$ promotion (labelled the $\tilde{E}'$ state [82]). Figure 5.6 shows an illustrative 2+1 REMPI spectrum, involving the $\nu_2'' = 5$ level of ND_3, together with a best-fit simulation.

The definitive identification of the $\tilde{C}'$ state necessitated some reappraisal of previous interpretations of the one photon absorption spectrum of ammonia, not least because a progression of parallel bands (labelled the $\tilde{C}$-$\tilde{X}$ system) had already been assigned [89] in terms of the $3pa_2''\leftarrow 3a_1$ electron promotion. Assuming that the complete set of excited (Rydberg) orbitals for ammonia can be derived by considering the orbitals of the appropriate united atom (neon) split in a field of D_{3h} symmetry, then we should expect just two singlet excited electronic states of ammonia at energies below that of the $\tilde{C}'$ origin. These are the $\tilde{A}^1A_2''$ and $\tilde{B}^1E''$ states; the mis-named '$\tilde{C}$ state' has since been shown to be a vibronic component of the $\tilde{B}$ state (see below). Vibronic progressions involving the ν_2' out-of-plane bending mode of both excited states had been identified in one photon absorption [80] but, once again, our knowledge of these excited states has been increased greatly by multiphoton studies. For example, the $\tilde{B}$-$\tilde{X}$ system of NH_3 and ND_3 (arising from the electronic promotion $3pe'\leftarrow 3a_1$) has been investigated by two photon resonant MPI both under beam conditions and, with sub-Doppler resolution, in the bulk [90,91]. Both types of experiment were crucial to the eventual spectroscopic assignment. The jet-cooled spectra, taken with a conventional nanosecond pulsed dye laser, provided unambiguous identification of the ν_2' vibronic band origins and of rovibronic transitions involving the low J',K' rotational states. However, full rotational analysis of the very congested room temperature spectra of the $ND_3(\tilde{B}$-$\tilde{X})$ bands, including measurements of the Zeeman splitting of selected lines, demanded much higher resolution than the 0.18 cm^{-1} (FWHM) 300K Doppler linewidth. Such could be achieved using a narrow bandwidth pulse amplified CW dye laser, the output of which is split into two beams and counterpropagated through the sample. This study serves to illustrate another of the well known advantages of two photon spectroscopy; namely, the way in which it is possible to obviate the first order Doppler shift by arranging for the molecules of interest to undergo a two photon transition by absorbing one photon from each of the two counterpropagating beams. Thus it was possible to measure molecule limited linewidths of individual $ND_3(\tilde{B}$-$\tilde{X})$ transitions and deduce a (rotational level independent) lifetime of *ca.* 0.25 ns for all ν_2' levels of the $ND_3(\tilde{B})$ state with bending vibrational quantum number $\nu_2' \leq 6$ [91]. As with the $\tilde{C}'$ state, the measured linewidths implied that $NH_3(\tilde{B})$ molecules predissociate at least an order of magnitude faster - a result that has since been confirmed by real time measurements involving two colour picosecond pump-probe ionization with detection of the resulting photoelectrons [92].

As mentioned in the Introduction, the ammonia molecule has been subjected to a number of doubly resonant multiphoton excitation studies. The aim, in most cases, has been to learn about excited state levels involving vibrational modes other than, or in addition to, ν_2'. One such class of experiments has involved two photon excitation to a selected ν_2' level of the $\tilde{A}$ state, followed by a one photon excitation to rovibronic levels of the $\tilde{B}$ or $\tilde{C}'$ state; absorption of one more photon leads to ionization [9,10]. In either case, the second step involves a transition between two Rydberg states both of which have planar equilibrium geometries. Such transitions will tend to be diagonal in ν_2, and thus offer a much better possibility of detecting weaker $\Delta\nu_1$, $\Delta\nu_3$ and/or $\Delta\nu_4\neq 0$ transitions. Experiments of this kind have allowed unique identification of the ν_1 symmetric stretching mode in the $\tilde{C}'$ state of NH_3 and ND_3 [9] - the former has also been identified in a careful REMPI-PES study of jet-cooled NH_3 [93]. They also enabled

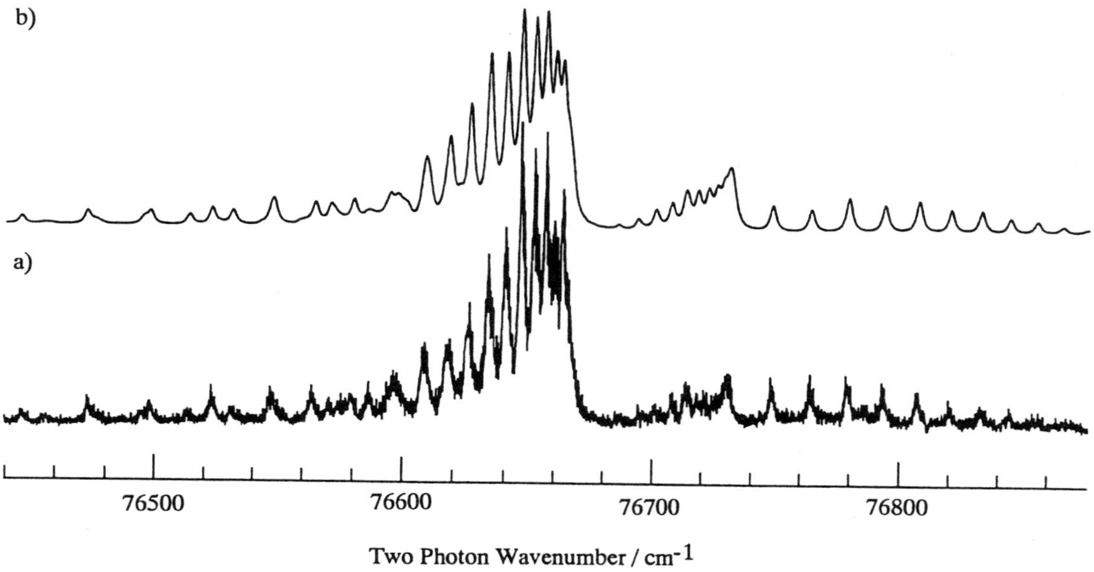

Fig. 5.6 (a) MPI spectrum of a room temperature sample of ND_3, resonance enhanced at the two photon energy by the v_2'=5 level of the $\tilde{E}'\,^1A_1'$ state [88]. The accompanying simulation (b) assumes contributions from both the T_0^0(A) and T_0^2(A) components of the two photon transition probability, with relative weights 1:2, the following excited state constants: ν_0 = 76667.5 cm^{-1}, B = 4.745 cm^{-1} and C = 2.762 cm^{-1}, a uniform Lorentzian linewidth of 3 cm^{-1} (fwhm) throughout and a rotational temperature of 300 K.

observation of some of the vibronic levels (of E"/E' symmetry according to whether v_2 = even/odd) associated with excitation of one quantum of each of ν_3 and ν_4 in the $\tilde{B}\,^1E''$ state of ammonia. Both ν_3 and ν_4 are degenerate (e') vibrations in planar NH_3 being, respectively, the antisymmetric stretch and in-plane bending modes. Either (or both) types of distortion could, in principle, be responsible for the small, but significant, Jahn-Teller distortion known to affect the $\tilde{B}$ state. Further clarification of the role of the Jahn-Teller effect in the $\tilde{B}$ state, and an interpretation of the '$\tilde{C}$ -$\tilde{X}$' system of ammonia, has been provided by rotationally resolved double resonance studies [12] in which $\tilde{B}$ state levels involving v_3'=1 and v_4'=2 have been populated by two photon excitation of molecules in selected excited rovibrational levels of the ground state (themselves prepared by tunable infrared laser excitation).

For completeness, we end this Section with mention of one other, somewhat different type of two colour REMPI experiment involving the ammonia molecule. As already discussed, two photon excitation at wavelengths (λ_1) *ca.* 300 nm results in quantum state selective population of the $\tilde{C}$' state, which can be conveniently monitored via the associated 2+1 REMPI. Introduction of a second laser pulse, tunable at wavelengths (λ_2) *ca.* 570 nm and timed so as to be essentially simultaneous with the λ_1 photons, can then reveal two distinct sets of double resonances. The first, which appear as broad 'dips' in the λ_1 induced REMPI signal, are attributable to stimulated emission pumping (SEP) to individual rovibronic levels of the first excited $\tilde{A}\,^1A_2''$ state. Careful measurements of the linewidths of these dips has provided quantitative insight into the quantum state dependent predissociation of $\tilde{A}$ state ammonia molecules [13,14]. The second, which appear as additional sharp peaks in the parent ion yield, have been observed in the specific case that λ_1 is fixed so as to excite the 2^1 level of the $\tilde{C}$' state and λ_2 is scanned so that the energy corresponding to absorption of two λ_1 photon + one λ_2 photon falls between the thresholds for forming the parent ion in its 0^0 and 2^1 levels. These sharp resonances have been attributed to autoionizing origins of the $nsa_1' \leftarrow 3a_1$ Rydberg series with n = 12-18, *i.e.* high lying members of the s Rydberg series, of which the predissociated $\tilde{A}$ state is the first (n=3) member [94].

5.3.3 Larger Polyatomic Molecules, Complexes and Clusters

As molecular size and mass increases, the spectroscopist's ambition becomes limited to vibronic, rather than full rovibronic, resolution, and structural inferences are necessarily based on, for example, observed Franck-Condon activity and the results of *ab initio* calculations or molecular simulations. Obviously, just as with the homonuclear diatomics discussed earlier in Section 5.3.1, two photon spectroscopy has played a major role in revealing *gerade* excited states of centrosymmetric species like the linear polyenes [20,95,96] and benzene [97]. Substituted benzenes, in particular, have proven popular systems for study by REMPI techniques because of their low ionization energies, the stability of their first excited (S_1) singlet states and the size of the oscillator strength linking these S_1 states to their respective ground (S_0) states; since the S_1-S_0 energy separation is generally more than half the first E_i, these transitions are well suited to study by 1+1 REMPI. In this Section we focus on a number of studies involving the aniline molecule.

The one colour two photon ionization spectrum of aniline, resonance enhanced at the one photon energy by the $S_1(^1B_2)$ state, was first reported in 1979 [98] but neither it, nor a later 2+2 REMPI study of the same transition [99], was able to add to the significant body of spectroscopic knowledge concerning the S_1 state already available from conventional absorption and emission studies [100,101]. The S_1-S_0 transition has been used as the first step in a number of *two* colour double resonance experiments involving jet-cooled aniline samples. For example, high vibrational levels of the S_0 state of aniline have been investigated using one laser (of frequency ν_1) to cause 1+1 REMPI via one of the well characterized levels of the S_1 state and then tuning a second (frequency ν_2) so as to cause stimulated emission pumping from this chosen S_1 level: the 'dump' transitions appear as dips in the ion yield caused by the first laser [102]. As in ammonia (above), a not dissimilar two colour scheme has also been used to probe the energy region around the first ionization limit. The 'observable' in these experiments was the total ion yield as a function of the sum frequency $\nu_1+\nu_2$, with ν_1 fixed so as to excite one selected vibronic level of the S_1 state. The S_1 levels involving only totally symmetric vibrational excitation (*e.g.*, ν_1', ν_{6a}') were found to show a marked propensity for direct Δv=0 photoionization; consequently it was possible to obtain a refined value for the adiabatic E_i of aniline, and to determine reasonably accurate wavenumbers for these particular vibrational modes in the cation [103]. Ion yield spectra excited via S_1 levels containing one or more quanta of a non-totally symmetric vibration (*e.g.*, ν_{10b}', ν_{15}') were found to much more structured. This structure arises as a result of vibrational autoionization (with a characteristic Δv=-1 propensity) of Rydberg states converging to the appropriate adiabatic E_i [104].

Such studies are complementary to REMPI-PES experiments, which have proved to be of considerable value both as a means of determining accurate ionization thresholds for polyatomic molecules and for establishing vibrational frequencies of the resulting cations. Even the earliest REMPI-PES studies of aniline sufficed to illustrate the benefits of the technique. The conventional gas phase HeI photoelectron spectrum of aniline is almost devoid of vibrational structure; in contrast, kinetic energy analysis of the photoelectrons ejected in the one colour two photon ionization of jet-cooled aniline, resonance enhanced at the one photon level by selected levels of the S_1 state, have yielded not just the first ($^2B_1$) [105] and second ($^2A_2$) [106] ionization thresholds but also wavenumbers for nine fundamental vibrational modes of the ground state aniline cation [105]. Subsequent ZEKE-PES studies have expanded this latter spectroscopic database to include 15 of the 36 normal modes of the cation, and yielded a refined value (62271 ± 2 cm^{-1}) for the adiabatic E_i [107,108].

Studies of the spectroscopy and dynamics of van der Waals complexes and of molecular clusters remain one of the vogue areas of gas phase chemical physics. Aromatic-rare gas complexes (*e.g.* aniline-Ar_n) have proved to be popular and convenient model systems for those interested in the electronic spectroscopy and the dynamics of van der Waals complexes. For example, one colour 1+1 REMPI spectroscopy with TOF mass analysis has shown the S_1–S_0 origins of the complexes aniline-Ar and aniline-Ar_2 to be red-shifted from that of bare aniline molecule by, respectively, 54 cm^{-1} and 112 cm^{-1} [105]. Vibronic structure observed close to the S_1–S_0 origin band of aniline-Ar and of related C_6H_5X-Ar complexes (where X is, for example, F, Cl or OH) has been assigned in terms of excitation of the (one stretch and two orthogonal bending) van der Waals modes in the S_1 state [109-111]. Given these assignments, it is possible to parameterise the van der Waals part of the potential energy surface of the complex. The fact that these two body terms allow almost quantitative prediction of the form of the vibronic structure around the S_1–S_0 origin of the aniline-Ar_2 complex, without any need for three body terms, has been used as evidence that the two Ar atoms in this complex must be on opposite sides of the aniline ring [110,111]. Two colour ZEKE-PES, resonance enhanced at the one photon energy by selected levels of the S_1 state, has enabled study of differences in the binding energies of the neutral and cationic complexes and allowed determination of the frequencies of the van der Waals modes in the ground state cation [112,113]. Observation of progressions in any of the low frequency van der Waals modes in the ZEKE photoelectron spectrum implies a significant change in the geometry of the complex upon ionization. Illustrative examples include (i) aniline-Ar, for which analysis of the ZEKE photoelectron spectrum indicates that photoionization of the S_1 complex is accompanied by an ~8° change in the angle between the line joining the Ar atom to the aniline centre of mass and the plane containing the carbon atoms [112,113], and (ii) the hydrogen bonded complexes phenol-methanol [114] and phenol-ethanol [115]. Given short laser pulses and a (variable) time delay between the S_1–S_0 excitation step and the subsequent photoionization, Knee and coworkers [113,116] have shown that it is possible to derive kinetic information about intramolecular vibrational redistribution (IVR) and vibrational predissociation in selected vibronic levels of the S_1 states of, amongst others, the aniline-Ar and aniline-CH_4 van der Waals complexes.

5.4 Excited State Dynamics Probed by REMPI

Molecular spectroscopy and structure are the central themes of this Chapter and, indeed, of this whole monograph. Nonetheless, this Chapter would be incomplete without some brief summary of the various ways in which REMPI spectroscopy is contributing to our knowledge of the dynamics of excited state molecules. Most have been mentioned in passing in the preceding text. As we have already seen, the REMPI technique is especially suited to studying the more long lived excited states of the neutral. This is because the signal (either an ion or a photoelectron current) depends upon at least some fraction of the resonant intermediate state population absorbing the necessary one or more additional photons to exceed E_i. Dissociative excited states, which in conventional absorption spectroscopy might reveal themselves as regions of continuous absorption, tend not to show up in REMPI spectroscopy because the up-pumping rate is usually much smaller than the rate of population loss via fragmentation. In favourable cases, however, repulsive states may reveal themselves in the vibrational structure apparent in a REMPI spectrum. For example, the vibrational structure evident in the one colour 2+2 REMPI spectrum of I_2, resonance enhanced at the two photon energy by the $E(0_g^+)$ ion-pair state, has been rationalized [117] by assuming that the first two photon step is actually a sequential absorption process, proceeding via the repulsive wall of the $B(0_u^+)$ state at the one photon energy, and that the second photon is absorbed (and the E state populated) only after some bond extension at the one photon energy. A two colour variant of the above, involving one ν_1 photon (to reach the B state above its dissociation limit) followed by $2\nu_2$ photons has recently been used to probe the ${}^3\Sigma_u^-(0_u^+)$ and ${}^3\Pi_u(0_u^+)$ ion pair states of Cl_2 at the extended geometries near their potential minima [118].

The Bristol group pioneered one of the first quantitative applications of REMPI spectroscopy in the study of excited state dynamics. By way of illustration, consider the REMPI spectrum of H_2O, resonance enhanced at the three photon energy by the $\tilde{C}\,^1B_1$ state origin. This shows resolvable rotational structure, even with a room temperature sample, but the number and strengths of the resolved lines were very different from that which is obtained from a calculation of the appropriate three photon rotational linestrengths [45,47]. This apparent discrepancy can be reconciled once it is appreciated that different rotational levels of the $\tilde{C}$ origin predissociate with different efficiencies. In this particular case, the excited state predissociation rate is found to scale with $\langle \hat{J}_a^2 \rangle$, the expectation value of the square of the *a*-axis rotational angular momentum in the excited state, implying that the predissociation involves coupling to the continuum associated with the $\tilde{B}\,^1A_1$ state. Similar analyses have been reported for a number of other Rydberg states of H_2O and D_2O [46] and for several other hydride molecules, notably H_2S and D_2S [47,48] and ammonia (see above). Studies involving the $\tilde{A}$ state of ammonia [13,14] have already been cited as one illustration of the use of double resonance methods (in this case SEP, in the form of 'ion dip' spectroscopy) as a means of probing the lifetimes of predissociated excited states. We have also highlighted examples of the use of pump-probe techniques to study excited state decay dynamics in real time (*e.g.*, predissociation of the $\tilde{C}'$ state of ammonia [63] and IVR and vibrational predissociation of aniline complexes in their S_1 state [116]). Given the increasing availability of reliable, tunable, ultrafast laser systems it is reasonable to anticipate that two colour pump-probe REMPI spectroscopy will find use in many more studies of excited state decay dynamics.

Finally, it is worth reiterating a point made in the Introduction - namely, that REMPI is a very sensitive, selective and convenient means of 'tagging' and detecting small localized concentrations of gas phase species, *e.g.*, atomic or molecular photofragments, or the products of cross-beam reactions. Obviously, an accurate knowledge of the energy disposal in such products, their recoil velocities and the angular distribution of these velocity vectors, the alignment of their rotational angular momenta, and the way all these quantities are correlated, can provide considerable insight into the detailed dissociation and/or reaction dynamics: Refs. [119] and [120] are two recent reviews which illustrate some of the ways in which REMPI spectroscopy is being put to good use by the reaction dynamics community.

5.5 Conclusions

This Chapter has concentrated on published work involving three representative systems to illustrate at least some of the many opportunities offered by REMPI methods. Without doubt, gas phase molecular spectroscopists will continue to be stimulated and rewarded by the measurement and analysis of wavelength resolved REMPI spectra of stable molecular species, of radicals and of molecular complexes. It seems inevitable that some of these REMPI excitation schemes will become accepted as the detection method of choice in, for example, reaction product ion imaging schemes [119,120]. Spectral complexity grows rapidly with molecular size; thus we can anticipate that future studies increasingly will take advantage of the benefits afforded by using (i) jet-cooled samples, (ii) mass and/or electron kinetic energy analysis and (iii) double (or even triple) resonance methods. Similarly, we can be assured that REMPI-PES methods and the ZEKE-PES technique will continue to provide a wealth of spectroscopic (and thus structural) information about molecular cations. Though not discussed in this Chapter, it seems probable that mass analysed threshold ionization (MATI) spectroscopy [121,122] will prove to be a viable alternative, high resolution route to measuring ionization thresholds and thus deriving information about the internal energy states of cations. As its name implies, MATI involves measurement of *ions* resulting from (delayed) pulsed field ionization of molecules in high-*n* Rydberg states. By a judicious arrangement of extraction fields it is possible to arrange for these 'threshold' ions to exhibit a TOF sufficiently different from that of all ions formed by direct one (or multi-) photon ionization. Since, in contrast to ZEKE-PES, this technique offers both wavelength and *mass* information, we may anticipate that MATI spectroscopy will find particular application in the study of cations of radicals, clusters and complexes where, because of the production method, species identification might be ambiguous.

Acknowledgements: Financial support from the Engineering and Physical Sciences Research Council, the Royal Society, NATO, the Nuffield Foundation and the Ciba Fellowship Trust has all been crucial to the multiphoton spectroscopy research program at Bristol. I am also enormously grateful to past and present colleagues in Bristol (notably R.N. Dixon, K.N. Rosser, C.M. Western, A.J. Orr-Ewing, J.M. Allen, J.M. Bayley, C.L. Bennett, S.G. Clement, J.D. Howe, R.A. Morgan, J.D. Prince, M.P. Puyuelo, R.J. Stickland and B. Tutcher), for their many and varied contributions to the Bristol multiphoton spectroscopy program, and to C.A. de Lange and W.J. Buma (Amsterdam) and J.W. Hudgens and R.D. Johnson III (NIST, Gaithersburg), and their respective colleagues, for their generous collaboration and friendship.

References

1. W.M. McClain and R.A. Harris, in *Excited States* (E.C. Lim, ed.), **3**, pp.1-56, Academic Press, New York (1978).
2. S.H. Lin, Y. Fujimura, H.J. Neusser and E.W. Schlag, *Multiphoton Spectroscopy of Molecules,* Academic Press, New York (1984).
3. M.N.R. Ashfold, *Mol. Phys.* **58**, 1 (1986).
4. M.N.R. Ashfold and J.D. Prince, *Contemp. Phys.* **29**, 125 (1988).
5. M.N.R. Ashfold and J.D. Howe, *Ann. Rev. Phys. Chem.* **45**, 57 (1994) and references therein.
6. N. Bjerre, R. Kachru and H. Helm, *Phys. Rev. A* **31**, 1206 (1985).
7. W.L. Glab and J.P. Hessler, *Phys. Rev. A* **35**, 2102 (1987); **42**, 5486 (1990).
8. E.Y. Xu, H. Helm and R. Kachru, *Phys. Rev. A* **39**, 3979 (1989).
9. J.M. Allen, M.N.R. Ashfold, C.L. Bennett and C.M. Western, *Chem. Phys. Lett.* **149**, 1 (1988).
10. X. Li, X. Xie, L. Li, X. Wang and C. Zhang, *J. Chem. Phys.* **97**, 128 (1992).
11. H. Rottke and K.H. Welge, *J. Chem. Phys.* **97**, 908 (1992) and references therein.
12. J.M. Allen, M.N.R. Ashfold, R.J. Stickland and C.M. Western, *Mol. Phys.* **74**, 49 (1991).
13. M.N.R. Ashfold, C.L. Bennett and R.N. Dixon, *Farad. Disc. Chem. Soc.* **82**, 163 (1986).
14. J. Xie, G. Sha, X. Zhang and C. Zhang, *Chem. Phys.Lett.* **124**, 99 (1986).
15. K. Kimura, *Adv. Chem. Phys.* **60**, 161 (1985).
16. K. Kimura, *Int. Rev. Phys. Chem.* **6**, 195 (1987).
17. S.T. Pratt, P.M. Dehmer and J.L. Dehmer, In *Advances in Multiphoton Processes and Spectroscopy*(S.H. Lin, ed.), **4**, pp.69-169, World Scientific, Singapore (1988).
18. R.N. Compton and J.C. Miller, in *Laser Applications in Physical Chemistry* (D.K. Evans, ed.), pp.221-306, Marcel Dekker, New York (1988).
19. K. Kimura and Y. Achiba, in *Advances in Multiphoton Processes and Spectroscopy* (S.H. Lin, ed.), **5**, pp.317-370, World Scientific, New Jersey (1989).
20. M.B. Robin, *Higher Excited States of Polyatomic Molecules,* Academic Press, New York, Vols. 1 and 2 (1974), Vol. 3 (1985).
21. H. Park and R.N. Zare, *J. Chem. Phys.* **99**, 6537 (1993) and references therein.
22. K. Müller-Dethlefs and E.W. Schlag, *Ann. Rev. Phys. Chem.* **42**, 109 (1991).
23. F. Merkt and T.P. Softley, *Int. Rev. Phys. Chem.* **12**, 205 (1993).
24. W.C. Wiley and I.H. McLaren, *Rev. Sci. Instrum.* **26**, 1150 (1955).
25. H. Kuhlewind, H.J. Neusser and E.W. Schlag, *J. Chem. Phys.* **82**, 5452 (1985)
26. P. Kruit and F.H. Read, *J. Phys. E* **16**, 313 (1983).
27. See, *e.g.*, N.P.L. Wales, E. de Beer, N.P.C. Westwood, W.J. Buma, C.A. de Lange and M.C. van Hemert, *J. Chem. Phys.* **100**, 7984 (1994) and references therein.
28. A.J.R. Heck, R.N. Zare and D.W. Chandler, *J. Chem. Phys.* **104**, 4019 (1996).
29. E.E. Marinero, C.T. Rettner and R.N. Zare, *J. Chem. Phys.* **80**, 4142 (1984).
30. T.N. Kitsopoulos, M.A. Buntine, D.P. Baldwin, R.N. Zare and D.W. Chandler, *Science* **260**, 1605 (1993) and references therein.
31. M.A. Buntine, D.P. Baldwin, R.N. Zare and D.W. Chandler, *J. Chem. Phys.* **94**, 4672 (1991).
32. A.G. Suits, R.L. Miller, L.S. Bontuyan and P.L. Houston, *J. Chem. Soc. Faraday Trans.* **89**, 1443 (1993)
33. S.M. Ball, G. Hancock, J.C. Pinot de Moira, C.M. Sadowski and F. Winterbottom, *Chem. Phys.Lett.* **245**, 1 (1995).
34. M.N.R. Ashfold, S.G. Clement, J.D. Howe and C.M. Western, *J. Chem. Soc. Faraday Trans.* **89**, 1153 (1993) and references therein.
35. R.D. Johnson III and J.W. Hudgens, *J. Chem. Phys.* **92**, 6420 (1990).
36. M.N.R. Ashfold, S.G. Clement, J.D. Howe and C.M. Western, *J. Chem. Soc. Faraday Trans.* **87**, 2515 (1991).
37. E. de Beer, M. Born, C.A. de Lange and N.P.C. Westwood, *Chem. Phys. Lett.* **186**, 40 (1991).
38. S.G. Clement, M.N.R. Ashfold and C.M. Western, *J. Chem. Soc. Faraday Trans.* **88**, 3121 (1992).

39. S.G. Clement, M.N.R. Ashfold, C.M. Western, E. de Beer, C.A. de Lange and N.P.C. Westwood, *J. Chem. Phys.* **96**, 4963 (1992).
40. S.G. Clement, M.N.R. Ashfold, C.M. Western, R.D. Johnson III and J.W. Hudgens, *J. Chem. Phys.* **96**, 5538 (1992).
41. N.P.L. Wales, E. de Beer, N.P.C. Westwood, W.J. Buma and C.A. de Lange, *J. Chem. Phys.* **100**, 7984 (1994).
42. M.N.R. Ashfold, B. Tutcher and C.M. Western, *Mol. Phys.* **66**, 981 (1989).
43. J.B. Milan, W.J. Buma, C.A. de Lange, C.M. Western and M.N.R. Ashfold, *Chem. Phys. Lett.* **239**, 326 (1995).
44. J.B. Milan, W.J. Buma, C.A. de Lange, K. Wang and V. McKoy, *J. Chem. Phys.* **103**, 3262 (1995).
45. M.N.R. Ashfold, J.M. Bayley and R.N. Dixon, *Chem. Phys.* **84**, 35 (1984).
46. M.N.R. Ashfold, J.M. Bayley and R.N. Dixon, *Can. J. Phys.* **62**, 1806 (1984).
47. M.N.R. Ashfold, J.M. Bayley, R.N. Dixon and J.D. Prince, *Ber. Bunsenges. Phys. Chem.* **89**, 254 (1985).
48. M.N.R. Ashfold, J.M. Bayley, R.N. Dixon and J.D. Prince, *Chem. Phys.* **98**, 289 (1985).
49. J.F. Black and I. Powis, *J. Chem. Phys.* **89**, 3986 (1988).
50. J. Xie and R.N. Zare, *J. Chem. Phys.* **93**, 3033 (1990).
51. G. Raseev and N. Cherepkov, *Phys. Rev. A* **42**, 3948 (1990).
52. K. Wang, and V. McKoy, *J. Chem. Phys.* **95**, 4977 (1991).
53. K. Wang, J.A. Stephens, V. McKoy, E. de Beer, C.A. de Lange and N.P.C. Westwood, *J. Chem. Phys.* **97**, 211 (1992).
54. E. de Beer, C.A. de Lange, J.A. Stephens, K. Wang and V. McKoy, *J. Chem. Phys.* **95**, 714 (1991).
55. K.S. Viswanathan, E. Sekreta, E.R. Davidson and J.P. Reilly, *J. Phys. Chem.* **90**, 5078 (1986).
56. K. Wang, J.A. Stephens and V. McKoy, *J. Chem. Phys.* **95**, 6456 (1991).
57. S.T. Pratt, P.M. Dehmer and J.L. Dehmer, *J. Chem. Phys.* **85,** 3379 (1986).
58. P.J. Miller, W.A. Chupka, J. Winniczek, and M.G. White, *J. Chem. Phys.* **89**, 4058 (1988).
59. G. Sha, D. Proch, C. Rose and K.L. Kompa, *J. Chem. Phys.* **99**, 4334 (1993).
60. H. Park and R.N. Zare, *J. Chem. Phys.* **99**, 6537 (1993) and references therein.
61. M.N.R. Ashfold, W.S. Hartree, A. Salvato, B. Tutcher and A. Walker, *J. Chem. Soc. Faraday Trans.* **86**, 2027 (1990).
62. G.P. Bryant, Y. Jiang, M. Martin and E.R. Grant, *J. Phys. Chem.* **96**, 6875 (1992); *J. Chem. Phys.* **101**, 7199 (1994) and references therein.
63. M.R. Dobber, W.J. Buma and C.A. de Lange, *J. Chem. Phys.* **101**, 9303 (1994) and references therein.
64. B. Yang, M.H. Eslami and S.L. Anderson, *J. Chem. Phys.* **89**, 5527 (1988).
65. R. Weinkauf and U. Boesl, *J. Chem. Phys.* **98**, 4459 (1993).
66. J. Baker and S. Couris, *J. Chem. Phys.* **103**, 4847 (1995) and references therein.
67. R.A. Morgan, M.A. Baldwin, A.J. Orr-Ewing, M.N.R. Ashfold, W.J. Buma, J.B. Milan and C.A. de Lange, *J. Chem. Phys.* **104**, 6117 (1996).
68. M.N.R. Ashfold, R.N. Dixon, J.D. Prince and B. Tutcher, *Mol. Phys.* **56**, 1185 (1988).
69. T.M. Orlando, S.L. Anderson, J.R. Appling and M.G. White, *J. Chem. Phys.* **87**, 852 (1987).
70. M.N.R. Ashfold, B. Tutcher, B. Yang, S.K. Jin and S.L. Anderson, *J. Chem. Phys.* **87**, 5105 (1987).
71. M. Takahashi, M. Fujii and M. Ito, *J. Chem. Phys.* **96**, 6486 (1992).
72. J.K. Lundberg, D.M. Jonas, B. Rajaram, Y. Chen and R.W. Field, *J. Chem. Phys.* **97**, 7180 (1992).
73. Y.F. Zhu, R. Shehadeh and E.R. Grant, *J. Chem. Phys.* **99**, 5723 (1993).
74. X.M. Song and T.A. Cool, *J. Chem. Phys.* **96**, 8664, 8675 (1992) and references therein.
75. E. Mayer and E.R. Grant, *J. Chem. Phys.* **103**, 10513 (1995).
76. K.K. Irikura and J.W. Hudgens, *J. Phys. Chem.* **96**, 518 (1992).
77. K.K. Irikura, R.D. Johnson III and J.W. Hudgens, *J. Phys. Chem.* **96**, 6131 (1992).
78. T.G. DiGuiseppe, J.W. Hudgens and M.C. Lin, *J. Phys. Chem.* **86**, 36 (1982); *J. Chem. Phys.* **79**, 571 (1983).
79. J.W. Hudgens, In *Advances in Multiphoton Processes and Spectroscopy* (S.H. Lin, ed.), **4**, pp. 171-296, World Scientific, Singapore (1988).
80. M.N.R. Ashfold, C.L. Bennett and R.J. Stickland, *Comm. At. Mol. Phys.* **19**, 181 (1987) and references therein.
81. G.C. Nieman and S.D. Colson, *J. Chem. Phys.* **68**, 5656 (1978); **71**, 571 (1979).
82. J.H. Glownia, S.J. Riley, S.D. Colson and G.C. Nieman, *J. Chem. Phys.* **72**, 5998 (1980); **73**, 4296 (1980).
83. M.N.R. Ashfold, R.N. Dixon and R.J. Stickland, *Chem. Phys.* **88**, 463 (1984).
84. M.N.R. Ashfold, R.N. Dixon, K.N. Rosser, R.J. Stickland and C.M. Western, *Chem. Phys.* **101**, 467 (1986).
85. J.H. Glownia, S.J. Riley, S.D. Colson, J.C. Miller and R.N. Compton, *J. Chem. Phys.* **77**, 68 (1982).
86. Y. Achiba, K. Sato, K. Shobotake and K. Kimura, *J. Chem. Phys.* **78**, 5474 (1983).
87. W.E. Conaway, R.J.S. Morrison and R.N. Zare, *Chem. Phys. Lett.* **113**, 429 (1985).
88. M.N.R. Ashfold, C.M. Western, J.W. Hudgens and R. D. Johnson III, *Chem. Phys. Lett. (in press).*
89. A.E. Douglas, *Disc. Faraday Soc.* **35**, 158 (1963).
90. M.N.R. Ashfold, R.N. Dixon, R.J. Stickland and C.M. Western, *Chem. Phys. Lett.* **138**, 201 (1987).
91. M.N.R. Ashfold, R.N. Dixon, N. Little, R.J. Stickland and C.M. Western, *J. Chem. Phys.* **89**, 1754 (1988).
92. M.R. Dobber, W.J. Buma and C.A. de Lange, *J. Phys. Chem.* **99**, 1671 (1995).
93. P.J. Miller, S.D. Colson and W.A. Chupka, *Chem. Phys. Lett.* **145**, 183 (1988).
94. D.T. Cramb and S.C. Wallace, *J. Chem. Phys.* **101**, 6523 (1994).
95. W.J. Buma, B.E. Kohler and T.A. Shaler, *J. Chem. Phys.* **96**, 399 (1992) and references therein.
96. H. Petek, A.J. Bell, Y.S. Choi and K. Yoshihara, *J. Chem. Phys.* **98**, 3777 (1993).
97. R.L. Whetten, S.G. Grubb, C.E. Otis, A.C. Albrecht and E.R. Grant, *J. Chem. Phys.* **82**, 1115, 1135 (1985) and

references therein.
98. J.H. Brophy and C.T. Rettner, *Chem. Phys.Lett.* **67**, 351 (1979).
99. L. Goodman and R.P. Rava, *J. Chem. Phys.* **74**, 4826 (1981).
100. J.C.D. Brand, D.R. Williams and T.J. Cook, *J. Mol. Spectrosc.* **20**, 359 (1966).
101. D.A. Chernoff and S.A. Rice, *J. Chem. Phys.* **70**, 2511 (1979) and references therein.
102. T. Suzuki, M. Hiroi and M. Ito, *J. Phys. Chem.* **92**, 3774 (1988).
103. M.A. Smith, J.W. Hager and S.C. Wallace, *J. Chem. Phys.* **80**, 3097 (1984).
104. J. Hager, M.A. Smith and S.C. Wallace, *J. Chem. Phys.* **84**, 6771 (1986).
105. J.T. Meek, E. Sekreta, W. Wilson, K.S. Viswanathan and J.P. Reilly, *J. Chem. Phys.* **82**, 1741 (1985).
106. B.J. Kim and P.M. Weber, *J. Phys. Chem.* **99**, 2583 (1995).
107. X. Zhang, J.M. Smith and J.L. Knee, *J. Chem. Phys.* **97**, 2843 (1992).
108. X. Song, M. Yang, E.R. Davidson and J.P. Reilly, *J. Chem. Phys.* **99**, 3224 (1993).
109. E.J. Bieske, M.W. Rainbird, I.M. Atkinson and A.E.W. Knight, *J. Chem. Phys.* **91**, 752 (1989)
110. E.J. Bieske, M.W. Rainbird and A.E.W. Knight, *J. Chem. Phys.* **94**, 7019 (1991).
111. E.J. Bieske, A.S. Vichanco, M.W. Rainbird and A.E.W. Knight, *J. Chem. Phys.* **94**, 7029 (1991).
112. M. Takahashi, H. Ozehi and K. Kimura, *J. Chem. Phys.* **96**, 6399 (1992).
113. Z. Xu, J.M. Smith and J.L. Knee, *J. Chem. Phys.* **97**, 2843 (1992).
114. T.G. Wright, E. Cordes, O. Dopfer and K. Müller-Dethlefs, *J. Chem. Soc. Faraday Trans.* **89**, 1609 (1993).
115. E. Cordes, O. Dopfer, T.G. Wright and K. Müller-Dethlefs, *J. Phys. Chem.* **97**, 7471 (1993).
116. X. Zhang and J.L. Knee, *Farad. Discuss.* **97**, 299 (1994) and references therein.
117. M.S.N. Al-Kahali, R.J. Donovan, K.P. Lawley and T. Ridley, *Chem. Phys. Lett.* **220**, 225 (1994).
118. M.S.N. Al-Kahali, R.J. Donovan, K.P. Lawley, T. Ridley and A.J. Yarwood, *J. Phys. Chem.* **99**, 3978 (1995).
119. P.L. Houston, *Acc. Chem. Res.* **28**, 453 (1995).
120. A.J.R. Heck and D.W. Chandler, *Ann. Rev. Phys. Chem.* **46**, 335 (1995).
121. L. Zhu and P. Johnson, *J. Chem. Phys.* **94**, 5769 (1991).
122. H.J. Dietrich, R. Lindner and K. Müller-Dethlefs, *J. Chem. Phys.* **101**, 3399 (1994).

6 Population Labelling Spectroscopy

Takayuki Ebata
Department of Chemistry, Graduate school of Science, Tohoku University, Sendai 980-77, Japan

6.1 Introduction

In general, electronic spectra of gas phase molecules are congested owing to the overlap of numerous bands and branches due to the transitions from thermally populated rovibrational levels. In addition, spectra often become very complicated due to perturbations among electronically excited states. They cause severe problems in analyzing the electronic spectrum of a large polyatomic molecule or a diatomic molecule consisting of heavy atoms, since its rotational constant is very small.

Population labelling spectroscopy enables the rotational structures of such congested electronic spectra to be resolved. This spectroscopy is essentially double resonance spectroscopy in which two tunable laser lights probe two different electronic transitions from a common level of the electronically ground state. Here, one of the electronic transitions is already analyzed by another means and the other transition is investigated. This spectroscopy was proposed by Schawlow *et al.* [1,2], who succeeded in analyzing the rovibrational structure of the congested $B^1\Pi_u \leftarrow X^1\Sigma_g^+$ electronic spectrum of gas phase Na_2, free from Doppler shifts or Doppler broadening. Population labelling spectroscopy is also used to separate a homogeneous line shape from an inhomogeneously broadened or congested spectrum. Population labelling spectroscopy has been successfully applied to analyze complicated molecular spectra of Cs_2 [3], BaI [4-7], NO_2 [8-10], CO_2^+ [11,12] and azulene [13].

There is another application of population labelling spectroscopy, which was proposed by Colson and coworkers [14]. They applied this spectroscopy to isolate an absorption band of a single species from the overlapping electronic spectra of molecules or a group of clusters with different conformations (isomers) and/or sizes which are generated in a supersonic jet. Atomic and molecular clusters, which are weakly bound by van der Waals forces or hydrogen-bonds, have been extensively studied by a number of researchers. The clusters are generated by an adiabatic expansion of gas phase molecules into vacuum through a pinhole, which forms a so-called supersonic jet. In the jet, a number of clusters of various sizes and different isomeric forms are simultaneously generated in the same region and their electronic states are located quite closely. Thus, when we measure an absorption spectrum by laser induced fluorescence (LIF) spectroscopy or resonance enhanced multiphoton ionization (REMPI) spectroscopy, the spectrum is overlapped by transitions of these species. Multiphoton ionization mass spectrometry can separate a species having a specific mass. However, it is sometimes difficult to avoid fragmentation of the clusters after the ionization and, moreover, it is impossible to discriminate between species having the same mass, such as isomers. Colson and coworkers called this spectroscopy as "Spectral hole burning spectroscopy" and first applied it to distinguish between the S_1 - S_0 electronic transitions of *cis* and *trans* isomers of *m*-cresol [14] and then extended their work to discriminate the S_1 - S_0 electronic spectrum of a specific species of phenol-$(H_2O)_n$ hydrogen-bonded clusters [15]. Since then, this technique has been successfully applied to distinguish the isomeric form of molecular clusters, such as the van der Waals clusters of benzene [16,17], hydrogen-bonded clusters of phenol derivatives [18-20], and large polyatomic molecules and clusters [21-23].

6.2 Principle and experimental setup

6.2.1 Population labelling spectroscopy for the analysis of complicated electronic spectrum

Population labelling spectroscopy is double resonance spectroscopy in which one measures the electronic spectrum as the spectrum of the population modulation. The modulation is induced by an intense scanning laser and is probed by a probe laser. Figure 6.1 shows the electronic energy levels of molecular system. The state $|0, J''\rangle$ is the lower electronic state, usually the electronically ground state, with the rotational level denoted by J''. The state $|1, v', J'\rangle$ is an electronically excited state with a rovibrational level denoted by v' and J'. We assume that the rotational structure of the state $|1, v', J'\rangle$ is already analyzed by another means. We also assume that the upper electronic state $|2, v, J\rangle$ has a complicated rovibrational structure and its rotational analysis is difficult by a conventional spectroscopic method such as LIF or REMPI. Population labelling spectroscopy is used to analyze the rotational structure of the $|2, v, J\rangle$ state. Schawlow *et al.* proposed four types of population labelling spectroscopy as shown in Fig. 6.1 [1]. In these schemes, two tunable lasers, a pump laser ($h\nu_1$) and a probe laser ($h\nu_2$), are used. The two lasers are either cw lasers or pulsed lasers. In the latter case, the pump laser is introduced prior to the probe laser.

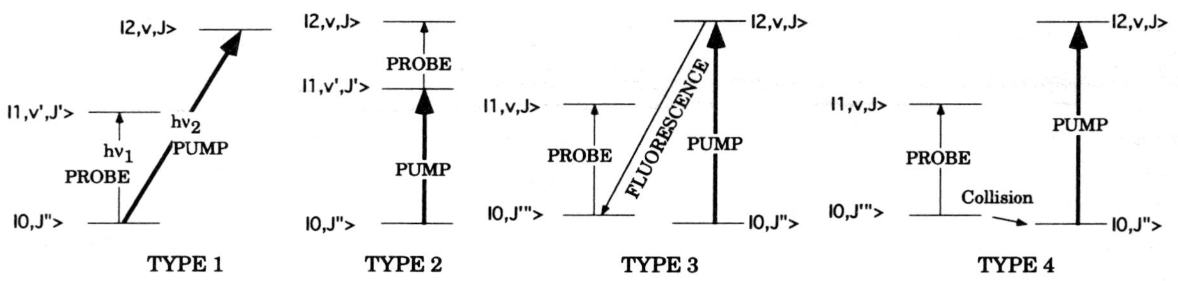

Fig. 6.1 Four excitation schemes of population labelling spectroscopy.

In type 1, the frequency of the weak probe laser ($h\nu_2$) is fixed to the specific $|1, v', J'\rangle \leftarrow |0, J''\rangle$ rotational transition and fluorescence (LIF) from the $|1, v', J'\rangle$ level is monitored. Thus, a single rotational level J'' of the state $|0\rangle$, that is $|0, J''\rangle$, is labelled by $h\nu_2$. Under this condition, the pump laser ($h\nu_1$) is scanned for the $|2, v, J\rangle \leftarrow |0, J''\rangle$ transition. The power of the pumping laser is set strong so that it depletes the population of each rotational level of the $|0\rangle$ state when its frequency is resonant with the $|2, v, J\rangle \leftarrow |0, J''\rangle$ transition. Since the fluorescence signal due to the probe laser is proportional to the population in $|0, J''\rangle$, a reduction of the fluorescence intensity occurs when $|2, v, J\rangle \leftarrow |0, J''\rangle$ transition shares the common J'' level of the $|0, J''\rangle$ state. Thus, by scanning the $h\nu_1$ laser frequency while monitoring the fluorescence from the $|1, v', J'\rangle$ level, a fluorescence-dip spectrum corresponding to the $|2, v, J\rangle \leftarrow |0, J''\rangle$ transition is obtained. The rotational selection rule for an electric dipole transition is $\Delta J = 0, \pm 1$, thus only three fluorescence dips corresponding to P, Q and R branches are observed from each $h\nu_2$ selected $|0, J''\rangle$ level, which greatly simplifies the $|2, v\rangle \leftarrow |0\rangle$ electronic spectrum. Finally, a set of the rotational energy levels of the $|2, v\rangle$ state can be obtained by measuring the $|2, v, J\rangle \leftarrow |0, J''\rangle$ fluorescence-dip spectrum for each J'' level, which is labelled by the $|1, v', J'\rangle \leftarrow |0, J''\rangle$ transition with $h\nu_2$. The advantages of the type 1 scheme are: (1) In this scheme, we observe the absorption spectrum as a depletion of the population by the sensitive LIF method instead of observing a direct absorption or an intensity modulation of irradiated laser light. Thus, one can measure the electronic spectrum even for the low density gas phase molecules whose direct absorption can not be observed. (2) The spectroscopy is applicable to the observation of a transition involving nonfluorescent state.

In type 3, the molecules excited to $|2, v, J\rangle$ emit fluorescence and return to the ground state levels, $|0, J'''\rangle$, with a rotational selection rule of $J''' - J = 0, \pm 1$. Thus, LIF intensities from these cascade-populated levels are strengthened and fluorescence enhancements are observed. Type 4 excitation scheme is used at a higher pressure. At high pressures, collisions transfer the molecules from nearby rotational levels, $|0, J'''\rangle$, to the level $|0, J''\rangle$ depleted by the intense $h\nu_1$. Therefore, the population of these neighboring rotational levels decrease when intense $h\nu_1$ laser is resonant on the $|2, v, J\rangle \leftarrow |0, J''\rangle$ transition and a dip is observed in the LIF signal monitored by $h\nu_2$. Type 2 is stepwise excitation, as described in Chapter 2, and we will not describe it in this chapter.

Figure 6.2(a) shows an experimental setup of population labelling spectroscopy with a cw-laser system used for the analysis of the C $^2\Pi_{3/2,1/2} \leftarrow$ X system of BaI [5]. The corresponding energy diagram is shown in Fig. 6.3. In Fig. 6.2(a), cw-laser beams 1 and 2 pass through the sample gas in a counterpropagated manner (In this case, the lasers cross a BaI molecular beam). Laser 2(L2) is fixed to the C $^2\Pi_{1/2} \leftarrow$ X transition and laser 1(L1) is scanned through the C $^2\Pi_{3/2} \leftarrow$ X transition. L1 and L2 are modulated at frequencies ω_1 and ω_2, respectively. Photomultiplier 1(PMT 1) monitors the fluorescence from C $^2\Pi_{3/2}$ through filter 1 (F1) and photomultiplier 2 (PMT 2) monitors the fluorescence from C $^2\Pi_{1/2}$ through filter 2 (F2). Lock-in 1 demodulates the excitation spectrum from the scanning laser L1, while lock-in 2 monitors the fluorescence signal from the fixed laser L2 modulated at the chopping frequency ω_1 of L1, thereby generates the double resonance signal. Figure 6.2(b) shows an experimental setup used for a pulsed laser system. A weak probe, pulsed laser 2, is fixed to the $|1\rangle \leftarrow |0\rangle$ transition, and the fluorescence emitted from $|1\rangle$ is monitored by a photomultiplier (PMT) through a filter. Photocurrent from the photomultiplier tube is processed by a boxcar integrator connected to a microcomputer. The pump, pulsed laser 1, is introduced prior to the laser 2 and is scanned to selectively depopulate ground state molecules through the $|2\rangle \leftarrow |0\rangle$ electronic transition. The double resonance spectrum can be obtained by monitoring the fluorescence from state $|1\rangle$ while scanning $h\nu_1$.

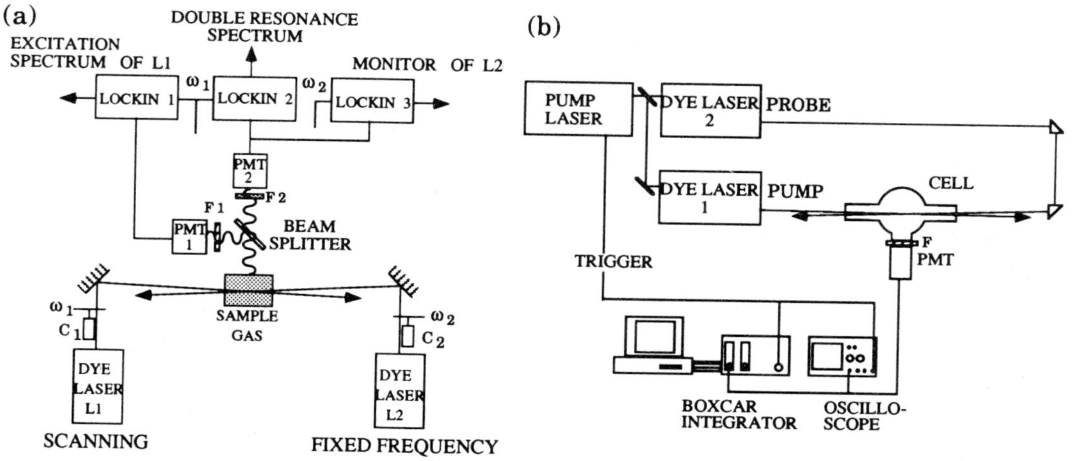

Fig. 6. 2 Experimental setup of population labelling spectroscopy by use of (a) cw-lasers and (b) pulsed lasers. F1, F2, F : Filter, PMT1, PMT2, PMT : Photomultiplier tube, C_1, C_2 : Chopper (frequency ω_1, ω_2).

Figure 6.3 shows an example of the type 1 and type 3 schemes applied for the BaI C $^2\Pi_{3/2,1/2}$ - X $^2\Sigma^+$ system [12]. Figure 6.3(a) shows the C $^2\Pi_{3/2}$ - X $^2\Sigma^+$ LIF spectrum and Fig. 6.3(b) shows the population labelling spectrum simultaneously observed. Here, laser 2 (L2) is tuned to a single rotational line, Q_{12}

($9^1/_2$), of the C $^2\Pi_{1/2}$ - X $^2\Sigma^+$ transition and laser 1(L1) is scanned through the C $^2\Pi_{3/2}$ - X $^2\Sigma^+$ subband. As seen in the figure, the LIF spectrum is heavily congested by overlaps of many rotational lines, while only four lines (two dips and two positive peaks) appear in the population labelling spectrum. In Fig. 6.3 (b), each dip occurs when L1 excites the molecules in the same rotational level with that monitored by L2, e.g., P_2 ($9^1/_2$) and Q_2 ($9^1/_2$). This scheme corresponds to type 1. On the other hand, each positive peak appears when the molecules excited by L1 emit fluorescence and return to the rotational level monitored by L2. When L1 is resonant with P_2 ($11^1/_2$) and P_{21} ($10^1/_2$) lines, it pumps the molecules to J = $10^1/_2$ and $9^1/_2$ levels in the C $^2\Pi_{3/2}$ state. The pumped molecules then emit fluorescence and return to J'' = $9^1/_2$ level via the R_2 ($9^1/_2$) or Q_2 ($9^1/_2$) transition, respectively. This repopulation results in the enhancement of fluorescence signal due to L2 which is fixed to Q_{12} ($9^1/_2$). A set of population labelling spectra measured at different rotational levels with fixed L2 give us detailed information of the rotational structure of the C $^2\Pi_{3/2}$ state.

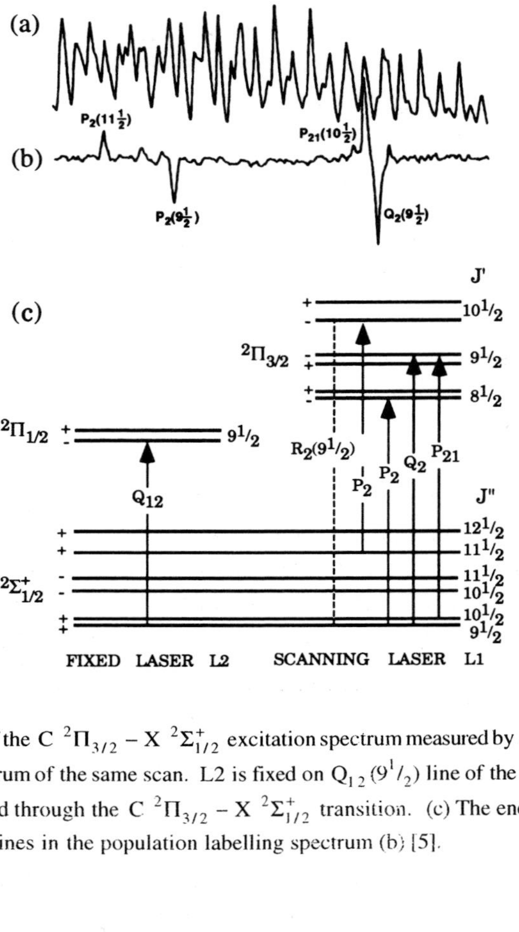

Fig. 6. 3 (a) A portion of the C $^2\Pi_{3/2}$ – X $^2\Sigma^+_{1/2}$ excitation spectrum measured by scanning Laser 1. (b) Population labelling spectrum of the same scan. L2 is fixed on Q_{12} ($9^1/_2$) line of the C $^2\Pi_{1/2}$ – X $^2\Sigma^+_{1/2}$ transition and L1 is scanned through the C $^2\Pi_{3/2}$ – X $^2\Sigma^+_{1/2}$ transition. (c) The energy level diagram identifying the modulated lines in the population labelling spectrum (b) [5].

6.2.2 *Doppler-free spectroscopy*

If the spectral bandwidths of the lasers are narrower than that of Doppler width, types 1 and 3 become Doppler-free spectroscopy. Figure 6.4 shows an example of the population labelling spectra obtained for different probe laser (L2) laser frequencies. We consider the case that L2 monitors the molecules having a velocity component of v_z in the opposite direction to the propagation of L2. If the pumping laser (L1) is introduced in the counterpropagated direction, the molecules having this velocity component are modulated when $\nu_1 = \nu_0 (1+ v_z / c)$, where ν_0 is the center frequency of the $|2\rangle \leftarrow |0\rangle$ transition. Thus, a sharp peak is observed at the frequency of $\nu_1 = \nu_0 (1+ v_z / c)$ in the population labelling spectrum. To obtain the

Doppler-free spectrum, the frequency of L2 (ν_2) must be fixed to the center of a Doppler broadened line of the $|1\rangle \leftarrow |0\rangle$ transition by using the Doppler-free spectroscopic technique described in Chapter 1 (see Fig. 1.18). In this case, the molecules with zero-velocity component along the beam direction are monitored by L2, and the intensity modulation by L1 is observed only for the molecules having a zero-velocity component and the spectrum free from Doppler broadening is obtained.

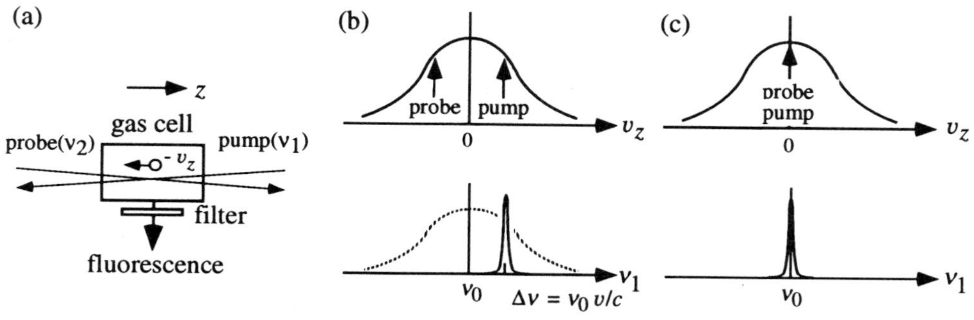

Fig. 6. 4 Schematic spectral profile with the counterpropagated laser beam arrangement. (a) Experimental setup. Here, fluorescence from the state pumped by ν_2 is monitored through a filter (see Fig. 6. 1). (b) (Top) shows the population distribution as a function of v_z. (Bottom) shows a profile of the population labelling spectrum when the probe laser monitors the molecules with the velocity component of - v_z. Dotted curve is the Doppler broadened spectrum. (c) (Top) shows that the probe laser monitors the molecules having the zero-velocity component. (Bottom) shows a profile of the population labelling spectrum.

6.2.3 Population labelling spectroscopy as a method of discriminating structural isomers

The excitation scheme is shown in Fig. 6.5. As was described previously, an application of the population labelling spectroscopy to distinguish the vibronic bands of single species from the electronic spectrum of several overlapping species or clusters formed in a supersonic jet was proposed by Colson and coworkers and is known as "spectral hole burning spectroscopy" [14].

In this method, both L1 and L2 are pulsed lasers and the pumping laser (L1) is introduced prior to the probe laser (L2). The L2 intensity is set to be weak, while that of L1 is strong enough to deplete the population of molecules in the ground state. Two schemes are shown in Fig. 6.5. In the scheme (A), L2 is fixed to an electronic transition of a specific species (isomer A). Either the fluorescence from the excited state or the ions generated by the resonance-enhanced multiphoton ionization (REMPI) is monitored. When the frequency of L1 is resonant with the transition of an isomer A, its population is depleted, that is a "hole" in the population of isomer (A) is created, and the signal due to L2 is weakened, while the signal is unchanged when L1 is resonant with the transition of isomer B. Therefore, by scanning L1 frequency while monitoring the fluorescence (ion) signal due to fixed L2, population labelling spectrum of isomer A is obtained (spectrum (b)). When L2 is fixed to the transition of isomer B (peak B) and L1 is scanned, a population labelling (dip) spectrum of isomer B is obtained. In this method, since L1 laser is intense, the obtained dip spectrum is often saturated and even a very weak transition can be observed. In scheme (B), on the other hand, L1 is fixed to the transition of isomer A and L2 is scanned. The repetition rate of L2 is set to twice that of L1 and two signals due to L2 with L1 off (corresponding to spectrum (a)) and on (corresponding to spectrum (b)) are recorded in an alternative manner. By subtracting spectrum (b) from spectrum (a), one can obtain the electronic spectrum of isomer A (spectrum (c)) alone.

Figure 6.6 shows an experimental setup. First, a supersonic free jet is formed by expanding a gaseous mixture of the sample gas and an inert gas into vacuum. The free jet is skimmed (Sk) to obtain a supersonic beam. An intense pump laser (L 1) crosses the supersonic beam. Then the probe pulse laser (L 2) beam crosses the molecular beam a few hundreds of μm downstream from the crossing point of the pumping laser. Since the beam velocity is estimated to be from 1000 to 1500 m s^{-1}, this spatial displacement of the two laser beams corresponds to a delay time of a few hundreds of ns between the two laser pulses. In this setup, the pumping laser frequency is fixed to the selected transition of a specific species in the molecular beam and the frequency of the probe laser is scanned through the S_1 - S_0 transition. The S_1 state molecules are further ionized by an ionization laser (L 3). The ions are mass-separated by a reflectron time-of-flight mass spectrometer (Re-TOF) and detected by a channel electron multiplier (D). To obtain the population labelling spectrum in a good signal-to-noise ratio, it is necessary to eliminate the intense signal (either the ion or fluorescence), generated by the pumping laser, from that generated by the probe laser. In this setup, a pulsed electric field opposite to the molecular beam direction is applied at a time between the two laser pulses, which removes the ions produced by L 1. In the case of fluorescence detection, it is easy to distinguish the two signals since the fluorescence lifetime of the S_1 state, which is typically a few tens of nanoseconds or less, is shorter than the time delay between the two laser pulses.

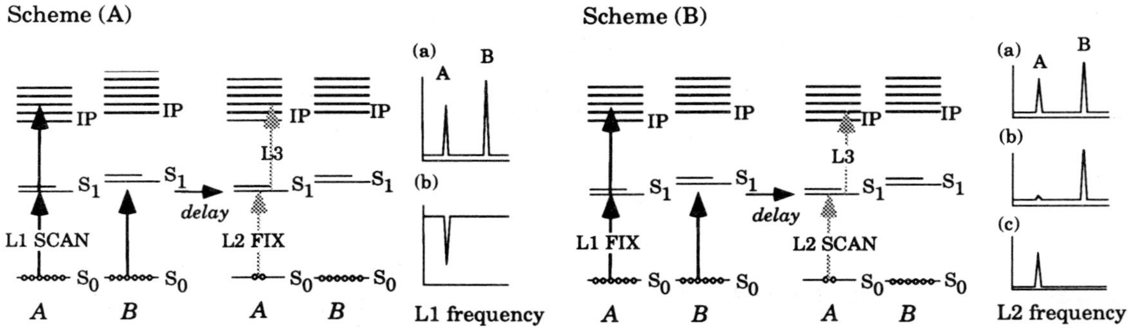

Fig. 6. 5 Two schemes of population labelling spectroscopy for the discrimination of structural isomers. **(Scheme A)** L2 is fixed to the band of isomer A and ion signal is monitored. L1 is introduced prior to L2 and is scanned. Spectrum (a) shows the S_1 - S_0 spectrum obtained by scanning L1. Spectrum (b) shows the population labelling spectrum of isomer A, which is obtained as a dip spectrum.
(Scheme B) L1 is fixed to the band of isomer A and L2 is scanned. Two spectra, L1 is off (spectrum (a)) and L1 is on (spectrum (b)), are observed in an alternative manner and the population labelling spectrum (spectrum (c)) is obtained by subtracting the spectrum (b) from the spectrum (a) [17].

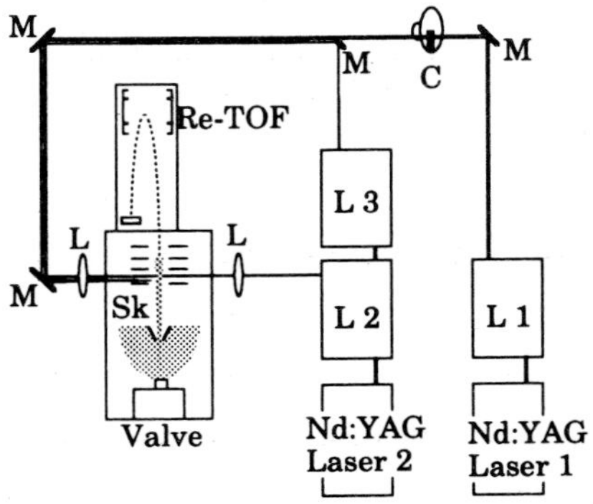

Fig. 6. 6 Experimental setup of population labelling spectroscopy applied to the discrimination of single isomer in a supersonic beam. L1: Pump laser, L2: Probe laser, L3: Ionization laser, L: Lens, M: Mirror, C: Chopper, Sk: Skimmer, Re-TOF: Reflectron time-of-flight mass analyzer [17].

Figure 6.7 shows an application of fluorescence detected population labelling spectroscopy to the S_1 - S_0 absorption spectrum of 2-naphthol [24]. Figure 6.7(a) shows the S_1 - S_0 LIF spectrum of bare 2-

naphthol and its hydrogen-bonded cluster with H_2O. In bare 2-naphthol there are two isomers, *cis* and *trans*, as are shown in the figure and the observed bands are due to the transitions of a mixture of the two species and their hydrogen-bonded clusters with H_2O. Figures 6.7(b)-(e) show the population labelling spectra obtained by monitoring the bands shown in Fig. 6.7(a), that is, (b) *cis*-2-naphthol monomer, (c) bare *trans*-2-naphthol monomer, (d) *cis*-2-naphthol-H_2O and (e) and *trans*-2-naphthol-H_2O. The spectra were obtained by using scheme (A) in Fig. 6.7. The distance between the pump laser and the probe laser was set to 10 mm, corresponding a delay time of 5.5 μs between the two laser pulses. It is obvious that each population labelling spectrum discriminates vibronic bands of each species from the LIF spectrum which is overlapped by the transitions of several species so that an analysis of the vibronic structure of each species is made very easy.

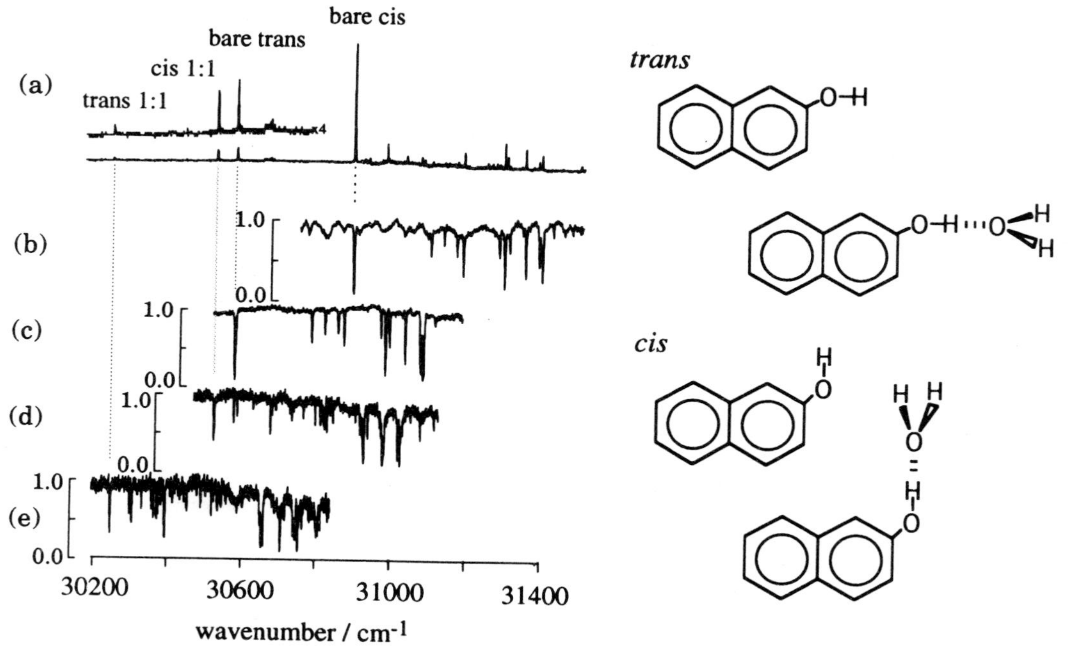

Fig. 6. 7 (a) S_1 - S_0 LIF spectrum of 2-naphthol and 2-naphthol-(H_2O) (1:1) cluster in a supersonic jet. Population labelling spectra of (b) *cis*-2-naphthol, (c) *trans*-2-naphthol, (d) *cis*-2-naphthol-(H_2O) (1:1) cluster and (e) *trans*-2-naphthol-(H_2O) (1:1) cluster [24].

6.3 Application of population labelling spectroscopy

A. Analysis of rotational structure of the $\tilde{B}^2\Sigma_u^+$ - $\tilde{X}^2\Pi_g$ system of CO_2^+ ion

The absorption spectrum of the $\tilde{B}\ ^2\Sigma_u^+(000)$ - $\tilde{X}\ ^2\Pi_g\ (000)$ band of gas phase CO_2^+ is known to show a complicated structure and only a part of the rotational branches could be assigned by conventional LIF spectroscopy [11]. The complexity of the rotational structure is thought to be due to the $\tilde{B}$ state being perturbed by nearby vibronic levels of the $\tilde{A}\ ^2\Pi_u$ state. Johnson *et al.* applied population labelling spectroscopy to analyze the rotational structure [11]. Figure 6.8(a) shows a typical population labelling spectrum, and the relevant energy level diagram is shown in Fig. 6.8(c). In this measurement, the probe laser frequency was fixed to the rotational line which was thought to be the $R_1(4^1/_2)$ line of the $\tilde{B}\ ^2\Sigma_u^+$ (000) - $\tilde{X}\ ^2\Pi_g$ (000) band, and the fluorescence from the upper level was monitored. The pump laser pulse was introduced 1.5 μs prior to the probe laser pulse and its frequency was scanned through the $\tilde{A}\ ^2\Pi_u$

(202) - $\tilde{X}$ $^2\Pi_g$ (000) transition whose rotational structure had already been analyzed. As can be seen in Fig. 6.8(a), three rotational lines, P, Q and R from the same rotational level of the $\tilde{X}$ state are observed and it was confirmed that the labelled level was $J'' = 4^1/_2$ and that of the $\tilde{B}$ $^2\Sigma_u^+$ (000) state was $J' = 5^1/_2$. The rotational structure of the $\tilde{B}$ $^2\Sigma_u^+$ (000) was analyzed by measuring the similar spectra from different labelled rotational levels and it was found that the $\tilde{B}$ $^2\Sigma_u^+$ (000) level was perturbed by more than two vibronic levels belonging to the $\tilde{A}$ $^2\Pi_u$ state.

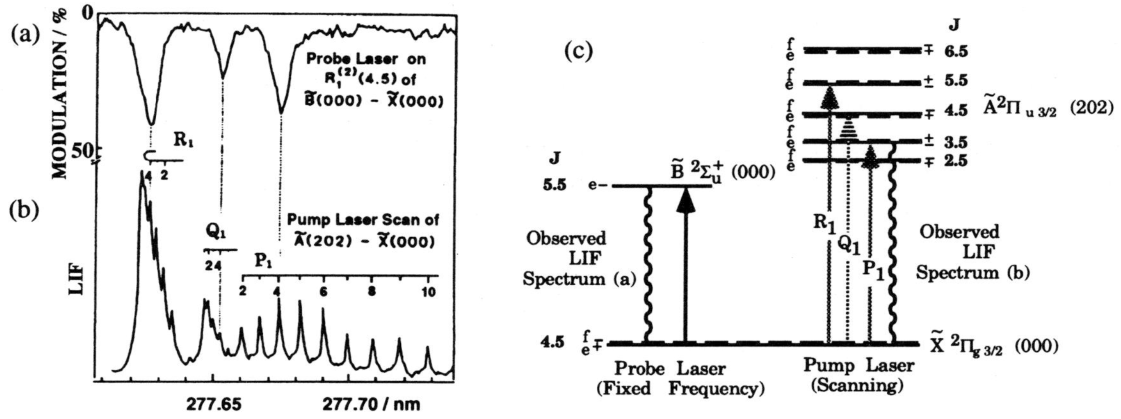

Fig. 6. 8 Population labelling spectroscopic measurement of CO_2^+. (a) LIF signal from probe laser fixed at $R_1(4.5)$ transition of the $\tilde{B}$ $^2\Sigma_u^+(000) - \tilde{X}$ $^2\Pi_{g\,3/2}(000)$ sub-band, while pump laser is scanned through the $\tilde{A}$ $^2\Pi_{u\,3/2}(000) - \tilde{X}$ $^2\Pi_{g\,3/2}(000)$ transition. (b) Excitation spectrum from the pump laser. (c) Level diagram indicating the origin of the transitions observed in (a) [11].

B. S_1 - S_0 transition of jet-cooled azulene For large polyatomic molecules, population labelling spectroscopy makes it possible to analyze a complicated vibronic structure and to investigate a broadening of an absorption band due to a short lifetime of the upper state. In addition, as described above, the type 1 scheme shown in Fig. 6.1 is very useful for observing the non-fluorescent state. A typical example is the observation of the first excited state (S_1) of jet-cooled azulene. The anomalous fluorescence characteristics of the S_1 and S_2 states of azulene are well known [25]. The S_2 state fluoresces even in the condensed phase, while the S_1 state is almost non-fluorescent due to fast internal conversion (IC). The fluorescence quantum yield of S_1 is reported to be 5 x 10^{-6} [26]. Though the IC rate constant in S_1 was measured in a mixed crystal with naphthalene [27], it was not clear whether the observed rate was equal to that under an isolated condition or not. One of the method was to measure the bandwidth of the S_1 - S_0 absorption spectrum of azulene cooled in a supersonic free jet. Rotational temperature of molecules in a supersonic jet was typically less than 10 K and so the transition occurred only from limited number of rotational levels. However, the measurement of the absorption spectrum would be very difficult because of a low molecular density in the jet and also of a very low S_1 - S_0 oscillator strength of azulene (f = ~ 0.009) [28]. The measurement of the S_1 - S_0 absorption spectrum of jet-cooled azulene was actually performed by Amirav and Jortner by using the slit jet absorption spectroscopy [29] and they reported the bandwidth of the 0,0 band to be 8.8 cm^{-1}. On the other hand, Suzuki *et al.* [13] applied population labelling spectroscopy to the same transition of the jet-cooled azulene to obtain the bandwidth. Figure 6.9 shows the energy level diagram of azulene and the obtained population labelling spectrum. In the experiment, the pump laser (ν_1)

was scanned through the S_1 - S_0 transition and the probe laser (ν_2), was fixed to the 0,0 band of the S_2 - S_0 transition to monitor the population of azulene in S_0 by 1 + 1 REMPI through S_2. As shown in Fig. 6.9(b), many strong dips corresponding to the S_1 - S_0 absorption appeared. Interesting point in the spectrum is that intensities of several bands exceed 50 %. The strong dip intensities are explained by that the molecules excited in the S_1 state decay so fast by IC that they have negligible chance to be dumped into the same S_0 state by stimulated emission within the laser pulse duration of ν_1. Thus, only $S_1 \leftarrow S_0$ absorption process occurs efficiently, resulting in a large reduction of the ground state population. This is the primary advantage of population labelling spectroscopy, that is, this spectroscopy is suitable for electronic states which have very short lifetimes. Accurate frequencies of vibronic bands of the S_1 state were determined from the population labelling spectrum. Under weak laser condition, the dip spectrum fitted well to a Lorentzian curve, and the bandwidth was determined to be 5.5 cm^{-1}, which corresponds to the IC rate constant of ~1 x 10^{12} s^{-1}. The bandwidth obtained by Suzuki *et al.* was narrower than that reported by Amirav and Jortner, and the difference was attributed to a low spectral resolution of the monochromator and also to higher rotational temperature of azulene in a slit jet in the absorption measurement. The fast IC of the S_1 sate was attributed to the large vibrational overlap integral in the matrix element of IC due to a drastic difference in the geometry of azulene between S_1 and S_0 [13].

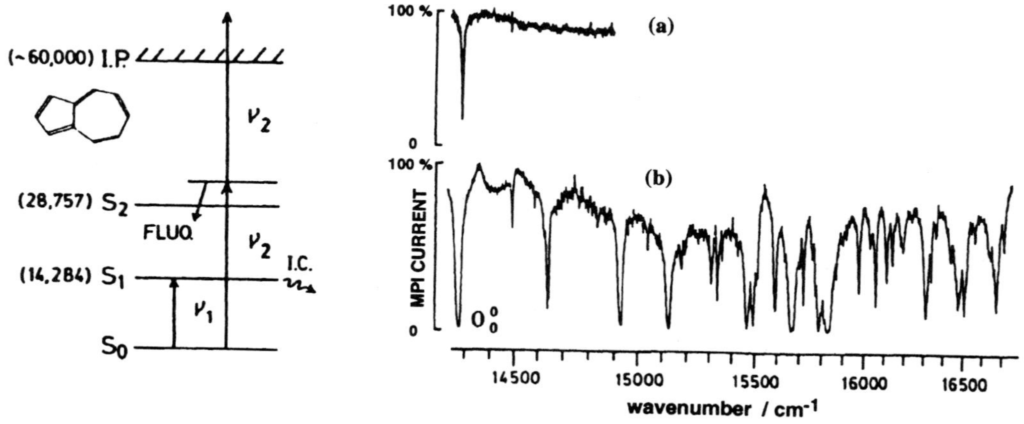

Fig. 6. 9 (Left) Energy level diagram of azulene. (Right) Population labelling spectrum of jet-cooled azulene measured under (a) low ν_1 laser power, and (b) high ν_1 laser power.

C. Conformational isomers of benzene dimer Benzene dimer is the simplest dimer of the aromatic molecular system and the determination of its stable structure is of fundamental importance both in theoretical and experimental points of view. The obtained result gives us a basic idea on the characteristics of intermolecular forces in the clusters consisting of aromatic molecules. Janda *et al.* showed that the dimer has a permanent dipole moment and suggested the structure in which two benzene molecules are

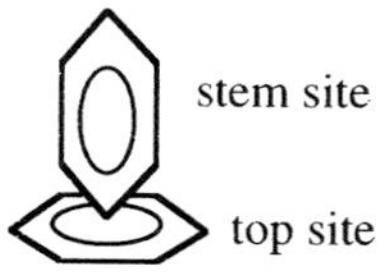

T-shaped

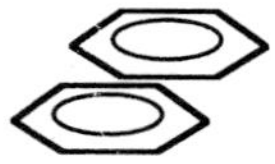

parallel displaced

perpendicular each other, a T-shaped structure [30]. On the other hand, Bernstein *et al.* proposed that a parallel displaced structure (sandwich structure) was more stable than the T-shaped structure based on a semi-empirical calculation for the intermolecular potential [31]. In the bottom part of Fig. 6.10 is shown the 6_0^1 band of the S_1 - S_0 transition of benzene dimer [32]. The 6_0^1 band of the dimer is red-shifted by

40 cm^{-1} from that of the monomer, and the band consists of a sharp doublet peak at low frequency side and a weak progression of ~ 17 cm^{-1} interval at high frequency side.

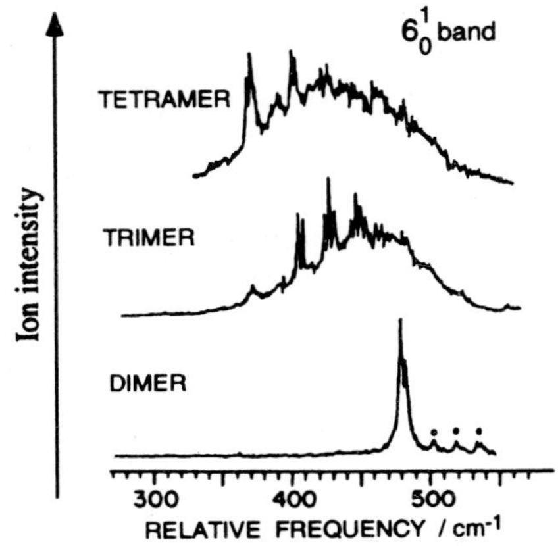

Fig. 6. 10 Resonance enhanced multiphoton ionization (REMPI) spectra of 6_0^1 band of benzene tetramer, trimer and dimer. The frequency scale is relative to the monomer origin at 38086.1 cm^{-1}. The dimer 6_0^1 band is red-shifted by 40 cm^{-1} from that of the monomer. Dotted bands are assigned to the intermolecular vibration [32].

If the dimer forms a parallel displaced structure, two benzene molecules are symmetrically equivalent. On the other hand, if the dimer forms a T-shaped structure, two molecules are nonequivalent. Since an environment on each molecule is different in the T-shaped dimer, the two molecules would have different S_1 - S_0 transition energies. Based on this idea, population labelling spectroscopy was applied by Scherzer *et al.* to investigate to which isomer the observed bands belonged or whether the bands were to be assigned to the transitions of the overlapping two isomers [16]. They measured the population labelling spectra of benzene dimer, $(C_6H_6)_2$, and isotopically substituted dimer, $(C_6H_6)(C_6D_6)$. Figure 6.11(a) shows the

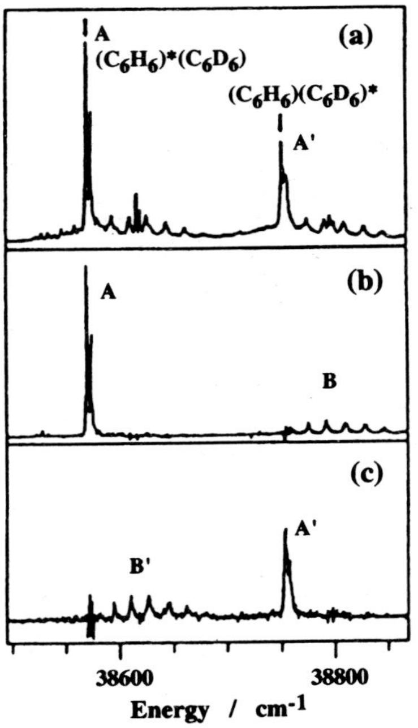

Fig. 6. 11 (a) 1+1' REMPI spectrum of $(C_6H_6)(C_6D_6)$ dimer. (b),(c) Population labelling spectra obtained by scanning monitoring laser frequency. The pump laser frequencies were fixed on the band A and A', respectively [16].

observed mass-separated two-color ionization spectrum of the 6_0^1 band of $(C_6H_6)(C_6D_6)$. Figure 6.11(b) and 11(c) show the population labelling spectra, where the excitation scheme (B) of Fig. 6.5 is applied. In Fig. 6.11(b), the frequency of the pump laser is fixed to band A in Fig. 6.11(a) and probe laser is scanned, and in Fig. 6.11(c), the pump laser is fixed to band A'. A complexity occurs in the $(C_6H_6)(C_6D_6)$ hetero dimer. If its structure is the T-shaped one, two isomers (1) and (2) are possible as shown below, and furthermore, two transitions corresponding to the electronic excitations localized on benzene, $(C_6H_6)^*(C_6D_6)$, and on deuterated benzene, $(C_6H_6)(C_6D_6)^*$, should be observed in each isomer. Actually, their transitions are clearly seen in the population labelling spectra. In Fig. 6.11(b), a sharp doublet band labelled by A is assigned to the electronic excitation localized on C_6H_6 on the stem site of the isomer (1),

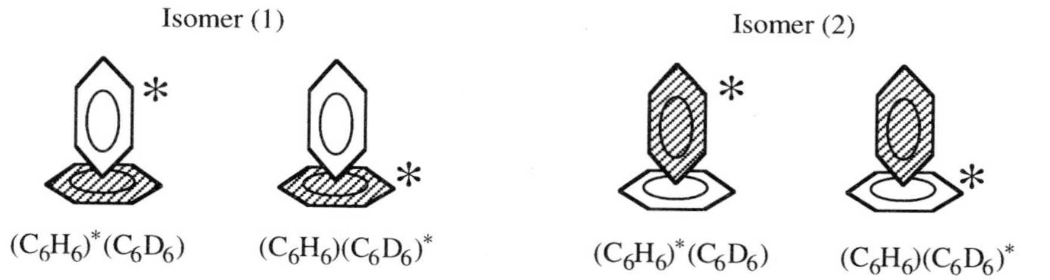

while a low-frequency vibrational progression labelled by B is assigned to the electronic excitation localized on C_6D_6 on the top site of the same isomer. On the other hand, in Fig. 6.11(c), sharp doublet band labelled by A' is assigned to the electronic excitation localized on C_6D_6 on the stem site of isomer (2), while a progression labelled by B' is assigned to that localized on C_6H_6 on the top site of the same isomer.

Figure 6.12(a) shows the two-color double resonant REMPI spectrum of $(C_6H_6)_2$. In the spectrum, a doublet peak labelled by A is assigned to the transition localized on C_6H_6 on the stem site, and a progression labelled by asterisks is assigned to C_6H_6 on the top site. Thus, the main feature observed in the electronic spectrum is attributed to that of the T-shaped dimer. However, in addition to the T-shaped dimer

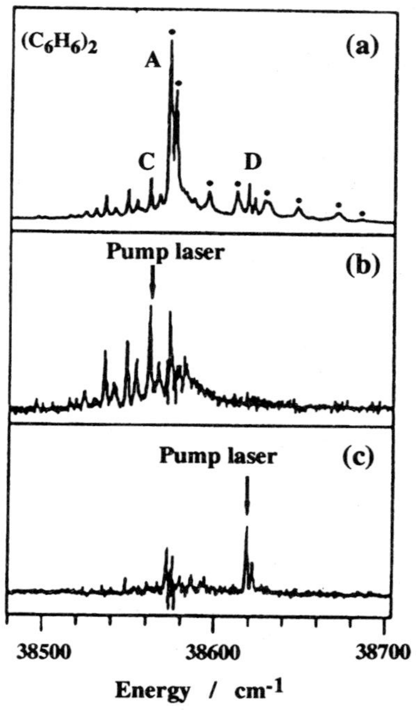

Fig. 6. 12 (a) 1+1' REMPI spectrum of the $(C_6H_6)_2$ dimer. Bands marked by dots are assigned to the T-shaped dimer. (b),(c)Population labelling spectra obtained by scanning the monitoring laser frequency. The pump laser frequencies were fixed to band C and D, respectively [16].

band, the other bands labelled by C and by D are observed ; the population labelling spectra obtained by fixing the laser frequency on C and D are shown in Fig. 6.12(b) and 12(c), respectively. Their spectral features are quite different from that of the T-shaped dimer, and they may be attributed to the parallel-displaced isomer , though more detailed experiments and theoretical works are necessar for firm conclusion.

As will be shown later, Raman spectroscopic studies combined with REMPI detection was performed by the group of Felker prior to the population labelling spectroscopic work and they confirmed that the two benzenes in the dimer are not symmetrically equivalent and proposed the T-shaped structure [33,34].

D. Hydrogen-bonded clusters of phenol In addition to clusters bound by van der Waals forces, hydrogen-bonded (H-bonded) clusters have also been extensively studied spectroscopically by many workers. H-bonded clusters of phenol with H_2O have attracted considerable attention [15,35-43], since phenol is a fundamental aryl alcohol and its H-bonding system is a prototype for other H-bonded clusters of aromatic molecules. Figure 6.13(a) shows the S_1 - S_0 LIF spectrum of phenol and phenol-$(H_2O)_n$ clusters in the band origin region. In the spectrum, band assignments with respect to sizes of the clusters had been performed by use of mass- analyzed REMPI spectroscopy [36,37]. However, as described above, it is impossible to distinguish isomers in the mass spectrum. It is seen from Fig. 6.13(a), the band origins of all the clusters are red-shifted from the band origin of bare phenol. This occurs because phenol is more acidic in S_1, thus the stabilization energy by H-bonding is larger in S_1 than S_0. Also the stabilization energy is different for different cluster size. Thus, each cluster has a different electronic transition energy. Among the bands assigned as vibronic bands of phenol-$(H_2O)_n$ clusters, the band of phenol-$(H_2O)_2$ having a peak at 36225 cm^{-1} is much broader (bandwidth is 25 cm^{-1}) than the bands of other clusters. Several explanations for the broadness were proposed: (1) The band is composed of transitions of several isomers of phenol-$(H_2O)_2$; (2) The spectrum is composed of a progression of low frequency vibrations of a single isomer; and (3) The spectrum is broadened by an anomalously short lifetime of the S_1 state. To investigate the anomalous band profile, Lipert *et al.* measured the population labelling spectrum of phenol-$(H_2O)_2$ by fixing the probe laser frequency on 36228 cm^{-1} [15]. Figure 6.13(b) shows the (Top) population labelling and (Bottom) REMPI spectra. As can be seen in the figure, both spectra are essentially identical and they concluded that the broad band is due to a single species of phenol-$(H_2O)_2$. They further investigated the band profile and reported that the band profile cannot be fitted by a single Lorentzian curve but is composed

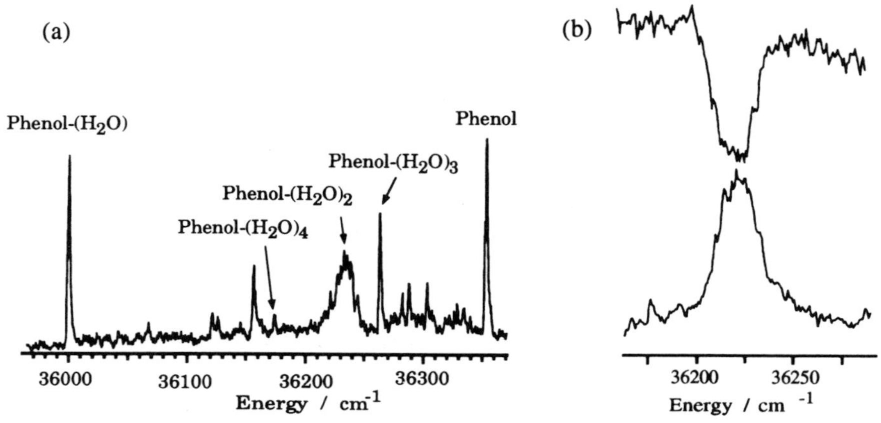

Fig. 6. 13 (a) S_1-S_0 LIF spectrum of phenol and phenol-$(H_2O)_n$ hydrogen-bonded clusters in the (0,0) band region. (b) (TOP) Population labelling spectrum of phenol-$(H_2O)_2$ cluster obtained by fixing probe laser frequency at 36228 cm^{-1} of the lower S_1-S_0 spectrum. Pump laser was fired 500 ns before the probe laser [15].

of a progression of 3 to 4 cm^{-1} or of two 6 to 8 cm^{-1} progressions. From the fact that only the phenol-$(H_2O)_2$ cluster exhibits this anomalous spectrum among all of the phenol-$(H_2O)_n$ clusters, they suggested that the geometry of phenol-$(H_2O)_2$ was different from others. As will be discussed later, very recently the IR spectrum of the OH stretching vibrations of phenol-$(H_2O)_n$ n = 1 - 4 clusters were measured and the structures of the clusters with n = 2 to 4 are determined to be a cyclic form in S_0 [44-46]. The anomalous vibronic structure of phenol-$(H_2O)_2$ is attributed to a large geometric change in S_1, which is induced by a drastic increase of acidity of phenol from S_0 (p*K*a = 10) to S_1 (p*K*a = 4) [47].

6.4 Other related spectroscopies

Though population labelling spectroscopy are generally applied to electronic transitions, it can be extended to vibrational spectroscopy. In this section, two forms of vibrational spectroscopy, which can also be classified as population labelling spectroscopy, are described; (1) infrared(IR) - UV double resonance spectroscopy and (2) stimulated Raman - UV double resonance spectroscopy. Vibrational spectroscopy has been extensively used to study the molecular structure and intermolecular interactions in the condensed phase, and the application to molecular clusters are also useful to determine their structures. There are two difficulties to overcome to perform vibrational spectroscopy for molecular clusters. First is that the concentration of clusters in a jet is too low to apply the conventional method. Second is the selection of single species (size and isomer) from the supersonic jet in which many species are simultaneously generated. Two spectroscopies described in this section have high sensitivity and selectivity sufficient to observe the vibrational spectra of the clusters.

Figure 6.14 shows the excitation schemes for IR-UV double resonance spectroscopy and stimulated Raman-UV double resonance spectroscopy. In both spectroscopies, a probe pulsed UV laser ($h\nu_{UV}$) monitors the population of a specific cluster in a supersonic jet by REMPI through the S_1 state. Under this condition, an intense tunable IR laser pulse ($h\nu_{IR}$) or two Raman pumping laser pulses ($h\nu_1$, $h\nu_2$) are introduced a few tens of ns prior to the UV laser pulse to pump the ground state cluster molecules to a vibrational level. A depletion in the REMPI signal is observed when the IR laser frequency, $h\nu_{IR}$, or the frequency difference of Raman pumping lasers, $h(\nu_1 - \nu_2)$, is resonant with the vibrational transition. Thus by scanning the laser frequency while monitoring the REMPI signal, an ion-dip spectrum which corresponds to the IR or Raman spectrum is obtained. These spectroscopies are called "ionization detected infrared spectroscopy (IDIRS)" and "ionization detected stimulated Raman spectroscopy (IDSRS)".

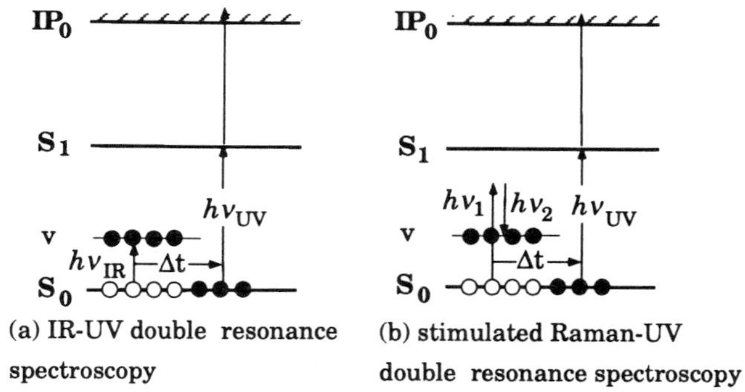

Fig. 6. 14 Population labelling spectroscopies for the observation of vibrational levels in the S_0 state.

Figure 6.15 shows IDIR spectra of jet-cooled phenol-$(H_2O)_n$, n = 0 - 4, in a region of the OH stretching vibration [46]. Here, $h\nu_{UV}$ was tuned to one of the bands of the S_1 - S_0 electronic transition of each cluster (see Fig. 6.13(a)), and the 1+1 REMPI signal through the S_1 state was monitored. A tunable

IR laser ($h\nu_{IR}$) generated by difference frequency mixing was introduced in a counterpropagated direction to $h\nu_{UV}$ and its frequency was scanned while monitoring the REMPI signal. In bare phenol, the phenolic OH stretching vibration is observed at 3657 cm^{-1}. In the phenol-H_2O 1:1 cluster, this vibration drastically shifts to the red, which indicates that the OH bond strength of the phenol site is drastically weakened upon the H-bond formation. In addition to the phenolic OH stretching vibration, OH stretching vibrations of the H_2O site are also observed. For larger size clusters, the observed spectra were compared with those obtained by *ab initio* calculations [48], and it was concluded that the structures of the clusters with n = 2 to 4 are ring-form as shown on the right side of Fig. 6.15. The bands which become broad and show a large red-shift with an increase of n are assigned to the stretching vibrations of the H-bonded OH groups in the cluster ring. The bands located at 3720 cm^{-1}, regardless of cluster size, are assigned to the stretching vibrations of the OH groups that protrude out of the ring.

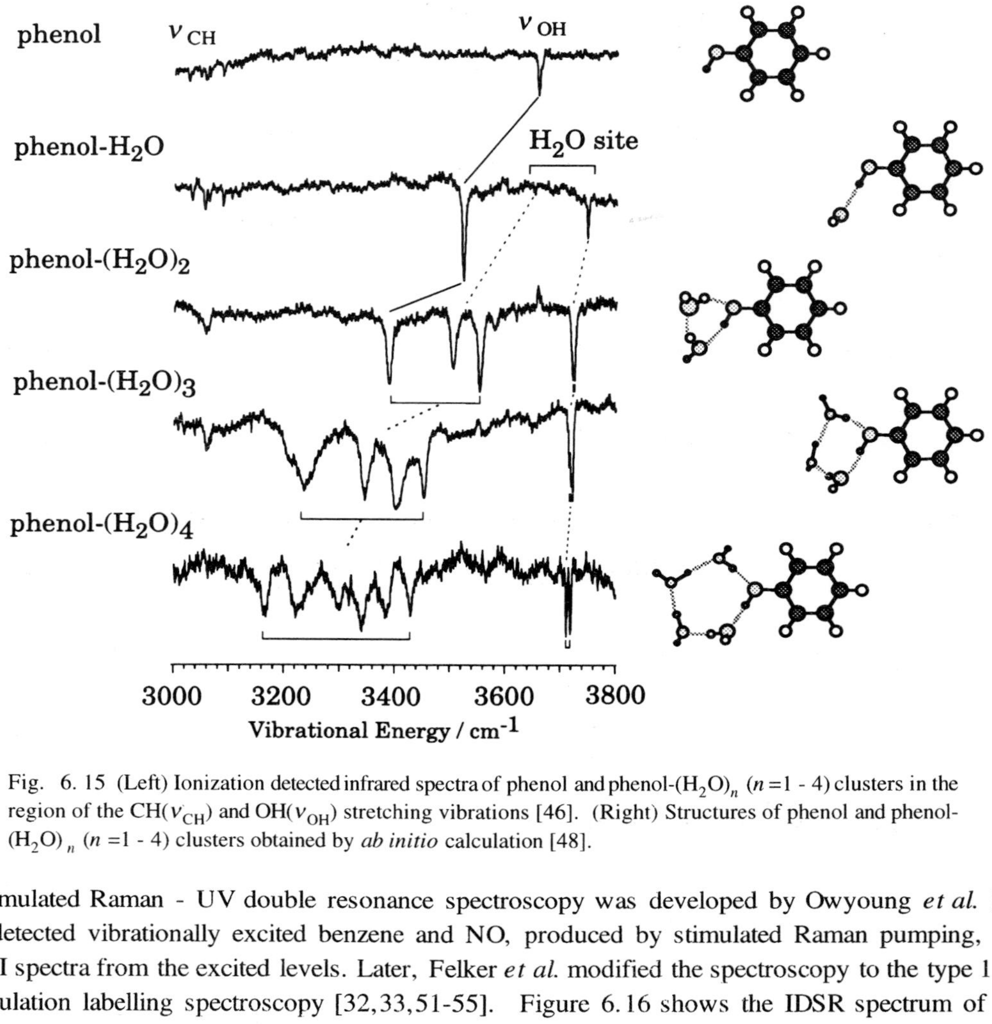

Fig. 6. 15 (Left) Ionization detected infrared spectra of phenol and phenol-$(H_2O)_n$ (n =1 - 4) clusters in the region of the CH(ν_{CH}) and OH(ν_{OH}) stretching vibrations [46]. (Right) Structures of phenol and phenol-$(H_2O)_n$ (n =1 - 4) clusters obtained by *ab initio* calculation [48].

Stimulated Raman - UV double resonance spectroscopy was developed by Owyoung *et al.* [49,50]. They detected vibrationally excited benzene and NO, produced by stimulated Raman pumping, by their REMPI spectra from the excited levels. Later, Felker *et al.* modified the spectroscopy to the type 1 scheme of population labelling spectroscopy [32,33,51-55]. Figure 6.16 shows the IDSR spectrum of mode 1 (ring breathing vibration) of benzene dimer. As described above, two stable isomers are proposed for the dimer, a parallel displaced (sandwich) dimer and a T-shaped dimer. The difference between their structures can be distinguished by Raman spectroscopy. In the dimer, the vibrational wavefunctions localized on the molecules on each site, are doubly degenerate. The wavefunction is expressed by a linear combination of excitations localized on either component (a) or component (b) [33],

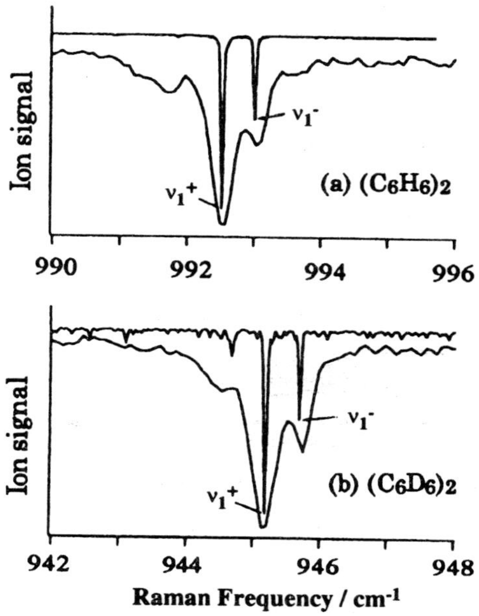

Fig. 6. 16 Ionization detected stimulated Raman spectrum of mode 1 vibration of (a) $(C_6H_6)_2$ and (b) $(C_6D_6)_2$ [33].

$$\left| \upsilon_j + \right\rangle = \alpha \left| \upsilon_j^a \right\rangle \left| 0^b \right\rangle + \beta \left| 0^a \right\rangle \left| \upsilon_j^b \right\rangle \tag{6.1}$$

$$\left| \upsilon_j - \right\rangle = \beta \left| \upsilon_j^a \right\rangle \left| 0^b \right\rangle - \alpha \left| 0^a \right\rangle \left| \upsilon_j^b \right\rangle . \tag{6.2}$$

Here, $\left| \upsilon_j^a \right\rangle$ and $\left| \upsilon_j^b \right\rangle$ are the vibrational wavefunctions of the υ-th quantum number of the j-th mode of monomer (a) and monomer (b) in the dimer, respectively. Since, $j = 1$ and $\upsilon = 1$ in the present case, Eqs. (6.1-2) become

$$\left| 1_1 + \right\rangle = \alpha \left| 1_1^a \right\rangle \left| 0^b \right\rangle + \beta \left| 0^a \right\rangle \left| 1_1^b \right\rangle \tag{6.3}$$

$$\left| 1_1 - \right\rangle = \beta \left| 1_1^a \right\rangle \left| 0^b \right\rangle - \alpha \left| 0^a \right\rangle \left| 1_1^b \right\rangle . \tag{6.4}$$

The Raman intensities to these levels may be expressed as,

$$I_1^+ = I_1 \left| \alpha + \beta \right|^2 \tag{6.5}$$

and

$$I_1^- = I_1 \left| \alpha - \beta \right|^2 \tag{6.6}$$

Here I_1 is the Raman transition intensity of the fundamental of mode 1 of the monomer. In the parallel displaced dimer, the two benzene molecules are symmetrically equivalent. In this case, $\alpha = \beta$, and only the $\left| 1_1 + \right\rangle$ level will be observed. In the T-shaped dimer, on the other hand, the two benzene molecules are not symmetrically equivalent. In this case, $\alpha \neq \beta$, and two levels, $\left| 1_1 + \right\rangle$ and $\left| 1_1 - \right\rangle$, can be observed in the

Raman spectrum. As can be seen in Fig. 6.16, mode 1 vibration is split into two with a separation of 0.6 cm^{-1}. This is a clear evidence that the two benzenes are symmetrically non-equivalent, that is the dimer forms a T-shaped structure.

6.5 Summary

Principles and applications of population labelling spectroscopy are described in this chapter. Development of a simplification technique is important to analyze a complicated absorption spectrum. The unique point of this spectroscopy is sensitivity and selectivity; we measure an absorption spectrum as a modulation of the population of the initial state, whose quantum number and parity are specified by another electronic transition, by use of very sensitive laser induced fluorescence (LIF) or resonance enhanced multiphoton ionization (REMPI) detection. We have seen how the rotational structures of complicated absorption spectra are simplified and also how electronic transitions of specific clusters are identified from overlapped absorption spectra of many clusters with different sizes and conformations. In addition to the selectivity with respect to the state and species, they have high sensitivity sufficient to measure the transitions of low density gas phase molecules. The high sensitivity owes to the use of LIF and REMPI method instead of monitoring a change of the absorption intensity. This spectroscopy can be applied not only to electronic spectroscopy but also to vibrational spectroscopy. The latter application enables us to observe infrared and Raman spectra of low-density gas phase molecules and molecular clusters.

References

1. M. E. Kaminsky, R. T. Hawkins, F. V. Kowalski and A. L. Schawlow, *Phys. Rev. Lett.* **36**, 671 (1976).
2. A. L. Schawlow, *J. Opt. Soc. Am.* **67**, 140 (1977).
3. M. Raab, G. Höning, W. Demtröder and C. R. Vidal, *J. Chem. Phys.* **76**, 4370 (1976).
4. R. B. Field, G. A. Capelle and M. A. Revelli, *J. Chem. Phys.* **63**, 3228 (1975).
5. M. A. Johnson, C. R. Webster and R. N. Zare, *J. Chem. Phys.* **75**, 5575 (1981).
6. M. A. Johnson, C. Noda, J. S. MacKillop and R. N. Zare, *Can. J. Phys.* **62**, 1467 (1984).
7. M. A. Johnson and R. N. Zare, *J. Chem. Phys.* **82**, 4449 (1985).
8. R. E. Teets, N. W. Carlson and A. L. Shawlow, *J. Mol. Spectrosc.* **78**, 415 (1979).
9. J. C. D. Brand, K. J. Cross and N. P. Ernsting, *Chem. Phys.* **59**, 405 (1981).
10. J. C. D. Brand and N. P. Ernsting, *J. Mol. Spectrosc.* **91**, 389 (1982).
11. M. A. Johnson, J. Rostas and R. N. Zare, *Chem. Phys. Lett.* **92**, 225 (1982).
12. M. A. Johnson, R. N. Zare, J. Rostas and S. Leach, *J. Chem. Phys.* **80**, 2407 (1984).
13. T. Suzuki and M. Ito, *J. Phys. Chem.* **91**, 3537 (1987).
14. R. J. Lipert and S. D. Colson, *J. Phys. Chem.* **93**, 3894 (1989).
15. R. J. Lipert and S. D. Colson, *Chem. Phys. Lett.* **161**, 303 (1989).
16. W. Scherzer, O. Trätzschmar, H. L. Selzle and E. W. Schlag, *Z. Naturforsch.* **47a**, 1248 (1992).
17. H. L. Selzle and E. W. Schlag, "*Reaction dynamics in clusters and condensed phase*" in *Proceedings of the 26th Jerusalem Symposia on Quantum Chemistry and Biochemistry.* (J. Jortner, R. D. Levine and B. Pullman ed.) Klumer academic publishers, Dordrecht/Boston/London (1993)
18. M. Pohl, M. Schmitt and K. Kleinermanns, *Chem. Phys. Lett.* **177**, 252 (1991).
19. M. Schmitt, H. Müller and K. Kleinermanns, *Chem. Phys. Lett.* **218**, 246 (1994).
20. M. Schmitt, U. Henrichs, H. Müller and K. Kleinermanns, *J. Chem. Phys.* **103**, 918 (1995).
21. H. Saigusa and E. C. Lim., *J. Phys. Chem.* **95**, 1194 (1991).
22. R. Takasu, N. Kizu and M. Itoh, *J. Chem. Phys.* **101**, 7364 (1994).
23. M. Kurono, R. Takasu and M. Itoh, *J. Phys. Chem.* **99**, 9668 (1995).
24. T. Ebata, Y. Matsumoto and N. Mikami, unpublished.
25. M. B. Beer and H. C. Longuett-Higgins, *J. Chem. Phys.* **23**, 1390 (1955).
26. D. Huppert, J. Jortner and P. M. Rentzepis, *J. Chem. Phys.* **56**, 4826 (1972).
27. R. M. Hochstrasser and T. -Y. Li, *J. Mol. Spectrosc.* **41**, 297 (1972).
28. G. R. Hunt and I. G. Ross, *J. Mol. Spectrosc.* **9**, 50 (1962).
29. A. Amirav and J. Jortner, *J. Chem. Phys.* **81**, 4200 (1984).
30. K. C. Janda, J. C. Hemminger, J. S. Winn, S. E. Novick, S. J. Harris and W. Klemperer, *J. Chem. Phys.* **63**, 1419 (1995).

31. M. Schauer and E. R. Bernstein, *J. Chem. Phys.* **82**, 3722 (1985).
32. J. P. Hopkins, D. E. Powers and R. E. Smalley, *J. Phys. Chem.* **85**, 3739 (1981).
33. B. F. Henson, G. V. Hartland, V. A. Venturo and P. M. Felker, *J. Chem. Phys.* **97**, 2189 (1992).
34. T. Ebata, S. Ishikawa, M. Ito and S. Hyodo, *Laser Chem.* **14**, 85 (1994).
35. H. Abe, N. Mikami and M. Ito, *J. Phys. Chem.* **86**, 1768 (1982).
36. R. J. Lipert and S. D. Colson, *J. Chem. Phys.* **89**, 4579 (1988).
37. R. J. Stanley and A. W. Castleman, Jr., *J. Chem. Phys.* **94**, 7744 (1991).
38. T. Ebata, M. Furukawa, T. Suzuki and M. Ito, *J. Opt. Soc. Am.* **B7**, 1890 (1990).
39. R. J. Lipert and S. D. Colson, *J. Phys. Chem.* **94**, 2358 (1990).
40. G. Reiser, O. Dopfer, R. Lindner, G. Henri, Müller-Dethlefs, E. W. Schlag and S. D. Colson, *Chem. Phys. Lett.* **181**, 1 (1991).
41. O. Dopfer, G. Lembach, T. G. Wright and K. Müller-Dethlefs, *J. Chem. Phys.* **98**, 1933 (1993).
42. T. Bürgi, M. Schütz and S. Leutwyler, *J. Chem. Phys.* **103**, 6350 (1995).
43. M. Gerhards and K. Kleinermanns, *J. Chem. Phys.* **103**, 7392 (1995).
44. S. Tanabe, T. Ebata, M. Fujii and N. Mikami, *Chem. Phys. Lett.* **215**, 347 (1993).
45. T. Ebata, N. Mizuochi, T. Watanabe and N. Mikami, *J. Phys. Chem.* **100**, 546 (1996).
46. T. Watanabe, T. Ebata, S. Tanabe and N. Mikami, *J. Chem. Phys.* **105**, 408 (1996).
47. E. L. Wehry and L. B. Rogers, J. Am. Chem. Soc. **87**, 4234 (1965).
48. H. Watanabe and S. Iwata, *J. Chem. Phys.* **105**, 420 (1996).
49. P. Esherick, A. Owyoung and J. Pliva, *J. Chem. Phys.* **83**, 3311 (1985).
50. J. Pliva, P. Esherick and A. Owyoung, *J. Mol. Spectrosc.* **125**, 393 (1987).
51. G. V. Hartland, B. F. Henson, L. L. Connell, T. C. Corcoran and P. M. Felker, *J. Phys. Chem.* **92**, 6877 (1988).
52. G. V. Hartland, B. F. Henson, V. A. Venturo and P. M. Felker, *J. Phys. Chem.* **96**, 1164 (1992).
53. V. A. Venturo and P. M. Felker, *J. Phys. Chem.* **97**, 4882 (1993).
54. M. Schaeffer, P. M. Maxton and P. M. Felker, *Chem. Phys. Lett.* **224**, 544 (1994).
55. M. W. Schaeffer, W. Kim, P. M. Maxton, J. Romascan and P. M. Felker, *Chem. Phys.* **242**, 632 (1995).

7 Zero Kinetic Energy (ZEKE) Photoelectron Spectroscopy

Klaus Müller-Dethlefs and Martin C.R. Cockett
Department of Chemistry, The University of York, Heslington, York YO1 5DD, UK

7.1 Introduction

The recent development and rapid growth of ZEKE (Zero Kinetic Energy) spectroscopy as a state of the art photoelectron spectroscopic technique have provided a means to explore the chemistry and spectroscopy of molecular ions at a level of detail and with a versatility of approach not previously available in other more traditional spectroscopic methods. Classical spectroscopic techniques such as microwave, IR, Raman, UV/Vis and NMR spectroscopies are often compromised by the need to generate the molecular ion first using methods such as microwave discharge. However, with ZEKE spectroscopy, which has at its heart the principles of photoelectron spectroscopy and photoionization, the ions are generated as a direct consequence of the method used to probe them and problems such as overlapping bands, molecule non-specificity and complex hot band structure can all be avoided in a single step. Since its development, the technique has been applied to a wide range of molecules, complexes, clusters and radicals and this review will attempt to illustrate the virtues of the technique by considering a cross section of some of these chemical systems.

As a preface to a detailed description of the ZEKE technique, and in order to set the development of the method in some historical context, it is helpful to start by considering the molecular orbital picture and its relationship to conventional photoelectron spectroscopy.

Our present understanding of chemistry and the chemical bond has been largely influenced by molecular orbital (MO) theory. The power of the molecular orbital approach in providing an understanding of the structure and reactivity of molecules lies in its description of chemical bonds. The legitimacy of the molecular orbital description and the resulting classification of molecular electronic states are borne out by the results of molecular electronic spectroscopy. In the one electron approximation, a bound-bound electronic transition can be viewed as the movement of a single electron between orbitals such as those shown schematically in Fig. 7.1. The excitation of an electron from an occupied molecular orbital to an unoccupied molecular orbital then corresponds to an electronic transition. However, one has to bear in mind that the orbital picture has limitations and electronic transitions really occur between states. Only transitions

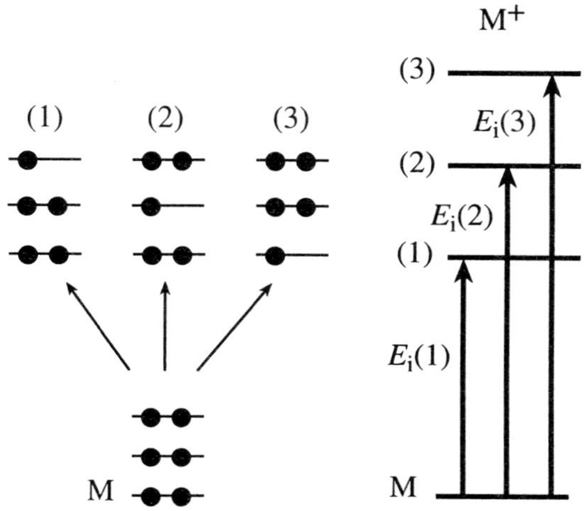

Fig. 7.1 Ionization from different orbitals with ionization energies $E_i(1)$, $E_i(2)$ and $E_i(3)$ leads to three different electronic cation states.

between states of correct symmetry species are allowed and it can be quite misleading to consider the corresponding orbital transition only. For example, in benzene the $S_1 \leftarrow S_0$ transition is in fact electronically forbidden, something not immediately apparent since a transition from an e_{1g} orbital to the excited e_{1u} orbital is formally allowed in the single electron picture. However, when the electronic symmetry selection rule for the total electronic symmetry (B_{1u}) of the excited state is taken into account, it is evident that the electronic $S_1 \leftarrow S_0$ transition in benzene is electronically forbidden: vibronic transitions can become allowed through vibronic coupling or more precisely Herzberg-Teller coupling [1].

The single electron molecular orbital picture is particularly useful when describing the photoionization of a molecule. In this case, an electron is excited from a bound orbital into the ionization continuum. For process (1) in Fig. 7.1 ionization takes place from the highest occupied molecular orbital (HOMO), thereby creating a molecular cation in its electronic ground state. The removal of an electron from an orbital lying below the HOMO results in the formation of an *excited* electronic state of the cation. When the photoionization energy is increased, lower lying orbitals can be ionized and higher excited states of the ion are formed, as depicted for processes (2) and (3) in Fig. 7.1. A photoelectron spectrum thus yields information about the ground state and electronically excited states of the cation. Fine structure will typically correspond to vibronic levels of the cation. A zero-order approximation in this case is that the energy of an electron in orbital n is equal to the negative of the ionization energy ($E_i(n)$ in Fig. 7.1). This approximation is the fundamental principle of *Koopmans*' theorem [2].

Although limited by its neglect of molecular orbital reorganization and electron correlation, Koopmans' theorem provides evidence that molecular orbitals are conceptually valid and has assumed a very important place in the development of our understanding of the electronic structure of molecules. If we examine the balance of energy expression associated with the photoelectric effect, which is the basis of photoelectron spectroscopy, we observe that

$$E_{\text{int}}^{\text{ion}}(n^+, S^+, v^+, N^+) = E_{\text{int}}^{\text{neutral}}(n, S, v, N) + h\nu - E_i - E_{\text{kin}}(e^-) \tag{7.1}$$

where n, S, v, N denote electronic, electronic spin, vibrational and rotational quantum numbers and the + superscript is used to indicate the ionic quantum numbers. E_i is the ionization energy, $E_{\text{kin}}(e^-)$ the kinetic energy of the departing photoelectron, and $h\nu$ the photon energy. It can be seen that the equation leads to populated electronic, vibrational and rotational states in the molecular cation.

The development in the early sixties of molecular *photoelectron spectroscopy* (PES) [3-5], provided a means to extract information about the electronic and vibronic states of molecular ions [6,7] through "back-titration" of the kinetic energy of the ejected photoelectron. Although there is no essential restriction on the resolution limit of PES (from energy conservation the energy of the ejected electron is quantized if we assume highly monochromatic photon energy and well defined states for the cation and the neutral molecule) the practical resolution limit of common geometrical or time-of-flight photoelectron analysers is about 10 meV (although recently in favorable cases 3 to 5 meV resolution has been demonstrated with a combination of He (I) radiation and a spherical analyser [8]).

When combined with *resonance enhanced multi-photon ionization* (REMPI), an improvement in resolution to around 3-5 meV [9,10] has also been achieved in some cases. Consequently, with the notable exceptions of hydrogen [11,12] and the high lying rotational states of NO^+ [9,10] where rotational structure has been resolved [11], conventional PES does not generally reveal the rotational state distribution of the resulting molecular ions. Similarly, the characteristically soft (low frequency) vibrations present in molecular clusters (*i.e.* the **inter**molecular vibrations) are very difficult to resolve. This limitation in resolution can be overcome by the use of what is now termed *ZEKE Spectroscopy* (Zero Kinetic Energy Photoelectron Spectroscopy), a method originally developed by one of us and further refined over the course of the last decade [12-23]. Many other groups have adopted the method and it is hoped that some flavour of the diversity of applications of the technique will be captured in the present review (for other recent reviews of the method see Refs. [19-23]).

In contrast to conventional photoelectron spectroscopy, the spectral resolution of ZEKE spectroscopy is limited only by the typical bandwidth of a tunable pulsed dye laser. This means that resolution of the order of 0.1 cm^{-1} is possible, which allows the prospect not only of resolving rotational states of larger molecular cations but also of preparing molecular ions in specific rovibronic states, thereby opening the way to new experiments in chemical reaction dynamics. The versatility of the ZEKE method in this respect can be more readily appreciated when one considers that in contrast to bound-bound transitions, for which the total angular momentum quantum number selection rule $\Delta J = \pm 1, 0$ is observed, angular momentum transfers between ion and molecule of more than ± 1 are possible for ionizing transitions.

A graphic illustration of the very much higher (about 3 orders of magnitude) resolving power of ZEKE spectroscopy can be seen in Fig. 7.2 in which the vibrationally resolved conventional photoelectron spectrum of NO excited by He(I) radiation [5] is compared with the very first ZEKE spectrum of NO recorded in 1984 [15] *via* the C $^2\Pi$ state in a 3+1 multi-photon ionization scheme [24]. The ZEKE spectrum shown in the lower part of Fig. 7.2 reveals the fully resolved rotational structure of the NO^+ cation while the conventional photoelectron spectrum is able only to resolve vibrational structure. The ZEKE spectrum also shows the advantage of combining ZEKE spectroscopy with a multi-photon ionization scheme in which an excited state of the neutral acts as a resonant intermediate (i.e. **R**esonance **E**nhanced **M**ulti-**P**hoton **I**onization, **REMPI**). REMPI spectroscopy is a well established and highly sensitive method of obtaining electronic excitation spectra with rovibronic state resolution. When combined with ZEKE-spectroscopy, one can achieve both very high sensitivity as well as molecule and state selectivity thereby extending the high energy resolution of optical spectroscopy to the ionic state. The high spectral resolution offered by the technique, coupled to the diversity of possible routes by which ionization can be achieved, has opened the way to studies of excited state and Rydberg state dynamics [16,25], photofragmentation, and the investigation of the dynamics of

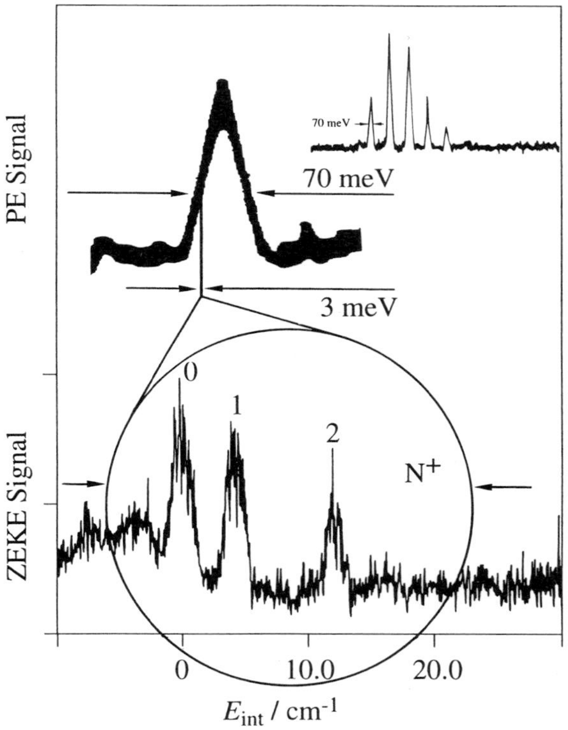

Fig. 7.2 The first VUV photoelectron spectrum of nitric oxide (Turner *et al.* [5]) (top), showing a vibrational progression from $v^+ = 0$ to $v^+ = 4$ of the NO^+ $^1\Sigma^+$ electronic ground state. The $v^+ = 0$ peak can be compared with the first ZEKE spectrum [14,15] (bottom) which shows the lowest rotational states in the $v^+ = 0$ vibrational level with total angular momentum N^+=0 to N^+=2.

photoionization apparent in the coupling between the molecular ion core and the ejected photoelectron [22,26].

One of the more important realizations in the development of ZEKE spectroscopy was the very long lifetime of very high-n Rydberg states within the "magic region" lying about 1 meV below each ion threshold [27]. The original concept behind ZEKE spectroscopy was to detect free photoelectrons at threshold by delayed pulsed field extraction techniques. This is still a valid description in the case of ZEKE photodetachment of anions (see below). The significance of the discovery of the long lived ZEKE Rydberg states was that it became clear that the ZEKE signal detected in a typical experiment was in most cases predominantly generated from *pulsed field ionization* of highly excited, long-lived Rydberg states with a principal quantum number $n > 150$. The form of ZEKE spectroscopy which makes use of this mechanism is known as ZEKE-PFI [27] spectroscopy and has become the most widespread experimental realization of the technique. The resolution of the method is largely unaffected by the fact that ionization is achieved in this case by field ionization and is effectively still determined mainly by the available laser bandwidth. The current resolution benchmark is the rotationally resolved ZEKE spectrum of benzene from which it was demonstrated that the benzene cation is planar and adequately described in the D_{6h} molecular point group, despite being subject to Jahn-Teller distortion [28]. ZEKE-spectroscopy has also been successfully applied to studies of the vibrational structure of large organic molecules [29], molecular [30] and metal clusters [31,32], and hydrogen-bonded systems [33]. One of the more routine strengths of the technique is that ionization energies can be determined with an accuracy comparable to that of Rydberg extrapolations but with less experimental effort.

Although rotationally resolved studies of large molecules provide compelling testimony to the power of the technique, ZEKE spectroscopy has been used to great effect in resolving vibrational structure in large polyatomic molecular systems in which vibronic congestion becomes high as internal energy is increased and in molecules characterized either by weak intramolecular or intermolecular bonding. As an example of the former we can compare the state-of-the-art time-of-flight (TOF) photoelectron spectrum [34] of para-difluorobenzene (top in Fig. 7.3), with the corresponding ZEKE spectrum [29], (bottom in Fig. 7.3). Clearly the ZEKE spectrum reveals a much more clearly resolved vibrational structure than the TOF photoelectron spectrum.

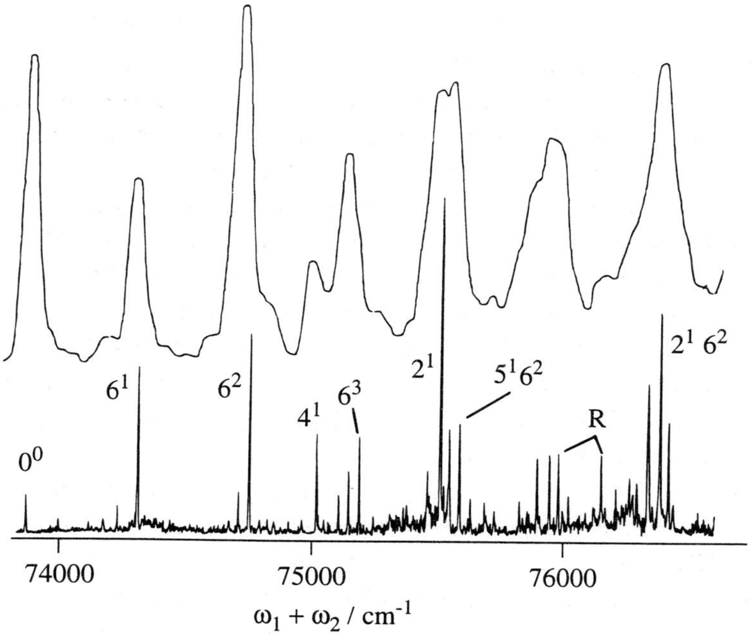

Fig. 7.3 Top: time-of-flight photoelectron spectrum [34] and (bottom) ZEKE spectrum [29] of para-difluorobenzene through the S_1 6^1 intermediate state.

7.2 Experimental ZEKE method

The term ZEKE spectroscopy is now commonly used to describe any experimental technique that detects electrons from a small, experimentally adjustable energy range around (i.e. above and below) a selected ionization threshold using delayed pulsed electric field extraction [14,15] (for a discussion of the older threshold techniques and their relation to ZEKE see refs. [16,21,22]). In contrast to PES, where a fixed photon energy is used for ionization, the ZEKE method involves the detection of electrons that are formed with zero kinetic energy when scanning the photon energy across the ionic threshold region. ZEKE electrons (within a certain bandwidth) are produced when the photon energy of the incident light matches the energy of a rovibronic transition between the initial state of the neutral molecule and the final state of the ion [16]. In the original ZEKE concept [14], as seen in Fig. 7.4, the free ZEKE electrons produced under (nearly) field-free conditions are extracted using a delayed pulsed electric field. Zero-field ionization and delayed pulsed-field extraction allow us to discriminate between ZEKE electrons and fast photoelectrons (by this we

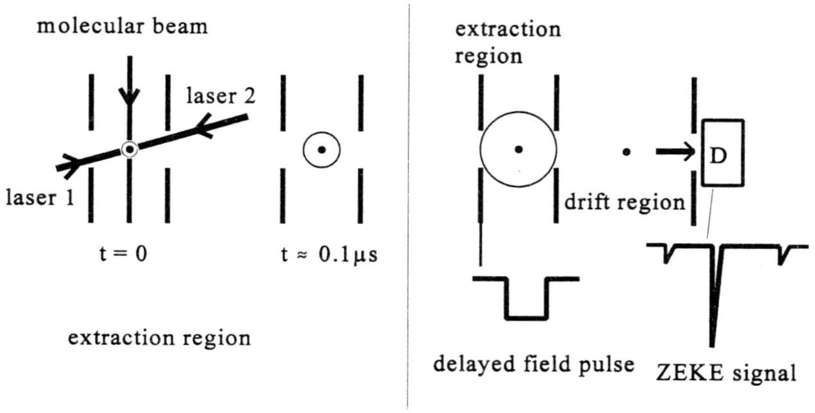

Fig. 7.4 Principle of delayed pulsed field extraction. Left:ZEKE electrons stay at their original position (assuming field free conditions) whereas kinetic electrons are successively diluted with increasing delay time. Laser 1 and laser 2 are the excitation and ionization lasers used in a typical two-colour ionization scheme. Right:The delayed field pulse extracts the ZEKE electrons from the extraction region through the field free drift region onto detector D.

mean having a kinetic energy greater than zero) by their respective times-of-flight (see Fig. 7.4). However, the electrons which contribute to the measured ZEKE signal in fact arise from two different sources. Firstly, the free ZEKE electrons (electrons of very low kinetic energy), which are generated by photoionization at threshold (a threshold implies, for molecules, a rovibronic eigenstate of the cation), are separated from the fast electrons using the steradiancy principle [35] and time-of-flight separation, and can be detected in a ZEKE experiment under certain exceptional experimental conditions. Secondly, as discovered by Reiser *et al.* [27], electrons emerging from pulsed-field ionization of the long-lived ZEKE Rydberg states of high principal quantum number n, constitute, at the very least, a significant proportion of the ZEKE signal (see Fig. 7.5). The free ZEKE electrons will usually only contribute significantly to the signal when the bandwidth of the ionizing radiation is comparable to or larger than the energy region occupied by the long-lived ZEKE Rydberg states. This is the case for typical VUV synchrotron radiation ZEKE experiments employing a normal incidence monochromator.

The form of ZEKE spectroscopy which utilizes pulsed field ionization of the metastable high-n ZEKE Rydberg states is known as ZEKE-PFI spectroscopy [27]. This method is generally applicable not only for small molecules like NO with their fewer degrees of freedom, but also for much larger molecules and molecular clusters where the lifetimes of the ZEKE Rydberg states are often found to exceed several tens of microseconds [20,36]. The pulsed field ionization mechanism is depicted in Fig. 7.5. The longevity of the ZEKE Rydberg states arises because the Rydberg electron is able to acquire non-penetrating character and no longer interacts with the positive ion core. Conversely, in the lower lying Rydberg states, as depicted in

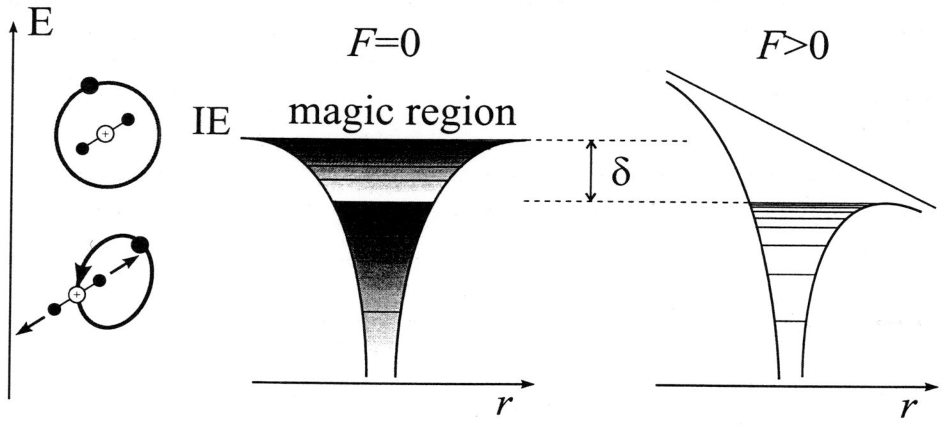

Fig. 7.5 Left: magic region of long lifetime ZEKE Rydberg states just below threshold with no electron core collisions, lower Rydberg states with core collisions. Right: pulsed field ionization.

Fig. 7.6, the Rydberg electron still undergoes regular collisions with the core, which leads to intramolecular relaxation processes such as predissociation into neutral fragments and autoionization. This decay of the lower Rydberg states during the typical delay times used in ZEKE experiments is the underlying reason why peak widths observed in ZEKE spectroscopy are generally no larger than 5 to 10 cm^{-1} even when high field strengths are used. From an experimental point of view this makes it particularly easy to obtain a ZEKE signal if a resolution of a few cm^{-1} is sufficient.

However, if laser limited resolution is required, one needs to design more sophisticated field ionization schemes. It has been demonstrated that in order to achieve rotational resolution in molecules such as benzene (*ca.* 0.2 cm^{-1}), a slowly rising electric field extraction pulse, either with a linear or multi-step slope, is highly effective [28,37,38]. Variation of the slope of the pulse allows the spectral resolution to be adjusted

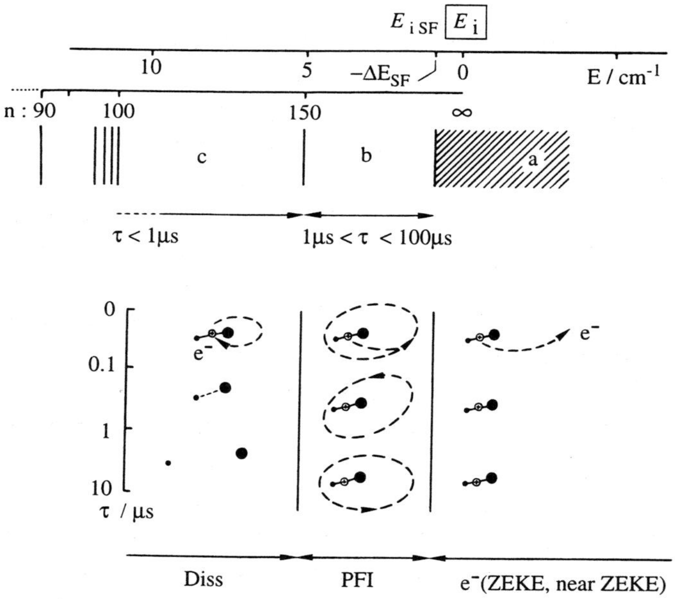

Fig. 7.6 Stabilization of ZEKE Rydberg states in the magic region, a few cm^{-1} below an ionization threshold. The Rydberg states below the magic region predissociate into neutral fragments since the Rydberg electron still undergoes collisions with the ion core. In the magic region the Rydberg electron no longer collides with the core and hence intramolecular relaxation processes are quenched. $E_{i\,SF}$ is the ionization energy red-shifted by an amount ΔE_{SF} by stray fields.

according to the laser bandwidth limitations and to the demands of the system under study (whether one requires vibrational or rotational resolution). Fig. 7.7 shows schematically the effect of pulse slope risetime on the time-of-flight (TOF) of the corresponding electrons produced by PFI. A fast pulse generates all the signal within a narrow TOF distribution whereas a slow pulse spreads the different "slices" of Rydberg states into a broader TOF distribution. Thus, for a particular TOF gate a smaller spectral "Rydberg slice" will be collected when the photon energy of the light source is scanned for a slow pulse than for a fast rise time pulse. A multi-step staircase-like extraction pulse provides a method for exactly determining the ionization energy under field free conditions [39,40].

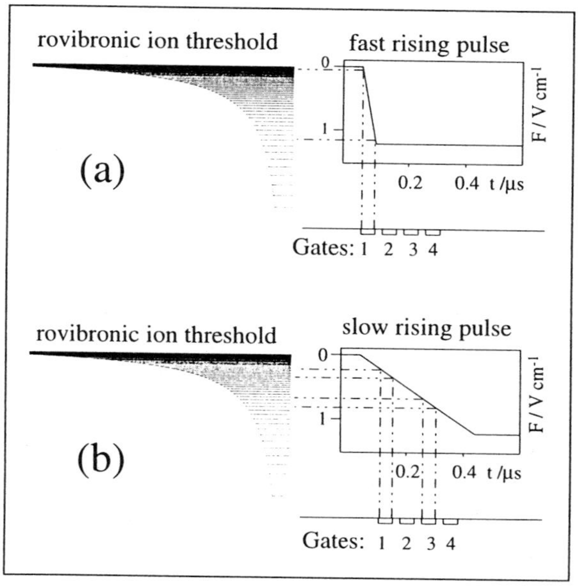

Fig. 7.7 Pulsed-field ionization of long-lived ZEKE Rydberg states using a fast-rising pulse (a, top) or slow-rising pulse (b, bottom) and detection within a certain time gate.

The origin of the long lifetime of the ZEKE Rydberg states has been a matter of some considerable debate and although the discussion is continuously evolving, a degree of consensus has been reached concerning the principal contributory effects. The longevity of the ZEKE Rydberg states is currently understood as being due to a combination of the effects of small homogeneous fields (which typically originate from electronic equipment in the laboratory) and inhomogeneous electric fields (associated with regions of localized charge in the spectrometer, *e.g.* ions). An alternative mechanism proposed that higher multipole moments of the core were responsible. The highly diffuse nature of high lying Rydberg states renders them susceptible to l and m_l mixing through external perturbation. Stray dc electric fields may cause substantial l mixing through the Stark effect [41,42], while inhomogeneous fields inducing m_l randomization arise from the presence of ions in low to medium concentrations [43-46].

While it is now generally accepted that the PFI mechanism is responsible for the ZEKE signal in experiments where the neutral molecule is ionized to give a cation, it is not applicable to photodetachment of mass selected anions to yield the vibronic structure of the corresponding neutral [31,47-49] because no Rydberg states exist for anions (in fact dipole bound states [50] can exist for anions but usually these will be very limited in number and cannot be compared to the huge manifold of Rydberg states which converge to each cation threshold). In this case, the only mechanism by which one can obtain a signal is the original ZEKE method involving the detection of free electrons with negligible kinetic energy. A particularly interesting application of this technique was to the study of the metastable activated complex IHI involved in the iodine hydrogen reaction accessed through the photodetachment of IHI^-. The ZEKE photodetachment spectrum of IHI^- is reproduced in Fig. 7.8 and shows resolved vibrational structure of the IHI transition state [48]. While this structure is easily resolved in the ZEKE experiment, it is clearly not resolved in the corresponding conventional photoelectron spectrum also shown superimposed in Fig. 7.8.

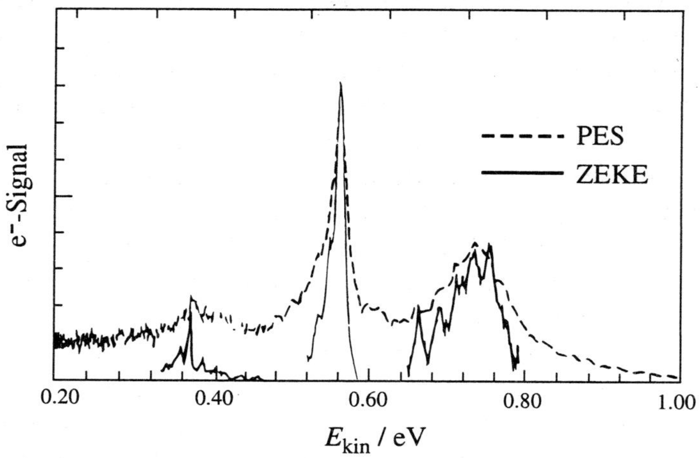

Fig. 7.8 Comparison of ZEKE (full line) and PES (dashed line) of IHI^-. From Refs. [48,49].

A typical experimental setup for a ZEKE experiment is shown schematically in Fig. 7.9. It consists of a laser system and a vacuum apparatus which includes the molecular beam source, the extraction plates and a μ-metal shielded flight tube with electron/ion detectors (dual multichannel plates) at each end. In a typical two-color experiment, both dye lasers (often frequency doubled) are pumped simultaneously by an excimer laser or a Nd:YAG laser. The first dye laser excites a specific vibronic or rovibronic level of the intermediate state and the second laser ionizes the molecules or promotes them into long-lived Rydberg states ($n > 150$) converging to (ro)vibronic levels of the electronic ground state or an electronically excited state of the cation. After a delay time of several μs, an extraction pulse is applied by either a simple electric pulsing device or by an arbitrary function generator. The electrons are detected with multichannel plates and their time-of-flight signal is recorded with boxcar integrators or a transient digitizer by setting narrow time gates (10-30 ns).

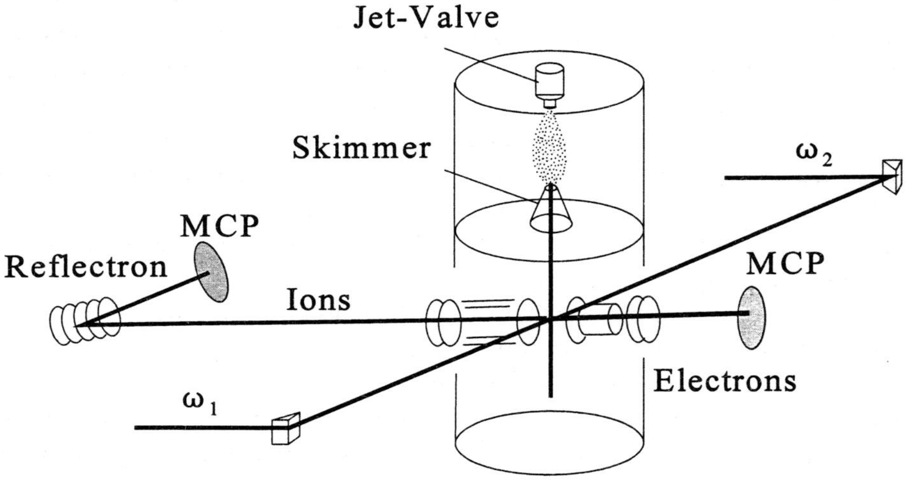

Fig. 7.9 Schematic ZEKE apparatus featuring molecular beam, two-colour laser excitation, linear electron extraction optics and reflectron time-of-flight mass spectrometer.

In any ZEKE experiment, some thought has to be given to the route by which photoionization is achieved. In many cases, this will be determined by experimental expediency and by the molecular system under study. The nature of modern commercially available dye lasers dictates the approach adopted by many groups in that the usable output normally covers the near infrared through soft ultraviolet region, which means that ionization has to be accomplished in more than one step. Consequently the most common approach is resonant enhanced multi-photon ionization in which an excited state of the neutral is used as a resonant intermediate state. The majority of ZEKE studies on molecular clusters and on larger molecular systems have been accomplished using REMPI schemes. However, while this approach is very useful for molecules having relatively low ionization energies and conveniently situated intermediate states, it is not necessarily the best approach for small lightweight molecules which often have somewhat higher ionization energies. In these cases, the use of tunable coherent vacuum ultraviolet or XUV light generated by frequency mixing techniques has proved very effective and a number of small molecules have been studied usually with rotational resolution. A third approach is the use of one-colour two-photon ionization in which the molecule is ionized directly from the ground state of the neutral. In contrast to REMPI, it works best when there are no bound electronically excited states lying in the region accessed by the first photon. For this reason it is sometimes known as non-resonant multi-photon ionization. While it is generally undesirable in this case to encounter a bound excited state at the one-photon level, it has been found that the presence of a repulsive intermediate state can provide positive benefits and open routes to regions of the ionic state potential not necessarily accessible in a one-photon experiment. In the following sections we shall present a cross section of the many applications of ZEKE spectroscopy, and will discuss examples of all three ionization schemes.

7.3. Small Molecules

7.3.1 Vibrational Resolution

Molecular Iodine The high resolution ZEKE-PFI spectrum of the ground electronic state of I_2^+ has been reported in three separate studies, each of which employed a different ionization route [51-53]. In the first study a (2+1') REMPI scheme was used in which a number of gerade Rydberg excited states acted as resonant intermediate states [51]. In the second, the valence $B^3\Pi 0_u^+$ state was used as an intermediate state in a (1+2') ionization scheme [52]. Both studies determined accurate vibrational constants and ionization energies and additionally both sets of spectra exhibited the effects of autoionization. In the (2+1') ZEKE spectra recorded for iodine in Ref. [51], vibrational autoionization resulted in markedly non-Franck-Condon behaviour with strong off-diagonal transition intensities in v^+,v'', while for the (1+2') ZEKE spectra reported in Ref. [52], the anomalous spin-orbit branching ratio reflected the considerable contribution of spin-orbit autoionization. In the third, most recent study [53], which was primarily concerned with electronically excited states of I_2^+, ionization was achieved directly from the electronic ground state of the neutral by coherent absorption of two UV photons in a one-colour ionization scheme. The spectra obtained in the latter study provide a good example not only of how ZEKE spectroscopy can be used to provide detailed vibrationally resolved spectra of the ion, but also of the ubiquitous nature of autoionization in the ZEKE spectroscopy of small molecules. In this case, the effect of the extensive autoionization is to greatly increase the amount of vibrational information obtained.

Deviations in intensity from those expected for direct ionization are often attributed in threshold photoionization and ZEKE spectroscopy to autoionization processes involving population of low-n Rydberg states which lie above the ionization threshold. In the conventional view of autoionization, these states may then spontaneously emit an electron which can be detected in the experiment. The kinetic energy of the emitted electron will depend upon which rovibronic state the ion is formed in, and its position relative to the autoionizing Rydberg state. Autoionization processes yielding a threshold photoelectron (E_{kin}=0) will only occur if there exists a rovibronic level in the autoionizing state which lies very close in energy to a similar level in the ionic state *i.e.* a discrete-discrete interaction. However, as we discussed earlier, ZEKE spectroscopy involves pulsed field ionization of so called ZEKE Rydberg states (high n (n>150), high l, high m_l) and so we have to modify our picture of how autoionization might occur. In this case, a high-n, low l Rydberg state is indirectly populated through an interaction with a low-n low l Rydberg state converging on some higher ionization threshold. The mixed character state undergoes rapid l and m_l mixing, which results in the electron being trapped in a metastable, non-penetrating orbital. The subsequent application of a pulsed electric field lowers the ionization threshold and ionization occurs. If the oscillator strength to the low-n Rydberg state is stronger than that into the manifold of high-n Rydberg states converging on the ionic level, then an enhancement of intensity may result. In effect, the transition into the ionic state will borrow intensity from the more strongly allowed transition into the low-n Rydberg state. However, when the situation is reversed, and the oscillator strength to the low-n Rydberg state is significantly less than that into the high-n Rydberg state, a window resonance can occur and a dip in intensity may be observed in the spectrum for a particular rovibrational line. Autoionization processes of this type will only affect intensities in the ZEKE spectrum if the coupling between the low-n Rydberg state and the ionization continuum is small. A large coupling to the continuum will result in a rapid decay of the state to yield a prompt electron through conventional fast autoionization. This may have the effect of depopulating the stabilized high-n (high-l high-m_l) Rydberg state and an overall (as opposed to localized) loss of intensity may result. In the case of iodine, the question of which mechanism describes the distribution of oscillator strength observed in the ZEKE spectrum is relatively straightforward.

The one-colour two-photon ZEKE spectrum of iodine [53] in the range 74750 to 82000 cm^{-1} is shown in Fig. 7.10. The spectrum shows vibrational progressions attributable to both spin orbit components of the $X^2\Pi_{\Omega,g}$ ground electronic state of the ion. Values for the adiabatic ionization energies for the two spin-orbit states (75069 ± 2 cm^{-1} for the $X^2\Pi_{3/2,g}$ state and 80266 ± 2 cm^{-1} for the $X^2\Pi_{1/2,g}$ state) can be readily derived from the origins of the two progressions. The most remarkable feature of this spectrum, however, is the highly extended vibrational progression observed for the lower spin orbit state. In the conventional photoelectron spectrum of iodine, the progression extends only as far as v^+=4, in accordance with what one might expect given the relatively small change in equilibrium geometry that accompanies removal of an electron from the predominantly non-bonding π_g orbital. This contrasts with the analogous progression in

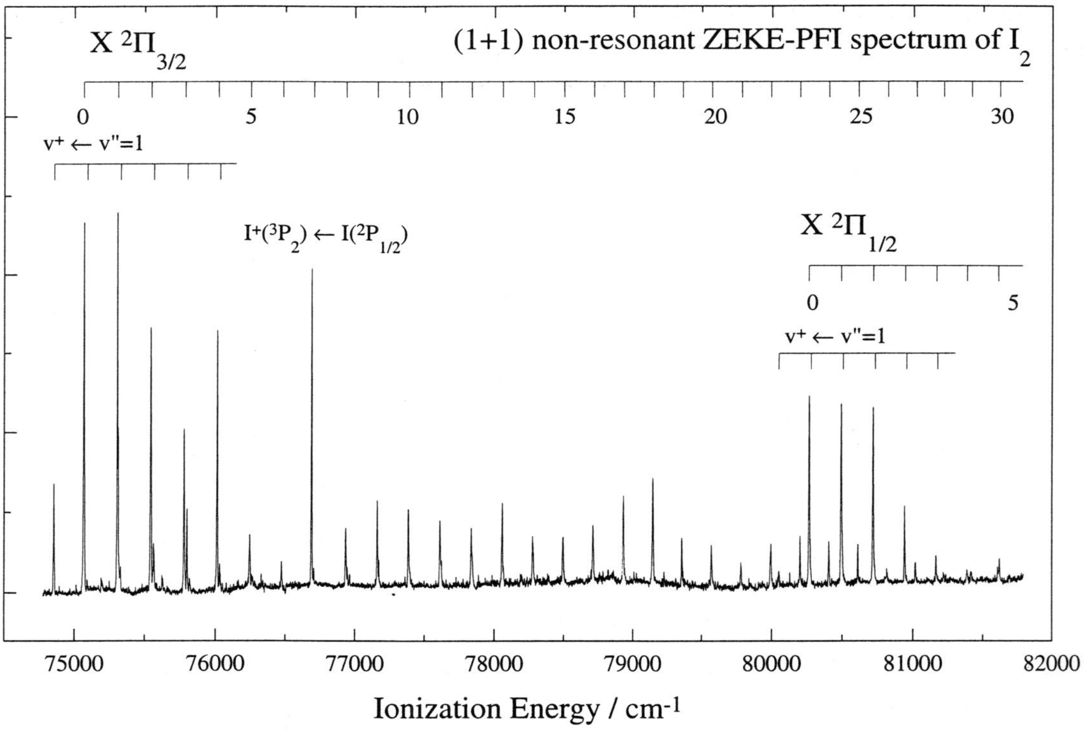

Fig. 7.10 The coherent one-colour (two-photon) ZEKE-PFI spectrum of the $X^2\Pi_{\Omega,g}$ state of I_2^+ in the range 74750 to 82000 cm^{-1}. From Ref. [53]. The appearance of such a long vibrational progression is due to electronic autoionization of Rydberg states based on higher electronic states. The sharp feature at 76687 ± 2 cm^{-1} is due to ionization of spin-orbit excited iodine atoms: $I^+(^3P_2) \leftarrow I(^2P_{1/2})$.

the ZEKE spectrum which can be followed as far as v^+=30 in the region shown in Fig. 7.10. The enhanced vibrational structure seen in the ZEKE spectrum is a classic example of intensity borrowing through electronic autoionization and can be explained in this case in terms of field induced electronic autoionization involving Rydberg states based on the $X^2\Pi_{1/2,g}$, $A^2\Pi_{\Omega,u}$, and the dissociative $B^2\Sigma_g^+$ state. An obvious benefit of the long progression is that accurate values for the vibrational constants can be derived from Birge-Sponer extrapolations. In this case values for ω_e and $\omega_e x_e$ of 240 ± 1 cm^{-1} and 0.71 ± 0.01 cm^{-1}, respectively, for the $X^2\Pi_{3/2,g}$ state and 229 ± 2 cm^{-1} and 0.75 ± 0.04 cm^{-1} for the $X^2\Pi_{1/2,g}$ state were determined.

7.3.2 Rotational Resolution

Nitric Oxide Nitric Oxide is the most extensively studied molecule in photoionization and serves as a model for the theoretical description of the photoionization process. The very first ZEKE spectrum recorded was that of NO ionized in a 3+1' ionization scheme *via* the $C^2\Pi$ state [15] and was successfully interpreted using multichannel quantum defect theory (MQDT) [54]. Ionization *via* the $A^2\Sigma^+$ state has been investigated by several groups. The A-state is a 3s Rydberg state with 94% s-, 5% d- and 0.2% p-contributions [55]. The ZEKE spectra recorded *via* N_A=0 in the A state [56] (see Fig. 7.11) show strong ΔN^+=±2 and ΔN^+=0 transitions with less intense ΔN^+=±1 and ±3 transitions. The molecular nature of the photoionization process is evident from the transitions with $\Delta N^+ = N^+ - N_A \neq 0$ (where N_A is the total angular momentum of the A state). In the simplest atomic like picture of the ionization process, a transition from a Rydberg state with electron angular momentum l will result in ejection into $l \pm 1$ continuum channels. The observation of the ΔN^+ =±2 transitions can be explained from the s and d contribution to the A state. However, the the p-contribution is far too small to explain the significant ΔN^+ =±1 contributions. An even l is required in the photoelectron continuum channel to explain an angular momentum transfer of $\Delta N^+ = N^+ - N$ = odd. *Ab-initio* calculations have shown that a scattering of the electron by the nonspherical core potential into continuum channels of opposite parity causes the appearance of signals forbidden in the atomic picture [56].

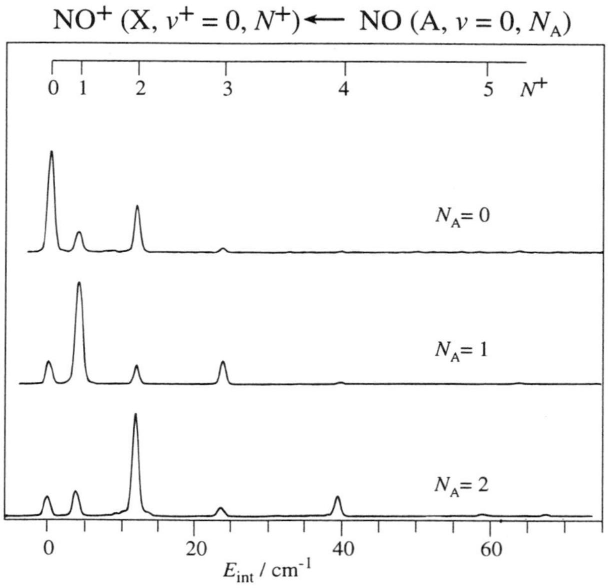

Fig. 7.11 Rotationally resolved ZEKE spectra of NO *via* different rotational levels N_A = 0 to 2 of the A $^2\Sigma^+$ (v=0, N_A) state (N: total angular momentum quantum number without electron spin). From Ref. [56].

The ZEKE spectra recorded for ionization through the N_A = 1 and 2 rotational states show a progressive decrease in intensity for ionizing transitions that involve a change in rotational quantum number (*i.e.* $\Delta N^+ \neq 0$) as the rotational quantum number N_A increases [57]. This effect has been explained in the theoretical work of Rudolph *et al.* [55] where the continuum partial wave channel matrix elements were obtained by an *ab initio* calculation. The observation that transitions involving a *negative* ΔN^+ are more intense compared to a *positive* ΔN^+, is observed in nearly all the rotationally resolved ZEKE measurements.

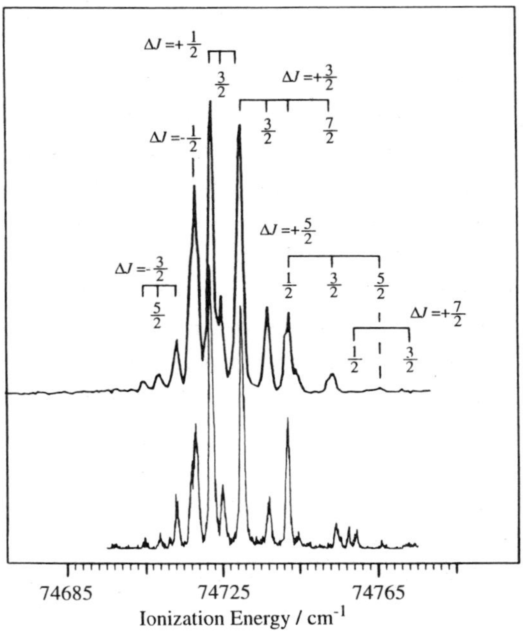

Fig. 7.12 A comparison of the two-photon $X^1\Sigma^+ (v^+ = 0) \leftarrow X^2\Pi_{1/2} (v = 0)$ spectrum [58] of NO with the one-photon $X^1\Sigma^+ (v^+ = 1) \leftarrow X^2\Pi_{1/2} (v = 0)$ spectrum [59].

More recently, ZEKE spectra from the $^2\Pi_{1/2}$ electronic ground state of NO have been recorded using both coherent two-photon [58] and one-photon ionization [59]. A comparison of the two-photon $X^1\Sigma^+ (v^+=0) \leftarrow X^2\Pi_{1/2} (v=0)$ spectrum with the one-photon $X^1\Sigma^+ (v^+=1) \leftarrow X^2\Pi_{1/2} (v=0)$ spectrum is shown in Fig. 7.12. The appearance of both spectra is remarkably similar. It should be noted that the two components of the Λ doublet in the electronic ground state are not resolved in either experiment and so parity conservation rules will not affect the observed distribution of rotational intensities. According to the *ab-initio* calculation, the highest occupied π molecular orbital is of predominant d-character, which means that emission into the odd l-continuum (p- and f- partial waves) is expected to dominate in the one-photon experiment while emission into even l-continuum (s-, d- and g- partial waves) is expected in the two-photon experiment. However, the maximum angular momentum transfer observed is ΔN=9/2 in both experiments, but both spectra are dominated by small ΔN differences up to 5/2. This indicates that emission into the s-channel is favoured in the two-photon experiment and emission into the p-channel in the one-photon experiment. Emission of electrons into continuum channels with larger l-values do not show a significant contribution. A possible interpretation of the similarity of the two spectra is that the electron acts as a spectator [21,60].

7.4. Larger Molecules

7.4.1 Vibrational Resolution: A breakdown of symmetry selection rules

The observation of vibronically forbidden band systems in ZEKE spectra has become something of a common phenomenon. Similar observations are frequently made in single photon absorption spectra of neutral molecules and generally occur through vibronic coupling. In the ZEKE spectroscopy of small molecules, a breakdown of the Franck-Condon principle is often attributed to autoionization and an example of this has already been discussed above for iodine. However, larger molecules appear to be far less susceptible to autoionization but are still frequently observed to break symmetry selection rules. In this section we shall consider two examples of aromatic molecules whose ZEKE spectra show vibronically forbidden band systems. The sensitivity and high resolution available in ZEKE spectroscopy reveal details in the ion spectrum which in many cases would not be seen in a conventional photoelectron spectrum. It is often these details which provide the means to elucidate the mechanism responsible for the observation of nominally forbidden transitions.

The process whereby electronic band systems can exhibit vibrationally induced bands involving non-totally symmetric vibrations, is based upon a breakdown of the Born-Oppenheimer approximation [61,62]. The generic term "vibronic coupling" is often used to describe any process where an interaction between a vibronic state of a given symmetry and a nearby electronic state of the same symmetry results in intensity borrowing. Of course, changes in point group between electronic states will inevitably result in a relaxation of vibronic selection rules and so it is important to have a reasonably thorough knowledge of the overall symmetries of each electronic state as well as positions of the electronic states involved in order to construct a viable vibronic coupling mechanism to account for the appearance of nominally forbidden vibronic bands. Naphthalene is a good example of a molecule of high symmetry whose neutral and low-lying ionic states are well characterized in terms of their symmetry and whose point group is the same in the neutral ground state as it is in the excited neutral and ionic states.

Naphthalene The $\tilde{A}^1B_{3u} \leftarrow \tilde{X}^1A_g$ absorption system ($S_1 \leftarrow S_0$) in naphthalene is electronically allowed but has a very small transition moment, directed along the long axis of the molecule. The most intense band in absorption (see the (1+1) REMPI excitation spectrum of the S_1 state in Fig. 7.13) [63] is not the 0^0_0 band but a band induced by the $\bar{8}$ mode vibration (the bar refers to a vibration having b_{1g} symmetry). The vibronic transition gains its intensity from the nearby $\tilde{B}^1B_{2u}$ (S_2) electronic state and has a transition moment directed along the short axis.

The first two doublet ionic states lie within 6000 cm^{-1} of each other and possess symmetries (A_u and B_{1u}) which might be expected to promote vibronic coupling, in a similar way to the first two neutral singlet states, through excitation of the $\bar{8}$ mode or indeed any other vibrational mode having b_{1g} symmetry.

The two-colour (1+1') ZEKE spectra [63], recorded *via* the S_1 band origin, as well as *via* three vibronic levels up to 700 cm^{-1} internal vibrational energy, are shown in Fig. 7.14. Each of the spectra shows a

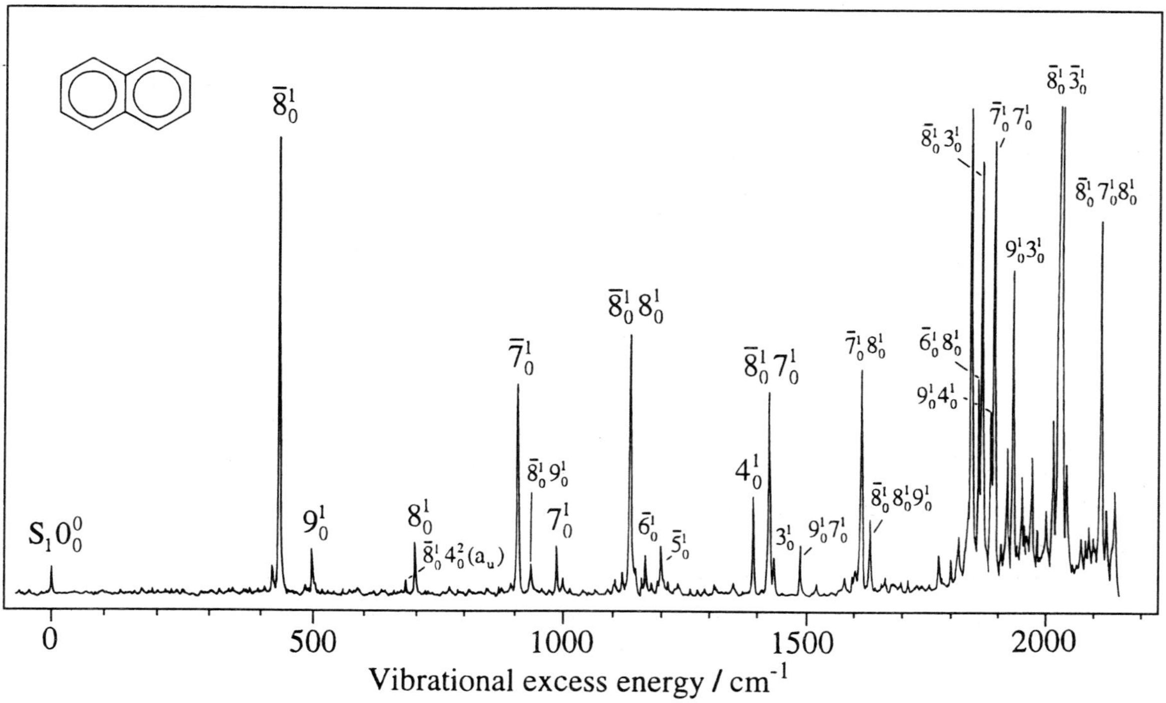

Fig. 7.13 The two-colour (1+1') REMPI ion current excitation spectrum of the S_1 state of naphthalene. From Ref. [63].

propensity for the diagonal transition (Δv=0), thereby demonstrating a clear correspondence between the vibrations excited in the S_1 state and those excited in the resulting ionic state. Consequently, it becomes straightforward to assign the peaks appearing at 455 and 911 cm^{-1} ion internal energy in Fig. 14 (b) to, respectively, one and two quanta excited in the non-totally symmetric $\bar{8}$ mode, and the peak at 507 cm^{-1} in Fig. 7.14(c) to one quantum in the totally symmetric 9 mode. Most of the peaks to higher energy appear as combination bands with one quantum in the respective dominant vibration excited in the ion. The spectrum in Fig. 7.14(d) recorded *via* the totally symmetric 8 mode in the S_1 state shows a deviation from the Δv=0 propensity rule, which suggests that a considerable displacement has occurred along the normal coordinate that describes the 8 mode upon ionization from the S_1 state. The peaks appearing at 763 and 1523 cm^{-1} can thus confidently be assigned to, respectively, one and two quanta excited in 8 mode in the ion.

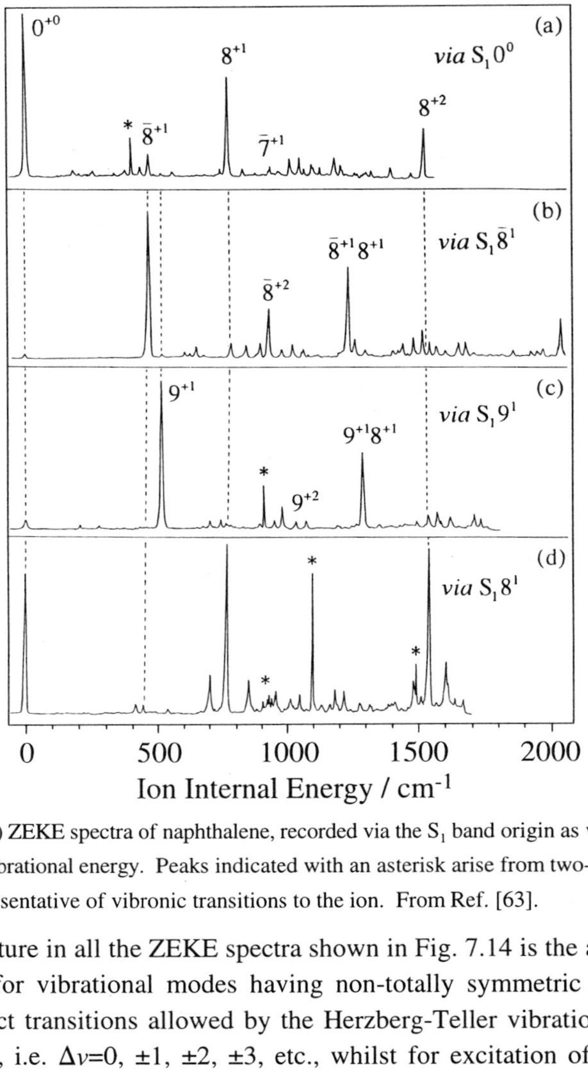

Fig. 7.14 The two-colour (1+1') ZEKE spectra of naphthalene, recorded via the S_1 band origin as well as *via* three vibronic levels up to 700 cm^{-1} internal vibrational energy. Peaks indicated with an asterisk arise from two-photon absorption of the ionizing laser and are not representative of vibronic transitions to the ion. From Ref. [63].

The most significant feature in all the ZEKE spectra shown in Fig. 7.14 is the appearance of relatively intense $\Delta v=\pm1$ transitions for vibrational modes having non-totally symmetric b_{1g} symmetry. For a_g vibrations, one would expect transitions allowed by the Herzberg-Teller vibrational selection rule for a totally symmetric vibration, i.e. Δv=0, ±1, ±2, ±3, etc., whilst for excitation of non-totally symmetric vibrations, in this case those having b_{1g} symmetry, one would expect transitions consistent with Δv=0, ±2, ±4, ±6 etc.

In proposing a possible vibronic coupling mechanism to account for the observation of $\Delta v=\pm1$ transitions in b_{1g} vibrational modes in the ZEKE spectra, it is helpful to start by considering the ground electronic state configuration of naphthalene as described by Åsbrink [64]. The ground neutral state configuration can be written in terms of its valence orbitals as

$$\cdots 9a_g^2\ 1b_{2g}^2\ 1b_{3g}^2\ 2b_{1u}^2 1a_u^2\ 2b_{2g}^0\ 2b_{3g}^0$$

The ground and first excited doublet cationic states arise from removal of the $1a_u$ and $2b_{1u}$ valence electrons to give states of symmetries A_u and B_{1u}, respectively. It is reasonably straightforward to envisage a vibronic coupling scheme between the two ionic states through excitation of a b_{1g} vibration in either state. For example, odd excitation of a b_{1g} mode in the $\tilde{X}^2A_u$ ionic state results in a vibronic state of B_{1u} symmetry. This vibronic state can interact with the excited $\tilde{A}^2B_{1u}$ ionic state which lies less than 6000 cm^{-1} to higher energy (see Fig. 7.15). Similarly, excitation of a b_{1g} mode in the B_{1u} state gives an A_u vibronic state which can interact with the ground ^2A_u ionic state.

If the vibronic wavefunctions for the initial $\tilde{A}^1B_{3u}$ (S_1) and final $\tilde{X}\,^2A_u$ (ground state of the ion) states are represented as ψ'_{ev} and ψ^+_{ev}, respectively, then the transition moment integral M, for the $\tilde{X}^2A_u \leftarrow \tilde{A}^1B_{3u}$ photoionization in naphthalene can be written as

$$M=\langle \psi'_{ev}|\boldsymbol{er}|\psi^+_{ev}\,\epsilon\rangle$$

where ***er*** is the electric dipole moment operator, and ϵ the free electron wavefunction. Vibronic coupling in the initial and final states implies that application of the Born-Oppenheimer approximation, in which the vibronic wavefunctions are written as products of the electronic and vibrational wavefunctions, is not appropriate in this case. In order for a transition to be allowed under the proposed vibronic coupling scheme, the direct product of the three terms involved in the matrix element must include the totally symmetric representation of the D_{2h} point group. Two ionization transitions can be considered to illustrate this point. If we first consider a $\Delta v=\pm1$ transition in the $\bar{8}$ mode for ionization from the S_1 band origin, then ψ'_{ev} will have $B_{3u} \times a_g = B_{3u}$ symmetry and ψ^+_{ev} will have $A_u \times b_{1g} = B_{1u}$ symmetry. The x, y and z components of ***er*** have b_{3u}, b_{2u} and b_{1u} symmetries, respectively, which means that the outgoing free electron wave in the case of a true ZEKE process or the Rydberg electron for PFI must correspondingly have either b_{1u}, a_u or b_{3u} symmetry. In this case, the symmetry of the final state wavefunction ψ^+_{ev} has changed from electronic A_u in the absence of vibronic coupling to vibronic B_{1u} on inclusion of vibronic coupling and consequently we can say that the transition gains intensity through vibronic coupling in the ionic state. If we now consider a $\Delta v=\pm1$ transition in the $\bar{8}$ mode for ionization *via* $S_1\bar{8}^1_0$, then ψ'_{ev} will have $B_{3u} \times b_{1g} = B_{2u}$ symmetry and ψ^+_{ev} will have $A_u \times a_g = A_u$ symmetry. In this case the symmetry of the initial state wavefunction ψ'_{ev} has changed from electronic B_{3u} in the absence of vibronic coupling to vibronic B_{2u} on inclusion of vibronic coupling and

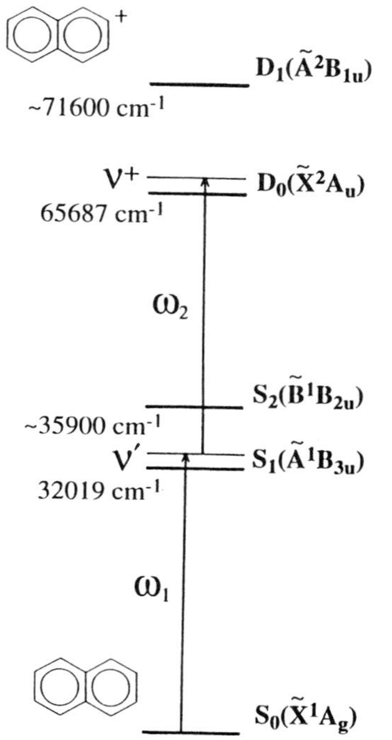

Fig. 7.15 Energy level diagram showing the excited states of the neutral and ionic states involved in vibronic coupling in naphthalene.

we can say that the transition has gained intensity through vibronic coupling in the initial neutral state. The conclusion in this case is that vibronic coupling in the initial and final states can allow Δv=odd components in b_{1g} vibrations to appear in ZEKE spectra recorded either *via* the S_1 state band origin or when one quantum in a b_{1g} vibration is excited in the S_1 state prior to ionization.

Para-Difluorobenzene Another example of a molecule which shows a breakdown of symmetry selection rules in its ZEKE spectrum is the para-difluorobenzene (p-DFB) cation. *Ab initio* calculations from von Niessen *et al.* [65] suggest a dominant electronic configuration for the valence molecular orbitals of the electronic groud state of p-DFB of ....$(b_{2u})^2(b_{3g})^2(b_{3u})^2(b_{1g})^2(b_{2g})^2(a_u)^0(b_{3u})^0$ within the D_{2h} symmetry group. The symmetry species for the electronic ground state of the cation has been determined to be $^2B_{2g}$ from He (I) and He (II) photoelectron spectra [5,66,67].

In the two-colour ZEKE spectra of p-DFB [68], the S_1 band origin, as well as a number of other S_1 vibronic states, were used as intermediate resonances ($S_1 0^0$, 3^1, 6^1, 9^2, 17^1, 27^1 and 30^1). The ZEKE spectrum recorded *via* the S_1 band origin is reproduced in Fig. 7.16. The energy region covered in the spectrum extends from the band origin (which gives the adiabatic ionization energy) to approximately 3000 cm^{-1} ion internal energy. The majority of the bands observed in this ZEKE spectrum were successfully assigned to vibrational states of the p-DFB cation. Other than the vertical transition to the band origin, the ZEKE spectrum is dominated by transitions into totally symmetric modes and combination bands thereof. The strongest transitions are to levels corresponding to one and two quanta excited in the ν_6, ν_5, ν_4 and ν_2 modes in the ion. However, transitions to non-totally symmetric modes ν_{30}, ν_{17} and ν_8 also appear. These modes are also observed in combination with a_g modes.

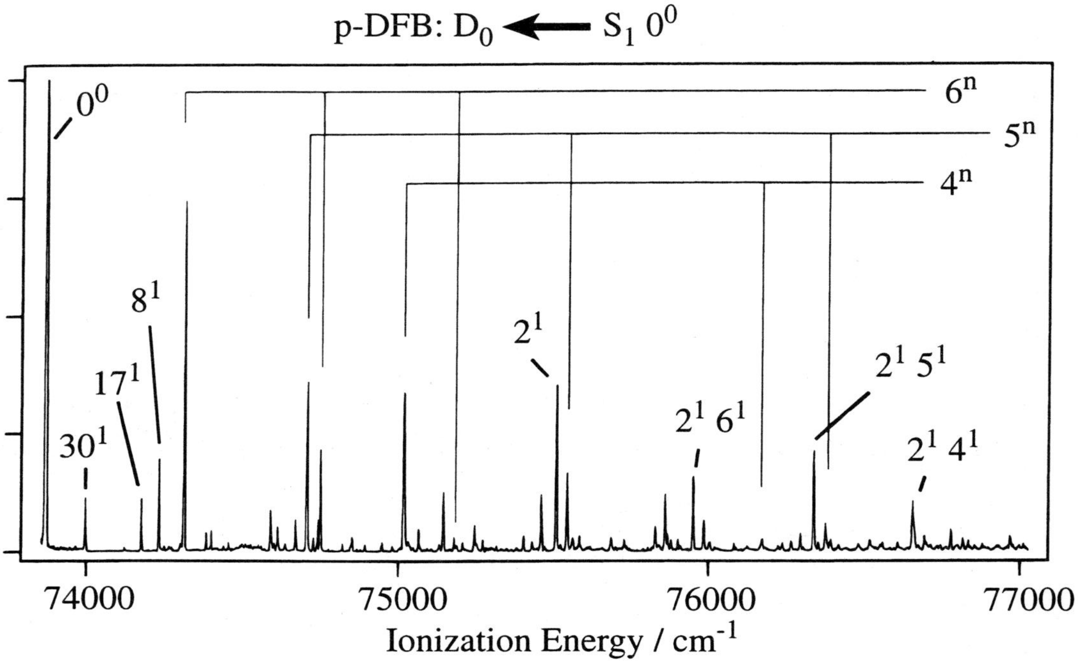

Fig. 7.16 Vibrationally resolved ZEKE spectrum of para-difluorobenzene *via* the S_1 vibrational origin (0^0) as intermediate resonance. From Ref. [68].

Figure 7.17 shows the ZEKE spectrum recorded *via* the S_1 6^1 state. The spectrum shows a progression in the ν_6 mode extending for 4 quanta and with a maximum intensity observed for the 6^2 level. Once again, the strongest transitions in Fig. 7.17 are to symmetric modes.

The most prominent feature of all the ZEKE spectra is the strong activity of the a_g modes, ν_2 to ν_6 (the a_g vibration ν_1 has not yet been seen in conventional photoelectron spectra and is, with the exception of the ZEKE spectrum from the S_1 0^0 level, outside the scanning range employed in the experiments reported here). The dominant activity of the ν_6 vibrational mode strongly suggests that a change in geometry occurs along the ν_6 normal coordinate upon ionization (although some account of Duschinsky rotation [69] of the normal coordinates within the molecular symmetry group D_{2h} may have to be taken into account).

As in the ZEKE spectrum recorded *via* the S_1 band origin, nominally forbidden transitions again occur *via* $S_1 6^1$ with the most prominent involving the ν_8, ν_{17} and ν_{30} vibrational modes. The appearance of these transitions cannot be explained within the constraints of the Born-Oppenheimer approximation and so, as with naphthalene, it is necessary to attempt to construct a vibronic coupling scheme involving other electronic states of the ion. However, if we consider as an example the ZEKE spectra recorded *via* the S_1 band origin, which show transitions to the 8^1 (a_u), 17^1 (b_{2g}) and 30^1 (b_{3u}) levels in the ion, the nearest electronic state with which the final vibronic state may interact is the D_5^+ state (B_{2u}) which lies nearly 5 eV above the electronic ground state of the ion. This state would only couple with the D_0 8^1 vibronic state and in any case such an interaction would necessarily be extremely weak and it is difficult to see how it could account for the observed activity of the ν_8 vibration. For transitions to the 17^1 (b_{2g}) and 30^1 (b_{3u}) levels, no known electronic state of the correct symmetry exists with which the vibronic state can interact. The implication is that other factors must be responsible for the activity of these vibrations in the ZEKE spectra.

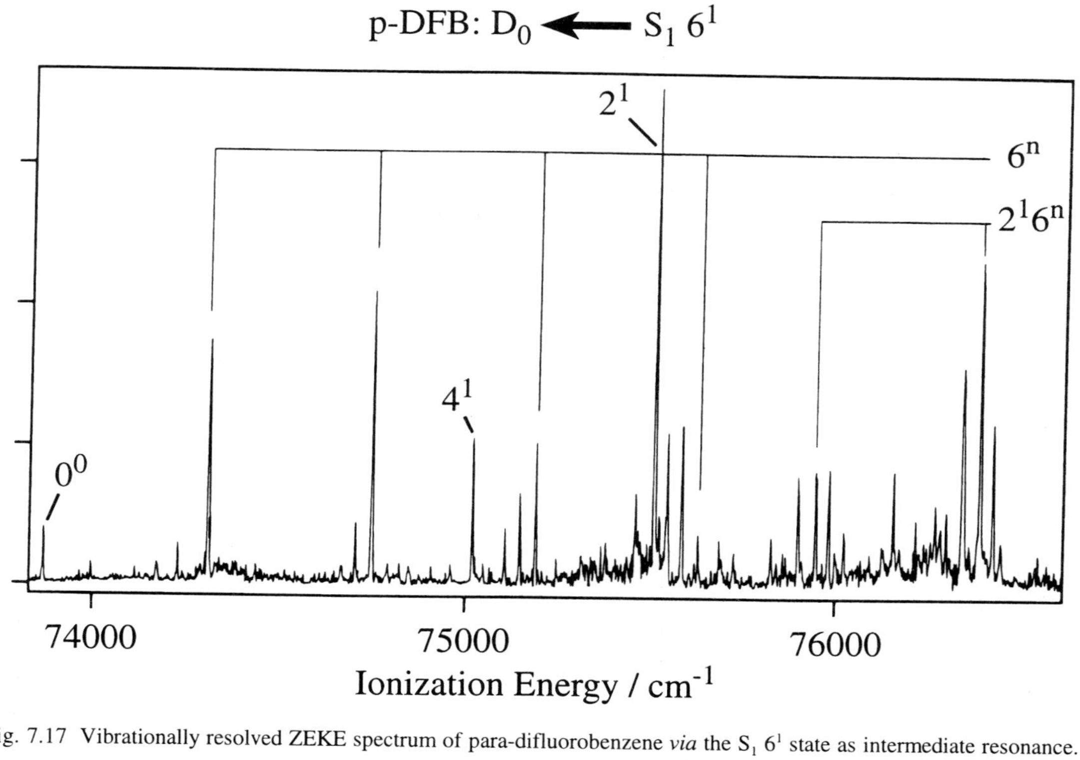

Fig. 7.17 Vibrationally resolved ZEKE spectrum of para-difluorobenzene *via* the S_1 6^1 state as intermediate resonance. From Ref. [68].

One possibility is a lowering of symmetry in the S_1 state combined with a breakdown of the Born-Oppenheimer approximation. Symmetry lowering in the S_1 state will allow a relaxation in the symmetry selection rules for some of the observed vibronic transitions, but for others a further reduction to C_2 would be necessary. The problem with assuming a lowering of symmetry is that one would then expect progressions in non-totally symmetric vibrations to dominate. The conclusion is that, in contrast to the situation encountered with naphthalene, invoking vibronic coupling to explain the appearance of non-totally symmetric vibrations in the ZEKE spectra is not straightforward. Furthermore, even when a lowering of symmetry is assumed in the S_1 state, it is still difficult to explain the appearance of all the transitions seen in the ZEKE spectrum.

In order to aid in the assignment of the observed vibrations in the p-DFB cation and to check the experimentally observed vibrational frequencies, *ab initio* calculations were performed [68]. The calculations support the experimental assignments for the ionic frequencies and lead to a D_{2h} structure for the S_0 and the ground ionic state. For the S_1 state, deviations from D_{2h} symmetry, mainly along the ν_8 coordinate are indicated, which goes some way to supporting the proposition that a lowering in symmetry in the S_1 state may be partly responsible for the observed breakdown of symmetry selection rules.

7.4.2 Rotational Resolution

Benzene Benzene is a benchmark molecule among the hydrocarbons and serves as an important prototype for aromatic systems in general. The electronic configuration of the neutral molecule is ...a_{2u}^2, e_{1g}^4. Upon ionization, one e_{1g} electron is removed from the highest molecular orbital, leaving one e_{1g} electron unpaired. This results in a doubly degenerate $^2E_{1g}$ electronic ground state of the cation. According to the Jahn-Teller theorem [70], for any non-linear polyatomic molecule in a degenerate electronic state, there exists a distortion of the nuclei along at least one non-totally symmetric normal coordinate that results in a splitting of the potential energy function so that the potential minimum is no longer at the symmetrical position [70-72]. Several authors have discussed structural distortions of the benzene cation [73,74]. According to quantum chemical *ab initio* calculations, in-plane distorted geometries with an elongated or compressed D_{2h} structure are more stable than the highly symmetrical D_{6h} configuration [73]. The experimentally determined value of the stabilization energy [75] of 266 cm^{-1} is about the half of the zero-point energy of the lowest frequency Jahn-Teller active normal vibration. The question arises whether the symmetry reduction deduced from the static Jahn-Teller distortion and predicted by theoretical calculations has an influence on the rotationally selective ionization dynamics [76] and on the rotational structure of the benzene cation. If the molecule was statically distorted to lower symmetry, transitions should appear in the rotationally resolved ZEKE spectra, which are forbidden in the undistorted, highly symmetrical D_{6h} configuration. Thus, the observed rotational transitions are a sensitive and clear indication of the symmetry of the cation.

If one quantum of a Jahn-Teller active normal vibration (for benzene normal modes with e_{2g} symmetry [70], ν_6-ν_9 in Wilson's notation [77]), is excited, linear dynamic Jahn-Teller coupling leads to a splitting into two vibronic states with $j=\pm 1/2$ and $j=\pm 3/2$ vibronic angular momentum [1,78]. The quadratic dynamic Jahn-Teller coupling further splits the $j=\pm 3/2$ state into two substates, but leaves the $j=\pm 1/2$ state doubly degenerate [1]. The excitation of an e_{2g} normal vibration in an E_{1g} electronic state leads to vibronic states with $B_{1g}\oplus B_{2g}\oplus E_{1g}$ vibronic symmetry species. First order interaction lifts the four-fold degeneracy and results in two doubly degenerate vibronic states with $B_{1g}\oplus B_{2g}$ and E_{1g} symmetry, respectively. Higher order interactions can only split the $B_{1g}\oplus B_{2g}$ state, but not the inherently doubly degenerate E_{1g} vibronic state.

Model calculations predict a D_{2h} symmetry for the cation, for which there are three equivalent structures [73] which can be envisaged as locally distorted elongated or compressed structures. This distortion to D_{2h} can be detected experimentally in a matrix at low temperatures [79]. The ESR spectra of $C_6H_6^+$ in a freon matrix, indicate an enhanced spin density at C1 and C4 at 4.2 K but equal spin densities on all C's at 100 K. Thus the question arises whether the D_{2h} structure is static and destabilized at higher temperatures by collisions with the matrix or whether a dynamic D_{2h} structure is stabilized by the matrix. For weak Jahn-Teller coupling, the stabilization energy for the distorted symmetry is smaller than the zero point energy of the Jahn-Teller active mode. Under collision-free conditions the three equivalent D_{2h} structures of the cation would dynamically interconvert rapidly, and the ground state of the cation would still be described in the D_{6h} symmetry group [75]. For strong Jahn-Teller coupling the cation would spend much time in one of the three structures, and would therefore be described in D_{2h}. A knowledge of the structure and the symmetry of the isolated benzene cation is desirable not only for testing quantum mechanical model calculations but also has a fundamental importance for organic chemistry. Rotationally resolved photoelectron spectroscopy allows an unambiguous determination of the symmetry of the cation. The molecular symmetry group can be deduced from the rotational transitions that occur, together with group theoretical considerations [1,80].

A low resolution scan of the ZEKE spectrum of benzene, recorded *via* the 6^1 level in the first singlet excited state of the neutral (S_1), shows in detail the vibronic structure of the ground electronic state of the ion [81]. Observable in the spectrum are bands at fundamental frequencies characteristic of the ν_6, ν_{16}, ν_4 and ν_1 vibrational modes. Harmonic progressions are completely absent and the higher energy portion of the spectrum exhibits a highly irregular and dense system of vibronically active states. The lowest active mode is the in-plane C-C-C ring bending mode, ν_6, which carries much of the vibronic intensity in the first 1000 cm^{-1} of the ZEKE spectrum. Distortion in this coordinate produces the planar acute and obtuse D_{2h} structures. A key pair of bands, which appear at about 350 cm^{-1} ion internal energy, are established in the vibronic analysis of the coarse ZEKE spectrum as the B_{1g} and B_{2g} components of the ν_6 fundamental, shifted to lower energy by linear Jahn-Teller coupling. However, it is not possible to establish the relative ordering of B_{1g} *vs.* B_{2g} from the vibrationally resolved ZEKE spectrum.

A high resolution ZEKE scan of this pair of bands, shown in Fig. 7.18, was recorded by exciting through the 6^1 $J' = 2$, $K' = 2$, $-l$ rotational level in the S_1 state [81] (here, J', K' and l refer to the total angular momentum, its molecule-fixed projection and the vibrational angular momentum, respectively [1]). Unique to this set of transitions, is the appearance of sub-bands terminating on states with rotational angular momentum projections $K^+ = 0$, which display only half the number of rotational lines as all $K^+ \neq 0$ sub-

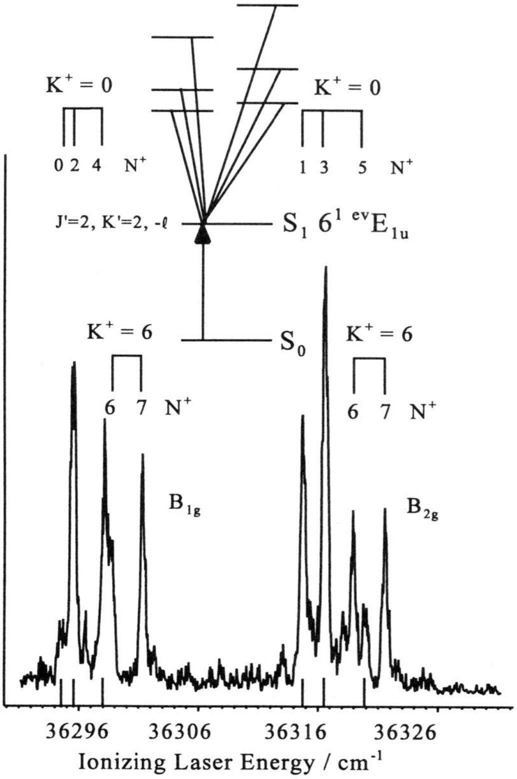

Fig. 7.18 A high resolution ZEKE scan of the benzene B_{1g} and B_{2g} components of the ν_6 fundamental recorded by exciting through the 6^1 $J' = 2$, $K' = 2$, $-l$ rotational level in the S_1 state [81] (here, J', K' and l refer to the total angular momentum, its molecule-fixed projection and vibrational angular momentum, respectively [1] and ${}^{ev}E_{1u}$ refers to the S_1 state vibronic symmetry).

bands. In the lower energy component of the doublet, only levels for which N^+ is even appear, whereas the upper component shows transitions only to levels for which N^+ is odd. The cause of this difference can be established by considering the nuclear-spin statistical weights of the individual vibronic levels in states with vibronic symmetry B_{1g} and B_{2g}.

In the molecular point group D_{6h}, the symmetry species of rovibronic states with projections K=0 show an alternating dependence on the total angular momentum quantum number. For a state with vibronic symmetry B_{1g}, the levels for which N is even have rovibronic symmetry B_{1g}, whereas those levels for which N is odd have rovibronic symmetry B_{2g}. In a state with vibronic symmetry B_{2g} this order is reversed. In order to conform with the requirements of nuclear-spin statistics, rovibronic states with symmetry species B_{1g} must have a nuclear-spin symmetry of A_1, whereas those with rovibronic symmetry B_{2g} must have a nuclear-spin symmetry of A_2. When nuclear-spin degeneracy factors are considered, states with rovibronic symmetry B_{1g} will be 13 times more abundant than states with rovibronic symmetry B_{2g}. This means that the rovibronic intermediate state selected in the two-colour ionization scheme will have predominantly B_{1u} symmetry and thus transitions to B_1 rovibronic symmetry states with a K=0 projection will be strongly favoured. Therefore, transitions from the 6^1 J' = 2, K' = 2, $-l$ rotational level in the S_1 state will terminate on even N^+ levels with K^+=0 for vibronic states of B_{1g} symmetry and on odd N^+ levels with K^+=0 for vibronic states of B_{2g} symmetry. The strong lines in the lower component shown in Fig. 7.18 corrrespond to even values of N^+ which unambiguously assigns this state the vibronic symmetry B_{1g}. The upper component on the other hand, shows greatest intensity in transitions to odd N^+ levels, which establishes this state as having B_{2g} vibronic symmetry. The rotational structure in the ZEKE spectrum has thus established unambiguously that the B_{2g} level in the quadratically split 6^1 (3/2) levels of the benzene cation, lies above the B_{1g} level.

The cation is clearly distorted by Jahn-Teller coupling in ν_6, but the effect of quadratic coupling is very small. It was concluded from the rotational intensities in the ZEKE spectrum that the wells in the pseudorotation coordinate correspond to local B_{1g} electronic configurations (the elongated structure) whereas the saddle points are locally B_{2g} (compressed). However, the coupling parameters that fit the vibronic structure in the coarse ZEKE spectrum establish that the energy difference between the stationary states is only 8 cm^{-1}, which is considerably less than the ν_6 zero point energy of 413 cm^{-1}. The conclusion is that the benzene cation is fluctuational, dynamically coupled and necessarily viewed in D_{6h} symmetry rather than in terms of D_{2h} structures with locally non-degenerate electronic configurations [81].

7.5. ZEKE spectroscopy of electronically excited cation states

The majority of studies in ZEKE spectroscopy have concentrated on the ground electronic state of the ion, due, in part, to concerns about how decay processes affecting the excited electronic states might compromise the technique. These concerns arise because, as the ion internal energy increases, the metastable high-n (n>150), high l, high m_l Rydberg states lying just below each ionization threshold are potentially subject to decay channels such as autoionization, predissociation and fluorescence. In spite of these concerns, Hepburn and coworkers have published a few recent examples of studies on electronically excited states of small molecules using coherent VUV and XUV light sources [82-84] and, significantly, have been able to apply the technique to the study of the strongly predissociated $A^2\Pi$ state of HBr^+ [82] as well as the fluorescing N_2O^+ $\tilde{A}^2\Sigma^+$ state [83]. With the exception of recent studies on the $a^1\Delta$ state of SH^+ [85] and the $a^4\Sigma_u^-$ state of I_2^+, all of the work so far published in this field on excited states involves direct ionization from the ground state of the neutral in a single step. A consequence of this is that the excited states accessed are, in principle, limited to those accessible by a one-electron excitation process and are further restricted by the spin multiplicity selection rules that accompany ionization from the neutral ground state.

An example of a molecule which shows a breakdown of this expectation is iodine for which the $A^2\Pi_{3/2,u}$ and $a^4\Sigma_u^-$ states have recently been probed by ZEKE spectroscopy [53]. The excited electronic states of the iodine cation, which should be accessible for direct one-photon ionization from the ground state, are the first excited $A^2\Pi_{3/2,u}$ and $A^2\Pi_{1/2,u}$ states ($\sigma_g^2\pi_u^3\pi_g^4$ (234)), and the $B^2\Sigma_g^+$ state ($\sigma_g^1\pi_u^4\pi_g^4$ (144)). Previous photoelectron studies have revealed bands attributable to all of these states [86,87] but there has been some additional speculation about the possible contribution of electronically excited states based on the 2421 ($\sigma_g^2\pi_u^4\pi_g^2\sigma_u^1$) electronic configuration to the region of the photoelectron spectrum lying between the two spin-orbit components of the $A^2\Pi_{\Omega,u}$ state [88]. These states might normally be considered one-electron forbidden states and are analogous to the bound 2431 ($^{1,3}\Pi_u$) states of the neutral molecule. The $\sigma_g^2\pi_u^4\pi_g^2\sigma_u^1$ configuration gives rise to the $^{2,4}\Sigma_u^-$, $^2\Delta_u$ and $^2\Sigma_u^+$ electronic states. Of these, the quartet state is additionally spin-forbidden (ΔS (core) = ½ for direct photoionization) for transitions from the ground state of the neutral molecule. It has also been proposed that these states are responsible for perturbations observed in the $A^2\Pi_{\Omega,u}$ states of Cl_2^+ [89,90] and Br_2^+ [90,91].

In all previous photoelectron studies on iodine, bands attributed to the $A^2\Pi_{\Omega,u}$ state were vibrationally unresolved whilst partially resolved structure for the ground state has only been observed in studies employing Ne (I) or H Lyman α radiation [87]. A recent threshold photoelectron study [92] was the first to partially resolve $A^2\Pi_{\Omega,u}$ state vibrational structure and the average vibrational spacings for the two spin-orbit components of approximately 120 cm^{-1} were measured.

In the present case, ionization is achieved directly from the ground state by coherent absorption of two UV photons in a one-colour ionization scheme. In principle, given sufficient photon energy, this scheme should enable us to access some of the excited states arising from the 2421 configuration, because absorption of two-photons allows us to simultaneously excite two electrons. However, the ΔS (core) = ½ spin-multiplicity selection rule should formally discriminate against transitions to the quartet state.

The $A^2\Pi_{3/2,u}$ and $a^4\Sigma_u^-$ excited states of I_2^+ The one-colour (two-photon) ZEKE-PFI spectrum [53] of I_2 in the range 86300 to 91250 cm^{-1} is presented in Fig. 7.19. The spectrum is dominated by a long progression with an average spacing of 120 cm^{-1} which is consistent with removal of a bonding π_u electron. On this basis, and from a comparison with previously recorded photoelectron spectra, this progression was assigned to the $A^2\Pi_{3/2,u}$ state. The observed onset of the progression appears at 86907 ± 2 cm^{-1} (corrected to account for field ionization shifts) and is labeled as peak 5 (the italics used in Fig. 7.19 indicate that no inference should be made from the numbering as to the absolute vibrational quantum number associated with a particular peak). This is the first fully vibrationally resolved photoelectron spectrum of the $A^2\Pi_{3/2,u}$ state of I_2^+. The onset can be compared with that measured in a previous, low resolution threshold photoelectron study [92] of 86228 ± 16 cm^{-1}. The apparent disparity arises from errors in determining the true band onset in the earlier much lower resolution threshold photoelectron spectrum.

It is apparent, that while the $A^2\Pi_{3/2,u}$ state progression dominates in this spectral region, there are three additional, extensive vibrational progressions, all of which appear with significantly lower intensity. Two of these progressions can be assigned to continuations of the two ground spin-orbit state progressions. They both appear here by virtue of autoionization. For the $X^2\Pi_{3/2,g}$ state, the progression can be followed from v^+=57 to v^+=90 while the $X^2\Pi_{1/2,g}$ state progression extends as far as v^+=40.

The fourth progression in the spectrum appears with an intensity about 20% of that of the $A^2\Pi_{3/2,u}$ state and covers an energy range of about 3400 cm^{-1} from an onset at 86403 cm^{-1} (corrected ionization energy). The measured average spacing of about 114 cm^{-1}, is similar to that measured for the $A^2\Pi_{3/2,u}$ state. The most reasonable assignment of this progression, given that ionization is achieved by absorption of two photons, is to one of the bound excited states of I_2^+ arising from the 2421 electronic configuration (i.e. one of the $^{2,4}\Sigma_u^-$, $^2\Delta_u$ and $^2\Sigma_u^+$ electronic states). The formal restriction imposed by the spin-multiplicity selection rule might at first appear to exclude the quartet state as a possible assignment. However, the quartet state, being the state of maximum multiplicity, will be the lowest lying of the four states. Furthermore, in the cases of Cl_2^+ [89,90] and Br_2^+ [90,91], the $^2\Sigma_u^-$, $^2\Delta_u$ and $^2\Sigma_u^+$ electronic states ought to lie to significantly higher energy than the $A^2\Pi_{3/2,u}$ state. For these reasons, the provisional assignment of this progression was to the $a^4\Sigma_u^-$ state (given that it is the lowest excited state with a multiplicity different from that of the ground $X^2\Pi_{\Omega,g}$ state of

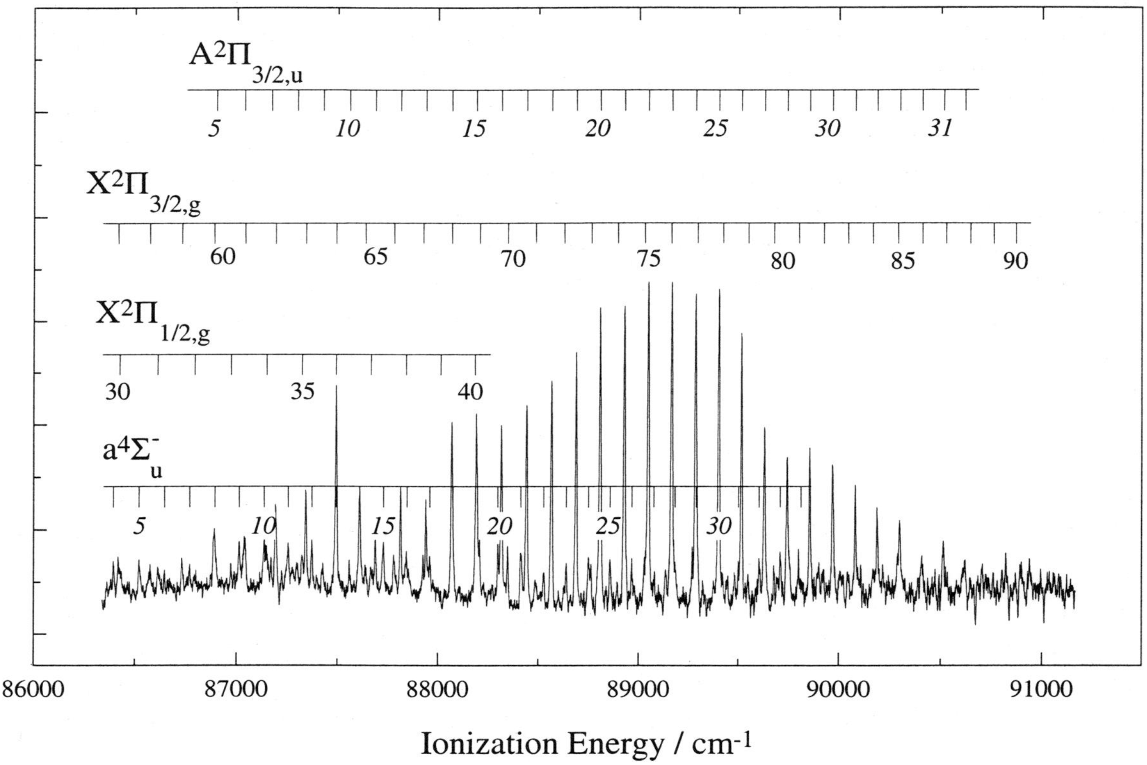

Fig. 7.19 The one-colour (two-photon) ZEKE-PFI spectrum [53] of the $A^2\Pi_{3/2,u}$ and $a^4\Sigma_u^-$ excited states of I_2^+ in the range 86250 to 91250 cm^{-1}. Note the continuation of vibrational progressions associated with the two spin-orbit components of the ground electronic state of the ion.

the ion). Further support for this assigment was provided by comparison with the ZEKE spectrum of the same states recorded *via* the valence B state of the neutral in a two-colour ionization scheme and for which a transition to the $a^4\Sigma_u^-$ state is fully allowed. Large changes in the relative intensities of the two states observed between the two ionization routes provided very strong additional evidence that the assignment was correct.

The appearance of the $a^4\Sigma_u^-$ state in the one-colour (two-photon) ZEKE spectrum is probably due to either configuration mixing with another state, or to spin-orbit interaction in the ion. In fact, recent relativistic Hartree Fock calculations [93] on I_2^+ show that the $A^2\Pi_{3/2,u}$ state has mixed $^2\Pi_{3/2,u}$ and $^4\Sigma^-_{3/2,u}$ character, which suggests one possible candidate for intensity borrowing. It was also suggested in Ref. [53] that autoionization may also be contributing to some extent to the appearance of this state.

7.6. ZEKE spectroscopy of complexes and clusters

The study of weakly bound clusters and complexes in the gas phase using high resolution laser spectroscopic techniques provides a direct experimental view of the perturbations induced in a solute molecule by the initial stages of solvation. While many of the more recent studies of van der Waals complexes in the neutral state have focussed on mono and polycyclic aromatic - rare gas complexes, the original pioneering work in this area was concerned with much smaller systems, concentrating initially on the iodine-rare gas van der Waals complexes [94]. The recent development of ZEKE spectroscopy has resulted in a rapid expansion in interest in the ionic state spectroscopy of both van der Waals and hydrogen bonded complexes. The inherently high spectral resolution of the technique, coupled with state-selective ionization through REMPI schemes enables ionization energies and hence binding energies to be accurately determined for individual complexes as well as enabling detailed vibrational analysis of the characteristically low-frequency intermolecular vibrational structure associated with these complexes.

The first ZEKE study of a van der Waals complex was carried out by Chewter *et al.* [95] on benzene argon. The ZEKE spectrum showed only one peak assigned to the ionization threshold, which was red shifted by 172 cm^{-1} with respect to the benzene origin. From that spectrum and the rather small change in ionization energy compared to the benzene monomer, the conclusion was drawn that the ionization of the complex does not lead to a drastic increase in binding energy of the complex. Since then, numerous other studies have been made on solvated aromatic complexes (*e.g.* aniline-Ar_n (n=1,2) [96], aniline-CH_4 [96,97], anthracene-Ar_n (n=1-5) [98], p-dimethoxybenzene-Ar_n (n=1,2) [99]) but only a few studies have been made on small van der Waals complexes. We shall consider here three examples of van der Waals complexes which fall into the latter category.

7.6.1 The iodine-argon van der Waals complex

Iodine-argon was one of the first complexes to be studied in the early jet spectroscopy experiments conducted by Levy and coworkers [94] in the 1970's. Most of the work carried out on this complex since then has been concerned with the $\tilde{B}^3\Pi 0_u^+ \leftarrow \tilde{X}^1\Pi 0_g^+$ system studied using laser induced fluorescence (LIF) spectroscopy. Unfortunately, the valence B state is not well suited as a resonant intermediate state because it lies only 15800 cm^{-1} above the electronic ground state. However, more recently, a number of gerade ns and nd Rydberg excited states, based upon both spin-orbit states of the $\tilde{X}\ ^2\Pi_{\Omega,g}$ I_2^+-Ar core, have been characterized using (2+1) mass-resolved REMPI spectroscopy. These Rydberg states lie between 53000 and

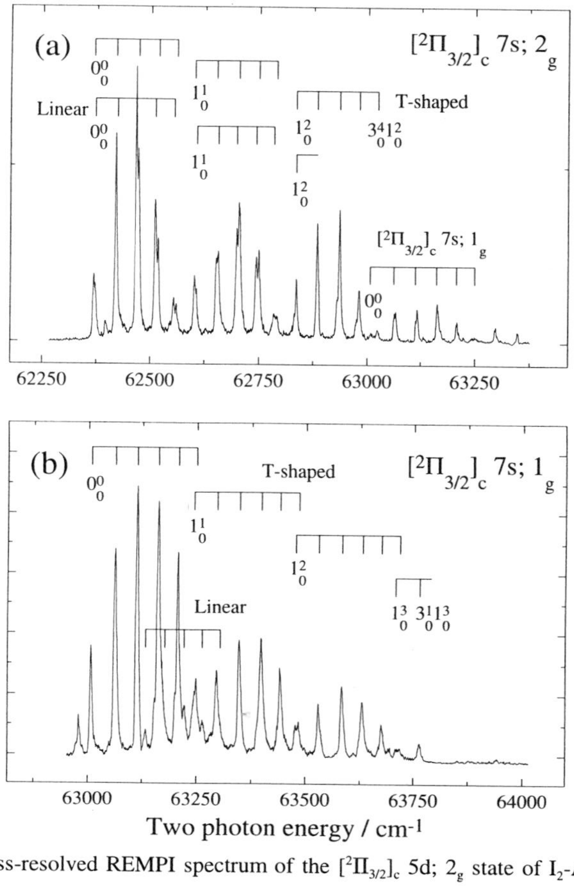

Fig. 7.20 (a) The (2+1) mass-resolved REMPI spectrum of the $[^2\Pi_{3/2}]_c$ 5d; 2_g state of I_2-Ar recorded with circularly polarized light in the range 62250 to 63250 cm^{-1}. (b) The (2+1) mass-resolved REMPI spectrum of the $[^2\Pi_{3/2}]_c$ 5d; 0_g^+ state of I_2-Ar recorded with linearly polarized light in the range 63000 to 64000 cm^{-1}. From Ref. [108].

69000 cm^{-1} and are much better suited as resonant intermediate states for ionization into the two spin-orbit components of the $\tilde{X}\ ^2\Pi_{\Omega,g}$ ionic ground state. Changes in equilibrium geometry along both the I···I and I_2···Ar coordinates of the T-shaped complex upon excitation to the respective Rydberg states result in extended vibrational progressions in both the I···I stretching mode and the I_2···Ar van der Waals stretching mode in the REMPI spectra. The substantial differences in binding energy observed for the 6s, 5d and 6d Rydberg states of the I_2-Ar complex correlate with the degree to which the positive charge of the core is shielded from the argon atom by the Rydberg electron. The more penetrating the Rydberg orbit, the smaller the effect of charge-induced dipole forces on the bonding. The observed binding energy increases accompanying increasing electronic excitation are manifested in the REMPI spectra as progressive increases in spectral red-shifts and intermolecular vdW stretching frequencies. A further issue which had been the subject of considerable debate since the early work of Levy was whether I_2-Ar adopts a T-shaped or linear geometry. Initial speculation suggested that it should be linear by analogy with the known linear geometry of the ClF-Ar complex [100]. However, the I_2-He complex had been found to be T-shaped [101] from a rotational analysis of the bands observed in the B-X fluorescence excitation spectrum and more recently the Br_2-Ne [102], Cl_2-Ne [103] and Cl_2-Ar [104] complexes have all been characterized as T-shaped from the B-X transition. The first direct experimental evidence for the geometry of I_2-Ar emerged from Burke and Klemperer's partially rotationally resolved B-X fluorescence excitation spectrum [105] which confirmed that I_2-Ar also adopts a T-shaped geometry. However, they also suggested that a linear isomer was responsible for an observed fluorescence excitation continuum underlying the discrete B-X transitions [106]. In the initial REMPI study of the Rydberg states of the complex [107], it was noted that many of the vibrational bands now assigned to the $[^2\Pi_{3/2}]_c$ 5d; 2_g Rydberg state were split into apparent doublets (in the notation used

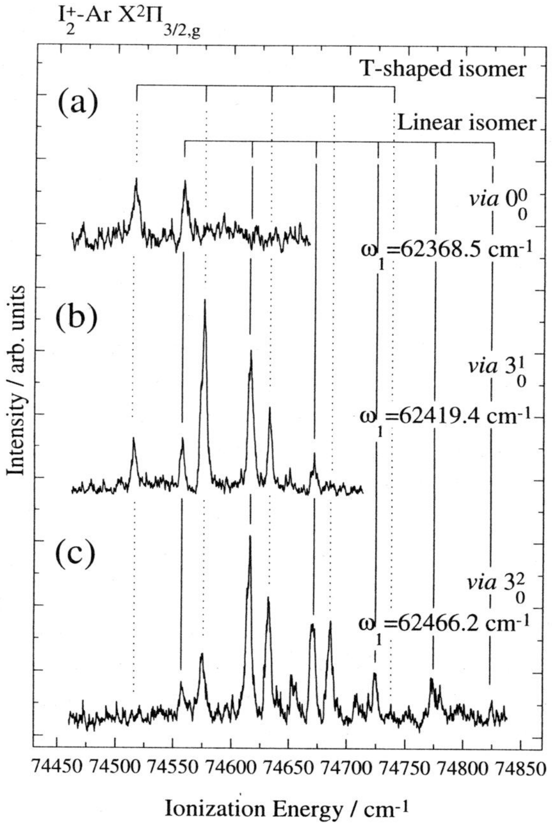

Fig. 7.21 The two-colour (2+1') ZEKE-PFI spectra of I_2^+-Ar ionized (a) *via* the overlapping band origins (0_0^0) of the $[^2\Pi_{3/2}]_c$ 5d; 2_g Rydberg state (b) *via* 3_0^1 and (c) *via* 3_0^2. From Ref. [108].

here, $[{}^{2S+1}\Lambda_{\Omega}]_c$ *nl*; Ω_g, the square brackets contain a description of the spin, oribital, and total angular momenta of the core on which the Rydberg state is based, *nl* refers to the principal quantum number *(n)* and angular momentum *(l)* of the Rydberg electron, and Ω_g gives the Hund's case (c) description of the system). It has subsequently emerged from the ZEKE spectra recorded *via* this state, that the splitting almost certainly arises from two partially overlapping vibrational progressions associated with two structural isomers [108] (In the original reference [107] the $[{}^2\Pi_{3/2}]_c$ 5d; 2_g state is referred to as the $[{}^2\Pi_{3/2}]_c$ 7s; 2_g state. However, the assignments given here conform to a recent reassignment of all states previously assigned to *ns* Rydberg states of I_2 for $n>6$ to $(n-2)$d Rydberg states).

The mass resolved (2+1) REMPI spectrum [108] recorded by monitoring the I_2^+-Ar mass channel is shown in Fig. 7.20. The spectrum is composed of partially overlapping vibrational progressions arising from the $[{}^2\Pi_{3/2}]_c$ 5d; 2_g (Fig. 7.20(a); recorded with circularly polarized light) and $[{}^2\Pi_{3/2}]_c$ 5d; 0_g^+ (Fig. 7.20(b); recorded with linearly polarized light) Rydberg states of I_2-Ar. The vibrational structure for both states essentially arises from simultaneous excitation of both the I-I stretch (ν_1) and the I_2-Ar van der Waals stretch (ν_3). For the $[{}^2\Pi_{3/2}]_c$ 5d; 0_g^+ state, the progression terminates abruptly at $3_0^1\,1_0^3$ (1 quantum in ν_3 plus 3 quanta in ν_1) which suggests that the complex dissociates at this point. The $3_0^1\,1_0^3$ band appears at an internal vibrational energy of 758 cm^{-1}, but we know from the spectral red shift of the band origin that the lower limit to the zero point dissociation energy for this state is 563 ± 5 cm^{-1}. Thus, it would appear that the coupling between the two vibrational modes is sufficiently weak to enable the complex to accommodate an excess 195 cm^{-1} of internal energy before it dissociates. This is consistent with the geometry of the complex being T-shaped. As mentioned above, the $[{}^2\Pi_{3/2}]_c$ 5d; 2_g Rydberg state progression shown in Fig. 7.20(a) reveals an additional complexity of structure when compared to the progression observed for the 0_g^+ state. For both v_1=0 and v_1=1, each vibrational band appears significantly broader than observed for the 1_g progression, and on most of the bands, an apparent splitting can be resolved. However, for v_1=2 the doublet structure has disappeared and the peaks have adopted the narrower profile seen in the 0_g^+ progression. In fact, the doublet structure arises, not from any splitting of the peaks, but from two partially overlapping vibrational progressions with near identical band origins. It appears from the spectrum that one of the progressions terminates at 1_0^2 while the other continues at least as far as $3_0^4 1_0^2$. If we assume, once again, that the point at which the progression terminates represents the point at which the complex dissociates, then we can conclude that for the shorter progression, the onset of dissociation occurs at 463 cm^{-1} internal vibrational energy (compared with a calculated value for D_0 of about 503 cm^{-1}), while for the longer progression, dissociation occurs at a lower limit of 655 cm^{-1}. On this basis, we can make a provisional assignment of the shorter progression to the linear isomer, for which the van der Waals stretch might be expected to couple more efficiently with the I_2 stretch, and the longer progression to the T-shaped isomer.

Although the REMPI spectrum certainly provides a great deal of circumstantial evidence for the existence of two isomers, the assignment can be greatly strengthened by extending the scope of this study to the ionic state by using ZEKE spectroscopy to probe the $[{}^2\Pi_{3/2}]_c$ 5d; 2_g Rydberg state in a two-colour (2+1') ionization scheme, and recording isomer specific ZEKE spectra. The ZEKE spectra of the ground electronic state of the ion recorded *via* several intramolecular vibrational levels in the $[{}^2\Pi_{3/2}]_c$ 5d; 2_g Rydberg state [108] show, in each case, two well separated vibrational progressions (Fig. 7.21). As the level of vibrational excitation is increased in the intermediate Rydberg state, so the vibrational activity in the resulting ZEKE spectra increases. The general propensity for diagonal transitions is consistent with the fairly small changes in geometry that occur on exciting from the Rydberg state, which is fundamentally ionic in character, to the ion. The experimental observations in this case are consistent with an assignment of the two overlapping Rydberg state progressions to two geometrical isomers. The measured difference in the ionization energies of 43 cm^{-1} for the two isomers provides an indication of their relative stabilities in the ion, with the linear isomer being the more weakly bound.

7.6.2 Ar-NO

Microwave and radiofrequency spectroscopy of the ground neutral state of the Ar-NO complex [109] has established the geometry of the molecule to be essentially T-shaped, with a vibrationally averaged deviation from a pure T-shape (where the apex is located at the NO centre of mass) of 5.175° with the Ar atom lying closer to the nitrogen than to the oxygen atom. The vibrationally averaged bond length between the Ar atom and the centre of mass of the NO moiety was derived as 3.711 Å.

Two-colour ZEKE spectra of Ar-NO were recorded in 1992 by Takahashi and Kimura [110,111]. The ionization threshold was reached by exciting through a number of vibrational levels in the $\tilde{C}^2\Pi$ state [112] of Ar-NO. The ZEKE spectra showed vibrational structure that was interpreted in terms of two vibrational series, one of which was assigned to an intermolecular bend and the other to an intermolecular stretch. The spectra shown in Fig. 7.22 were recorded *via* the $\tilde{C}$ state with zero, one and two quanta of the intermolecular stretch excited, respectively (only the intermolecular stretch appears in the (2+1′) REMPI spectrum [113,114]). The vibrational spacing for the intermolecular bend series was measured as ≈ 79 cm^{-1}, while that of the stretch was measured as ≈ 94 cm^{-1}. A more accurate analysis led to values for the harmonic vibrational frequencies of 80.3 cm^{-1} and 99.6 cm^{-1} with anharmonic constants [115] of 1.0 cm^{-1} and 3.55 cm^{-1} for the intermolecular bend and stretch, respectively. Franck-Condon calculations were performed in that work which made two significant approximations. Firstly, harmonic oscillator potentials were used and secondly, no account of the Duschinsky rotation [69] of the normal coordinates was taken. Within these assumptions, the intensities of the vibrational components in the (2+1′) REMPI spectrum were used to derive the changes in the geometry of the $\tilde{C}$ state. Although the signs of these changes were not known, it was reasonably assumed that the change would be negative, *i.e.* the bond length shortened. Since no bending mode structure was seen for the $\tilde{C}$←$\tilde{X}$ transition it was assumed that the Ar-N-O bond angle did not change. The geometric

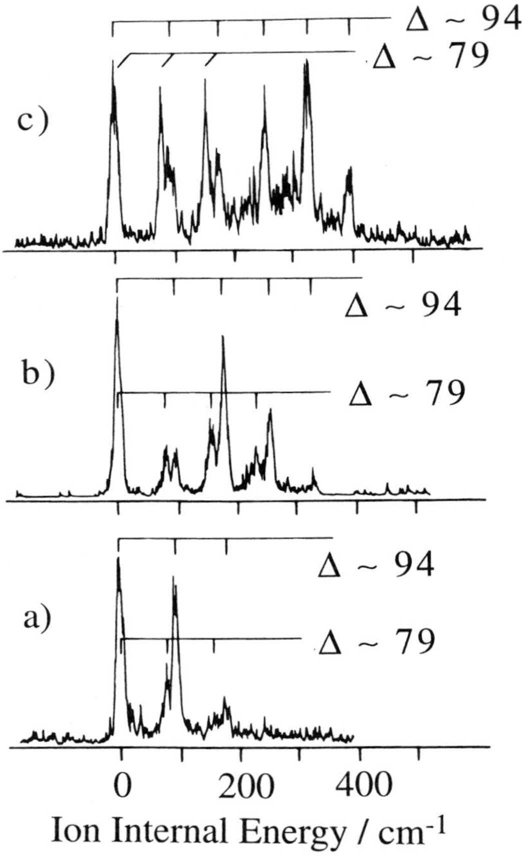

Fig. 7.22 ZEKE spectra of Ar-NO with zero (a), one (b) and two (c) quanta of the intermolecular stretch excited in the intermediate $\tilde{C}$ state. From Ref. [110].

parameters for the $\tilde{C}$ state were then used, together with the experimental intensities in the ZEKE spectrum, to derive geometric parameters for the ground cationic state. Once again, since the absolute values of the changes could not be determined, the change in the bond length was assumed (again reasonably) to be negative. It should be noted that some errors in the use of the complex geometry were made in references [110, 111] which have subsequently been outlined and corrected [116]. Two *ab initio* calculations on the Ar-NO^+ complex have been published recently [116,117]. The earlier calculation [117], which used a size consistent "modified CIPSI" approach (configuration interaction with multiconfigurational zeroth order wavefunction selected by iterative process) [118], generated contour plots of the potential energy surface, which seemed to suggest that the minimum energy of the Ar-NO^+ complex corresponded to a bond angle of *ca.* 90°. This result was in agreement with the more recent calculation [116] that used large basis sets and the MP2 level of theory [119]. The experimentally derived bond angle would agree well with the calculated values, if the derived change in bond angle was taken to be positive rather than negative. The calculated bond length derived in the earlier *ab-initio* study [117] disagreed with the experimental result, whereas the more recent study supported it [116].

The difference between the ionization energy of the Ar-NO complex and the NO molecule is equal to the difference in zero point dissociation energies of the Ar-NO neutral and cation ground states. The ionization energy of NO^+ is known very accurately from both Rydberg series extrapolations [120] and ZEKE spectroscopic studies [27]. The D_0 value for the Ar-NO^+ ion was re-evaluated as 920±20 cm^{-1} [116], using the adiabatic ionization energy of Ar-NO (73869±6 cm^{-1}) [110].

7.6.3 Xe_2

The rare gas dimers represent one of the simplest van der Waals species and have consequently been the subject of many experimental and theoretical studies [121,122]. The most recent theoretical study, by Ma *et al.* [123] calculated the ionization energies of all R_1R_2 species (R_1, R_2=He, Ne, Ar and Kr), using the *Gaussian 2* [124] approach. The only ZEKE spectroscopic study has come from the White group [125]. In that study, one-photon ionization of Xe_2, formed by supersonic expansion of Xe through a 20 μm nozzle, was used. The ionizing radiation was produced by four-wave sum mixing [126]. The ZEKE spectrum showed a long progression in the internuclear vibration of the Xe_2^+ $A^+\,{}^2\Sigma_u^+$ state, but, unfortunately, the adiabatic ionization energy was not determined due to poor Franck-Condon factors in that energy region. Studies with isotopically-enriched samples were also performed. Analysis of the spectrum led to an assignment of the vibrational structure, which indicated that the lowest vibrational feature corresponded to v^+=56 in the ion. The harmonic vibrational frequency ω_e was determined as 114.7 cm^{-1}, with $\omega_e x_e$ equal to 0.417 cm^{-1}. An extrapolation led to a value of 90312 cm^{-1} being determined for T_e. The adiabatic ionization energy was thus derived as 90360 ± 70 cm^{-1} by adding the zero point energy to the T_e value. It was suggested that the Franck-Condon factors for excitation to the ionization threshold could be greatly improved by exciting *via* a suitable intermediate state in a two-colour ZEKE experiment, thereby leading to a more accurate determination of the ionization threshold. Finally, the dissociation energy of the A^+ state, D_0, was determined to be 7660 ± 70 cm^{-1}. As well as the A^+ state, the B^+ and C^+ states were also investigated in the ZEKE study. For the B^+ state although it was not possible to completely unravel the vibrational assignment, a value for the dissociation energy, D_0, was estimated as ≥ 1230 cm^{-1} while the vibrational frequency ω_e was found to be ≥ 45 cm^{-1}. For the C^+ state, the dissociation energy, D_0, was accurately determined as 442 ± 2 cm^{-1} with a value determined for the vibrational frequency ω_e of 23.1 cm^{-1}.

7.7. Hydrogen bonded clusters

The enormous amount of attention paid to hydrogen bonding over the years stems from its ubiquitous presence in biological systems [121]. The detailed study of hydrogen bonding in such systems is not straightforward, however, and attention has been focused on rather smaller systems in environments that are more amenable to interpretation by the spectroscopist. The central role played by hydrogen bonding in the extraordinary physical properties of solid and liquid water provides an additional incentive to look at smaller systems. The biggest successes have come from the study of 1:1 bonded complexes in molecular beams [127,128].

Although numerous studies have been performed on neutral complexes, there are very few studies that have concentrated on ionic species. These species are of obvious importance because fundamental processes, such as solvation, depend on ionic species interacting with neutral molecules. The paucity of such studies is due mainly to the difficulty of producing significant quantities of ionic complexes in the gas phase (although infrared spectroscopy of ions is an extremely active field [129,130]). Moreover, conventional PES does not offer sufficient resolution to reveal the low-frequency intermolecular vibrational structure which characterizes the spectroscopy of these complexes. By way of an illustrative example of this problem the photoelectron spectra of phenol, phenol-H_2O and phenol-$(H_2O)_n$ are shown in Fig. 7.23.

The hydrogen bonded complexes studied so far using ZEKE spectroscopy have all involved phenol as the proton-donating moiety and include phenol-water [33,131,132] (Ph-H_2O), phenol-methanol [133] (Ph-MeOH), phenol-ethanol [134] (Ph-EtOH), the phenol-dimer [135] (Ph_2) and phenol-dimethylether [136] (Ph-DME). The ZEKE spectra of each complex were recorded by tuning the first dye laser to an S_1 vibronic state of the complex under study, and then scanning the second dye laser through the ionization thresholds to populate ZEKE Rydberg states. These were subsequently field-ionized by a delayed extraction pulse of 0.7 V/cm (delay 2 μs).

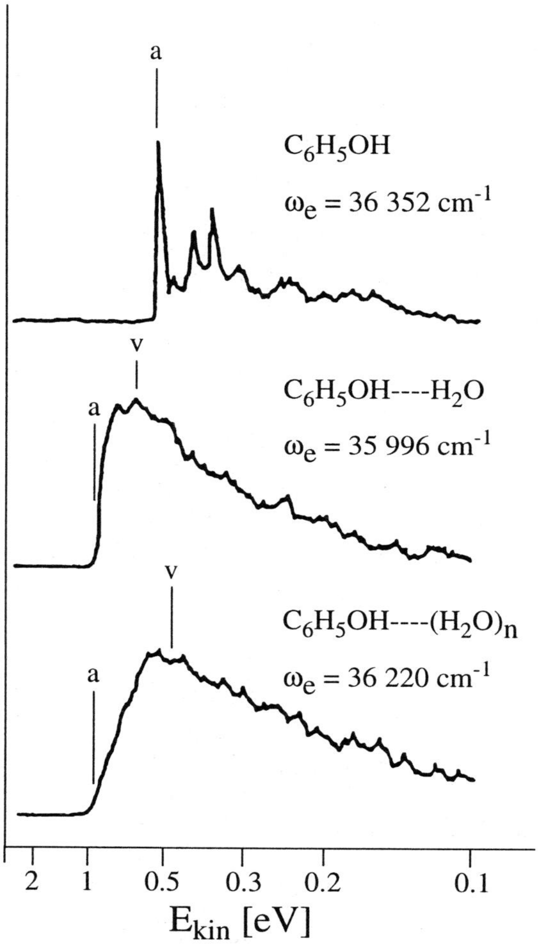

Fig. 7.23 Photoelectron spectra of phenol (top), phenol-water (middle) and higher phenol clusters (bottom). From Ref. [141].

7.7.1 Phenol-Water

This complex, together with the series phenol-$(H_2O)_n$ (n=1-4), has received much attention over the last 10-15 years, with several experimental techniques being employed. A number of *ab initio* studies on the neutral ground electronic state have also been published [137-140], all of which agreed that the complex has C_s symmetry with the protons on the water oriented perpendicular to the plane of the phenol ring (*i.e.* symmetric with respect to reflection in the plane). The experimental studies on the 1:1 complex have mainly concentrated on the S_1 state (formed by a $\pi^* \leftarrow \pi$ excitation on the phenol ring), although some information is available on the S_0 state. The 1+1′ REMPI spectrum of the Ph-H_2O complex shows very little in the way of vibrational structure. The transition into the vibrationless state is the most intense transition, suggesting that the geometries of the S_1 and S_0 states are fairly similar along the hydrogen-bonding coordinates. Two other intermolecular vibrations appear, which have been assigned with a high degree of confidence to the stretching vibration (σ) at 156 cm^{-1} and the weak in-plane wagging vibration (γ') at 121 cm^{-1}.

In contrast to the neutral states of the 1:1 phenol-water complex, information about the low-frequency intermolecular vibrations of the *ionic* complex, was much less comprehensive. The conventional (one-colour, two-photon) time-of-flight photoelectron spectrum (REMPI-PES) [141] *via* the vibrationless S_1 state did not show resolved structure. Two-colour photoionization efficiency (REMPI-PIE) [142] measurements *via* the vibrationless S_1 origin and the S_1 level with one quantum of the intermolecular stretch excited showed the presence of at least one intermolecular vibration, which was observed as a progression of steps in the ion

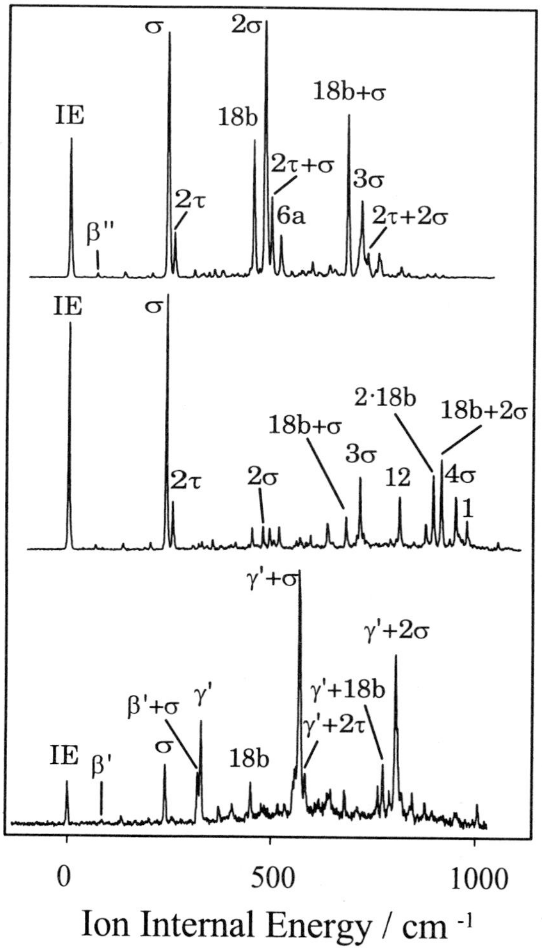

Fig. 7.24 ZEKE spectrum of phenol-water *via* vibrational origin of S_1 (0^0, top), *via* intermolecular stretch excited in S_1 (σ^1, middle) and *via* intermolecular in-plane wag excited in S_1 (γ', bottom). From Ref. [132].

yield spectra. This vibration, with a spacing of 242 cm^{-1}, was assigned to the intermolecular stretching mode of the ionic complex. However, no additional intermolecular vibrational modes were resolved.

Very recently, the 1:1 phenol-water complex has been investigated in detail by ZEKE spectroscopy [132]. The study included the threefold deuterated complex (where the two water hydrogens and the phenolic hydrogen have been substituted by a deuterium atom). For the protonated (h_3) complex, ZEKE spectra were obtained by exciting through three intermediate vibronic states; S_10^0 (the band origin), $S_1\sigma^1$ (one quantum excited in the intermolecular stretch) and $S_1\gamma'^1$ (one quantum excited in the intermolecular in-plane wag). The resulting spectra are shown in Fig. 7.24. From the ZEKE experiment, the (field-free) ionization energy (including the correction for the extraction field) was determined to be 64027 ± 4 cm^{-1}. The ionization energy red shift of the h_3 complex is determined to be 4601 ± 8 cm^{-1} with respect to the ionization energy of the isolated phenol molecule [143] (68626 ± 4 cm^{-1}).

Five out of the six possible intermolecular vibrations were observed and measured. As noted above, the most obvious feature is the progression of the intermolecular stretch (see Fig. 7.24), which also appears in combination with other intermolecular modes. The long progression (consisting of five quanta) in the intermolecular stretch is indicative of the expected large change in the hydrogen bond length upon ionization. This observation has also been made in the photoionization efficiency measurements [113], where a bond decrease of 0.018 Å was estimated from a one-dimensional Franck-Condon factor calculation. The progression itself was found to be quite anharmonic which is a general observation in the ZEKE spectra of the different hydrogen-bonded complexes. The increase in frequency of the intermolecular stretch of the cation (240 cm^{-1}) over that in the S_1 state (157 cm^{-1}) [132] and the S_0 state (155 cm^{-1}) [137] illustrates the large increase in binding energy upon ionization. The other intermolecular vibrations that were observed are the out-of-plane bend (β'') at 67 cm^{-1}, the in-plane bend (β') at 84 cm^{-1}, the first overtone of torsional mode (2τ) at 257 cm^{-1} and the in-plane wag (γ') at 328 cm^{-1}. In assigning each of these vibrational modes, a number of arguments were used including the magnitude of vibrational frequency shifts on deuteration, comparison with the *ab initio* calculations [140] carried out as part of the study, and observation of the changes in intensity as the intermediate S_1 vibronic level was changed. A good example of the latter may be seen in Fig. 7.24. In the ZEKE spectrum recorded *via* the intermolecular $S_1\gamma'^1$ level (bottom), the ionic intermolecular in-plane wag (γ') is strongly enhanced compared to the other two ZEKE spectra recorded *via* the $S_1\sigma^1$ (middle) and S_1 origin level (top).

A further point of interest was that the intramolecular ν_{18b} vibration of the complex at 450 cm^{-1} was found to couple strongly with the intermolecular stretch, which explained why the ν_{18b} mode vibration appeared so prominently in the ZEKE spectrum of this complex, while it was seen only weakly in the ZEKE spectrum of the isolated phenol molecule. This coupling was identified from normal mode pictures obtained *via* the *ab initio* calculations [140] of the frequencies.

7.7.2. Phenol-Methanol

The study of the Ph-MeOH complex was carried out in a similar way to the Ph-H_2O complex. The (1+1′) REMPI spectrum was difficult to interpret owing to the comparatively dense vibrational structure. Vibrational levels in the S_1 state were used as intermediate states on the way to ionization. The ZEKE spectrum obtained by exciting *via* the S_1 vibrationless level [133] is particularly striking (see Fig. 7.25), with progressions of *ca.* ten quanta in a low-frequency vibrational mode of 34 cm^{-1}, denoted η_1 [144], appearing in combination with components of an anharmonic progression of the intermolecular stretch of 278 cm^{-1}. This pattern suggests a rather substantial change of geometry upon ionization. The adiabatic ionization energy was derived as 63207 ± 4 cm^{-1} (field-free value) which represented an increase of bonding over that in the S_0 state of 5421 ± 8 cm^{-1}, again indicating a large increase in bond strength. The latter point is also exemplified by the large increase in the intermolecular stretch over the values of 176 cm^{-1} in the S_1 state [133] and the value of 162 cm^{-1} in the S_0 state obtained by dispersed fluorescence spectroscopy [145]. Additionally, between the latter components, another set of progressions of the intermolecular mode of 34

cm^{-1} were seen, this time in combination with a third intermolecular mode of 52 cm^{-1}. The ZEKE spectrum obtained *via* the S_1 state, with one quantum of the lowest frequency intermolecular mode excited, showed the same vibrations but with a substantially changed Franck-Condon envelope which allowed the identification of a fourth intermolecular mode, denoted η_4, at 153 cm^{-1}. A slightly different envelope for the 34 cm^{-1} vibration was also obtained when exciting through the S_1 state with one quantum of the intermolecular stretch (σ) excited. The other two intermolecular modes of the Ph-MeOH cationic complex were identified from the ZEKE spectrum obtained *via* a combination band; their values being 76 cm^{-1} and 158 cm^{-1}. So, as with Ph-H_2O, where five intermolecular vibrations were assigned, for phenol-methanol all six intermolecular modes of the cationic complex were identified by using different intermolecular vibrational S_1 levels as intermediate resonances.

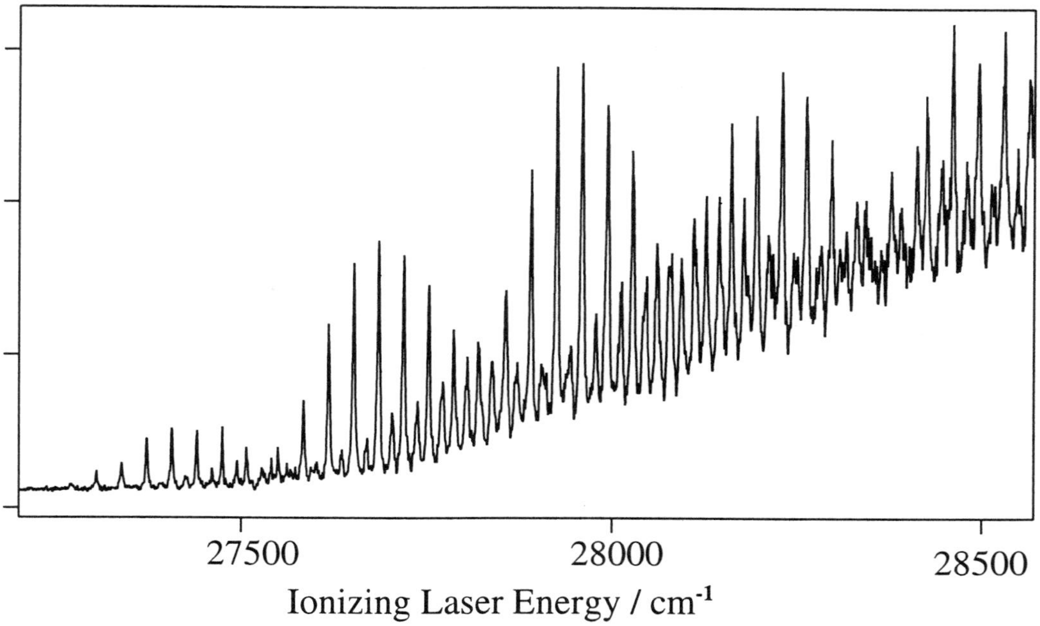

Fig. 7.25 ZEKE spectrum of phenol-methanol *via* vibrational origin of S_1 0^0. From Ref. [133].

7.8. Conclusions

The excellent resolution of the ZEKE method coupled to a wide variety of ionization schemes employing non-linear resonant and non-resonant multi-photon ionization as well as single photon vacuum ultraviolet (VUV) and soft X-ray (XUV) ionization schemes has resulted in applications to a diverse range of molecules, complexes, clusters and radicals. The examples presented in this chapter are just a small selection of the systems which have been studied so far, but it is hoped that they have provided an indication of the special strengths and features of ZEKE spectroscopy. In almost every example, the unique properties of this method are responsible for making new and important discoveries which would have been inaccessible using earlier forms of photoelectron spectroscopy or other more traditional spectroscopic techniques. In many cases, the major strength of ZEKE spectroscopy lies in the unprecedented levels of resolution available. This may be applied to resolve low frequency intermolecular vibrations in van der Waals or hydrogen bonded complexes or to revealing rotational structure of molecules such as the benzene cation from which the true nature of Jahn-Teller distortion has been determined for the first time.

A continually developing area has been the study of photoionization dynamics [26] and in achieving an undertanding of the properties of the ZEKE Rydberg states with particular emphasis on the origin of their long lifetimes. Other new areas of activity include femtosecond ZEKE experiments [146,147] where the ZEKE channel is used as a probe of chemical reaction dynamics and the study of radicals [148] and anions [49] from which the first direct measurements of transition states in chemical reactions have been made [57].

Another aspect important for chemistry concerns the dynamics of hydrogen-bonded clusters. Attempts have been made to probe proton transfer in such systems by short pulse lasers in the femto- to picosecond regime [149]. The combination of this pump-probe technique with ZEKE detection should be particularly suitable to project out the proton transfer coordinate from the excited S_1 state into the ion. A particularly interesting development is mass-selected ZEKE detection [150], also termed MATI (mass analysed threshold ionization), which can be made fully equivalent to ZEKE electron detection in terms of resolution and signal strength [151,152] and provides a means to study fragmentation pathways in cation clusters [153]. MATI can also be used to follow the proton transfer channel in hydrogen bonded cluster ions, something which is of particular interest for larger clusters such as hetero trimers [154].

Acknowledgements: Support from the *Deutsche Forschungsgemeinschaft*, the *Bundesministerium für Forschung und Technologie*, the *Fonds der Chemischen Industrie*, the *Commission of the European Union* and the *Engineering and Physical Sciences Research Council* is gratefully acknowledged.

References

1. G. Herzberg, *Electronic Spectra of Polyatomic Molecules*, Van Nostrand, New York (1966).
2. T.A. Koopmans, *Physica*, **1**, 104 (1933).
3. F. I. Vilesov, B. I. Kurbatov and A. N. Terenin, *Sov. Phys. Dokl.* **6**, 490 (1961).
4. D. W. Turner and M. I. Al Joboury, *J. Chem. Phys.* **37**, 3007 (1962).
5. D. W. Turner, C. Baker, A. D. Baker and C. R. Brundle, *Molecular Photoelectron Spectroscopy*, Wiley, London (1970).
6. J. Berkowitz, *Photoabsorption, Photoionization and Photoelectron Spectroscopy*, Academic Press, New York (1979).
7. K. Kimura, S. Katsumata, Y. Achiba, T. Yamazaki and S. Iwata, *Handbook of HeI Photoelectron Spectra of Fundamental Organic Molecules*, Japan Sci. Soc. Press, Tokyo (1981).
8. P. Baltzer, L. Karlsson, M. Lundqvist and B. Wannberg, *Rev. Sci. Instrum.* **64**, 2179 (1993).
9. K. S. Viswanatan, E. Sekreta, E. R. Davidson and J. P. Reilly, *J. Phys. Chem.* **90**, 5078 (1986).
10. D. J. Leahy, K. L. Reid, H. Park and R. Zare, *J. Chem. Phys.* **97**, 4948 (1992).
11. L. Åsbrink, *Chem. Phys. Lett.* **7**, 549 (1970).
12. A. Niehaus and M. W. Ruf, *Chem. Phys. Lett.* **11**, 55 (1971).
13. S. Southworth, C. M. Truesdale, P. H. Kobrin, D. W. Lindle, W. D. Brewer and D. A. Shirley, *J. Chem. Phys.* **76**, 143 (1982).
14. K. Müller-Dethlefs, M. Sander and E. W. Schlag, *Z. Naturforsch.* ***A* 39**, 1089 (1984).
15. K. Müller-Dethlefs, M. Sander, and E. W. Schlag, *Chem. Phys. Lett.* **112**, 291 (1984).
16. K. Müller-Dethlefs and E. W. Schlag, *Annu. Rev. Phys. Chem.* **42**, 109 (1991).
17. F. Merkt and T. P. Softley, *Int. Rev. Phys. Chem.* **12**, 205 (1993).
18. T. G. Wright, G. Reiser and K. Müller-Dethlefs, *Chem. Britain*, p128, Feb. 1994.
19. I. Fischer, R. Lindner and K. Müller-Dethlefs, *J. Chem. Soc. Faraday Trans.* **90**, 2425 (1994).
20. K. Müller-Dethlefs, O. Dopfer and T. G. Wright, *Chem. Rev.* **94**, 1845 (1994).
21. K. Müller-Dethlefs, *High Resolution Spectroscopy with Photoelectrons*, in: *High Resolution Laser Photoionization and Photoelectron Studies*, (I. Powis, T. Baer and C.-K. Ng ed.) Chapter II, p.21, Wiley, Chichester (1995).
22. *High Resolution Laser Photoionization and Photoelectron Studies*, (I. Powis, T. Baer and C.-K. Ng ed.) Wiley, Chichester (1995).
23. K. Müller-Dethlefs, E.W. Schlag, E. Grant, K. Wang and B. V. McKoy, *Adv. Chem. Phys.* **90**, 1 (1995).
24. REMPI (resonance-enhanced multi-photon ionization) spectroscopy accesses an electronically-excited state of a molecule or complex using m photons and then absorption of a further n photons of the same color causes ionization. The process is then known as an $m + n$ REMPI process; if the colours of the photons are different, then the process is known as an $m + n'$ REMPI process.
25. E. R. Grant and M. G. White, *Nature* **354**, 249 (1991).
26. K. Wang and V. McKoy, *Ann. Rev. Phys. Chem.* **46**, 274 (1995).
27. G. Reiser, W. Habenicht, K. Müller-Dethlefs and E. W. Schlag, *Chem. Phys. Lett.* **152**, 119 (1988).
28. R. Lindner, B. Beyl, H. Sekiya and K. Müller-Dethlefs, *Angew. Chem.* **105**, 631 (1993); *Angew. Chem. Int. Ed. Engl.* **32**, 603 (1993).
29. G. Reiser, D. Rieger, T. G. Wright, K. Müller-Dethlefs and E. W. Schlag, *J. Phys. Chem.* **97**, 4335 (1993).
30. I. Fischer, A. Strobel, J. Staeker, G. Niedner-Schatteburg, K. Müller-Dethlefs and V. E. Bondybey, *J. Chem. Phys.* **96**, 7171 (1992).

31. T. Kitsopoulos, C. J. Chick, Y. Zhao and D. M. Neumark, *J. Chem. Phys.* **95**, 1441 (1991).
32. G. Ganteför, D. M. Cox and A. Kaldor, *J. Chem. Phys.* **93**, 8395 (1990).
33. G. Reiser, O. Dopfer, R. Lindner, G. Henri, K. Müller-Dethlefs, E. W. Schlag and S. D. Colson, *Chem. Phys. Lett.* **181**, 1 (1991).
34. E. Sekreta, K.S. Visvanathan and J.P. Reilly, *J. Chem. Phys.* **90**, 5349 (1989).
35. W. B. Peatman, T. B. Borne, and E. W. Schlag, *Chem. Phys. Lett.* **3**, 492 (1969).
36. T. G. Wright, E. Cordes, O. Dopfer and K. Müller-Dethlefs, *J. Chem. Soc. Faraday Trans.* **89**, 1609 (1993).
37. H. Sekiya, R. Lindner and K. Müller-Dethlefs, *Chem. Lett.* p.485 (1993).
38. R. Lindner, H. Sekiya and K. Müller-Dethlefs, *Angew. Chem.* **105**, 1384 (1993); *Angew. Chem. Int. Ed. Engl.* **32**, 1364 (1993).
39. R. Lindner, H.-J. Dietrich and K. Müller-Dethlefs, *Chem. Phys. Lett.* **228**, 417 (1994).
40. R. Lindner, H. Sekiya and K. Müller-Dethlefs, *J. Chem. Phys.* to be submitted.
41. W. A. Chupka, *J. Chem. Phys.* **98**, 4520 (1993); ibid. **99**, 5800 (1993).
42. M. Bixon and J. Jortner, *Mol. Phys.* **89**, 373 (1996); *J. Chem. Phys.* **105**, 1363 (1996); *J. Phys. Chem.* **100**, 12866 (1996).
43. D. Bahatt, U. Even and R. D. Levine, *J. Chem. Phys.* **98**, 1744 (1993).
44. U. Even, M. Ben-Nun and R. D. Levine, *Chem. Phys. Lett.* **210**, 416 (1993).
45. F. Merkt and R. N. Zare, *J. Chem. Phys.* **101** 3495 (1994).
46. F. Merkt, H. Xu and R. N. Zare, *J. Chem. Phys.* **104**, 950 (1996).
47. T. N. Kitsopoulos, I. M. Waller, J. G. Loeser and D. M. Neumark, *Chem. Phys. Lett.* **159**, 300 (1989).
48. I. M. Waller, T. N. Kitsopoulos and D. M. Neumark, *J. Phys. Chem.* **94**, 2240 (1990).
49. D. M. Neumark, *Annu. Rev. Phys. Chem.* **43**, 153 (1992); R. B. Metz, S. E. Bradforth, D. M. Neumark, *Adv. Chem. Phys.* **81**, 1 (1992).
50. C. E. H. Dessent, C. G. Bailey and M. A. Johnson, *J. Chem. Phys.* **102**, 6335 (1995).
51. M. C. R. Cockett, J. G. Goode, K. P. Lawley and R. J. Donovan, *J. Chem. Phys.* **102,** 5226 (1995).
52. M. C. R. Cockett, *J. Phys. Chem.* **99**, 16228 (1995).
53. M.C.R. Cockett, R. J. Donovan and K. P. Lawley, *J.Chem. Phys.* **105**, 3347 (1996).
54. S. Fredin, D. Gauyacq, M. Horani, Ch. Jungen, G. Lefevre and F. Masnou-Seeuws, *Mol. Phys.*, **60**, 825 (1987).
55. H. Rudolph, V. McKoy and S.N. Dixit, *J. Chem. Phys.* **90**, 2570, (1989).
56. G. Reiser and K. Müller-Dethlefs, *J. Phys. Chem.* **96**, 9 (1992).
57. M. Sander, L.A. Chewter, K. Müller-Dethlefs and, E.W. Schlag, *Phys. Rev.* **A 36**, 4543 (1987).
58. A. Strobel, I. Fischer, J. Staecker, G. Niedner-Schatteburg, K. Müller-Dethlefs and V. E. Bondybey, *J. Chem. Phys.* **97**, 2332 (1992).
59. R. T. Wiedmann, M. G. White, K. Wang and V. McKoy, *J. Chem. Phys.* **98**, 7672 (1993).
60. K. Müller-Dethlefs, *J. Chem. Phys.* **96**, 4821 (1991).
61. G. Herzberg and E. Teller, *Z. Phys. Chem.* **21**, 410 (1933).
62. G. Orlandi and W. Siebrand, *J. Chem. Phys.* **58**, 4513 (1973).
63. M. C. R. Cockett, H. Ozeki, K. Okuyama and K. Kimura, J. Chem. Phys. **98**, 7763 (1993).
64. L. Åsbrink, C. Fridh and E. Lindholm, *Z. Naturforsch.* ***A*** **33,** 172 (1978).
65. W. von Niessen, G. H. F. Diercksen and L. S. Cederbaum, *Chem. Phys. Lett.* **45**, 295 (1977).
66. M. H. Palmer, W. Moyes, M. Spiers and J. N. A. Ridyard, *J. Mol. Struct.* **49**, 105, (1978).
67. G. Bieri, L. Åsbrink and W. von Niessen, *J. Elec. Spectrosc. Rel. Phen.* **23**, 281 (1981).
68. G. Reiser, D. Rieger, T. G. Wright, K. Müller-Dethlefs and E. W. Schlag, *J. Phys. Chem.* **97**, 4335 (1993).
69. (a) F. Duschinsky, *Acta Physiochem. U.R.S.S.* **1**, 551 (1937); (b) see also: G. J. Small, *J. Chem. Phys.* **54**, 3300 (1971).
70. H. A. Jahn and E. Teller, *Proc. Roy. Soc. London* **A 161**, 220 (1937).
71. U. Öpik and M. H. L. Pryce, *Proc. Roy. Soc. London* ***A*** **238**, 425 (1956).
72. I. B. Bersuker, *The Jahn-Teller Effect and Vibronic Interactions in Modern Chemistry*, Plenum Press, New York (1984).
73. K. Raghavachari, R. C Haddon, T. A. Miller and V. E. Bondybey, *J. Chem. Phys.* **79**, 1387 (1983).
74. T. A. Miller and V. E. Bondybey, *Molecular Ions: Spectroscopy, Structure and Chemistry*, North-Holland, Amsterdam (1983).
75. S. R. Long, J. T. Meek and J. P. Reilly, *J. Chem. Phys.***79**, 3206 (1983).
76. K. Wang and B. V. McKoy, *Ann. Rev. Phys. Chem.* **46**, 275 (1995).
77. E. B. Wilson Jr., *Phys. Rev.* **45,** 706 (1934).
78. H. C. Longuet-Higgins, U. Öpik, M. H. L. Pryce, F. R. S. Sack and R. A. Sack, *Proc. Roy. Soc. London* ***A*** **244**, 1 (1958).
79. M. Iwasaki, K. Toriyama and K. Nunome, *J. Chem. Soc. Chem. Comm.*320 (1983).
80. P.R. Bunker, *Molecular Symmetry and Spectroscopy*, Academic Press, New York (1979).
81. R. Lindner, K. Müller-Dethlefs, C. Wedum, K. Haber and E. R. Grant, *Science* **271**, 1698 (1996).
82. A. Mank, T. Nyugen, J. D. D. Martin and J. W. Hepburn, *Phys. Rev.* ***A*** **51,** R1 (1995).
83. W. Kong, D. Rogers and J. W. Hepburn, *Chem. Phys. Lett.* **221,** 301 (1994).
84. W. Kong, D. Rogers and J. W. Hepburn, *J. Chem. Phys.* **99,** 8571 (1993).
85. J. B. Milan, W. J. Buma and C. A. De Lange, *J. Chem. Phys.* **104**, 521 (1996).
86. A. B. Cornford, D. C. Frost, C. A. McDowell, J. L. Ragle and I. A. Stenhouse, *J. Chem. Phys.* **54,** 2651 (1971); A. W. Potts and W. C. Price, *Trans. Faraday Soc.* **67,** 1242 (1971).
87. B. R. Higginson, D. R. Lloyd and P. J. Roberts, *Chem. Phys. Lett.* **19,** 480 (1973); H. Van Lonkhuyzen and C. A. Lange,

Chem. Phys. **89,** 313 (1984).
88. M. Jungen, *Theoret. Chim. Acta* **27,** 33 (1972).
89. R. P. Tuckett and S. D. Peyerimhoff, *Chem. Phys.* **83,** 203 (1984).
90. P. M. Boerrigter, M. A. Buijse and J. G. Snijders, *Chem. Phys.* **83,** 203 (1987).
91. T. Harris, J. H. D. Eland and R. P. Tuckett, *J. Mol. Spectrosc.* **98,** 269 (1983).
92. A. J. Yencha, M. C. R. Cockett, J. G. Goode, R. J. Donovan, A. Hopkirk and G. C. King, *Chem. Phys. Lett.* **229,** 347 (1994).
93. W.A. de Jong, L. Visscher and W.C. Nieuwpoort, submitted to *J. Chem. Phys.* (1997).
94. R. E. Smalley, D. H. Levy and L. Wharton, *J. Chem. Phys.* **64,** 3266 (1976); *ibid.* **67,** 671 (1978); G. Kubiak, P. S. H. Fitch, L. Wharton and D. H. Levy, *J. Chem. Phys.* **68,** 4447 (1978).
95. L. A. Chewter, K. Müller-Dethlefs and E. W. Schlag, *Chem. Phys. Lett.* **135,** 219 (1987).
96. X. Chang, J. Smith and J. L. Knee, *J. Chem. Phys.* **97,** 2843 (1992).
97. J. Smith, X. Chang and J. L. Knee, *J, Chem. Phys.* **99,** 2550 (1993).
98. M. C. R. Cockett and K. Kimura, *J. Chem. Phys.* **100,** 3429 (1994).
99. M. C. R. Cockett, K. Okuyama and K. Kimura, *J. Chem. Phys.* **97,** 4679 (1992).
100. S. J. Harris, S. E. Novick, W. Klemperer and W. E. Falconer, *J. Chem. Phys.* **61,** 193 (1974).
101. R. E. Smalley, L. Wharton and D. H. Levy, *J. Chem. Phys.* **68,** 671 (1976).
102. F. Thommen, D. D. Evard and K. C. Janda, *J. Chem. Phys.* **82,** 5259 (1985).
103. D. D. Evard, F. Thommen, K. J. Janda, *J. Chem. Phys.* **84,** 3630 (1986).
104. D. D. Evard, J. I. Cline and K. C. Janda, *J. Chem. Phys.* **88,** 5433 (1988).
105. M. L. Burke and W. Klemperer, *J. Chem. Phys.* **98,** 6642 (1993).
106. M. L. Burke and W. Klemperer, *J. Chem. Phys.* **98,** 1797 (1993).
107. M. C. R. Cockett, J.G. Goode, K. P. Lawley and R. J. Donovan, *Chem. Phys. Lett.* **214,** 27 (1993); M. C. R. Cockett, J. G. Goode, R. R. J. Maier, K. P. Lawley and R. J. Donovan, *Chem. Phys. Lett.* **101,** 126 (1994).
108. M. C. R. Cockett, D. A. Beattie, R. J. Donovan and K. P. Lawley, *Chem. Phys. Lett.* **259**, 554 (1996).
109. P.D.A. Mills, C.M. Western and B.J. Howard, *J. Phys. Chem.* **90**, 4961 (1986); *ibid.* **90**, 3331 (1986).
110. M. Takahashi, *J. Chem. Phys.* **96**, 2594 (1992).
111. K. Kimura and M. Takahashi, Optical Methods for Time- and State-Resolved Chemistry, *SPIE Proceedings Series*, **1638**, 216 (1992). *NOTE*: Fig. 10 of that work was attributed to the ZEKE spectrum obtained when exciting through the vibrationless $\tilde{C}$ state - however, reference [110] indicates that this was in fact the $\tilde{C}$ state, but with one quantum of the intermolecular stretch excited.
112. The notation of the term symbol for the NO moiety, with a tilde over it, for the molecular complex, follows the nomenclature used in laser-induced fluorescence - see the two recent reviews by Heaven: M.C. Heaven, *Annu. Rev. Phys. Chem.* **43**, 283 (1992); M.C. Heaven, *J. Phys. Chem.* **97**, 8567 (1993).
113. K. Sato, Y. Achiba and K. Kimura, *J. Chem. Phys.* **81**, 57 (1984).
114. J.C. Miller and W.-C. Cheng, *J. Phys. Chem.* **89**, 1647 (1985).
115. Note that in reference [110], an error has occurred and the anharmonicities reported are actually twice the value they should be.
116. T.G. Wright, V. Špirko and P. Hobza, *J. Chem. Phys.* **100**,5403 (1994).
117. J.-M. Robbe, M. Bencheikh and J.-P. Flament, *Chem. Phys. Lett.* **210**, 170 (1993).
118. Size-consistency in an *ab initio* calculation implies that the energy of the complex, when there is infinite separation between the components, is equal to the sum of the energy of the separate components.
119. The abbreviation HF implies the use of Hartree-Fock theory with the basis set indicated: D.R. Hartree, *Proc. Camb. Phil. Soc* R. **24**, 89 (1928). V. Fock, *Z. Physik* **61**, 126 (1930). C.C.J. Roothaan, *Rev. Mod. Phys.* **23**, 161 (1951). The abbreviation MP2 implies the use of Møller-Plesset perturbation theory to second order: C. Møller and M.S. Plesset, *Phys. Rev.* **40**, 618 (1934). The abbreviation QCISD(T) implies the use of the quadratic configuration method, including single and double excitations with a perturbative correction for triple excitations: J.A. Pople, M. Head-Gordon and K. Raghavachari, *J. Chem. Phys.* **87**, 5968 (1987).
120. E. Miescher, *Can. J. Phys.* **54**, 2074 (1976).
121. P. Hobza and R. Zahradník, *Intermolecular Complexes: The Role of van der Waals Systems in Physical Chemistry and in the Biodisciplines*, Elsevier, Amsterdam (1988).
122. C.Y. Ng, *Adv. Chem. Phys.* **52**, 263 (1983).
123. N.L. Ma, W.-L. Li and C.Y. Ng, *J. Chem. Phys.* **99**, 3617 (1993).
124. L.A. Curtiss, K. Raghavachari, G.W. Truck and J.A. Pople, *J. Chem. Phys.* **94**, 2221 (1991).
125. R.G. Tonkyn and M.G. White, *J. Chem. Phys.* **95**, 5582 (1991).
126. Descriptions of the experimental procedure to produce this radiation for use in ZEKE experiments are given in: R.G. Tonkyn and M.G. White, *Rev. Sci. Inst.* **60**, 1245 (1989); H.H. Fielding, T.P. Softley and F. Merkt, *Chem. Phys.* **155**, 257 (1991).
127. M. Ito, *J. Molec. Struct.* **177**, 173 (1988).
128. See, for example, and references therein: M. Ito, In *Vibrational Spectra and Structure*, (J. R. Durig ed.) Vol. 15, Elsevier, Amsterdam (1986). M. Ito, T. Ebata and N. Mikami, *Annu. Rev. Phys. Chem.* **39**, 123 (1988). M. Ito, S. Yamamoto, T. Aoto and T. Ebata, *J. Molec. Struct.* **237**, 105 (1990). M. Gerhards, B. Kimpfel, M. Pohl, M. Schmitt and K. Kleinermanns, *J. Molec. Struct.* **270**, 301 (1992).
129. T. Oka, *Rev. Mod. Phys.* **64**, 1141 (1992).

130. Hirota, E. *Chem. Rev.* **92**, 141 (1992).
131. O. Dopfer, G. Reiser, R. Lindner and K. Müller-Dethlefs, *Ber. Bunsen-Ges. Phys. Chem.* **96**, 1259 (1992).
132. O. Dopfer, G. Reiser, K. Müller-Dethlefs, E. W. Schlag and S. D. Colson, *J. Chem. Phys.* **101**, 974 (1994).
133. T. G. Wright, E. Cordes, O. Dopfer and K. Müller-Dethlefs, *J. Chem. Soc. Faraday Trans.* **89**, 1609 (1993).
134. E. Cordes, O. Dopfer, T. G. Wright and K. Müller-Dethlefs, *J. Phys. Chem.* **97**, 7471 (1993).
135. O. Dopfer, G. Lembach, T. G. Wright and K. Müller-Dethlefs, *J. Chem. Phys.* **98**, 1933 (1993).
136. T. G. Wright, O. Dopfer, E. Cordes and K. Müller-Dethlefs, *J. Am. Chem. Soc.* **116**, 5880 (1994).
137. M. Schütz, T. Bürgi, Leutwyler, S. and T. Fischer, *J. Chem. Phys.* **98**, 3763 (1993).
138. M. Schütz, T. Bürgi and S. Leutwyler, *J. Mol. Struct. (THEOCHEM)* **276**, 117 (1992).
139. D. Feller and M. W. Feyereisen, *J. Comp. Chem.* **14**, 1027 (1993).
140. P. Hobza, R. Burcl, V. Spirko, O. Dopfer, K. Müller-Dethlefs and E. W. Schlag, *J. Chem. Phys.* **101**, 990 (1994).
141. K. Fuke, H. Yoshiuchi, K. Kaya, Y. Achiba, K. Sato and K. Kimura, *Chem. Phys. Lett.* **108**, 179 (1984).
142. R. J. Lipert and S. D. Colson, *J. Chem. Phys.* **89**, 4579 (1988).
143. G. Lembach, *Diplomarbeit* Technische Universität, München (1992); G. Lembach, *et al.* to be published.
144. For the hydrogen-bonded complexes Ph-MeOH, Ph-EtOH, Ph-DME and Ph_2 following nomenclature has been used for the six intermolecular vibrations: ξ_i for bends and the torsion (i=1-5) and σ for the stretch in the S_1 state; η_i for bends and the torsion (i=1-5) and σ_+ for the stretch in the ionic ground state.
145. H. Abe, N. Mikami, M. Ito and Y. Udagawa, *J. Phys. Chem.* **86**, 2567 (1982).
146. T. Bäumert, R. Thalweiser and G. Gerber, *Chem. Phys. Lett.* **209**, 29 (1993).
147. I. Fischer, M. J. J. Vrakking, D. M. Villeneuve and A. Stolow, *Chem. Phys.* **207**, 331 (1996).
148. R. T. Wiedmann, M. G. White, K. Wang and V. McKoy, *J. Chem Phys.***100**, 4738 (1994)
149. J. A. Syage and J. Steadman, *J. Chem. Phys.* **95**, 2497 (1991); J. A. Syage, *J. Phys. Chem.* **97**, 12523 (1993).
150. L. Zhu and P. Johnson, *J. Chem. Phys.* **94**, 5769 (1991).
151. H. J. Dietrich, R. Lindner and K. Müller-Dethlefs, *J. Chem. Phys.* **101**, 3399 (1994).
152. H. J. Dietrich, K. Müller-Dethlefs and L. Ya Baranov, *Phys. Rev. Lett.* **76**, 3531 (1996).
153. H. J. Neusser and H. Krause, *Int J. Mass Spectrom. Ion Proc.* **131**, 193 (1994).
154. O. Dopfer, M. Melf and K. Müller-Dethlefs, *Chem. Phys.* **207**, 437 (1996).

8 Nonlinear Raman Spectroscopy

Hiro-o Hamaguchi
Department of Chemistry, School of Science,
The University of Tokyo, 7-3-1 Hongo, Tokyo 113, Japan

8.1 Spontaneous and Nonlinear Raman Scattering

Raman scattering in a narrow sense means spontaneous Raman scattering which was discovered by C. V. Raman in 1928 and which bears his name. Raman scattering in a broader sense refers to a class of one-step multi-photon processes in which a Raman resonance condition is satisfied; of the n angular frequencies of the n photons involved, two satisfy the relation $\omega_i - \omega_j = \Omega$, where ω_i and ω_j are the angular frequencies of the ith and jth photons, and Ω (>0) is the angular frequency of a Raman active excitation of a molecule. In **spontaneous Raman scattering**, two photons with angular frequencies ω_1 and ω_2 are involved. While the molecule undergoes a transition from the initial state i to the final state f via the virtual state *v*, one ω_1 photon is absorbed and one ω_2 photon is emitted by spontaneous emission (Fig. 8.1). Having a clear phase relation between states i and f, Raman scattering is a one-step two-photon process and not a combination of the two consecutive one-photon processes. Two Raman resonance conditions are possible; the **Stokes** Raman resonance $\omega_1 - \omega_2 = \Omega$ and the **anti-Stokes** Raman resonance $\omega_2 - \omega_1 = \Omega$. The same ω_1, ω_2 pair of photons in a Raman resonance are involved in **Stimulated Raman Scattering (SRS),** in which the ω_2 light field is so intense that stimulated emission dominates spontaneous emission (Fig. 8.1-b). In SRS the ω_2 photons are emitted coherently in the form of a beam, in contrast to spontaneous Raman scattering in which the ω_2 photons are emitted in all directions. **Coherent anti-Stokes Raman scattering (CARS)** and **Coherent Stokes Raman scattering (CSRS)** involve a third photon ω_3 in addition to the two Raman resonant photons, ω_1 and ω_2 ($\omega_1 - \omega_2 = \Omega$). In CARS, two ω_1 photons are converted to one ω_2 and one ω_3 photon; $\omega_3 = 2\omega_1 - \omega_2$ (Fig. 8.1-c). In CSRS, two ω_2 photons are converted to one ω_1 and one ω_3 photon; $\omega_3 = 2\omega_2 - \omega_1$. Both CARS and CSRS are parametric processes in which no real transition of the molecule is involved. The ω_3 photons are coherently emitted in the form of a beam. **Raman resonant four-wave mixing** is a more general case of CARS and CSRS in which four photons, ω_1, ω_2, ω_3, ω_4, are involved and in which one ω_1 and one ω_3 photon are converted to one ω_2 and one ω_4 photon with no real transition of the molecule. Still higher-order Raman processes are conceivable based on the same principle, but they are yet to be experimentally established.

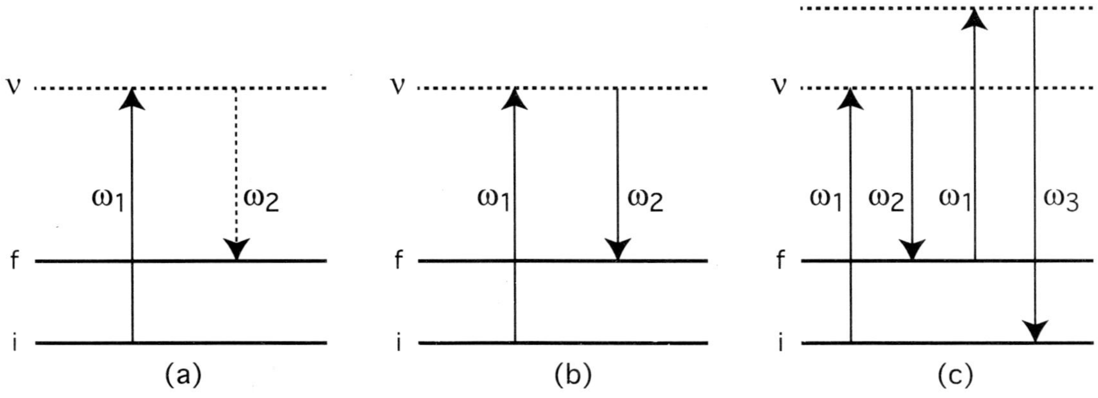

Figure 8.1 Schematic diagram for spontaneous Raman scattering (a), SRS (b), and CARS (c). Arrows directed upward represent absorption and those downward emission (full line stimulated and dotted line spontaneous emission).

There is not much difference between spontaneous Raman scattering and stimulated Raman scattering except that the black-body radiation field has to be considered for the former process [1]. However, the convention is to separate spontaneous Raman scattering from other Raman scattering processes which are categorized as **nonlinear Raman scattering**. This is because the signal intensity of spontaneous Raman scattering is proportional to the intensity of the incident light, while those of the other Raman scattering processes are not linearly dependent on the incident light intensities. Hyper Raman scattering is often categorized as a nonlinear Raman scattering process. However, it is not a Raman process according to the definition given above, because no Raman resonance is involved. Note that, in hyper Raman scattering, the angular frequency ω_2 of the scattered photon satisfies the relation $2\omega_1-\omega_2=\Omega$ or $\omega_2-2\omega_1=\Omega$.

In this Chapter, we focus on SRS and CARS, and spectroscopies based on them. The readers are recommended to refer to existing textbooks and reviews [2-4] for more general treatment of nonlinear Raman processes including hyper Raman scattering.

8.2 Principle of Nonlinear Raman Spectroscopy

8.2.1 Coherent Excitation of Molecules and Nonlinear Raman Scattering

If two beams of monochromatic radiation with angular frequencies ω_1 and ω_2 ($\omega_1>\omega_2$) are incident on a material, a beat electric field having the difference angular frequency $\omega_1-\omega_2$ is generated. If the two angular frequencies satisfy a Raman resonance condition, $\omega_1-\omega_2=\Omega$, this beat electric field resonantly excites a material excitation which is a coherent collective excitation of molecules. In many instances, a molecular vibration (with angular frequency Ω) is coherently excited for a number of molecules within a macroscopic region of the material that is large compared with the wavelengths of the incident radiations. This coherent excitation of molecules produces a periodic variation of the dielectric property of the material that oscillates with the angular frequency Ω. If the ω_1 field interacts with this oscillation, the mixing results in two angular frequencies, $\omega_1+\Omega$ and $\omega_1-\Omega$. The angular frequency $\omega_1+\Omega$ corresponds to CARS, accounting for the generation of one photon with angular frequency $\omega_3=\omega_1+\Omega$. The angular frequency $\omega_1-\Omega=\omega_2$ corresponds to **stimulated Raman gain (SRG),** in which stimulated Raman scattering manifests as an increase in the number of ω_2 photons. The mixing of the material oscillation with the ω_2 field produces $\omega_2+\Omega$ and $\omega_2-\Omega$. The angular frequency $\omega_2+\Omega=\omega_1$ represents **stimulated Raman loss (SRL)** in which stimulated Raman scattering manifests as a decrease in the number of the ω_1 photons. The angular frequency $\omega_2-\Omega$ corresponds to CSRS, in which one $\omega_3=\omega_2-\Omega$ photon is generated.

The qualitative description of nonlinear Raman processes given above is more quantitatively developed using the following susceptibility formalism. By the two incident monochromatic radiations ω_1 and ω_2 , a polarization P is induced in the material;

$$\begin{aligned} \boldsymbol{P} &= \boldsymbol{P}^{(1)} + \boldsymbol{P}^{(2)} + \boldsymbol{P}^{(3)} + \ldots.. \\ &= \chi^{(1)} \boldsymbol{E}_i + \chi^{(2)} \boldsymbol{E}_i\boldsymbol{E}_j + \chi^{(3)} \boldsymbol{E}_i\boldsymbol{E}_j\boldsymbol{E}_k + \ldots.. \end{aligned} \tag{8.1}$$

where $\chi^{(1)}$, $\chi^{(2)}$, and $\chi^{(3)}$ are the linear, the second-order, and the third-order susceptibilities which are tensors of rank two, three, and four, respectively. $\boldsymbol{E}_i$ is the electric field of the ith incident radiation. Note that the second line of Eq. (8.1) is only conceptual and not mathematically exact; polarizations $\boldsymbol{P}^{(i)}$ contain different Fourier components having different angular frequencies and the susceptibility is defined for each of these Fourier components. The linear polarization $\boldsymbol{P}^{(1)}$ is responsible for the propagation of the incident radiations. The real part of $\chi^{(1)}$ corresponds to the refractive index and the imaginary part to the absorption coefficient. The second-order polarization $\boldsymbol{P}^{(2)}$ accounts for the frequency mixing processes including the sum and the difference frequency generation, the second harmonic generation, and the optical rectification. $\boldsymbol{P}^{(2)}$ vanishes for a system with a center of symmetry. Nonlinear Raman processes are associated with the third-order polarization $\boldsymbol{P}^{(3)}$. In order for $\boldsymbol{P}^{(3)}$ to be appreciable, intense optical electric fields have to be applied. A rough estimate of the ratio $\chi^{(n+2)}/\chi^{(n)}$ is 10^{-22} V^2 m^{-2} [5]. By focusing a picosecond laser pulse with 10 μJ energy into a 0.1 mm^2 spot, we obtain a laser field of ~10^9 V m^{-1}. With this electric field, $\boldsymbol{P}^{(3)}/\boldsymbol{P}^{(1)} \sim 10^{-4}$, which means that the amplitude of the nonlinear Raman signal is 10^{-4} of the amplitude of the propagating incident laser fields. It is possible to generate a still more intense field by using a higher laser power. However, with such an extremely intense field, other undesired nonlinear processes like multiphoton ionization are induced concomitantly and mask the nonlinear Raman signal. Nonlinear Raman processes became experimentally accessible only after the development of reliable high power laser sources.

In practical experiments with lasers, the incident radiation fields can be considered, to a good approximation, as monochromatic plane waves whose electric fields are given as complex forms,

$$\boldsymbol{E}_i = 1/2\{\underline{\boldsymbol{E}}_i\exp[i(\boldsymbol{k}_i\boldsymbol{r}-\omega_i t)]+\underline{\boldsymbol{E}}_i{}^*\exp[-i(\boldsymbol{k}_i\boldsymbol{r}-\omega_i t)]\}, \tag{8.2}$$

where $\underline{\boldsymbol{E}}_i$ is the complex amplitude and $\boldsymbol{k}_i$ the wave vector of the ith incident radiation, $\boldsymbol{r}$ is the spatial coordinate, and t is time. The third-order polarization $\boldsymbol{P}^{(3)}$ is expressed as a sum of Fourier components $\boldsymbol{P}^{(3)}(\omega_n)$ having angular frequency ω_n,

$$\begin{aligned}\boldsymbol{P}^{(3)} &= \Sigma\, \boldsymbol{P}^{(3)}(\omega_n)\\ &= \Sigma\, 1/2\{\underline{\boldsymbol{P}}^{(3)}(\omega_n)\exp[i(\boldsymbol{k}_n\boldsymbol{r}-\omega_n t)] + \underline{\boldsymbol{P}}^{(3)*}(\omega_n)\exp[-i(\boldsymbol{k}_n\boldsymbol{r}-\omega_n t)]\},\end{aligned} \tag{8.3}$$

where the wave vector $\boldsymbol{k}_n$ is determined by the phase matching condition that is discussed in the following section of CARS spectroscopy (8.4.1). The complex amplitude $\underline{\boldsymbol{P}}^{(3)}(\omega_n)$ of the nth component of $\boldsymbol{P}^{(3)}$ is related to the product of three field amplitudes using the third-order susceptibility $\chi^{(3)}(\omega_n=\omega_i+\omega_j+\omega_k)$ as,

$$\underline{\boldsymbol{P}}^{(3)}(\omega_n) = D\chi^{(3)}(\omega_n=\omega_i+\omega_j+\omega_k)\underline{\boldsymbol{E}}_i\underline{\boldsymbol{E}}_j\underline{\boldsymbol{E}}_k\,. \tag{8.4}$$

The degeneracy factor D represents the D equivalent terms that are obtained by the permutation of the three fields. D=6 if the three incident radiations have three different angular frequencies, D=3 if any two of the three are the same, and D=1 if the three angular frequencies are identical. Angular frequencies with minus signs can appear in (8.4) with the complex conjugates of the field amplitudes $\underline{\boldsymbol{E}}^*$.

For CARS, we put $\omega_i=\omega_j=\omega_1$ and $\omega_k=-\omega_2$ with D=3. Then the amplitude of the third-order CARS polarization $\underline{\boldsymbol{P}}^{(3)}{}_{\mathrm{CARS}}$ is given as,

$$\underline{\boldsymbol{P}}^{(3)}{}_{\mathrm{CARS}} = \chi^{(3)}{}_{\mathrm{CARS}}\underline{\boldsymbol{E}}_1\underline{\boldsymbol{E}}_1\underline{\boldsymbol{E}}_2{}^*, \tag{8.5}$$

where $\chi^{(3)}{}_{\mathrm{CARS}} = 3\chi^{(3)}(2\omega_1-\omega_2=\omega_1+\omega_1-\omega_2)$. For SRG, $\omega_i=\omega_1$, $\omega_j=-\omega_1$, $\omega_k=\omega_2$ with D=6. The amplitude of the SRG polarization $\boldsymbol{P}^{(3)}{}_{\mathrm{SRG}}$ is expressed using the SRG susceptibility $\chi^{(3)}{}_{\mathrm{SRG}}=6\chi^{(3)}(\omega_2=\omega_1-\omega_1+\omega_2)$,

$$\underline{\boldsymbol{P}}^{(3)}{}_{\mathrm{SRG}} = \chi^{(3)}{}_{\mathrm{SRG}}\underline{\boldsymbol{E}}_1\underline{\boldsymbol{E}}_1{}^*\underline{\boldsymbol{E}}_2\,. \tag{8.6}$$

Equivalent expressions are obtained for SRL and CSRS as,

$$\underline{\boldsymbol{P}}^{(3)}{}_{\mathrm{SRL}} = \chi^{(3)}{}_{\mathrm{SRL}}\underline{\boldsymbol{E}}_1\underline{\boldsymbol{E}}_2\underline{\boldsymbol{E}}_2{}^*, \tag{8.7}$$

$$\underline{\boldsymbol{P}}^{(3)}{}_{\mathrm{CSRS}} = \chi^{(3)}{}_{\mathrm{CSRS}}\underline{\boldsymbol{E}}_1{}^*\underline{\boldsymbol{E}}_2\underline{\boldsymbol{E}}_2, \tag{8.8}$$

where $\chi^{(3)}_{SRL}=6\chi^{(3)}(\omega_1=\omega_1+\omega_2-\omega_2)$ and $\chi^{(3)}_{CSRS}=3\chi^{(3)}(2\omega_2-\omega_1=-\omega_1+\omega_2+\omega_2)$, respectively. Note that a field product having the angular frequency $\omega_1-\omega_2$ ($\underline{\boldsymbol{E}}_1\underline{\boldsymbol{E}}_2{}^*$) or $\omega_2-\omega_1$ ($\underline{\boldsymbol{E}}_1{}^*\underline{\boldsymbol{E}}_2$) is involved in (8.5)-(8.8), which allows a Raman resonance condition to be satisfied. The analytical expressions for $\chi^{(3)}$ including the four nonlinear Raman susceptibilities, $\chi^{(3)}_{CARS}$, $\chi^{(3)}_{SRG}$, $\chi^{(3)}_{SRL}$, and $\chi^{(3)}_{CSRS}$, are given in Ref. [4].

The field amplitude $\underline{\boldsymbol{E}}_R(\omega_R)$ of the nonlinear Raman signal at angular frequency ω_R is obtained by introducing the amplitude of the third-order nonlinear Raman polarization $\underline{\boldsymbol{P}}^{(3)}{}_R(\omega_R)$ into the wave equation,

$$\nabla\times\nabla\times\underline{\boldsymbol{E}}_R(\omega_R)\exp[i\boldsymbol{k}_R\boldsymbol{r}] - (n\omega_R/c)^2\underline{\boldsymbol{E}}_R(\omega_R)\exp[i\boldsymbol{k}_R\boldsymbol{r}] = 4\pi(\omega_R/c)^2\underline{\boldsymbol{P}}^{(3)}{}_R(\omega_R)\exp[i\boldsymbol{k}_R\boldsymbol{r}], \tag{8.9}$$

where $\boldsymbol{k}_R$ is the wave vector of the signal radiation, n the refractive index at the angular frequency ω_R, and c the speed of light in vacuum. Note that ω_R is equal to $2\omega_1-\omega_2$ for CARS, ω_1 for SRL, ω_2 for SRG, and $2\omega_2-\omega_1$ for CSRS. The expressions for the signal intensities are different among the four nonlinear Raman processes and are discussed separately in the following sections.

8.2.2 General properties of nonlinear Raman susceptibilities

Being a tensor of rank four, a nonlinear Raman susceptibility χ^R has 81 elements some of which may be related to each other by symmetry requirements. Here, χ^R represents one of $\chi^{(3)}_{CARS}$, $\chi^{(3)}_{SRG}$, $\chi^{(3)}_{SRL}$, and $\chi^{(3)}_{CSRS}$. The number of independent elements becomes smaller for a medium possessing a higher symmetry. The independent elements of χ^R for 32 crystal groups are tabulated in Ref. [2]. For isotropic media like gases, liquids, and solutions, only three elements are independent; χ^R_{1111}, χ^R_{1122}, χ^R_{1212}, and χ^R_{1221} with $\chi^R_{1111}=\chi^R_{1122}+\chi^R_{1212}+\chi^R_{1221}$, where 1,2 stand for x,y,z.

In the vicinity of a Raman resonance, the nonlinear Raman susceptibility has the general form,

$$\chi^R = \chi^{NR} + \Sigma\, \chi^Q/(\Omega_Q-\omega_1+\omega_2+i\Gamma_Q), \tag{8.10}$$

where χ^{NR} is the non-resonant term whose amplitude is not dependent on $\omega_1-\omega_2$, and χ^Q is the amplitude of the Raman term resonant with the Qth excitation of the molecule. In most cases, Q stands for a Raman active vibration of the molecule having the angular frequency Ω_Q. The damping constant Γ_Q corresponds to the homogeneous width of the Qth Raman transition.

If both of the incident angular frequencies ω_1 and ω_2 are much smaller than those of any of the electronic transitions of the molecule (off-resonance condition), χ^{NR} and χ^Q are both real and their tensor elements satisfy the following relations,

$$\chi^{NR}_{1122} = \chi^{NR}_{1212} = \chi^{NR}_{1221} = 1/3\chi^{NR}_{1111}\ , \tag{8.11}$$

$$\chi^Q_{1212} = \chi^Q_{1221} = \rho_Q\chi^Q_{1111} \text{ and } \chi^Q_{1122} = (1-2\rho_Q)\chi^Q_{1111}, \tag{8.12}$$

where ρ_Q is the depolarization ratio of the Qth Raman transition. Equation (8.10) is most commonly used in the simulation analysis of nonlinear Raman band shapes, which often show complicated behavior due to interferences between the resonant and non-resonant terms and also interferences among the resonant terms.

If, on the other hand, either ω_1 or ω_2 is in resonance with a one-photon allowed electronic transition of the molecule (one-photon resonance), the virtual state becomes a real intermediate state and the relaxation process in this internediate state must be taken into account. As a result, χ^Q becomes complex and Eq. (8.11) is no longer valid. Furthermore, if $2\omega_1$ is resonant with a two-photon allowed electronic transition of the molecule (two photon resonance), χ^{NR} becomes complex and Eq. (8.11) becomes invalid. Though these electronic resonances make the observed nonlinear Raman spectra much more difficult to analyze, they provide extra information on the electronic properties of the molecule. For example, the two-photon resonance condition can be used to study the two-photon absorption spectrum of a molecule for which the detection of two-photon absorption is otherwise difficult [6].

8.2.3 Nonlinear Raman spectroscopies

Spectroscopy utilizing a nonlinear Raman process is called nonlinear Raman spectroscopy. Here we discuss SRG, SRL, CARS, and CSRS spectroscopies. These are all performed with two incident laser beams having angular frequencies ω_1 and ω_2 ($\omega_1 > \omega_2$). The ω_1 laser beam is often called the anti-Stokes beam and the ω_2 beam the Stokes beam. SRG/SRL spectroscoppy detects the gain/loss in the Stokes/anti-Stokes laser intensity induced by a Raman resonance (Fig. 8.2). SRL spectroscopy is sometimes called **inverse Raman spectroscopy** (IRS). SRG spectroscopy and SRL spectroscopy are looking at the same physical phenomenon (stimulated Raman scattering) from two opposite sides. SRG and SRL spectroscopies are therefore referred to as SRS spectroscopy. CARS/CSRS spectroscopy detects a coherent signal generated at the anti-Stokes/Stokes side of the ω_1/ω_2 laser (Fig. 8.2). CARS and CSRS are analogous but physically distinct processes.

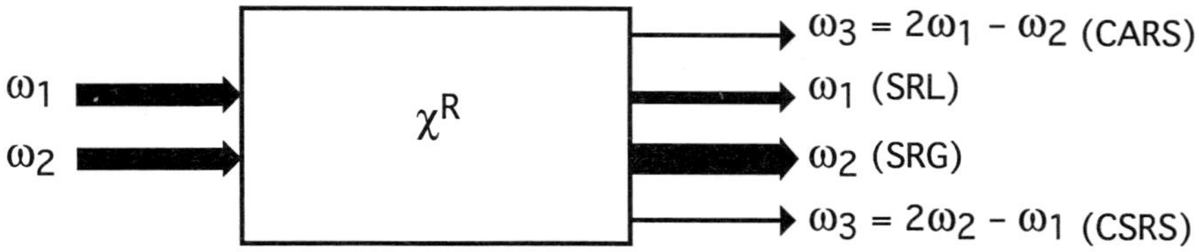

Figure 8.2 Schematic diagram showing the CSRS, SRG, SRL, and CARS processes.

Two different schemes are used for detecting a Raman resonance. The **scanning scheme** uses one monochromatic laser beam with a fixed angular frequency and one monochromatic laser beam with a tunable angular frequency. In scanning CARS spectroscopy, for example, ω_1 is fixed and ω_2 is tunable (Fig. 8.3a). If the angular frequency ω_2 is scanned, a CARS signal with angular frequency $\omega_3 = 2\omega_1 - \omega_2$ is generated when $\omega_1 - \omega_2$ satisfies a Raman resonance condition $\omega_1 - \omega_2 = \Omega$. In the **multiplex scheme**, one monochromatic laser beam with a fixed angular frequency and one broad-band laser beam with continuously distributed angular frequencies are used. In multiplex CARS spectroscopy, ω_1 is fixed and ω_2 is distributed (Fig. 8.3-b). Instead of scanning ω_2, a monochromator or polychromator is used to separate different CARS signals at different angular frequencies ω_3 and they are detected simultaneously by an optical multichannel detector.

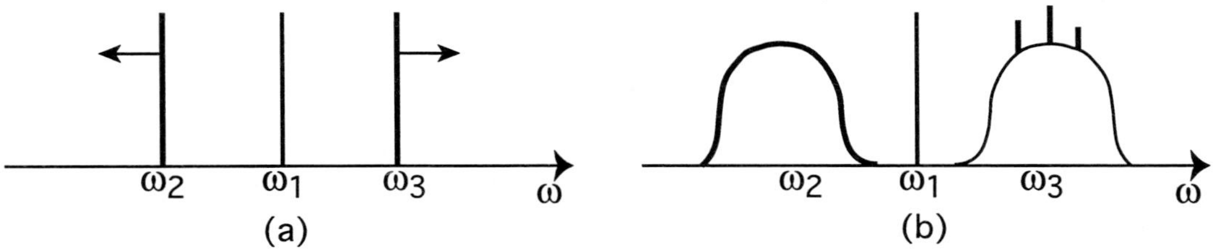

Figure 8.3 Scanning CARS spectroscopy (a) and multiplex CARS spectroscopy (b).

There are a few distinct advantages of nonlinear Raman spectroscopy as compared with ordinary spontaneous Raman spectroscopy. First, frequency resolution of the order of 10 MHz is achievable, which far exceeds the resolution (typically 1 cm^{-1}=30 GHz) of spontaneous Raman spectroscopy. An example of high resolution SRS spectroscopy is given in Section 8.3.2. Note, however, that the high resolution is achieved only with scanning nonlinear Raman spectroscopies. Second, because the nonlinear Raman signal is obtained in the form of a beam, the spatial filtering of the signal from isotropically emitted background radiations is very efficient. This makes nonlinear Raman spectroscopies extremely useful for applications to highly fluorescent molecules or high-temperature radiative materials, for which spontaneous Raman spectroscopy is not applicable. SRS spectroscopy for condensed-phase studies including fluorescence elimination is described in Section 8.3.3. Third, CARS has the extra freedom to control the polarizations of the three laser beams involved. As shown in Section 8.4.3, selective elimination of signals with a particular value of depolarization ratio is possible in polarization sensitive CARS (PSCARS) spectroscopy. Fourth, being a kind of laser spectroscopy, nonlinear Raman spectroscopy is readily applicable to fast and ultrafast time-resolved measurements. Recent applications include femtosecond time-resolved SRG spectroscopy (Section 8.3.4) and picosecond time-frequency two-dimensional multiplex CARS spectroscopy (Section 8.4.4). Finally, in CARS spectroscopy, spatially-resolved Raman information is obtainable by properly focusing the ω_1 and ω_2 beams at the sample point of interest. This makes CARS very suitable for combustion diagnostics [7], which is out of the scope of the present article, and also for the study of molecules in supersonic jet beams. CARS spectroscopy of molecular clusters in a supersonic jet is described in section 8.4.4.

The obvious drawback of nonlinear Raman spectroscopy concerns the fact that it requires a rather complicated and often very expensive optical set-up using lasers. This has hindered the development of a "turn-key" nonlinear Raman spectrometer for general use, in contrast to the recent development of optical-fiber based versatile spontaneous Raman spectrometers. Nonlinear Raman spectroscopy requires further development before it will serve as a fully complementary alternative of spontaneous Raman spectroscopy.

8.3 SRS Spectroscopy

8.3.1 SRG and SRL signal intensities and polarizations

In the standard configuration of SRS spectroscopy (Fig. 8. 4), the incident ω_1 and ω_2 beams, travelling in the z direction, are colinearly superimposed in the sample and the gain/loss in the ω_2/ω_1 intensity after traversing path length L is measured.

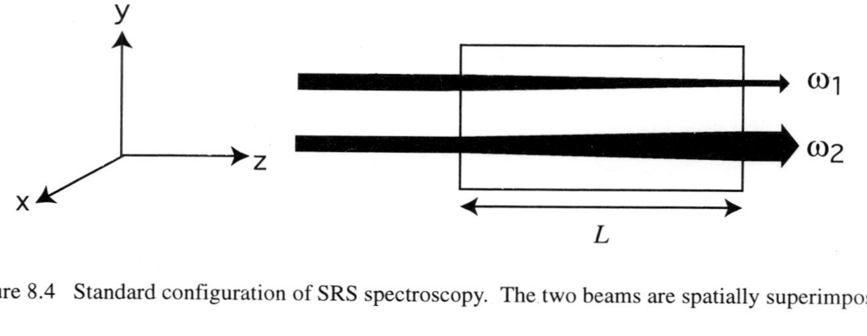

Figure 8.4 Standard configuration of SRS spectroscopy. The two beams are spatially superimposed.

The expression for the SRG/SRL signal amplitude is obtained by introducing $\boldsymbol{P}^{(3)}{}_{SRG}/\boldsymbol{P}^{(3)}{}_{SRL}$ into the right hand side of Eq. (8.9). For SRG spectroscopy, the product of this signal amplitude with the incident ω_2 field gives the following expression for the SRG signal intensity,

$$I_2(\mathrm{L}) = I_2\exp[-(32\pi^2/n_1n_2c^2)\mathrm{Im}\chi^{(3)}{}_{SRG}I_1L] \tag{8.13}$$

where I_1 and I_2 are incident intensities of the ω_1 and the ω_2 beams; $I_2(L)$ is the intensity of the ω_2 beam after passing through the interaction length L; n_1 and n_2 are the refractive indices at ω_1 and ω_2. The imaginary part of $\chi^{(3)}_{SRG}$, $\text{Im}\chi^{(3)}_{SRG}$, is negative and this makes ΔI_2 positive (gain). For a small intensity change $\Delta I_2 = I_2 - I_2(L)$, Eq. (8.13) is reduced to:

$$\Delta I_2/I_2 = -(32\pi^2\omega_2/n_1n_2c^2)\text{Im}\chi^{(3)}_{SRG}I_1L. \tag{8.14}$$

The plot of $\Delta I_2/I_2$ vs $\omega_1-\omega_2$ gives a Raman spectrum with Lorentzian resonances that are derived from the expression of χ^R given in Eq. (8.10). The corresponding expression for SRL spectroscopy is obtained as,

$$\Delta I_1/I_1 = -(32\pi^2\omega_1/n_1n_2c^2)\text{Im}\chi^{(3)}_{SRL}I_2L, \tag{8.15}$$

where $\text{Im}\chi^{(3)}_{SRL}$ is positive and ΔI_1 is negative (loss).

In the standard configuration of SRS spectroscopy (Fig. 8.4), two different polarization settings are possible. In the parallel setting, where both the ω_1 and the ω_2 beams are polarized in the same direction x, the intensity change $\Delta I_{//}$ is derived from χ^R_{xxxx}. For the perpendicular setting, in which the ω_1 and ω_2 polarizations are perpendicular to each other, the intensity change $\Delta I_\perp$ is associated with χ^R_{xyyx}. It follows from Eq. (8.12) that the ratio $\Delta I_{//}$ $\Delta I_\perp$ gives the depolarization ratio ρ which is exactly the same as that for spontaneous Raman scattering.

8.3.2 High resolution quasi-cw SRS spectroscopy

The field of high resolution rovibrational spectroscopy of gases has benefitted greatly from the development of new SRS techniques since the late 1970's. Figure 8.5 shows the scheme of quasi-cw SRS spectroscopy that is most often employed for high resolution studies [8]. Quasi-cw SRS spectroscopy uses a pulsed pump beam with a tunable wavenumber and a quasi-cw probe beam with a fixed wavenumber. In the SRG mode, the pump beam is in the anti-Stokes side and the probe beam in the Stokes side. In the SRL mode, the reverse is the case. The intensity change in the probe beam, which appears as a dip on a quasi-cw contour in the case of SRL (Fig. 8.5), is detected by a photodiode and processed by a boxcar integrator. A high pass filter is used to eliminate the dc component in the signal, so that only the change in the probe intensity (the dip) is detected and thereby the dynamic range of the detecting electronics is effectively extended. Quasi-cw gating of the probe beam is necessary to avoid the saturation of the detector. The quasi-cw detection scheme utilizes the high peak power of a pulsed laser to induce maximum gain/loss signals and, simultaneously, makes the full use of the ultra high frequency and amplitude stability of a single mode cw laser, to assure the maximum detectivity of small signals.

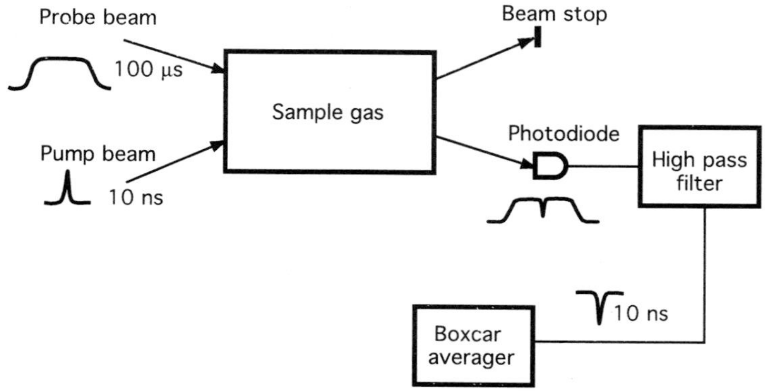

Figure 8.5 The principle of quasi-cw SRS spectroscopy [8].

Figure 8.6 shows a typical set up for quasi-cw SRS spectroscopy [8]. The pumping beam is obtained by amplifying the output of an Ar^+-laser pumped single-mode dye laser by a three stage dye amplifier that is pumped by a frequency-doubled Q-switched Nd:YAG laser. The final output from the amplifier (1 cm^{-1} tunable, 60 MHz (=0.002 cm^{-1}) line width, 12 ns pulse width, 10 Hz repetition rate, 1 MW peak power) is used as the pump (ω_1 for SRG and ω_2 for SRL) beam. Three separate lasers are used as probe sources (ω_2 for SRG and ω_1 for SRL). They include a ring dye laser (575–620 nm tunable), a Kr^+ laser (647.1 nm), and an Ar^+ laser (488.0 or 514.5 nm). In either case, single-mode operation is necessary to reduce the amplitude noise due to the mode hopping. The power level of the probe beam is around 350 mW, which is adjusted to maximize the signal while avoiding population saturation and also detector saturation. The probe beam is modulated to a quasi-cw pulse train of 10 Hz repetition rate by a chopper. The pump and probe beams are focused and crossed in a sample cell with a typical crossing angle of 1.4°. The probe beam is spatially separated from the pump beam and then filtered once again by a dispersive filter consisting of a 1200 lines mm^{-1} grating. The change in the probe intensity is detected by a photodiode and processed by a boxcar integrator. The signal is normalized to the pump intensity that is simultaneously monitored by a photodiode and the second channel of the boxcar integrator. The absolute wavenumbers of the pump and probe beams are measured by a wavemeter to ±0.001 cm^{-1} in the wavenumber difference. The wavemeter consists of a modified Michelson interferometer, having two corner cubes, of which one can move mechanically on a 1m ball bearing track, and a counter that counts the number of fringes produced. A stabilized single mode He-Ne laser is used as a standard.

By changing the dye, the wavelength region of 550–620 nm is covered by the amplified pump beam. Combination of this pump beam with the Kr^+ 647.1 nm line covers the Stokes wavenumber region of 680–2000 cm^{-1} in the SRG mode. The anti-Stokes wavenumber region of 2000–4360 cm^{-1} is accessible with either the Ar^+ 488.0 nm or 514.5 nm line in the SRL mode. The lower wavenumber region is covered by the combination of thepulsed pump beam with the ring dye laser. The wavenumber resolution of this SRS spectrometer is limited by the width of the pump beam, 0.002 cm^{-1}, which is only 60 % larger than the theoretical Fourier transform limit of a 12 ns Gaussian pulse.

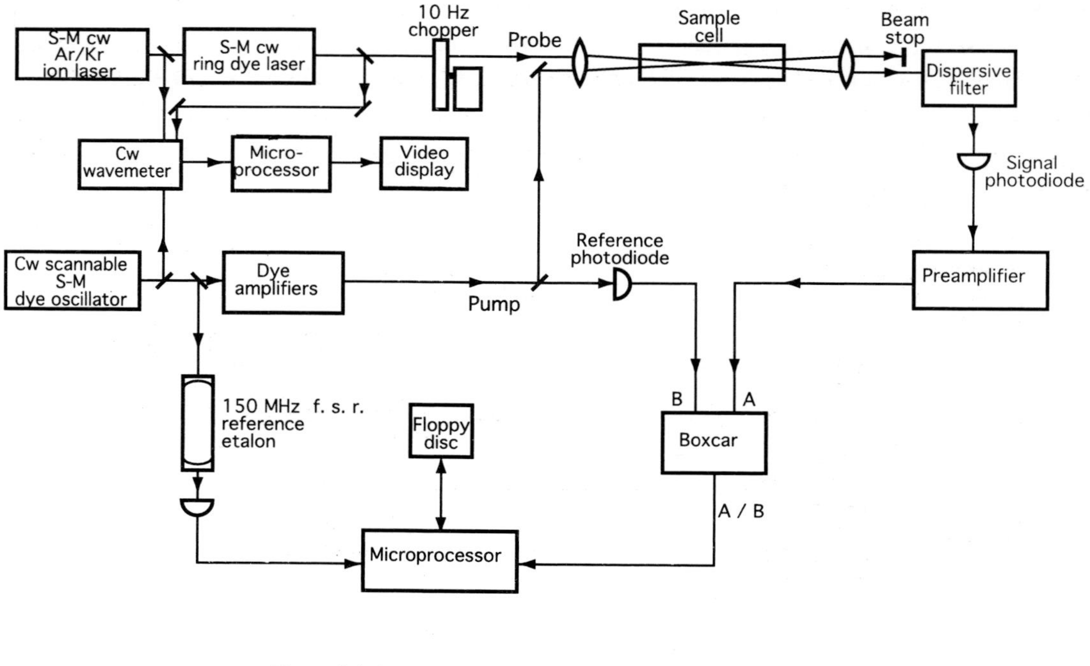

Figure 8.6 A representative quasi-cw SRS spectrometer [8].

As a representative high resolution SRS spectrum, the SRG spectrum of SF_6 is shown in Fig. 8.7. The Kr^+ 647.1 nm line is used as the probe beam. The sample pressure is 3.8 Torr. A 1 cm^{-1} region from 773.6–774.6 cm^{-1} is displayed. This region is dominated by the Q branches of the ν_1 fundamental overlapped with the $\nu_1+\nu_6 \leftarrow \nu_6$ hot band. Analysis including a Coriolis interaction in the ν_6 (F_{2u}) level has been carried out to yield high accuracy rovibrational constants [8].

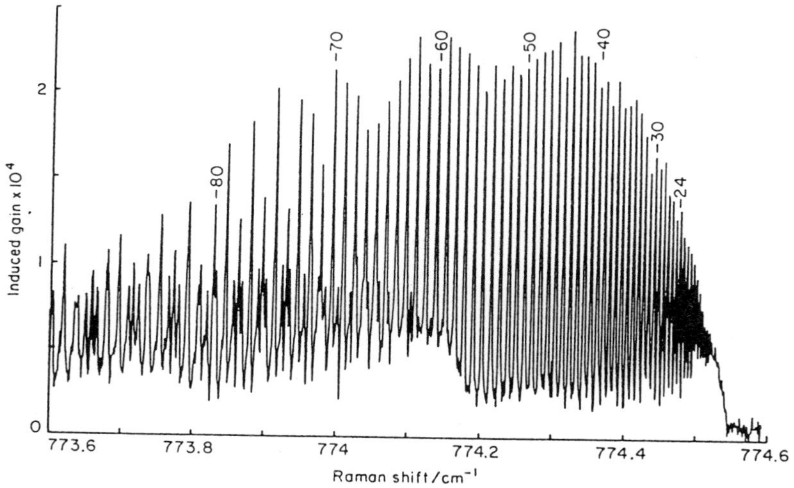

Figure 8.7 High resolution SRG spectrum of SF_6. Reproduced with permission from Ref. 8.

8.3.3 SRS spectroscopy for condensed-phase studies

High repetition rate mode-locked picosecond lasers are suitable sources for medium-low resolution SRS spectroscopy for condensed-phase studies. A typical set up is shown in Fig. 8.8 [9]. A mode-locked Ar^+ laser pumps two dye lasers A and B. Dye laser A is operated as a synchronously-pumped mode-locked laser and its output (570–630 nm tunable, 5–20 ps pulse width, 81.8 MHz repetition rate, 30–100 mW average power) is used as the probe beam. The pump beam (600–700 nm, 20 ps, <4 MHz, 30 mW) is obtained from Dye laser B which is operated in the mode–locked cavity dumped mode. The pump and probe beams are collimated and focused through a sample cell onto the entrance slit of a double monochromator. The probe beam is selected by the double monochromator and detected by a photodiode. A lock-in amplifier is used to pick up the signal that is induced synchronously with the pump laser pulses. By properly combining the dye lasers, the wavenumber difference region 0–3300 cm^{-1} can be scanned.

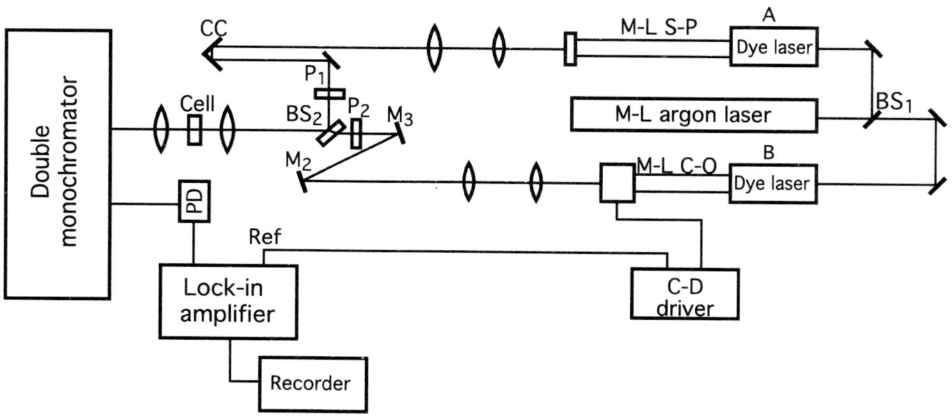

Figure 8.8 A high repetition rate picosecond SRS spectrometer [9].

The SRL/SRG spectrum of liquid carbon tetrachloride obtained with this set up is shown in Fig. 8.9 [9]. The symmetric relationship between SRL and SRG is very clearly demonstrated. The spectrum also indicates the low wavenumber feasibility of SRS spectroscopy, though an intense artifact masks the region of 0 cm^{-1} in this particular example. With the use of a more sophisticated optical set up, this feature can be totally removed. The same set up has been successfully applied for fluorescence elimination. The SRS spectra of highly fluorescent dyes including Rhodamine 6G have been reported [9].

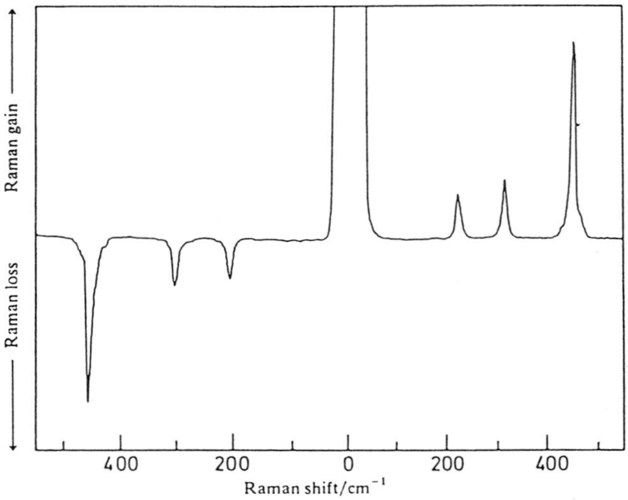

Figure 8.9 SRL/SRG spectrum of carbon tetrachloride. Reproduced with permission from Ref. 9. Note the fact that the signal is positive (gain) in the Stokes region and is negative (loss) in the anti-Stokes region. The artifact around 0 cm^{-1} is due to stray light that is not removed by the monochromator.

8.3.4 Femtosecond time-resolved multiplex SRG spectroscopy

Using three femtosecond laser pulses, the photoexcitation pulse, and the ω_1 and ω_2 pulses, femtosecond time-resolved pump-probe SRG spectroscopy can be performed. Figure 8.10 shows the principle of the experiment [10]. The photoexcitation pulse (630 nm, 100 fs pulse width), which precedes the ω_1 (700 nm, 200 fs) and ω_2 pulses (400–1000 nm), drives the molecule to an excited state. The photophysical and photochemical processes following photoexcitation are monitored by multiplex SRG spectroscopy. The three laser beams are focused and superimposed spatially on a sample and, after passing through the sample, the ω_2 beam is separated and spectrally resolved by a polychromator. The SRG signal on the ω_2 beam is detected by a CCD detector.

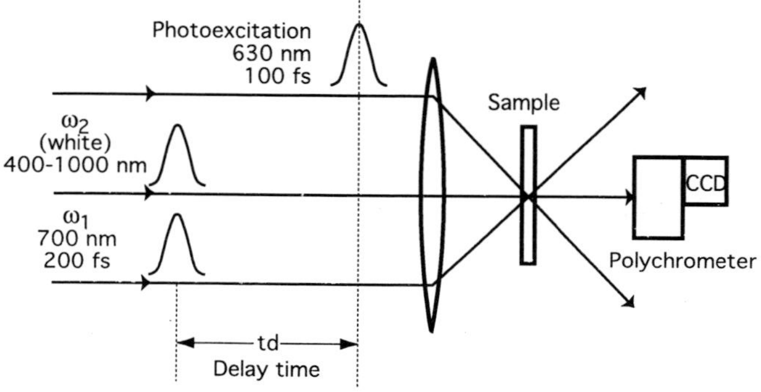

Figure 8.10 Femtosecond pump-probe multiplex SRG spectroscopy.

Figure 8.11 shows a set of time-resolved SRG spectra obtained after the photoexcitation of a kind of polydiacetylene, PDA-3BCMU, deposited on a KCl crystal [10]. The spectral change around 1200 cm^{-1} is discussed in terms of a geometrical change from an acetylene-like structure, $(=CR\text{-}C\equiv C\text{-}CR=)_x$, to a butatriene-like structure, $(\text{-}CR=C=C=CR\text{-})_x$, where R stands for a substituent. Note the fact that all the observed bands are broadened due to the intrinsic broadening in the ω_1 pulse; a 200 fs sech^2-shaped laser pulse results in a spectral width of 50 cm^{-1} by the Fourier transform relationship.

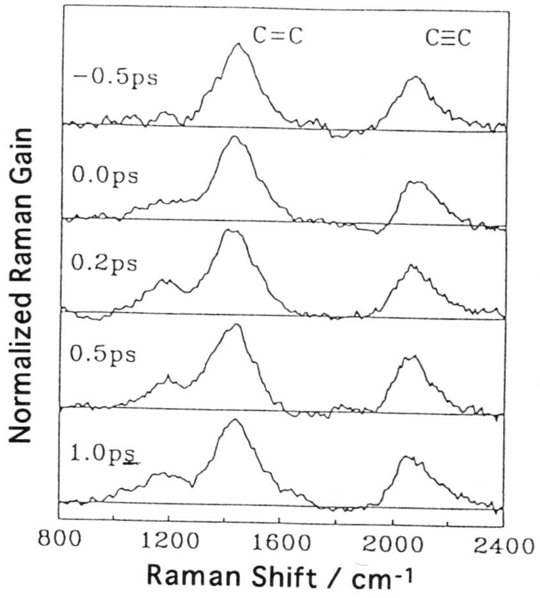

Figure 8.11 Femtosecond time-resolved SRG spectra of PDA-3BCMU (a kind of polydiacetylene) deposited on a KCl crystal. Reproduced with permission from Ref. [10].

8.3.5 Other Variations of SRS spectroscopy

In addition to the standard SRG and SRL spectroscopies described above, there are many other ways of performing SRS spectroscopy. **Raman induced Kerr effect (RIKE)** spectroscopy probes the birefringence that is induced in the material by a Raman resonance [11]. A small rotation in the plane of polarization of either the Stokes or anti-Stokes beams is detected by a crossed polarizer/analyzer. RIKE spectroscopy enables a background free detection of a stimulated Raman signal, if the birefringence in the optics is properly compensated for. In SRG and SRL spectroscopies, a small change in the probe laser intensity has to be measured. This often limits the detectivity of the SRG and SRL signals. **Photoacoustic Raman (PAR)** spectroscopy detects the acoustic wave that is generated during the non-radiative relaxation of the Raman-resonantly induced material excitations [12]. PAR spectroscopy does not require the separation of the ω_1 and ω_2 beams and therefore is suitable for low-frequency Raman spectroscopy. Note, however, that the PAR signal depends monotonically upon the excitation energy $h\Omega$ of the material and goes to zero as Ω approaches zero. **Ion detected stimulated Raman (IDSR)** spectroscopy probes the excited-state population in the material that is produced by Raman pumping [13]. A third laser beam is used to selectively photoionize the molecules in the excited state and not those in the ground state. The IDR method is background free and is highly sensitive because of the high sensitivity of the ion detection. Though these variations of SRS spectroscopy played certain roles in the past development of nonlinear Raman spectroscopy, their practical use has been limited. This is primarily because of complex signal detection schemes as compared to SRS. Now that we can accurately measure light intensities over an extensive dynamic range, the simple detection scheme in the standard SRS spectroscopy is favorably used in the majority of practical applications.

8.4 CARS Spectroscopy

8.4.1 CARS signal intensities and polarizations

In the standard CARS experiment (Fig. 8. 12), the incident ω_1 and ω_2 beams are configured to satisfy the phase matching condition so that the signal is generated in the form of a new beam at $\omega_3=2\omega_1-\omega_2$. The phase matching condition of CARS is given as,

$$\boldsymbol{k}_3 = 2\boldsymbol{k}_1 - \boldsymbol{k}_2, \tag{8.16}$$

where $\boldsymbol{k}_1$, $\boldsymbol{k}_2$, and $\boldsymbol{k}_3$ are the wave vectors of the ω_1, ω_2, and ω_3 beams, respectively. The crossing angle α is determined by the dispersion of the refractive index of the sample and is usually small (Fig. 8.12 is exaggerated). The incident beams are nearly parallel to the z axis which is taken parallel to the ω_3 beam.

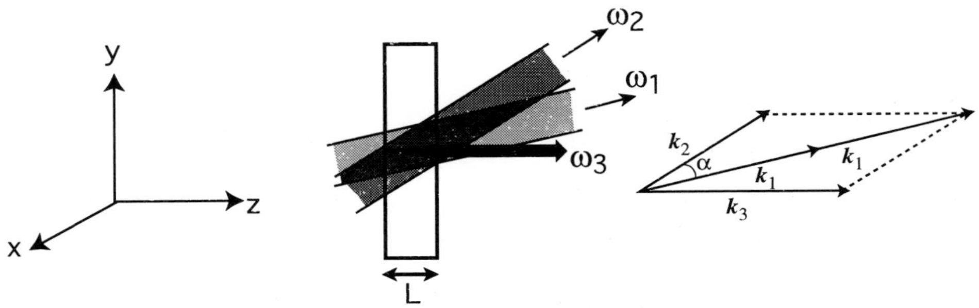

Figure 8.12 Phase matching condition of CARS.

The expression for the CARS intensity is derived from Eq. (8.9) by introducing $\boldsymbol{P}^{(3)}_{\mathrm{CARS}}$ in the right hand side and taking the square modulus,

$$I_3(L) = 256\pi^4\omega_3^2/n_1^2n_2n_3c^4\,|\chi^{(3)}_{\mathrm{CARS}}|^2 I_1^2 I_2 L^2[\mathrm{sinc}(\Delta kL/2)]^2 , \tag{8.17}$$

where I_1 and I_2 are the incident intensities of the ω_1 and ω_2 beams; $I_3(L)$ is the intensity of the CARS ω_3 beam after traversing an interaction length L; n_1, n_2, and n_3 are the refractive indices at ω_1, ω_2, and ω_3, respectively; sinc(x)=(sin x)/x; Δk is the z component of the phase mismatch vector $\Delta\boldsymbol{k} = 2\boldsymbol{k}_1 - \boldsymbol{k}_2 - \boldsymbol{k}_3$. If the phase matching condition (8.16) is satisfied, $\Delta k=0$ and $\mathrm{sinc}(\Delta kL/2)=1$. Then, the CARS intensity is determined by three factors; i) the square modulus of $\chi^{(3)}_{\mathrm{CARS}}$, ii) the intensity product $I_1^2I_2$ of the incident beams, and iii) the square of the interaction length L.

Some complexity arises from the first factor i). In the first place, the CARS intensity is not proportional to the molecular number density but it is proportional to the square. This contrasts with the SRS intensity which is proportional to the number density (though not explicitly stated in the previous section). The difference is due to the fact that the SRS intensity is related to Im $\chi^{(3)}_{\mathrm{SRS}}$ but the CARS intensity is related to $|\chi^{(3)}_{\mathrm{CARS}}|^2$. Note that $\chi^{(3)}$ is proportional to the number density. Second, interference between the non-resonant and the Raman-resonant terms and that among the Raman-resonant terms occur even under off-resonance conditions. As a consequence, the CARS band shape can vary between a positive Lorentzian to a negative Lorentzian through dispersive forms, depending on the relative magnitudes of the non-resonant and the Raman resonant terms. It is therefore necessary to simulate the entire CARS spectrum before obtaining any spectroscopic parameters like the band center, the band width, and the intensity. Much precautions must be taken before using CARS spectra for quantitative as well as qualitative analysis of materials.

The phase matching condition of CARS can be significantly relaxed by reducing the interaction length L. Figure 8.13 shows the theoretical phase mismatch factor f=[sinc($\Delta kL/2$)]^2 calculated for liquid carbon tetrachloride [14].

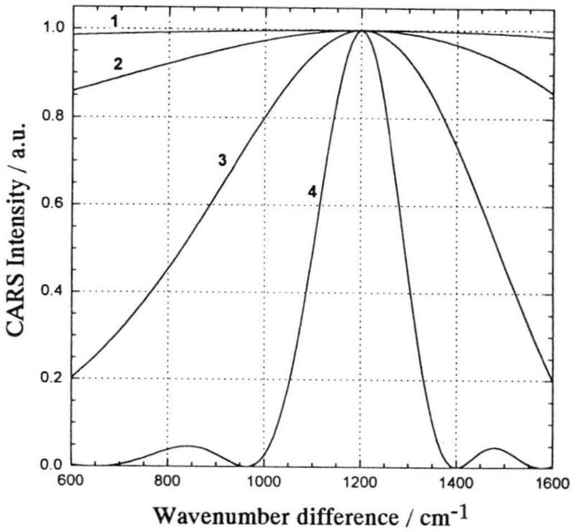

Figure 8.13 Calculated CARS phase mismatch factor for carbon tetrachloride. The interaction length L=300 μm (curve 1), 1 mm (2), 5 mm (3), and 10 mm (4). Reproduced with permission from Ref. [14].

In this calculation, the crossing angle α is fixed at 2.2° so that f=1 at 1200 cm^{-1}. For a large interaction length (L=10 mm), f drops sharply as $\omega_1-\omega_2$ is detuned from 1200 cm^{-1} and becomes zero at 1000 and 1400 cm^{-1}. This means that, as $\omega_1-\omega_2$ is scanned, the crossing angle must be continuously readjusted so that f remains nearly equal to unity. For a small interaction length (L=300 μm), on the other hand, the factor f drops by less than 5 % even at 600 cm^{-1} detuning. There is no need of readjusting the crossing angle if a 300 μm thick sample is used. Broadband multiplex CARS spectroscopy that covers a full 1000 cm^{-1} region thus becomes feasible with the use of a short path length (see Section 8.4.3). Note that the CARS intensity is proportional to L^2 and the magnitude of the signal decreases by a factor of 10^3 on going from L=10 mm to L=300 μm. However, this decrease of the signal intensity is very well compensated for in the multiplex CARS scheme, if a highly sensitive multichannel detector, such as a CCD, is used.

As three beams with different angular frequencies are involved, there is a three-dimensional freedom of manipulating the light polarizations in CARS spectroscopy. In polarization sensitive CARS (PSCARS) spectroscopy, the polarizations of the ω_1 and ω_2 beams as well as that of the ω_3 beam are properly controlled so that the resultant spectrum is given a specificity concerning these polarization characteristics. If the angle between the directions of polarization of the linearly polarized ω_1 and ω_2 beams is fixed at β, the polarization direction of the signal is determined with [15],

$$\tan\theta = \rho \tan\beta, \tag{8.18}$$

where θ is the angle between the directions of polarization of the ω_1 and the signal beam, and ρ is the Raman depolarization ratio. As a representative setting, the case of β = 60° is shown in Fig. 8.14, where the direction of polarization of ω_1 is taken as the x axis. If the transmission axis of the analyzer $\boldsymbol{P}_a$ is set at a particular angle φ_a, the CARS signal with the polarization direction perpendicular to $\boldsymbol{P}_a$ is completely eliminated. Therefore, in PSCARS spectroscopy, one can selectively eliminate Raman bands with arbitrary depolarization ratios by merely adjusting the angle φ_a. For example, the nonresonant background with ρ = 1/3 is eliminated by setting φ_a= 120 °. A weak Raman band that is overlapped by another stronger band may be detected by suppressing the latter by setting φ_a to a proper angle, provided that the two bands have different depolarization ratios. This polarization feasibility of CARS spectroscopy is lacking in spontaneous Raman and SRS spectroscopies in which elimination of a particular depolarization ratio is possible if and only if ρ=0.

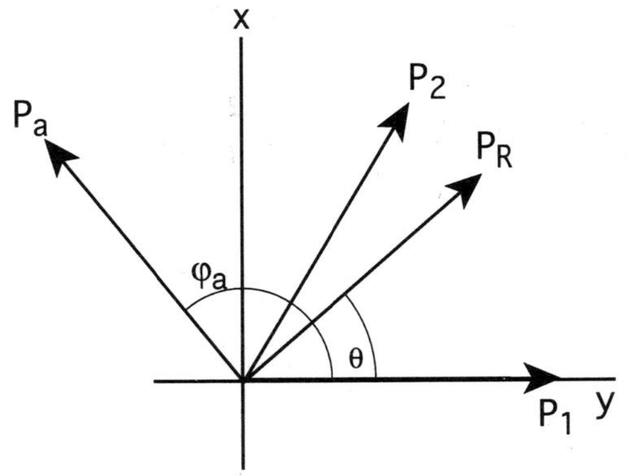

Figure 8.14 A representative polarization setting in PSCARS spectroscopy.

At the end of this section, it is appropriate to note that the above discussion of CARS holds for CSRS as well, if ω_1 and ω_2 are interchanged. CSRS gives much the same spectral information as CARS does, except for the case of "extra resonance" in which a Raman resonance with a transition between non-populated excited states appears in the CSRS spectrum [16]. Extra resonance does not occur in CARS.

8.4.2 CARS spectroscopy in supersonic jet beams

The high sensitivity of CARS spectroscopy makes it possible to observe low density gaseous molecules that exist only in a low temperature supersonic jet beam. Figure 8.15 shows a CARS set up for jet spectroscopy [17]. The system is based on a pulsed Q-switch Nd:YAG laser. The second harmonic of this laser is used as the ω_1 beam (532 nm, 10 ns pulse width, 10 Hz repetition rate, 10 mJ/pulse energy). The ω_2 beam is obtained from a dye laser which is pumped either by the second or the third harmonic of the same Nd:YAG laser (wavelength tunable, 0.25 cm^{-1} line width with an air-spaced etalon, 10 ns, 10 Hz, 2–10 mJ/pulse).

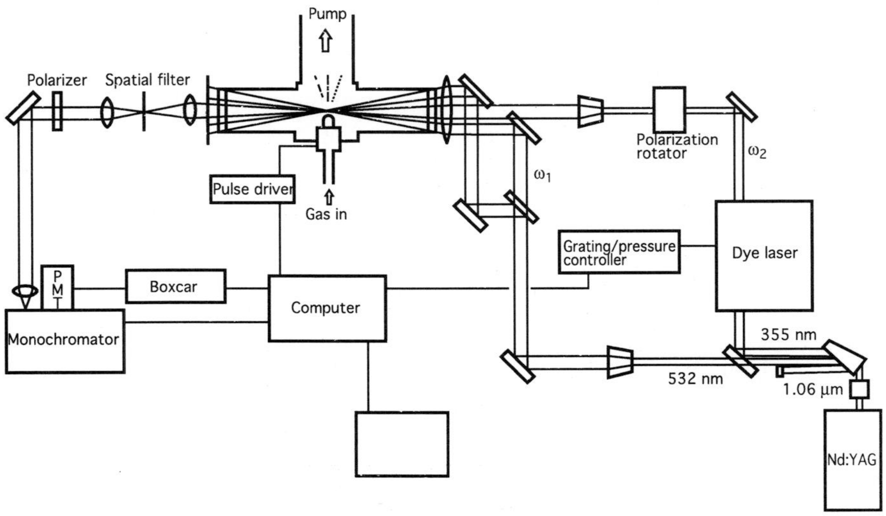

Figure 8.15 A CARS set up for supersonic jet spectroscopy [17].

The ω_1 and ω_2 beams are crossed in the supersonic jet in the so called BOXCARS geometry (Fig. 8.16), in which a three-dimensional phase-matching condition is satisfied and in which the spatial separation of the CARS beam from the ω_1 and ω_2 beams is better performed. The crossed geometry in BOXCARS is also advantageous for jet spectroscopy because it provides better spatial resolution. The signal beam is separated first by a spatial filter and then filtered by a monochromator and detected by a photomultiplier. The signal is processed by a boxcar integrator.

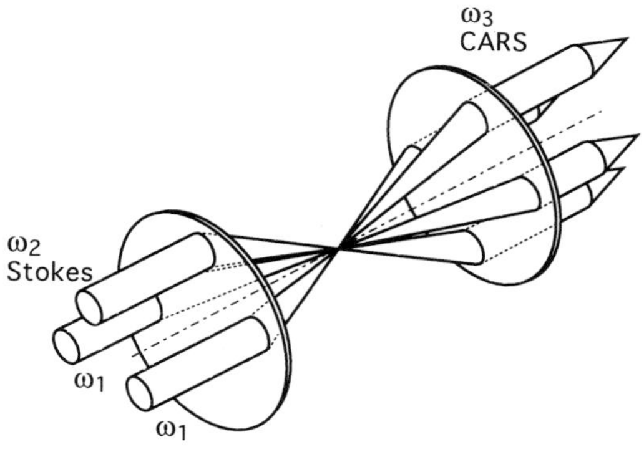

Figure 8.16 BOXCARS beam configuration of CARS spectroscopy [17].

Van der Waals clusters of CO_2 have been studied with this CARS set up [17]. Figure 8.17 shows CARS spectra in the region of the ν_1 Q-branch of CO_2 obtained from the jet expansion of the CO_2/He mixture with different concentrations and pressures. For a low concentration (1 %, Fig. 8.16-c), the spectrum is dominated by a strong Q-branch of the monomer at 1285 cm^{-1}. For a high concentration (5 %, Fig. 8.16-a), on the other hand, extra peaks marked D (dimer), T (trimer), and P (polymer) are observed in addition to the strong peak of the monomer. These extra bands are attributed to the van der Waals clusters of CO_2 . For the dimer, the observed frequency is 4.2 cm^{-1} red shifted from the monomer band, while the infrared frequency of the dimer is known to be 0.7 cm^{-1} blue shifted from the monomer. This finding has been used to suggest that the structure of the dimer has a C_{2h} symmetry.

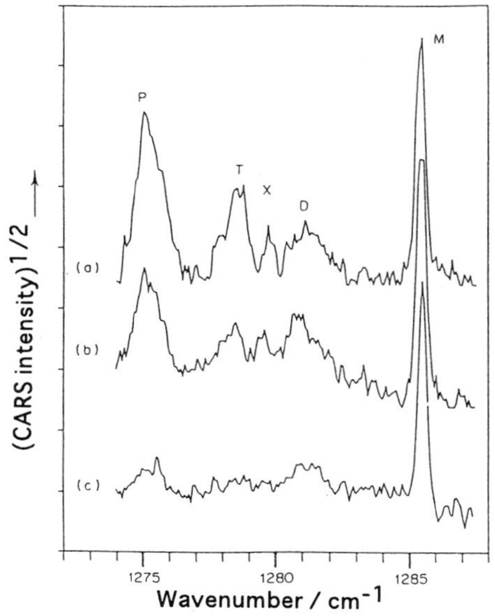

Figure 8.17 CARS spectra of CO_2 in jet expansions of CO_2/He mixtures with different concentrations and pressures. (a) 5 % CO_2 and 12 atm pressure, (b) 2 % and 14.6 atm, (c) 1 % and 18 atm. Reproduced with permission from Ref. [17].

8.4.3 Polarization Sensitive Multiplex CARS Spectroscopy

The multiplex detection scheme, which enables to measure a large number of CARS spectra with different polarization settings, is most suitable for PSCARS spectroscopy. Figure 8.18 shows a broadband multiplex CARS system for PSCARS measurements [14]. The frequency doubled output of a Nd:YLF laser (523. 6 nm, 7 ns duration, 2 kHz repetition rate, 250 mW average power) is split into two. One half pumps a dye laser whose cavity is optimized for broadband lasing. A mixture of fluorescein 548 and rhodamine 6G in ethylene glycol is used to obtain broadband output that covers the wavelength region 543 nm to 575 nm with an average power of up to 10 mW. This broadband output is used as the ω_2 radiation. The other half of the doubled output is used as the ω_1 radiation. This combination of ω_1 and ω_2 covers the Raman shift ($\omega_1-\omega_2$) region of 700 to 1700 cm^{-1}. The ω_1 and ω_2 beams are collimated to diameters of about 1 mm on the sample, which is in the form of a flowing thin liquid film of 300 μm thickness. The timing between the two beams is adjusted by an optical delay line. The generated CARS signal is focused onto the entrance slit of a single polychromator and detected by an intensified photodiode array after being dispersed. A retardation plate (λ/2) and three Glan-Thomson prism polarizers (P1, P2, PA) are used to perform the polarization sensitive measurements.

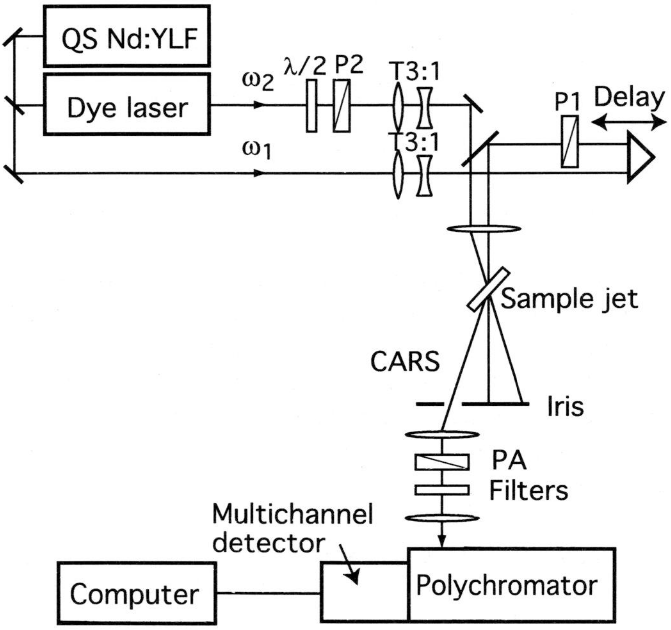

Figure 8.18 A broadband multiplex CARS spectrometer for polarization sensitive measurements [14].

Figure 8.19 shows PSCARS spectra of liquid cyclohexane measured with $\beta = 60°$ and $\varphi_a = 0°$ (a), 143° (b), 130° (c), 120° (d), 100° (e), and 90° (f). The exposure time needed to measure one CARS spectrum is 30 seconds. The noisy lines represent the measured spectra and the smooth lines the calculated spectra. The measured raw spectra are corrected for the intensity distribution of the ω_2 radiation, the wavelength dependence of the spectrometer and detector response, and the phase mismatch, by referring to the nonresonant background spectra of carbon tetrachloride which does not show any strong Raman resonances in the region 700 cm^{-1} to 1700 cm^{-1}. The following three points are noteworthy in these spectra. First, the nonresonant background is totally eliminated in spectrum d which is measured with $\varphi_a = 120°$. This polarization setting eliminates the CARS signal with $\rho = 1/3$ including the nonresonant background. As a result, all the observed bands in spectrum d have Lorentzian shapes arising solely from the Raman resonance terms.

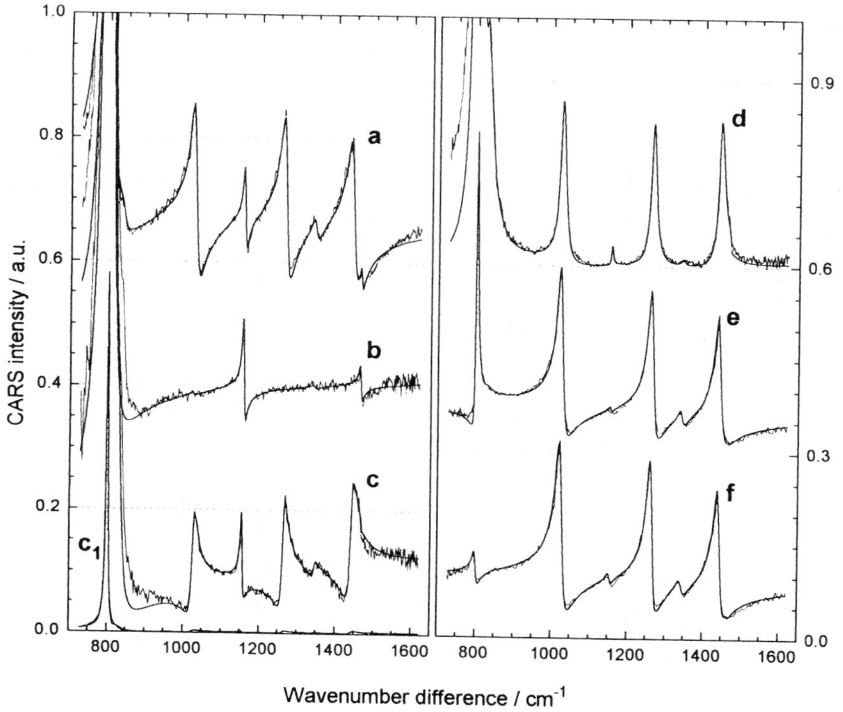

Figure 8.19 PSCARS spectra of cyclohexane. Reproduced with permission from Ref. [14].

Second, all the depolarized bands with $\rho = 3/4$ are eliminated in spectrum b with $\varphi_a = 143°$. Equation 8.18 tells that the depolarized bands have the direction of polarization 52.4°. The polarization setting $\varphi_s = 143°$ therefore eliminates the depolarized bands. A very weak polarized band at 1466 cm^{-1} is clearly seen in spectrum b, which is hidden by the strong depolarized 1443 cm^{-1} band in the other polarization settings. Third, the polarized bands are effectively suppressed in spectra e and f depending upon their depolarization ratios. The band at 1157 cm^{-1} is the maximally suppressed in spectrum e, while the intensity of the 801 cm^{-1} band becomes minimum in spectrum f.

All the spectra in Fig. 8.19 can be fitted to theoretical curves using the following equation with the same set of band parameters;

$$I_{\mathrm{PSCARS}}(\Delta\omega,\varphi_a) = \left|\chi^{\mathrm{NR}} g^{\mathrm{NR}}(\varphi_a) + \Sigma\chi_Q g^Q(\varphi_a)/(\Omega_Q - \Delta\omega + i\Gamma_Q)\right|^2, \tag{8.19}$$

where $\Delta\omega = \omega_1 - \omega_2$, and χ^{NR} and χ^Q are the third-order CARS susceptibilities corresponding to the nonresonant background and the Qth Raman resonance, $g^{\mathrm{NR}}(\varphi_a)$ and $g^Q(\varphi_a)$ are the angle factors for the nonresonant background and the Qth Raman resonance, Ω_Q is the Qth Raman frequency, and Γ_Q is the damping constant (band width) for the Qth Raman transition. Equation (8.19) is readily derived from the general form of the nonlinear Raman susceptibility in Eq. (8.10). The theoretical curves shown in Fig. 8.19 were obtained with a total of 7 Raman resonances being taken into account. The parameters thus determined give accurate band centers, band widths, band intensities, and depolarization ratios for the seven Raman transitions including the very weak one at 1466 cm^{-1}. The accuracy of these band parameters, especially that of the depolarization ratio, far exceeds what is obtained by ordinary spontaneous Raman spectroscopy. Note that, by changing φ_a in a smaller steps, the number of CARS spectra to be fitted can be increased and therefore the fitting analysis based on Eq. (8.19) can be applied to cases with a still larger number of Raman resonances. The limitation of Eq. (8.19) is that it is not applicable to one-photon resonant CARS.

8.4.4 Picosecond Time-frequency Two-Dimensional Multiplex CARS Spectroscopy

By combining the multiplex CARS technique with a streak camera, a new type of time-resolved CARS spectroscopy, termed two-dimensional multiplex CARS (2-D CARS) spectroscopy [18], is performed. The principle of 2-D CARS spectroscopy is shown in Fig. 8.20. Two CARS probing beams, ω_1 (monochromatic, nanosecond duration) and ω_2 (broadband, nanosecond duration), are used with the third pico/femtosecond laser beam for the photoexcitation. The generated CARS signal, which is distributed both in the frequency and time domains, is first frequency-resolved by a spectrograph and then time-resolved by a streak camera. The resultant 2-D CARS image is detected by a CCD detector. The frequency resolution in 2-D CARS spectroscopy is determined by the polychromator and the time resolution by the streak camera. Intersection of the obtained 2-D image at a specific time gives the time-resolved CARS spectrum at that time, while intersection at a specific frequency gives the time dependence of the CARS intensity at that frequency. 2-D multiplex CARS spectroscopy enables one to collect all the CARS spectral and temporal information with only a single measurement. The efficiency of the 2-D CARS method, if compared against the time needed to obtain the same set of time-resolved CARS spectra, is more than two orders of magnitude higher than the conventional time-resolved scanning CARS technique using two picosecond lasers. Such a high efficiency has made it practical to study the structure and dynamics of highly fluorescent photoexcited molecules in condensed phases by using CARS spectroscopy.

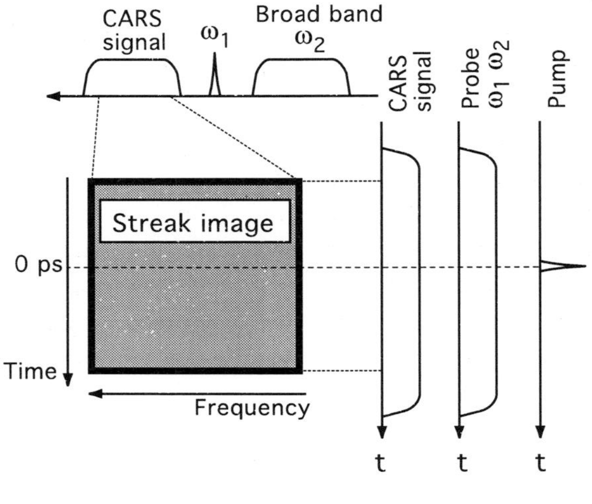

Figure 8. 20 The principle of time-frequency 2-D CARS spectroscopy [18].

Figure 8.21 shows a setup of picosecond 2-D multiplex CARS spectroscopy [18]. The same ω_1 and ω_2 beams as described in Section 8.4.3 are used. The photoexcitation beam, which is used for pumping the molecule to the excited electronic state, is obtained from a self-mode-locked Ti:sapphire oscillator/regenerative amplifier system. The output of the regenerative amplifier is divided into two portions. One portion (5 %) is used for triggering a streak camera. The other portion (95 %) is guided to an optical delay line to generate a time delay that is needed to compensate for the electronic delay in the streak camera. It is then frequency doubled and used as the photoexcitation beam (390 nm, 500 fs duration, 1 kHz repetition rate, 50 μJ/pulse). The three beams, ω_1, ω_2, and photoexcitation, are focused and spatially and temporally superimposed on a sample of 300 μm thick liquid film. The generated CARS signal is collimated and focused onto the entrance slit of an astigmatism corrected single polychromator. The signal is frequency resolved by the spectrograph and time-resolved by a streak tube and finally detected as a 2-D image by a CCD camera. The temporal response of this system is 15 ps if a proper aperture is inserted in order to limit the solid angle of the signal light entering the detector; angular spread results in a temporal broadening of the detected signal.

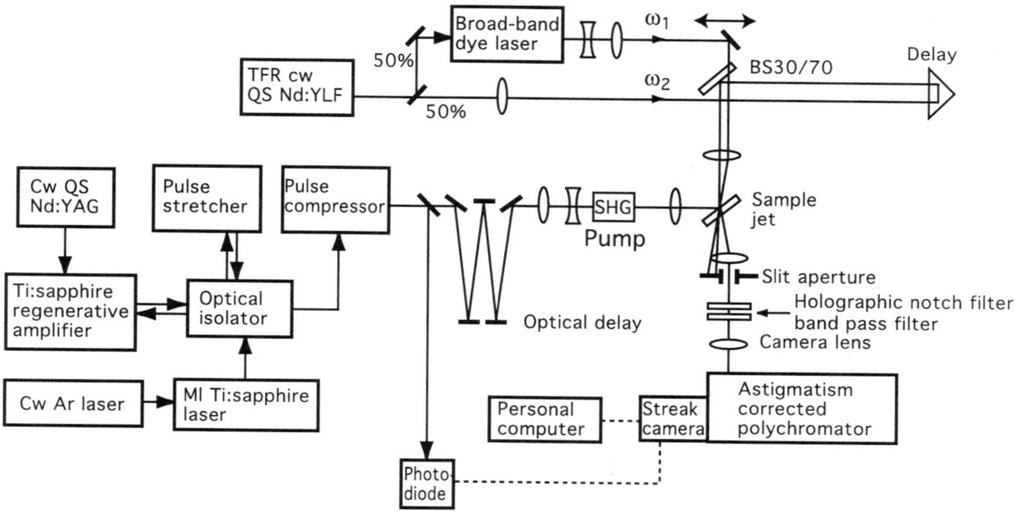

Figure 8. 21 A set up for picosecond 2-D multiplex CARS spectroscopy [18].

A 2-D CARS image obtained after the photoexcitation of all-trans retinal in cyclohexane is shown in Fig. 8. 22. The ordinate corresponds to time (increasing from up to down) and the abscissa the wavenumber shift $\omega_1 - \omega_2$ (increasing from right to left). The horizontal line marked "0 ps" shows the very short-lived fluorescence emitted from photoexcited all-trans retinal. The four vertical lines which are observed throughout the time course and which are marked "Solvent" are due to the Raman bands of cyclohexane. The three vertical lines marked "Transient" are ascribed to the CARS signals from the lowest excited triplet (T_1) state of all-trans retinal. This experiment is carried out under a one-photon resonance with the strong $T_n \leftarrow T_1$ absorption at 460 nm and therefore the CARS bands due to the T_1 state are strongly enhanced. The three transient CARS signals rise after the photoexcitation with a delay time constant of a few tens of picoseconds, which corresponds very well to the S_1 to T_1 intersystem crossing time constant. A similar 2-D CARS experiment on 9-cis retinal has proved that the 9-cis to all-trans isomerization takes place on the T_1 potential surface with a time constant of 900 ps [18]. Picosecond time-frequency 2-D multiplex CARS spectroscopy will find many more useful applications in the field of photoinduced dynamics of excited molecules.

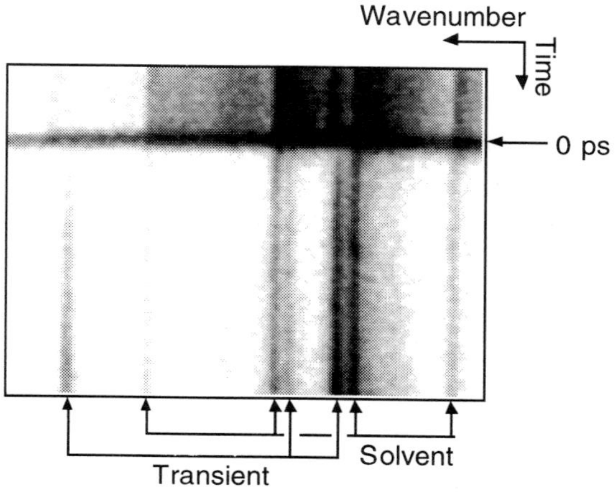

Figure 8. 22 2-D CARS image after the photoexcitation of all-trans retinal.

8.5 Future Perspective

Though the history of nonlinear Raman spectroscopy extends for more than three decades, its primary contribution to molecular spectroscopy has so far been limited to high resolution studies of gases. Extremely high spectral resolution, which is never attainable with conventional spontaneous Raman spectroscopy, has given nonlinear Raman spectroscopy a unique status.

Other aspects of molecular spectroscopy in which nonlinear Raman techniques may play significant roles are the time-resolved and/or polarization sensitive measurements of condensed phase materials including fluorescent molecules. These measurements are extremely difficult, if not impossible, to perform with the scanning scheme, because they require a number of similar experiments with different time delays and/or polarizations to be repeated in a reasonable span of time. The recent development of highly efficient multiplex techniques has opened a way to many exciting, new, and practical applications of time-resolved and/or polarization sensitive nonlinear Raman spectroscopy.

High resolution measurements of gases with the scanning scheme and **time-resolved and/or polarization sensitive measurements** of condensed-phase materials using the multiplex scheme will act as the two driving wheels of nonlinear Raman spectroscopy in the near future.

References

1. D. Lee and A. C. Albrecht, In *Advances in Infrared and Raman Spectroscopy* (R. J. H. Clark and R. E. Hester, eds.), Vol. 12, p. 179, Wiley-Heiden, Chichester (1985).
2. M. D. Levenson and S. S. Kano, *Introduction to Nonlinear Spectroscopy*, Revised Version, Academic Press, Orlando (1988).
3. Y. R. Shen, *The Principle of Non-linear Optics*, John Wiley & Sons, New York (1984).
4. *Advances in Non-linear Spectroscopy*, (R. J. H. Clark and R. E. Hester, eds.) John Wiley & Sons, Chichester (1988).
5. R. Loudon, *The Quantum Theory of Light,* Oxford University Press, Oxford (1973).
6. R. T. Lynch, Jr. and H. Lotem, *J. Chem. Phys.*, **66**, 1905 (1977).
7. A. C. Eckbreth and P. W. Schreiber, in *Chemical Applications of Nonlinear Raman Spectroscopy* (A. B. Harvey, ed.), Chapter 2, p. 27, Academic Press, New York (1981).
8. P. Esherick and A. Owyoung, in *Advances in Infrared and Raman Spectroscopy*, (R. J. H. Clark and R. E. Hester, eds.), Vol. 9, p. 130, Heiden, London (1982).
9. J. Baran, D. Elliot, A. Grofcsik, W. J. Jones, M. Kubinyi, A. J. Langley, and V. U. Nayar, *J. Chem. Soc., Faraday Trans. 2*, **79**, 865 (1983).
10. M. Yoshizawa, Y. Hattori, and T. Kobayashi, *Phys. Rev. B*, **49** 13259 (1994).
11. D. Heiman, R. W. Hellwarth, M. D. Levenson, and G. Martin, *Phys. Rev. Lett.*, **36**, 187 (1976).
12. G. A. West, D. R. Siebert, and J. J. Barrett, *J. Appl. Phys.*, **51**, 2823 (1980).
13. P. Esherick and A. Owyoung, *Chem. Phys. Lett.*, **103**, 235 (1983).
14. B. N. Toleutaev, T. Tahara, and H. Hamaguchi, *Appl. Phys. B*, **59**, 369 (1994).
15. S. A. Akhmanov and N. I. Koroteev, *Methods of Nonlinear Optics in Light Scattering Spectroscopy*, Nauka, Moscow (1981).
16. Y. Prior, A. R. Bogdan, M. Dagenais, and N. Bloembergen, *Phys. Rev. Lett,* **46**, 111 (1981).
17. J. W. Nibler and G. A. Pubanz, in Ref. 4, Chapter 1, p. 1.
18. T. Tahara, B. N. Toleutaev, and H. Hamaguchi, *J. Chem. Phys.*, **100**, 786 (1994); T. Tahara and H. Hamaguchi, *Rev. Sci. Instru.* **65**, 3332 (1994).

9 Quantum Beat Spectroscopy

H. Bitto and J. Robert Huber

Physikalisch-Chemisches Institut der Universität Zürich, Winterthurerstrasse 190, CH-8057 Zürich, Switzerland

9.1 Introduction

The enormous success of nuclear magnetic resonance spectroscopy in the last decade has been the story of the pulsed Fourier transform technique or, in other words, that of time-resolved magnetic spectroscopy [1]. The resonances in a complex NMR spectrum cover a broad frequency range which in conventional spectroscopy is slowly scanned by a monochromatic radio-frequency source to locate the nuclear spin transitions. In contrast, the pulse technique uses a short radio-frequency pulse, the time-energy uncertainty of which is sufficiently broad as to cover all the NMR resonances of the system. The subsequent time evolution of the excited nuclear spin states, more specifically, the free precession of all the coherences, is detected and provides the free induction decay (FID). A Fourier transformation brings the time-resolved signal into the frequency domain and reveals - with a high signal/noise ratio - the contributions of the individual resonances. Time domain spectroscopy in the optical regime follows a similar procedure. Here a short light pulse generates within its energy uncertainty a superposition of coherently excited molecular states. Their time evolution, more directly, their relaxation to a common ground state is superposed by interferences and monitored to yield the 'time domain spectrum' similar to the FID in NMR spectroscopy. A Fourier transformation converts this signal finally to the frequency domain and thus recovers the excited state spectrum.

Before going into the details of time-resolved optical spectroscopy some introductory remarks are in order. The frequency domain spectra of atoms and molecules are usually interpreted in terms of transitions between stationary states or eigenstates $(|n\rangle)$. Time domain spectroscopy, however, deals with non-stationary states which are described as linear combinations or coherent superpositions of eigenstates as given by

$$|\psi(t)\rangle = \sum_{n} a_n e^{-iE_n t/\hbar} |n\rangle \ . \tag{9.1}$$

Hence, following excitation of a non-stationary state $|\psi(t)\rangle$ the contribution of each eigenstate, which is subject to a periodic phase change, introduces oscillations into the time evolution which are manifested as an intensity modulation in the time-resolved emission decay. This phenomenon is called quantum beats. Since the modulation pattern is determined by the energies E_n of the superposed eigenstates, or more specifically by their energy differences, the oscillatory fluorescence decay can be utilized for a form of time domain spectroscopy called quantum beat spectroscopy.

Though the basic concept is simple and coherent superposition is at the heart of quantum mechanics, it was not until the early sixties that the theory and experimental techniques were established for time domain spectroscopy in the optical region. In the first experimental demonstration Alexandrov [2] and, independently, Dodd, Kaul and Warrington [3] used short light pulses generated by shuttered spectral lamps to prepare atoms in coherent superposition states showing quantum beats. Since the quantum beat frequencies are orders of magnitude smaller than the optical transition frequencies, quantum beats are essentially unaffected by the Doppler effect. It is this very property which makes quantum beats an excellent tool for high-resolution spectroscopy.

The aim of early quantum beat experiments was mainly to demonstrate the phenomenon. Only the advent of tunable dye lasers, making available intense and narrow-band light pulses, intensified the interest in quantum beats and many studies on atoms and diatomics have since been published. For a review of these experiments the reader is referred to the articles by Alexandrov [4], Dodd and Series [5], and Haroche [6]. With the introduction of supersonic beam techniques in molecular spectroscopy, spectral congestion in polyatomic molecules could be reduced to such a degree that McDonald and coworkers were able to detect the first quantum beats in a polyatomic molecule in 1978 [7]. Soon after, several research groups applied this method mainly to explore intra-molecular state couplings and the dynamics of energy redistribution in molecules [8-12]. In this context we refer in particular to the work of Zewail and coworkers [9], who used picosecond laser pulses to produce coherences between rotational or vibrational states, and to the investigations of Lim *et al.* [10] and Kommandeur *et al.* [13] on heterocyclic molecules. The work in Zürich was started about 15 years ago [14,15]. Inspired by the quantum beat experiments on atoms mentioned above and on NO_2 by Brucat and Zare [16], we began to develop the quantum beat technique for high resolution spectroscopy of polyatomic molecules with the goal to study not only dynamics but also molecular structure [17-27]. By high resolution we refer to the frequency domain; this should not be confused with high temporal resolution obtained with pico- or femtosecond pulses. Under the latter conditions the beat frequencies exceed the Doppler width so that the corresponding energy differences can usually be measured by conventional spectroscopy.

9.2 Theoretical aspects

9.2.1 Two level quantum beats

We start this introduction with the simplest case of a quantum beat and consider the four-level system depicted in Fig. 9.1. The molecular levels $|a\rangle$ and $|b\rangle$ are taken here as electronically excited states, $|g\rangle$ is the initial state from which excitation occurs, and $|f\rangle$ is the final state to which the molecule relaxes on emission of a photon. The states are assumed to be eigenstates of the molecular Hamiltonian including a possible external magnetic or electric field. States $|a\rangle$ and $|b\rangle$ are split by an energy E_{ab} which is orders of magnitude smaller than their energy spacings to the other states.

If the system is exposed to a short resonant laser pulse, the Fourier bandwidth of which exceeds the energy spacing E_{ab}, the molecules are prepared in a superposition state

$$|\psi(t=0)\rangle = c_a|a\rangle + c_b|b\rangle, \tag{9.2}$$

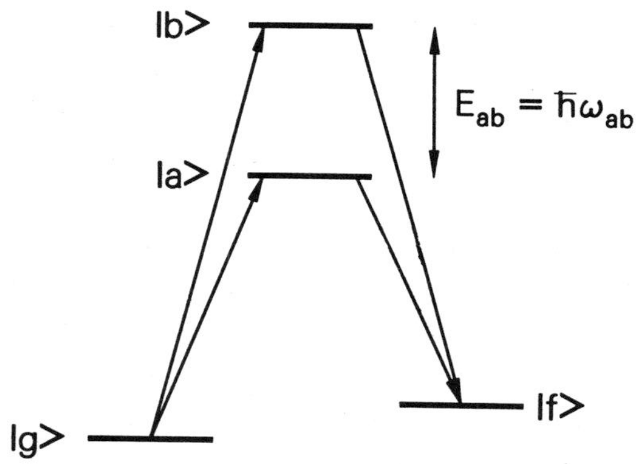

Fig. 9.1 Excitation and detection scheme of two level quantum beat.

where the excited state lifetime and the period of the precession frequency $\hbar/E_{ab}$ are much longer than the laser pulse (delta pulse excitation). While such an excitation creates a fixed phase relation among the excited states - a Hertzian coherence [28] - no net optical coherence $|a\rangle \leftrightarrow |g\rangle$ and $|b\rangle \leftrightarrow |g\rangle$ is present in the molecular ensemble after the excitation process. The destruction of the latter coherence, often referred to as optical dephasing, is caused by the motion of the molecules in the gas phase or by destructive interference of the light waves if the excitation is broadband [6,29,30].

The time evolution of the superposition state Eq. (9.2) can be described in the eigenstate basis by

$$|\psi(t)\rangle = c_a e^{-i\omega_a t}|a\rangle + c_b e^{-i\omega_b t}|b\rangle, \tag{9.3}$$

where $\omega_a = E_a/\hbar$ and $\omega_b = E_b/\hbar$ represent the respective eigenstate energies. Assuming a constant laser pulse intensity over the energy range E_{ab}, the coefficients c_a and c_b are proportional to the transition dipole matrix elements $\langle a|\mu|g\rangle = \mu_{ag}$ and $\langle b|\mu|g\rangle = \mu_{bg}$, respectively. After introduction of the phenomenological decay constants γ_a for $|a\rangle$ and γ_b for $|b\rangle$, we obtain

$$|\psi(t)\rangle \sim \mu_{ag} e^{-(i\omega_a + \gamma_a/2)t}|a\rangle + \mu_{bg} e^{-(i\omega_b + \gamma_b/2)t}|b\rangle. \tag{9.4}$$

The time evolution of this superposition state is monitored most conveniently by the fluorescence to a ground state $|f\rangle$ whose time-resolved intensity $I_{fl}(t)$ is given by the expression

$$I_{fl}(t) \sim |\langle f|\mu|\psi(t)\rangle|^2. \tag{9.5}$$

After inclusion of Eq. 9.4 and taking the modulus, the resulting terms may be arranged as follows

$$I_{fl}(t) \sim |\mu_{ag}|^2 |\mu_{fa}|^2 e^{-\gamma_a t} + |\mu_{bg}|^2 |\mu_{fb}|^2 e^{-\gamma_b t} + 2|\mu_{ag}\mu_{bg}\mu_{fa}\mu_{fb}| e^{-(\gamma_a+\gamma_b)t/2} \cos(\omega_{ab}t + \theta), \tag{9.6}$$

where we have introduced the phase angle θ to maintain the generality of the expression (*vide infra*). The first two terms describe the independent (incoherent) decays of the eigenstates $|a\rangle$ and $|b\rangle$ while the last term - the cross or coherence term - describes the oscillatory decay referred to as the *quantum beat* since it is a manifestation of the quantum mechanical interference.

Equation (9.6) reveals the physics of the decay process and shows that time-resolved measurements of the emission permit us to determine small energy differences from the beat frequency $\omega_{ab} = E_{ab}/\hbar$. In Fig. 9.2 an example is depicted which was observed in the fluorescence of the molecule CS_2 [31]. The real part of its Fourier transform plotted on the right side shows a Lorentzian line at a frequency of $\omega_{ab}/2\pi = 1.37$ MHz beside the dominating Lorentzian line at zero frequency due to the exponential decays. The width of the lines is given by the natural linewidth which is 66 kHz for the 2.4 μs lifetime. This represents indeed excellent resolution for optical spectroscopy. The high resolution is due to the fact that the beat frequencies lie in the radio-frequency range and are subject to a Doppler broadening negligible compared to the natural linewidth.

At this stage, it is instructive to address the fundamental quantum mechanical nature of the quantum beat phenomenon. In the language of quantum mechanics, the situation of exciting a superposition state $|\psi\rangle$ is equivalent to that of two states, $|a\rangle$ and $|b\rangle$, sharing a photon. If the path of the photon is known, i.e. from which of the two states it is emitted, the interference pattern is lost and if the interference is observed, the information about the path is lost. In other words, the operators associated with the path information and the interference phenomenon do not commute [32]. Quantum beats are the effect of a *single* particle (atom or molecule) and a *single* photon scattered by two indistinguishable channels. In order to observe quantum beats, an experimentalist must ensure that both channels remain indistinguishable during the measurement

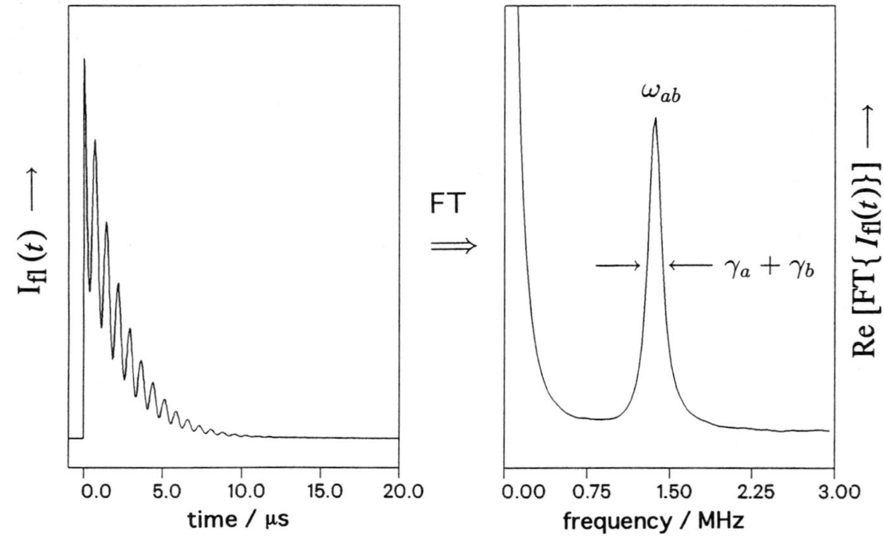

Fig. 9.2 Fluorescence decay and real part of its Fourier transform showing a two level quantum beat.

process. In an actual experiment an ensemble of particles is excited and the signals of the independently emitted photons are accumulated with respect to the time after the laser pulse in order to reveal the measured oscillatory decay signal. Importantly, each photon carries the information about both the emission channels. In Eq. (9.6) the extent of interference is reflected in the magnitude of the dipole transition moments involved. If the two moments of absorption are equal, $\mu_{ag} = \mu_{bg}$, and also those of emission, $\mu_{fa} = \mu_{fb}$, the beat amplitude is maximum. Whereas if one of the moments is zero - which would make the channels distinguishable - the coherent term vanishes. Non-equal moments lead to an intermediate beat amplitude.

9.2.2 Quantum beats in multilevel systems

In extending the above treatment to a realistic but still isolated molecule, we consider the level scheme of Fig. 9.3. As long as the coherences among the set of ground states $\{|f\rangle\}$ are absent - an assumption which holds for all the measurements we shall discuss (see below) - the intensity of the spectrally integrated, total fluorescence is the sum of the signals of the individual emission channels given by Eq. (9.6). Consequently only one beat frequency is observed. Nonetheless, the ground states are relevant for the quantum beat sig nal in so much as different emission channels may have differently polarized dipole transition moments and therefore different phases θ. Hence, the beat amplitude of the resulting total fluorescence may be reduced compared to that of the individual signals.

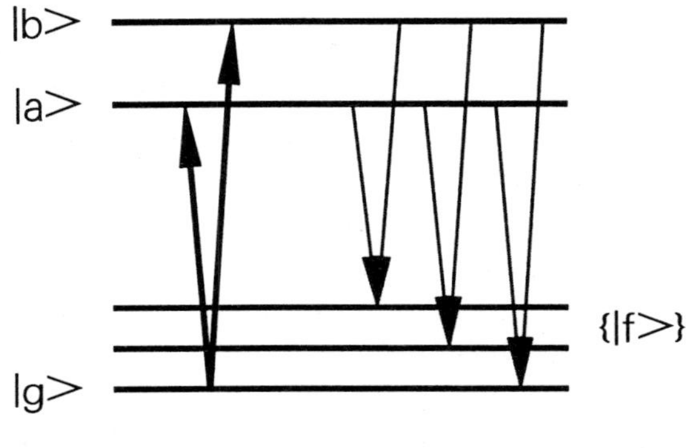

Fig. 9.3 Quantum beats in many level systems in molecules

It should be noted that excitation with strong laser pulses, in particular with femtosecond pulses, may give rise to stimulated emission and thereby produce also coherence among ground state levels within the coherence width of the excitation pulse, *i.e.* its Fourier bandwidth [33]. Since under nanosecond excitation conditions the coherence width is considerably smaller than the energy interval spanned by the states $\{|\mathrm{f}\rangle\}$ to which emission occurs, ground state coherence can safely be neglected.

Next we generalize the expression for the time-resolved fluorescence intensity Eq. (9.5) to the case of many-level systems in the ground as well as the excited state. In addition the polarization of the excitation ε^{L} and that of the detected fluorescence ε^{D} will be taken into account. As with the expectation value of any observable, the time-resolved fluorescence can be expressed as the trace of the product of the excited state density matrix ρ and of the operator of the observable

$$I_{\mathrm{fl}}(t) \sim \mathrm{Tr}\big(\rho(t) L_{\mathrm{F}}\big). \tag{9.7}$$

In this case the operator is the "fluorescence monitoring operator" given by

$$L_{\mathrm{F}} = \sum_{\mathrm{f}} \varepsilon^{\mathrm{D}*}\mu|\mathrm{f}\rangle\rho_{\mathrm{ff}}\langle\mathrm{f}|\mu\varepsilon^{\mathrm{D}}, \tag{9.8}$$

where ρ_{ff} absorbs the spectral response function of the detection system and the frequency dependence of the emission rate. In analogy, the excited state density matrix after the laser pulse reflects the excitation process

$$\rho(t=0) = \sum_{\mathrm{g}} \varepsilon^{\mathrm{L}*}\mu|\mathrm{g}\rangle\rho_{\mathrm{gg}}\langle\mathrm{g}|\mu\varepsilon^{\mathrm{L}}. \tag{9.9}$$

The summation over the diagonal ground state density matrix accounts for the incoherent superpositions of the respective quantum beat signals. In the most general case, the time evolution of the density matrix ρ is obtained by solving the Liouville or the master equation. Allowing now for a set of coherently excited eigenstates $\{|\mathrm{n}\rangle\}$, the free evolution of the density matrix in the eigenstate basis $\rho_{\mathrm{nn'}}$ is simply

$$\rho(t)_{\mathrm{nn'}} = \rho(0)_{\mathrm{nn'}} \exp\big(-((\gamma_{\mathrm{n}} + \gamma_{\mathrm{n'}})/2 + i(\omega_{\mathrm{n}} - \omega_{\mathrm{n'}}))t\big). \tag{9.10}$$

Using Eq. (9.7) - Eq. (9.10) the fluorescence signal of an ensemble of excited molecules is expressed by

$$I_{\mathrm{fl}}(t) \sim \sum_{\mathrm{g}}\rho_{\mathrm{gg}}\sum_{\mathrm{f}}\rho_{\mathrm{ff}}\sum_{\mathrm{nn'}}\langle\mathrm{n}|\varepsilon^{\mathrm{L}*}\mu|\mathrm{g}\rangle\langle\mathrm{g}|\mu\varepsilon^{\mathrm{L}}|\mathrm{n'}\rangle\langle\mathrm{n'}|\varepsilon^{\mathrm{D}*}\mu|\mathrm{f}\rangle\langle\mathrm{f}|\mu\varepsilon^{\mathrm{D}}|\mathrm{n}\rangle \\ \exp\big(-((\gamma_{\mathrm{n}} + \gamma_{\mathrm{n'}})/2 + i(\omega_{\mathrm{n}} - \omega_{\mathrm{n'}}))t\big). \tag{9.11}$$

If $\{|\mathrm{n}\rangle\}$ consists of N eigenstates and if they are coherently excited from one ground state $|\mathrm{g}\rangle$, a fluorescence decay is observed which exhibits $N(N-1)/2$ superimposed quantum beats. In many cases this general expression can be simplified. For example if the eigenstates are angular momentum states that can be described in a limiting coupling case, closed form expressions can be derived for the sum of the products of transition moments using angular momentum theory [34,35]. In particular, dealing with polarized excitation and detection, the dependence of the signal on the molecular dynamics and on the excitation/detection geometry can be separated and the powerful formalism of the density matrix theory applied [36,37].

Although a quantum electrodynamic treatment is required for a full understanding [6,29,30], this simple description clearly illustrates the strength of the quantum beat phenomenon when applied to molecular spectroscopy. In the diluted gas phase, this time-resolved technique provides (i) Doppler-free spectroscopy, since it rests on a single particle effect and only very small energy differences are measured, (ii) high energy resolution that is limited only by the lifetime of the coherently excited states, and (iii) a method that is essentially insensitive to saturation effects [38]. In summary quantum beat spectroscopy is a method by which energy splittings of coherently excited levels can be measured with high accuracy; the method is fundamental, simple, direct, and sensitive.

9.3 The quantum beat experiment

Quantum beats of molecules are measured under collisionless conditions in a cold molecular beam using pulsed laser excitation and time-resolved detection. Figure 9.4 shows the "classical" orthogonal experimental arrangement, in which the molecular beam, the laser beam and the detector axis are mutually perpendicular. The individual components will be described below. Further technical details of our apparatus, which has been designed for quantum beat spectroscopy on the nanosecond and microsecond time scales, are found in a number of articles [14,20,22].

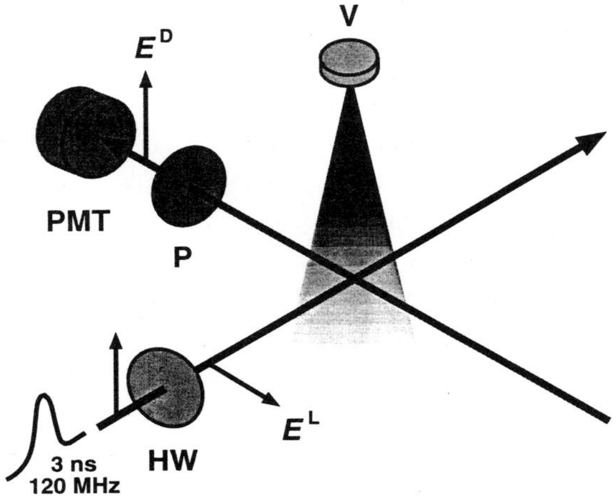

Fig. 9.4 Experimental set-up (valve V, photomultiplier tube PMT, polarizer P, half-wave retarder HW).

9.3.1 Preparation of cold molecules

Quantum beats of atoms and of molecules are based on the same physical principles but the experimental conditions to detect this phenomenon in polyatomic molecules are more demanding due simply to the increased number of degrees of freedom. The rotational and vibrational degrees of freedom dramatically increase the density of states and in turn the number of transitions. As a consequence of the optical selection rules, many of these states cannot be excited coherently so that a mixture of incoherent and coherent states is initially prepared. Since incoherently excited states contribute only to the non-modulated fluorescence, the relative modulation depth of the quantum beats is reduced, often to such a degree that quantum beat measurements are not feasible. In particular, broad-band excitation provided by electron bombardment or sudden impact as used in atomic quantum beat spectroscopy or in beam foil spectroscopy [39] can therefore not be employed for molecules.

The selectivity of the excitation process and hence the relative intensity of the beat amplitude is, however, greatly increased by cooling the molecules to low temperatures and by using narrow-band laser pulses. By seeding the molecules in a noble gas they can be effectively cooled to a rotational temperature of a few K in a supersonic jet expansion. In combination with pulsed laser excitation, a pulsed valve is most appropriate for this purpose. It produces short gas pulses (~ 200 μs) of high particle density (10^{15} per cm^3) and, operating at moderate repetition frequencies (*e.g.* 20 Hz), requires only small pumps to maintain vacuum ($\leq 10^{-4}$ mbar). Moreover, a pulsed jet offers the benefit of small sample consumption which is particularly important when expensive isotopic species are investigated. To avoid molecular collisions during the time evolution of the superposition state, the laser crosses the molecular beam about 100 to 200 nozzle diameters (Ø = 0.3 mm) below the orifice, where the mean distance between particles has become larger than 1000 Å.

The beam source can easily be modified such that transient molecules, in particular radical species, can be generated [40]. A KelF fixture can be attached to the face plate of the valve. Within the fixture two 1 mm diameter copper rod electrodes are positioned centrally and perpendicularly to the 20 mm long channel of 1 mm diameter. Discharge is initiated by a $\approx 10\mu s$, ≤ 1000 V electrical pulse with variable current limitation. The discharge pulse is delayed to coincide with the arrival at the electrodes of the leading edge of the gas pulse which contains the precursor in 1 % concentration.

9.3.2 Excitation and detection

Commercially available dye lasers, which deliver tunable radiation of about 1 GHz bandwidth in pulses of ~ 5 ns, are suitable for selective excitation of single rotational states in small polyatomic molecules. In principle their bandwidth is adequate for quantum beat experiments, although it does not provide the ultimately achievable resolution of 88 MHz (FWHM) given by a 5 ns (FWHM) Fourier transform-limited (FTL) pulse with a Gaussian profile. Such pulses are required for larger molecules to ensure clean coherent excitation and were used in most of the experiments presented below. They were generated by amplification of the injected mode of a ring dye laser in a pulsed dye amplifier, a schematic of which is shown in Fig. 9.5. The amplification takes place in three dye cuvettes transversally pumped by a XeCl excimer laser. The gain in the last cuvette is most efficiently used in a double pass arrangement. Feedback of pulses scattered back into the ring dye laser is suppressed with an optical diode. In most experiments, UV radiation was required and was generated by frequency doubling of the amplified pulses in a KDP crystal.

Under typical operating conditions the injected power was up to 500 mW, the UV pump laser provided pulse energies of 150 mJ, and the amplified pulses had energies of 5 mJ to 10 mJ, depending on the dye and the wavelength. Frequency doubling yielded UV pulses with an energy of 500 μJ. The bandwidth of the fundamental was measured to be 85 ± 10 MHz. Assuming a Gaussian frequency profile, the frequency doubled pulses were estimated to have a 120 MHz bandwidth in agreement with the measured pulse duration of 3.5 ± 0.5 ns.

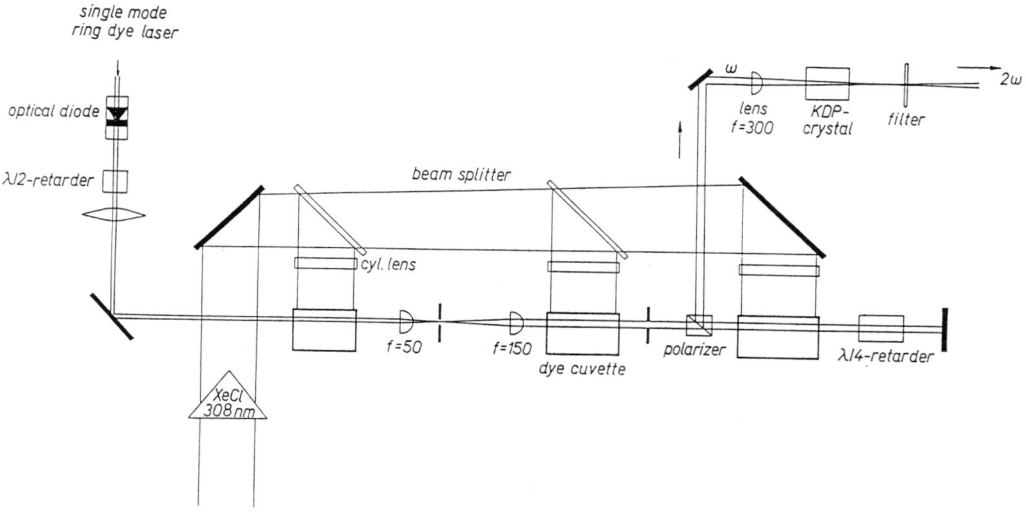

Fig. 9.5 Schematic of pulsed dye amplifier

The emission of the excited molecules was collected with a $f/1$ collimator and detected with a photomultiplier tube connected to a transient digitizer or a digital storage oscilloscope. The molecular beam was imaged onto a slit aperture oriented parallel to the molecular beam, which restricts the observation to excited molecules travelling on the beam axis. Assuming straight trajectories of the molecules starting at the orifice of the nozzle, their velocity component perpendicular to the beam axis, *i.e.* parallel to the laser beam, depends on the distance from the axis. Doppler broadening is, therefore, reduced according to the width of the slit without the necessity to skim the molecular beam and to set up expensive, differential pumping. Under typical operating conditions (slit width 2 mm, magnification of image 1.3), Doppler-limited linewidths of 200 to 250 MHz were achieved in the excitation spectra.

Some experiments require specific excitation and detection polarizations ($\boldsymbol{E}^{\mathrm{L}}$, $\boldsymbol{E}^{\mathrm{D}}$, in Fig. 9.4) which are selected with a half wave retarder (HW) and a polarizer (P), respectively. In order to avoid undesired Zeeman splitting, the experiment must be carried out in zero magnetic field by compensating the Earth's field with three pairs of Helmholtz coils. The coils are also used to generate magnetic fields of up to 1.5 mT which, in most experiments, were oriented coaxial to the molecular beam axis [20]. Radio-frequency magnetic fields for the double resonance experiments described in section 9.4.3 were generated by a pair of Helmholtz coils (Ø = 60 mm) mounted in front of the detection optics in the vacuum chamber [31,41]. A plate capacitor is introduced in place of the coils when Stark quantum beat experiments are performed. The capacitor was oriented such that the plates were parallel to the plane defined by the laser and the molecular beam axes. The fluorescence was observed through a transparent electrode made out of an indium coated quartz plate. Electric field strengths of up to 2500 V/cm could be generated for an electrode spacing of 20 mm [24,42].

9.3.3 Data acquisition and analysis

A computer controls the laser scanning, data acquisition and signal averaging. In taking spectra the laser wavenumber is tuned in discrete steps and at each step a fluorescence decay is measured and stored. As illustrated in Fig. 9.6 the measured spectra are, therefore, three-dimensional. They represent the fluorescence intensity as a function of the laser wavenumber and of the time after the laser pulse. For the measurement of fluorescence decays with improved sampling and signal/noise ratio, the laser is tuned directly to the resonances and the signal of the appropriate number of laser pulses is averaged. Such decays are usually analyzed by Fourier transformation using a fast Fourier transform routine. The Fourier transform of a real function, such as the fluorescence decay, is in general a complex function which can be displayed in different ways. We use the real part of the quantum beat signal since it provides narrower lines than power or amplitude spectra. Typically 2048 data points per fluorescence decay were sampled at sampling rates of up to 400 MHz. Thus frequencies up to a Nyquist frequency of 200 MHz can be measured reliably. The sampling interval in the frequency domain is determined by the total observation time T as $\Delta f = 1/T \approx 200$ kHz and limits the resolution in the frequency domain spectra. By reducing the sampling rate and concomitantly the Nyquist frequency, the observation time can be extended and thereby the resolution improved.

The resolution is limited by the natural linewidth of the excited state, but from an experimental point of view, it is limited by the actual observation time of the fluorescence. In fact the resolution can even be pushed beyond the limit of the natural linewidth (see Fig. 9.7) when the emission of the excited molecules is biased in favor of the long-lived species in the data analysis [5,44-46]. This combined deconvolution/convolution procedure removes the exponential damping and gives the fluorescence a Gaussian shape by multiplying the experimental fluorescence decay with an offset Gaussian function $g(t) = \exp\left(-((t-t_0)/a)^2\right)$. The resulting lines in the real part of the Fourier transform show a Gaussian line shape of a width reduced by a factor r if the parameters $t_0 = 2\ln(2)r^2\tau$ and $a = 2\sqrt{\ln 2}\, r\tau$ are chosen, where $\tau = \gamma^{-1}$ is the fluorescence lifetime. The gain in resolution is in accordance with Heisenberg's uncertainty principle because the natural linewidth is valid only for a non-selected ensemble of excited molecules decaying by spontaneous emission. However, the improved resolution is gained only at the expense of the signal/noise ratio.

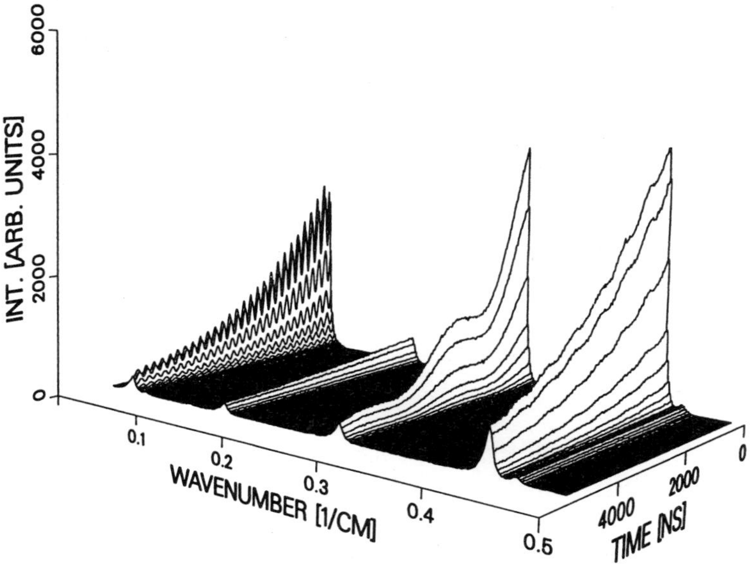

Fig. 9.6 Time-resolved fluorescence excitation spectrum. The spectrum is a portion of the 10V band of $^{13}CS_2$ [43].

9.3.4 Pump-probe quantum beat experiment

Time-resolved pump-probe schemes are widely employed in experiments which require high temporal, *e.g.* femtosecond, resolution. We have chosen this method in order to extend quantum beat spectroscopy to vibrationally excited states of the electronic ground state, the IR emission of which is difficult to detect with sufficient time resolution and signal/noise ratio. An initial laser pulse, the pump pulse, prepares the molecules in the excited states of interest. With a probe pulse the time evolution of the excited molecules can be investigated, *e.g.* via the induced emission, if the delay of the probe pulse with respect to the pump pulse

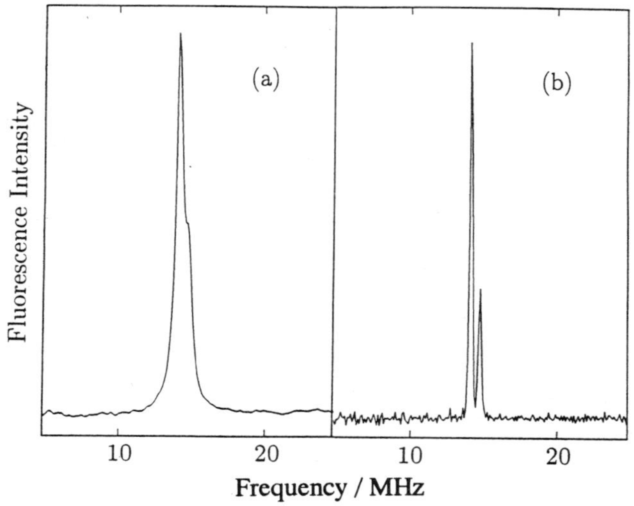

Fig. 9.7 Fourier transforms of an experimental fluorescence decay. Sub-natural linewidth deconvolution of the Fourier transform in (a) resolves two hyperfine components (b) [44].

is scanned. The duration of both laser pulses and their relative delay jitter should be much shorter than the periods of the characteristic oscillations of the time evolution in order to guarantee a high temporal resolution. In our experiment, molecules are coherently pumped with an IR laser pulse and the time evolution of the vibrationally excited superposition state is probed by the absorption of a variably delayed UV pulse, $|g; S_0, v=1\rangle \xrightarrow{IR} |\psi(t); S_0, v=1\rangle \xrightarrow{UV} |f; S_1, v=1\rangle$. The time-integrated fluorescence of the state excited by the UV pulse $|f; S_1, v=1\rangle \xrightarrow{FL} |S_0\rangle$ is observed as a measure of the UV absorption [17,47]. While a commercial dye laser with a 1.2 GHz bandwidth was used for the generation of the UV pulses, a home-built optical parametric oscillator (OPO) pumped by a injection-seeded Nd:YAG laser, delivered IR pulses of 5 ns duration. By inserting an intracavity etalon the bandwidth could be reduced from 0.8 cm^{-1} to 0.03 cm^{-1} and even single mode operation (< 0.007 cm^{-1}) could be achieved for short scans. The experimental set-up was as described in Fig. 9.4, the additional probe laser beam travelling coaxially to the pump beam.

9.4 Examples of molecular quantum beat spectroscopy

9.4.1 Excitation of sublevels of rovibronic states. Zeeman, hyperfine and Stark quantum beats

Zeeman quantum beats arise from a superposition of magnetic sublevels of a single molecular state split by an external magnetic field $\boldsymbol{B}$. These beats constitute the time-resolved analogue of the famous Hanle effect, in which the polarized fluorescence emitted by an atom or molecule is depolarized under the influence of a magnetic field. In fact the understanding of the Hanle effect made a treatment necessary based on the concept of excited state coherence and inspired Aleksandrov and, independently, Dodd, Kaul and Warrington to experimentally demonstrate quantum beats for the first time [2,3].

Figure 9.8 shows a Zeeman quantum beat observed in a rotational level of the triplet state $^3A_2$ in the CS_2 molecule. The rotational state with angular momentum J = 1 is split into 3 equally spaced sublevels due to the linear Zeeman interaction

$$H_{mag} = -g_J \mu_B \boldsymbol{J}\boldsymbol{B}, \tag{9.12}$$

where g_J is the Landé factor and μ_B the Bohr magneton. Taking the direction of $\boldsymbol{B}$ as the quantization axis

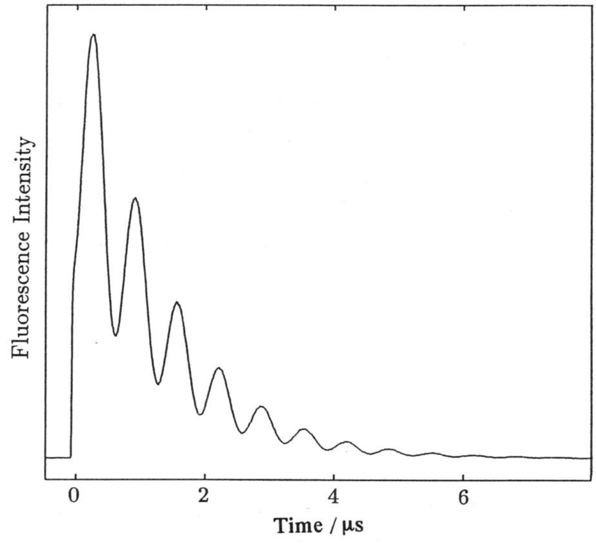

Fig. 9.8 Zeeman quantum beat of the $J=1$ rotational level in the 9U band of CS_2 [48].

and starting from the ground state $|J = 0, M = 0\rangle$, a laser of σ polarization $(\varepsilon^{L} \perp \boldsymbol{B})$ excites a superposition of the two states $|J = 1, M = +1\rangle$ and $|J = 1, M = -1\rangle$ according to the optical selection rules $\Delta M = \pm 1$ and thus creates a $\Delta M = 2$ coherence $|\psi\rangle = \sum_{M=-1}^{M=1} c_M |J = 1, M\rangle$. The emission is modulated at a frequency $\omega = 2\omega_L = 2g_J\mu_B B/\hbar$, which is twice the Larmor frequency. The relatively long emission lifetime of this triplet state provides such a high resolution that only a very small magnetic field is required to observe the effect.

In the general case, a rotational level with total angular momentum J shows $2J + 1$ equally spaced sublevels which give rise to $2J - 1$ coherences when excited by σ polarized light (see Fig. 9.9). As all the coherences oscillate at the same frequency, only one beat frequency is observed and the fluorescence decay can be described by

$$I_{fl}(t) = I_{fl}(0)\exp(-\gamma t)[A_0(\phi) + A_1(\phi)\cos(\omega t)]. \tag{9.13}$$

Here A_0 represents the unmodulated part and A_1 the modulation amplitude both of which depend on the

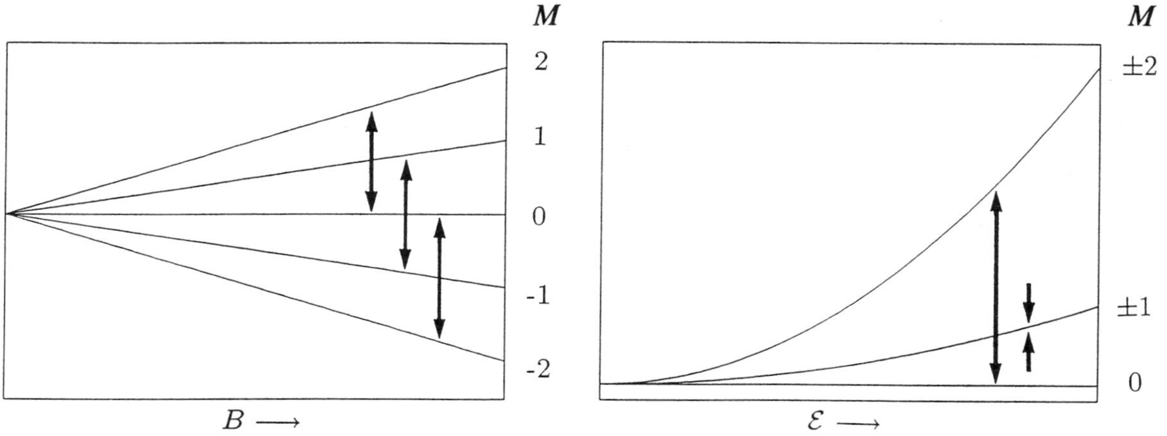

Fig. 9.9 Coherences in field-induced quantum beats of a $J = 2$ state in the presence of (a) linear Zeeman splitting and (b) quadratic Stark splitting. The laser polarization is perpendicular to the field axes.

angle between the magnetic field and the detection polarization. The beat amplitude A_1 assumes its maximum negative value for $\phi = \pi/2$ (σ polarized detection) and vanishes for $\phi = 0$ (π polarized detection). These polarization conditions are realized with our excitation detection geometry shown in Fig. 9.4. In another common experimental arrangement, the magnetic field is directed along the detection axis whereby the linear detection polarization is always perpendicular to the magnetic field irrespective of the direction of the polarization. Under these conditions the signal has an amplitude which is independent of the direction of the detection polarization

$$I_{fl}(t) = I_{fl}(0)\exp(-\gamma t)[A_0 + A_1\cos(\omega t + 2\phi)] \tag{9.14}$$

but the phase of the oscillation is determined by the angle ϕ between the detection polarization and the σ-polarized excitation.

The geometry dependence of the Zeeman beat signals which has been worked out in detail for the example of $J = 1$ can be found in the textbook by Corney [28]. Closed form expressions for the Zeeman beat signals in molecules are found in Ref. [23] and [16]. From pure Zeeman quantum beat measurements the Landé factor of the excited rotational state can be extracted. For instance in the $^3A_2$ state in the CS_2 molecule

Landé g factors were determined which are independent of J. This observation indicates that the magnetic moment of the spin component excited in this triplet state is due to spin-uncoupling [48,49].

A rotational state is also split into sublevels if the molecule contains nuclei with non-zero magnetic moments. The interaction of the electron and nuclear spins lifts the degeneracy of the nuclear states which creates the hyperfine structure of the rovibronic state. As hyperfine interaction is usually the weakest interaction in the molecule, the nuclear spin I is coupled last to the angular momenta resulting in a total angular momentum $\boldsymbol{F} = \boldsymbol{J} + \boldsymbol{I}$. The excitation with polarized light produces an anisotropic distribution of molecules in the excited state. Because the total angular momentum $\boldsymbol{F}$ is a constant of motion, $\boldsymbol{J}$ and $\boldsymbol{I}$ precess around $\boldsymbol{F}$, causing a decrease of the rotational alignment and a periodic transfer of molecular polarization to nuclear spin polarization. This change in alignment is observed as a modulation in the anisotropy of the emitted fluorescence which oscillates at frequencies determined by the hyperfine (hf) splittings. In other words, following coherent excitation of the hf levels $\{|J, F\rangle\}$ of the molecular state $|J\rangle$, hf quantum beats are observed [34]. The time-resolved emission intensity $I_{\text{fl}}(t)$ can be expressed in terms of a geometrical factor which depends on the angle θ between the laser ε^{L} and the detection ε^{D} polarization, and dynamical factors which describe the molecular angular momentum coupling as well as absorption and emission [21]:

$$I_{\text{fl}}(t) \sim \exp(-\gamma t)\left[1 + A_0 P_2(\cos\theta) + \sum_j A_j \cos(\omega_j t) P_2(\cos\theta)\right]. \tag{9.15}$$

The decay constant γ is assumed to be the same for all the hf components of a particular state, $P_2(\cos\theta)$ denotes the second order Legendre polynomial and the sum runs over all possible quantum beats j. Since the optical selection rule $\Delta F = 0, \pm 1$ applies to excitation and detection, coherent superpositions can only be created and observed between hf levels which differ by $\Delta F = 0, \pm 1, \pm 2$. The amplitudes A_j are largest if the molecules are excited via R branch transitions and decrease with increasing J. Therefore, exciting $J = 1$ levels via the R(0) transition as shown in Fig. 9.2 and 9.10 provides the deepest modulation. In these examples we excited $J = 1$ rovibronic levels of the $^3A_2$ state in $^{13}CS_2$. As ^{13}C has a spin $I = 1/2$ the rotational level is split into two hf components $|J = 1, F = 1/2\rangle$ and $|J = 1, F = 3/2\rangle$ and a single beat frequency is observed.

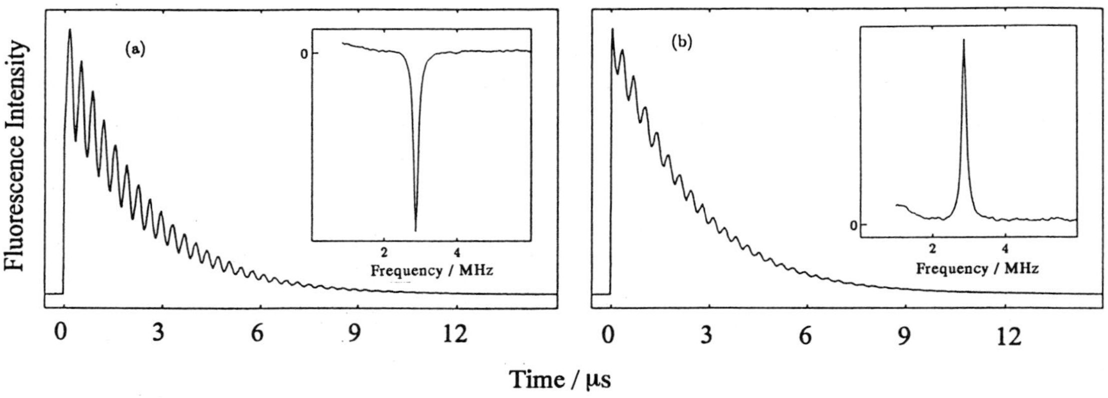

Fig. 9.10 Hyperfine quantum beats observed with unpolarized detection in the (0,4,2) band of the B_2 state of $^{13}CS_2$ [48]. Since in (a) the excitation polarization ε^{L} is parallel to the detection axis both polarization components of the emission are perpendicular to ε^{L} and consequently both components contribute a $\theta = 90°$ signal. In contrast, for an excitation polarized perpendicular to the detection axis as in (b), one emission component is polarized parallel whereas the other one is polarized perpendicular. The corresponding beat signals, having opposite phase, partially cancel so that only a quarter of the maximum amplitude remains.

According to Eq. (9.15) the fluorescence consists of three parts, an unmodulated isotropic part, an unmodulated anisotropic part and a modulated anisotropic part. Due to the $P_2(\cos\theta)$ dependence, the modulation is deepest if the excitation and detection polarizations are parallel, vanishes at the magic angle $\theta = 54.7°$, and reaches half of the maximum amplitude for perpendicular polarizations but with negative phase. The quantum beats can also be observed by unpolarized detection though with reduced amplitude as illustrated in Fig. 9.10.

Quantum beats induced by an external electrical field are in many respects related to Zeeman quantum beats. The interaction of the electrical field $\boldsymbol{E}$ with the permanent electric dipole moment μ of the molecule splits the Zeeman components of the rotational states $|J, M\rangle$ and, after coherent excitation, quantum beats can be measured if the excitation and detection polarizations are perpendicular to the electric field. Although Stark quantum beats can be observed in molecules excited to states of any multiplicity, we restrict the following discussion to pure singlet states. The Stark effect of a non-degenerate angular momentum state is of second order and quadratic in the field strength $\boldsymbol{E}$ as opposed to the linear, first-order Zeeman effect. The energy shift E_{el} of a Zeeman level $|J, M\rangle$ due to the interaction

$$H_{\text{el}} = -\mu \boldsymbol{E} \tag{9.16}$$

can generally be expressed by [50]

$$E_{\text{el}}(J, M) = \sum_i \left(A_i(J) + B_i M^2\right)\mu_i^2 \boldsymbol{E}^2 \tag{9.17}$$

where μ_i is the projection of the dipole moment vector μ onto the i'th principal axis of the molecule. The factors A_i and B_i, expressed as perturbation series, contain the energetic separations of states coupled by μ_i and hence depend on the molecular structure parameters such as the rotational constants. Owing to the fact that states with equal $|M|$ remain degenerate and tune quadratically with the field strength $\boldsymbol{E}$, we observe $J-1$ different $\Delta M = 2$ Stark quantum beat frequencies $\left(\varepsilon^{\text{L}} \perp \boldsymbol{E}\right)$ given by

$$\omega_{M+2,\,M} = \sum_i B_i 4(M+1)\mu_i^2 \boldsymbol{E}^2 \qquad 0 \le M \le J-2. \tag{9.18}$$

As an example we consider excitation of the rotational state $J = 2$ illustrated in Fig. 9.9. The excitation gives rise to a beat frequency ω_{2-0} and ω_{0--2}, which, due to the $|M|$ degeneracy, have the same absolute value. The third frequency ω_{1-1} is zero and, therefore, this coherence contributes to the unmodulated anisotropic

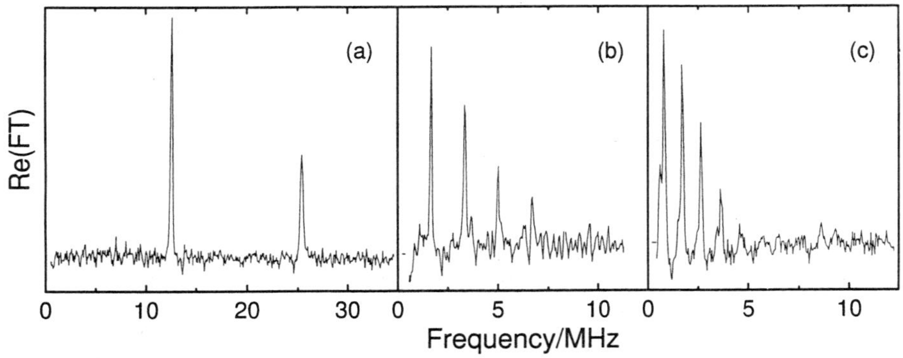

Fig. 9.11 Fourier spectra of Stark quantum beats showing regularly spaced multiplets of ΔM=2 coherences. The measurements were performed exciting a J = 3 (a), J = 5 (b), and J = 6 (c) rotational state of the 12^1_0 band in the S_1 state of HC≡CCDO [42].

emission. Thus for $J > 2$ a regularly spaced $(J-1)$-multiplet of lines appears in the Fourier transform spectrum as displayed in Fig. 9.11. It should be noted that the high resolution of quantum beat spectroscopy permits the measurement of Stark splittings at relatively low electric field strengths. Given the experimental arrangement described in 9.3.2, the fluorescence signal is of the form of Eq. (9.13) where ϕ represents here the angle between the detection polarization and the σ-polarized excitation [42].

Stark quantum beat spectroscopy, introduced to measure atomic polarizabilities [51], has been applied to determine the permanent electric dipole moment of molecules in different vibrational states. Diatomic molecules, having only one dipole component μ and symmetric top molecules in $^1\Sigma$ states lend themselves for a direct analytic evaluation [50,52]. Based on a perturbative treatment, corresponding analytic expressions for the tuning coefficients of slightly asymmetric top molecules have been given by Hack et al. [42], but, in general, the Stark tuning coefficients are calculated numerically. Dipole moments of electronically excited molecules with C_{2u} symmetry were measured by Vaccaro et al. (H_2CO), by Brucat and Zare (NO_2), and by Ohta and Tanaka (pyrimidine) [16,53,54]. In order to determine a dipole moment vector that is not directed along one of the principal axes of the molecule, the dependence of the Stark tuning of different rotational states on the dipole vector components μ_i through the functions $B_i(J)$ in Eq. (9.17) is exploited. Therefore, it is possible to determine magnitude *and* orientation of the dipole moment vector by analyzing a set of selected rotational states. This approach was successfully employed in the work on propynal in the S_1 electronic state [24]. Furthermore, the rotational constants of vibronically excited, short-lived species can be determined together with the dipole moment because the Stark shifts depend on the energy separation of the rotational states and, hence, on the rotational constants. Such a self-consistent determination of structure parameters has been demonstrated on vibrationally excited S_1 propynal [25,42].

9.4.2 Coherent excitation of eigenstates. Molecular quantum beats

In the examples considered above sublevels of unperturbed rovibronic levels were excited coherently. In polyatomic molecules, particularly if they are excited to electronically excited states, the interaction of the rovibronic states to background states is commonly observed. The resulting spectral perturbations are of general interest since they are a manifestation of intramolecular energy flow. Quantum beat spectroscopy provides us with the means to characterize the eigenstates and to elucidate the coupling as well as the nature and properties of the perturbing states. In the following we first present an example of resolved eigenstates carried out in the manner described above and then proceed to quantum beats of coherently excited eigenstates. All examples involve the perturbations of rovibronic singlet states caused by spin-orbit interaction. For this purpose we briefly introduce the interaction of a pure singlet state $|s\rangle$ and a pure triplet state $|l\rangle$ giving rise to two mixed states

$$|n\rangle = c_{sn}|s\rangle + c_{ln}|l\rangle \qquad |n'\rangle = c_{sn'}|s\rangle + c_{ln'}|l\rangle \tag{9.19}$$

The states $|s\rangle$ and $|l\rangle$ are often referred to as Born-Oppenheimer (BO) states and $|n\rangle$ and $|n'\rangle$ as eigenstates of the molecular hamiltonian $\hat{H}_{MOL} = \hat{H}_{BO} + \hat{H}_{SO}$. The operator $\hat{H}_{SO}$ describes the spin-orbit interaction. The energy spacing of the two eigenstates expressed in terms of the energy difference of the BO states ΔE_{sl} and the spin-orbit matrix element υ_{sl} is [20]

$$\Delta E_{nn'} = \left(\Delta E_{sl}^2 + 4|\upsilon_{sl}|^2\right)^{1/2}. \tag{9.20}$$

Non-linear polyatomic molecules in singlet electronic states possess only a negligible magnetic moment in contrast to molecules excited to a triplet electronic state. The eigenstates acquire a magnetic moment that reflects the admixture of the triplet states. Determination of both Landé g factors and hyperfine splittings by quantum beat spectroscopy can be employed to quantify this triplet admixture and, in turn, to elucidate the

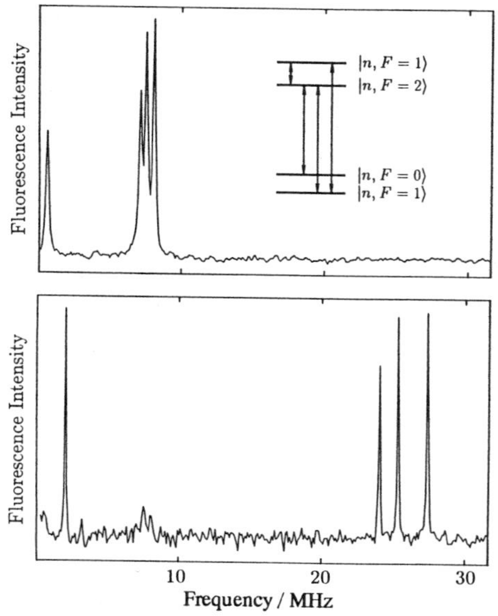

Fig. 9.12 Hyperfine quantum beats in the eigenstates of a rotational level $J = 1$ in the 9^1_0 band of propynal S_1. In the top panel the Fourier transform of the singlet eigenstate and in bottom panel that of the triplet eigenstate is shown.

underlying coupling mechanism [55-58]. This, for instance, has been the aim in the investigation of the extensive perturbations present in the $^1B_2$ state of $^{12}CS_2$ and $^{13}CS_2$ [43,59,60]. In the spectrum of $^{13}CS_2$, a portion of which has already been presented in Fig. 9.6, the hf quantum beats of eigenstates are clearly evident. Here, we focus on a more pedagogical example which has been studied in propynal (HC ≡ CCHO) excited to the S_1 electronic state [22]. In one particular vibronic band a series of rotational levels are split into a doublet of eigenstates due to spin-orbit interaction. Since propynal has two inequivalent protons with nuclear spin $I = 1/2$, each of the eigenstates is further split into four hf levels with $F = J \pm 1/2 \pm 1/2 = J - 1, J, J, J + 1$. Coherent excitation of the hf levels of an eigenstate gives rise to quantum beats, examples of which are plotted in Fig. 9.12. The splitting of the hf levels in the singlet state $\{|s, F\rangle\}$ is negligibly small whereas that in the triplet state $\{|l, F\rangle\}$ is considerable. This hf splitting is transferred to the eigenstates as shown in the diagram of Fig. 9.13. As a consequence of the different admixture of the same zero-order triplet state, the spectra of eigenstates belonging to one doublet are scaled images of each other from which the eigenstate hf structure can be reconstructed. From the spectroscopic point of view the knowledge of the hf splittings of the zero-order triplet state are more interesting because they are directly related to the electronic wavefunction. Fortunately, the spectrum of the zero-order triplet state can easily be recovered from the eigenstate spectra simply by adding the corresponding beat frequen-

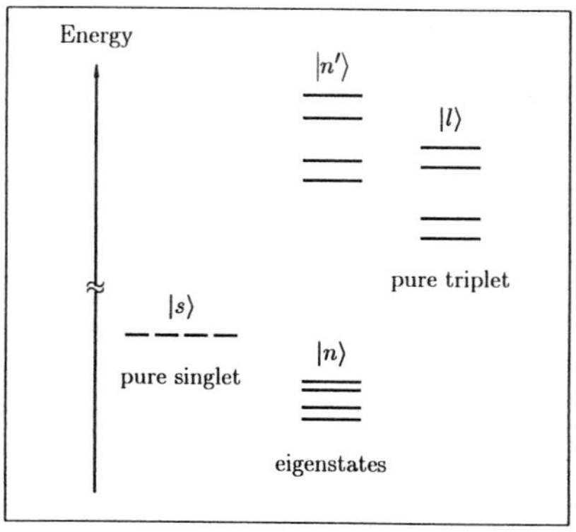

Fig. 9.13 Interacting singlet and triplet rovibronic level split by hyperfine interaction

cies $\omega_{FF'}(l) = \omega_{FF'}(\mathrm{n'}) + \omega_{FF'}(\mathrm{n'})$. A detailed analysis of such quantum beat spectra yielded the Fermi contact and dipole-dipole constants of the lowest triplet electronic state T_1 of propynal and, with those at hand, insight into the spin distribution, more generally, into the electronic structure of a polyatomic molecule in an electronically excited state.

If the Fourier bandwidth of the laser pulse exceeds the energy spacing between the eigenstates, quantum beats at the frequency $\omega_{\mathrm{nn'}} = \Delta E_{\mathrm{nn'}}/\hbar$ are observed. According to Eq. (9.6) and Eq. (9.11) the time-dependent emission intensity $I_{\mathrm{fl}}(t)$ of the coherently excited states $|\mathrm{n}\rangle$ and $|\mathrm{n'}\rangle$ is then given by

$$I_{\mathrm{fl}}(t) = C\left\{|c_{\mathrm{sn}}|^4 \mathrm{e}^{-\gamma_{\mathrm{n}} t} + |c_{\mathrm{sn'}}|^4 \mathrm{e}^{-\gamma_{\mathrm{n'}} t} + 2|c_{\mathrm{sn}}|^2 |c_{\mathrm{sn'}}|^2 \mathrm{e}^{-(\gamma_{\mathrm{n}}+\gamma_{\mathrm{n'}})t/2} \cos(\omega_{\mathrm{nn'}} t)\right\}, \tag{9.21}$$

where γ_{n} and $\gamma_{\mathrm{n'}}$ are the decay constants of $|\mathrm{n}\rangle$ and $|\mathrm{n'}\rangle$. In the derivation we considered that only the zero-order singlet state is optically accessible, i.e. $\langle \mathrm{g}|\mu\varepsilon^{\mathrm{L}}|\mathrm{s}\rangle \neq 0; \langle \mathrm{f}|\mu\varepsilon^{\mathrm{D}}|\mathrm{s}\rangle \neq 0$, whereas the triplet state is dark, $\langle \mathrm{g}|\mu\varepsilon^{\mathrm{L}}|l\rangle \langle \mathrm{f}|\mu\varepsilon^{\mathrm{D}}|l\rangle = 0$. Consequently, the dipole transition moments of an eigenstates reflect the admixture of $|\mathrm{s}\rangle$, $\langle \mathrm{g}|\mu\varepsilon^{\mathrm{L}}|\mathrm{n}\rangle = c_{\mathrm{sn}} \langle \mathrm{g}|\mu\varepsilon^{\mathrm{L}}|\mathrm{s}\rangle$, and the transition moments of the singlet state can be factored out and incorporated in the constant *C*. Since these moments contain the polarization of excitation and detection, the modulation depth, $2|c_{\mathrm{sn}}|^2|c_{\mathrm{sn'}}|^2$, is independent of these polarizations. The quantum beats are therefore isotropic in contrast to the beats discussed so far. Due to the fact that for this two-level system [20]

$$|c_{\mathrm{sn}}|^2 |c_{\mathrm{sn'}}|^2 = (\upsilon_{\mathrm{s}l}/\hbar\omega_{\mathrm{nn'}})^2, \tag{9.22}$$

the emission decay provides not only the energy spacings of the two eigenstates from the beat frequency $\omega_{\mathrm{nn'}}$, but also the coupling strength in terms of the spin-orbit matrix element $\upsilon_{\mathrm{s}l}$ from the beat amplitude.

Figure 9.14 shows an example of isotropic quantum beats obtained after excitation of propynal into its S_1 state [20]. As discussed above, each eigenstate is split into 4 hf components. The Fourier spectrum in Fig. 9.14a reveals 4 different beat frequencies which are due to coherences between the hf components of the states $|\mathrm{n}, F\rangle$ and $|\mathrm{n'}, F\rangle$ with the same total angular momentum *F*. Consequently these quantum beats are isotropic. Quantum beats of this type, although without resolved hf components, were the first ones measured in a polyatomic molecule by J.D. McDonald and coworkers in their pioneering work [7]. In general, both isotropic and anisotropic quantum beats may be present simultaneously. As the amplitude of the anisotropic quantum beats decreases rapidly with increasing angular momentum *J*, we have observed in a number of molecules that for $J \geq 3$ only isotropic quantum beats remain. In the other cases, the two types of beats can be separated by utilizing their polarization behaviour.

In measuring the quantum beats displayed in Fig. 9.14a, a perfect compensation of the Earth's magnetic field was required to avoid additional beats due to Zeeman splitting of the hf levels. On the other hand a spectroscopic assignment of the individual hf states is conveniently achieved by application of a controlled magnetic field *B* that splits each hf beat signal into $2F + 1$ Zeeman components M_F [20]. Under these conditions, the isotropic quantum beats are due to $\Delta M_F = 0$ coherences. Figure 9.14b shows the eigenstate hf components $|\mathrm{n}, F\rangle$ and $|\mathrm{n'}, F'\rangle$ being split into the Zeeman sublevels $|\mathrm{n}, F, M_F\rangle$ and $|\mathrm{n'}, F', M'_F\rangle$. Owing to the high resolution only a very weak magnetic field is needed, which allows us to treat the field effect by first order perturbation theory. The results is the simple eigenvalue expression for the state $|\mathrm{n}, F, M_F\rangle$

$$E_{\mathrm{n}} = E_{\mathrm{n}}(B=0) + \mu_{\mathrm{B}} g_{\mathrm{n}} M_{F_{\mathrm{n}}} B, \tag{9.23}$$

where the Landé factor of this state $g_{\mathrm{n}} = |c_{l\mathrm{n}}|^2 g_l$ is given in terms of the BO triplet state *g* factor. From the

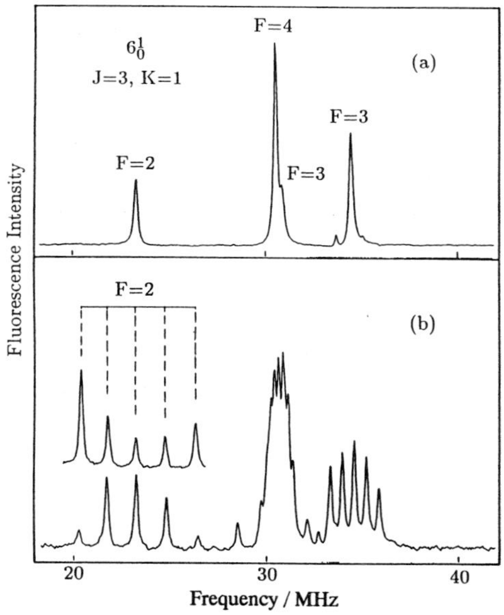

Fig. 9.14 Fourier spectra of quantum beats observed in the $J = 3$, $K = 1$ rotational state of the 6^1_0 band of S_1 propynal in (a) zero field and in (b) a magnetic field of 0.15 mT. Coherences between the eigenstates $|n, F\rangle$ and $|n', F\rangle$ of the four hf states $F = 2, 3, 3, 4$ give rise to four quantum beats, two of which are overlapped in the line at 30.5 MHz [20]. Each beat is split into $2F + 1$ components by the magnetic field. Changing the laser polarization from $\boldsymbol{B}\|\varepsilon^L$ to $\boldsymbol{B}\perp\varepsilon^L$ (inset of (b)) provides a characteristic change of the intensity pattern of the Zeeman components.

energy difference $\omega_{nn'}$ between $|n, F, M_F\rangle$ and $|n', F, M_F\rangle$ and for $\Delta M_F = 0$ we obtain the frequency splitting of the Zeeman-split isotropic quantum beats

$$|\omega_{nn'}(M_F + 1) - \omega_{nn'}(M_F)| = |g_n - g_{n'}|\mu_B B/\hbar. \tag{9.24}$$

The splittings of the beats shown in Fig. 9.14b are thus proportional to the difference in the g factors of the participating eigenstates. The Landé factor g_l of the pure BO triplet state $|l\rangle$, an important quantity for the description of the molecular structure, is easily found with Eq. (9.19) and yields

$$|\Delta E_{sl} g_l| = |\hbar\omega_{nn'}(g_n - g_{n'})|. \tag{9.25}$$

Unless the detuning ΔE_{sl} of the BO states $|s\rangle$ and $|l\rangle$ is accidently zero, the value of the Landé factor g_l can also be determined. However, for the signs of the g factors we need the results of anisotropic Zeeman beat experiments described in 9.4.1.

Figure 9.14b shows the Fourier spectrum of Zeeman-split quantum beats of propynal when a magnetic field of a mere 0.15 mT is applied. Based on the splitting into $2F + 1$ lines the four coherences with $F = 2$, 3, 3, 4, which are expected for the excited $J = 3$ rovibronic level, are identified. In the case of spectral overlap, a simple polarization change of the pump laser with respect to the magnetic field axis $\boldsymbol{B}$ ($\|$, $\perp$, or the magic angle 54.7°) facilitates the assignment because this polarization change causes a characteristic change in the intensity pattern of the M_F components within a given $\boldsymbol{F}$ coherence.

By means of the Zeeman-split quantum beats shown in Fig. 9.14b, an elegant level anticrossing experiment can be carried out which yields directly the magnitude of the spin-orbit matrix elements v_{sl} given in Eq. (9.20). By varying B, a single Zeeman component M_F of the triplet eigenstate $|n'\rangle$ is brought into resonance with that of the corresponding singlet state $|n\rangle$. At this field strength the beat frequency has

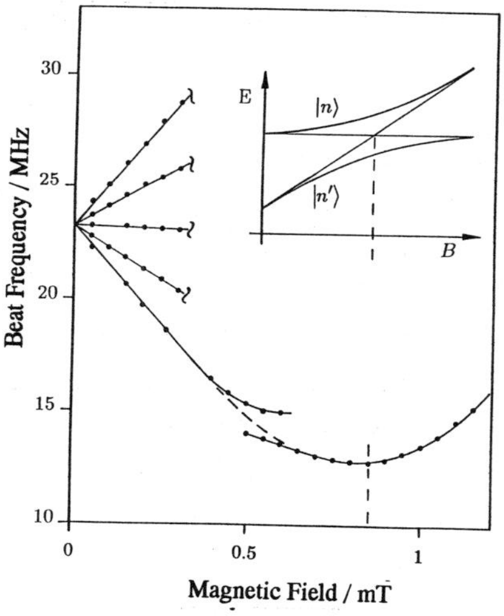

Fig. 9.15 Zeeman tuning and level anti crossing of the $F = 2$ quantum beat frequencies shown in Fig. 9.14. At low magnetic field strengths a linear Zeeman tuning is observed. In the 0.5 to 1.0 mT range the singlet and triplet eigenstates $|n, F = 2, M_F = -2\rangle$ and $|n', F = 2, M_F = -2\rangle$ show anticrossing. Weak interaction of a third state causes a local anticrossing at 0.5 mT.

reached a minimum value $\omega_{nn'}(\min) = 2\, v_{sl}/\hbar$. Since such level anticrossing experiments require variation of B over only a small low field range (0 - 3 mT in the case of propynal) they can be performed without experimental difficulties [19]. The result of such a level anticrossing experiment is displayed in Fig. 9.15.

The applications of quantum beats discussed so far were of a spectroscopic nature. They were aimed at the spectroscopist's goal to determine molecular structure parameters and to elucidate molecular state couplings. In this context we considered quantum beats of coherently excited two level systems or of incoherently superposed sets thereof. On the other hand, quantum beat spectroscopy as a time-resolved method is also a valuable tool for *directly* studying the energy flow among excited molecular states [61]. While Felker and Zewail studied the energy redistribution among vibronic states on a picosecond time scale, excellently reviewed in [61], we applied this method to investigate the dynamics of molecules on a longer time scale. We were particularly interested in the intersystem crossing dynamics which are intermediate between the "small molecule case" [62-65], where a single excited state decays radiatively, and the "large molecule case" [62-65], where the excited state decays non-radiatively into a quasi-continuum formed by a dense manifold of overlapping states. As an extension of Eq. (9.19) the intermediate case is then described by a BO state $|s\rangle$ coupled to a manifold of N-1 states $\{|l\rangle\}$ which results in a set of N eigenstates

$$|n\rangle = c_{sn}|s\rangle + \sum_{l=1}^{N-1} c_{ln}|l\rangle \tag{9.26}$$

In the following example $|s\rangle$ represents a BO singlet state, $\{|l\rangle\}$ a manifold of dark BO triplet states, and the initial and final states $|g\rangle$ and $|f\rangle$ are BO singlet states. Consequently, the time evolution of the fluorescence intensity after coherent excitation of the eigenstates $\{|n\rangle\}$ becomes in analogy to Eq. (9.21)

$$I_{fl}(t) = C\left\{\sum_{n=1}^{N} |c_{sn}|^4 \exp(-\gamma_n t) + \sum_{n=1,\, n'>n}^{N} 2|c_{sn}|^2 |c_{sn'}|^2 \exp\left(-(\gamma_n + \gamma_{n'})t/2\right)\cos(\omega_{nn'}t)\right\}, \tag{9.27}$$

where the second term is often referred to as the coherent decay term while the first one is called the incoherent decay term. The fluorescence thus consists of a multiexponential decay with $n_{QB} = N(N-1)/2$ superimposed quantum beats. Furthermore, when the excitation involves different sets $\{m\}$ of coherently excited states, the total fluorescence is simply the sum of the coherent contributions over all sets, *i.e.* $I_{fl}(t)^{tot} = \sum_{\{m\}} I_{fl}(t)_m$. For instance, in the presence of hf interaction a set of eigenstates of equal F gives rise to a coherent signal while the individual signals of sets of different F are incoherently superimposed.

As long as the quantum beat spectra are not too congested, quantum beats provide a simple but elegant method for the determination of the state density, which is a salient parameter for characterizing multilevel systems. This quantity is obtained by counting the number of quantum beat frequencies in a Fourier spectrum, $n_{QB} \approx N^2/2$, and relating $\sqrt{2n_{QB}}$ to the Fourier bandwidth of the laser pulse. Figure 9.16 shows the results of such a state count by which the number of eigenstates was determined for a series of rotational levels in the 2_0^1 vibronic band of S_1 propynal [66]. A clear increase of the number of eigenstates, and hence of the number of coupling triplet states, with increasing rotational angular momentum J of the excited levels is apparent. This effect could be attributed to a breakdown of the selection rules of spin-orbit interaction due to mixing of rovibronic states in the triplet manifold [66].

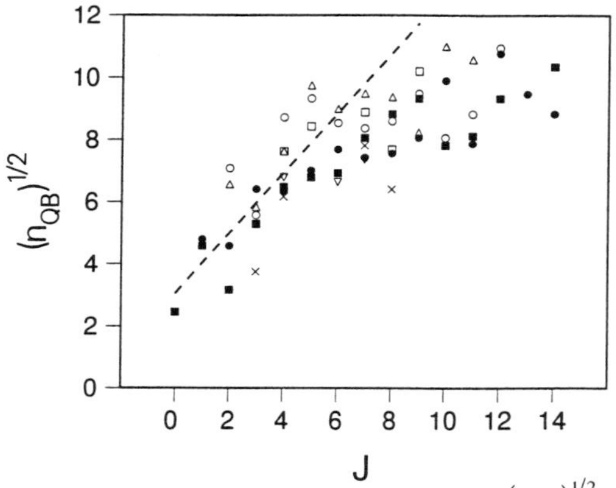

Fig. 9.16 Results of quantum beat state counting $(n_{QB})^{1/2}$ in the 2_0^1 band of S_1 propynal as a function of the excited rotational levels J. The different symbols represent data of different K states [66].

If a large number of eigenstates is excited coherently, the fluorescence decay assumes a new appearance [64] as illustrated by the "biexponential" fluorescence decays of butynal and pyrimidine shown in Figs. 9.17 and 9.18. At time $t = 0$ all oscillations are in phase (second term in Eq. (9.27)), contributing constructively to the time-resolved fluorescence. Shortly afterwards, the oscillations of different frequency begin to interfere destructively giving rise to a fast decay referred to as the dephasing of the coherence. The time scale of the dephasing is determined by the coherence width of the laser pulse or the width of the spectral envelope of the eigenstates whichever is narrower. At longer times the fluorescence becomes dominated by the incoherent emission from the individual eigenstates (first term in Eq. (9.27)). It represents an average exponential decay with superimposed recurrences which arise when oscillations come back into phase. The presence of recurrences indicates, as is generally true, that dephasing has not yet reached the limiting large molecule case.

Phenomenologically, an average fluorescence decay can be described by two exponential decays

$$I_{fl}(t) = A^+ e^{-\gamma_+ t} + A^- e^{-\gamma_- t} \tag{9.28}$$

The slow decay rate is given by the average eigenstate linewidth $\gamma_- = \langle \gamma_n \rangle$ while the fast decay rate is determined by the width of the spectral envelope $\gamma_+ = \Delta$ (expressed in angular frequencies). On the basis of a model previously invoked to describe radiationless transitions [62] this behaviour can be further analyzed [63,64]. In particular, the ratio of the pre-exponential factors

$$A_{fs} = A^+/A^- \simeq \pi \Delta \rho \tag{9.29}$$

was shown to be related to the effective number of coupled states and provides an estimate of their density ρ. The parameter A_{fs} is widely used to determine the density of coupling background states [65]. In the examples shown in Figs. 9.17 and 9.18 the measured density of coupling triplet states, exhibiting a rotational state dependence, indicates mixing of rovibronic states in the triplet state. The extent of the rotational state dependence is related to the pertinent coupling mechanism [67,68]. However, a warning is in order. A fast decay component does not necessarily originate from dephasing as has been found in S_1 pyrazine under saturation conditions with nanosecond laser pulses. The misinterpretation of this decay as dephasing caused a long-lasting debate on the dynamics of S_1 pyrazine until it was shown to be due to a saturation effect [13,69].

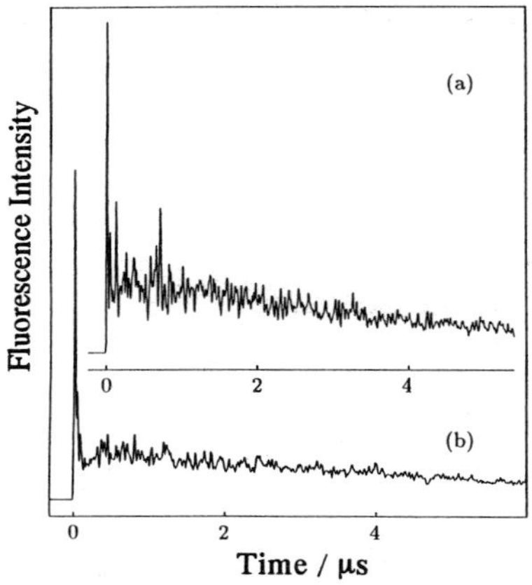

Fig. 9.17 Fluorescence decays of rovibronic levels (a) $J = 1$ and (b) $J = 2$ in the 4^1_0 band of S_1 butynal [67]. The decays illustrate the features of a large multilevel system discussed in the text.

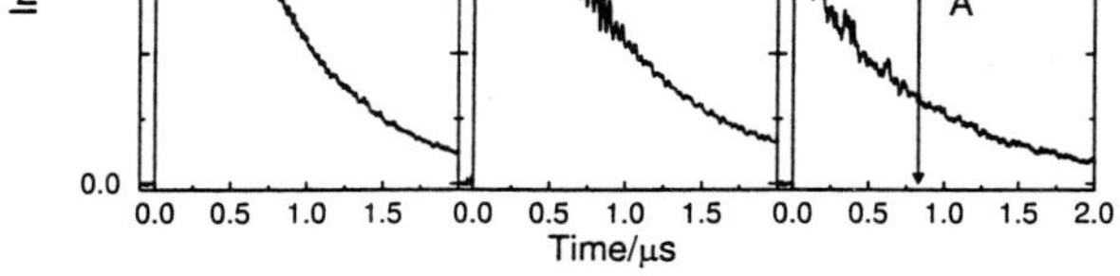

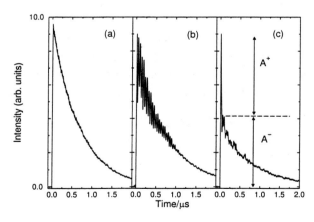

Fig. 9.18 Fluorescence decays of rovibronic levels (a) $J = 0$, (b) $J = 2$ and (c) $J = 5$ in S_1 pyrimidine exhibiting the transition from exponential to biexponential behaviour within a vibronic band [68].

Quantum beats of multilevel systems also contain information about the level structure in terms of ergodicity or chaos. An irregular or chaotic level structure exhibits a nearest neighbor spacing distribution which is well described by a Wigner distribution whilst a regular structure is characterized by a Poisson distribution. For Wigner distributed spacings, the fluorescence intensity is predicted to show a depression immediately following the fast decay [64,70]. This "correlation hole" observed (Fig. 9.17) in the time-resolved fluorescence of S_1 butynal [27,67] is the effect of level repulsion in an irregular level structure which causes small level spacings to disappear.

9.4.3 Time-resolved double resonance techniques

The application of quantum beat spectroscopy illustrated above was confined to electronically excited states. Its application to the electronic ground state should be as profitable, since IR emission lifetimes promise sub-kHz resolution. The detection of IR emission with sufficient time resolution and sensitivity is, however, difficult. Consequently quantum beats in the IR emission of a polyatomic molecule have never been reported and more elaborate methods have to be applied. The IR-pump UV-probe technique introduced in section 9.3.4 is a simple and convenient method as is demonstrated by the two examples in Figs. 9.19 and 9.20 [47,71]. The former shows a ground state Stark quantum beat observed in the aldehydic C-H stretch $v_2 = 1$ of propynal. Since the state excited by the σ polarized IR pulse $\left(\varepsilon^{\mathrm{IR}} \perp \boldsymbol{E}\right)$ and probed by the σ polarized UV pulse $\left(\varepsilon^{\mathrm{UV}} \perp \boldsymbol{E}\right)$ is a $J = 2$ rotational state, the coherently excited Zeeman levels $M = 0$ and

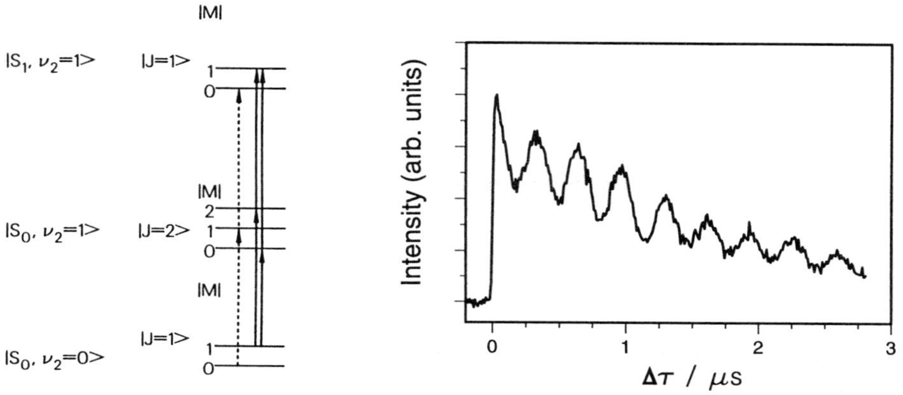

Fig. 9.19 Pump-probe scheme (left) and ground state Stark quantum beat of a $J = 2$ level (right)

$|M| = 2$ give rise to a $\Delta M = 2$ coherence (see Fig. 9.19 left) manifested as a single beat frequency (Fig. 9.19 right). The overall decay of the signal is due to the drift of the excited molecules out of the probing zone in the molecular beam. This "decay" limits the resolution in the present case to 300 kHz. In principle, however, the resolution can be extended if the laser beams are chosen to be coaxial with the molecular beam.

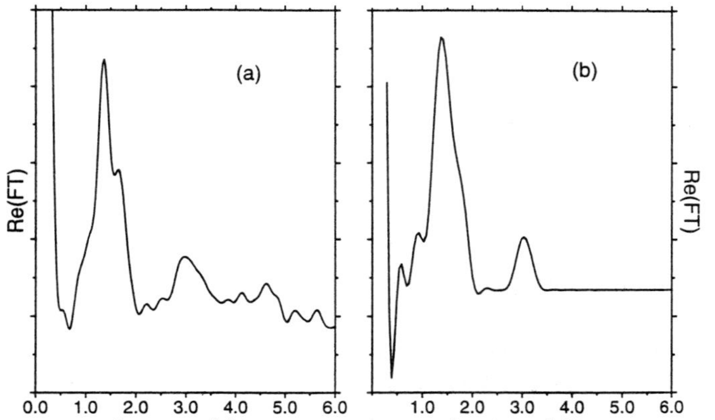

Fig. 9.20 Fourier transform of (a) experimental and (b) simulated ground state nuclear quadrupole quantum beat of a $J = 1$ state in a C-H stretch mode of S_0 pyrimidine.

The pump-probe technique provides a modulation depth which is more pronounced than that of Stark quantum beats observed in the S_1 fluorescence of the same molecule (see section 9.4.1). This can be attributed to the increased selectivity of the detection of the anisotropic quantum beats. Furthermore, the modulation depth can be maximized by reducing the incoherent contribution to the signal if the pump occurs *via* R branch excitation and the probe *via* P branch detection. Other types of anisotropic quantum beats also profit from this improved modulation depth as is shown for the hf quantum beats observed after excitation of a C-H stretch vibrational level of pyrimidine in Fig. 9.20 [71]. This type of quantum beat measured in a number of rotational states yielded the nuclear quadrupole coupling constants of the two nitrogen nuclei. Typical hf splittings were between 1 and 3 MHz and would be difficult to measure by conventional IR spectroscopy due to the residual Doppler broadening in a molecular beam.

A second kind of double resonance experiments deals with the manipulation of coherences among Zeeman sublevels, bringing time-domain magnetic resonance to electronically excited states. Following the optical pulse, the Fourier bandwidth of which covers the energy range of the pertinent molecular sublevels, a continuous or pulsed rf field drives the system. Simultaneously its time evolution is monitored by time-resolved emission whereby the optical detection upgrades the detection efficiency by orders of magnitude over that of rf photon detection. Such a double resonance experiment has been demonstrated with the $J = 1$ rotational level in the 17U band of the $^3A_2$ electronic state of CS_2 [41]. Employing, on one hand, σ polarized excitation and detection, a Zeeman quantum beat of the coherently excited $|J=1, M=+1\rangle$ and $|J=1, M=-1\rangle$ levels is observed, where the $\Delta M = 2$ coherence oscillates at twice the Larmor frequency ω_L as previously shown in Fig. 9.8. Choosing, on the other hand, π polarized excitation and keeping σ polarized detection, a non-modulated exponential decay is produced which is entirely due to incoherent emission to $J \geq 1$ states. The decay changes drastically if the optically excited molecules are subjected to a linearly polarized rf field, $\boldsymbol{B}_1 = 2\boldsymbol{B}_1^0 \cos(\omega_0 t + \phi)$, oriented parallel to the excitation axis and perpendicular to the static magnetic field $\boldsymbol{B}$. The transitions involved in the presence of a resonant rf field, $\omega_0 = \omega_L$, are depicted in the scheme of Fig. 9.21. Since the laser pulse excites exclusively the $|J=1, M=0\rangle$ state, the beat pattern evident in the decays (Fig. 9.21) is due to the coherences induced by the rf field between Zeeman sublevels. The coherences are manifested as $\Delta M = 2$ and $\Delta M = 1$ beats, of which the former are selectively detected under pure σ polarization.

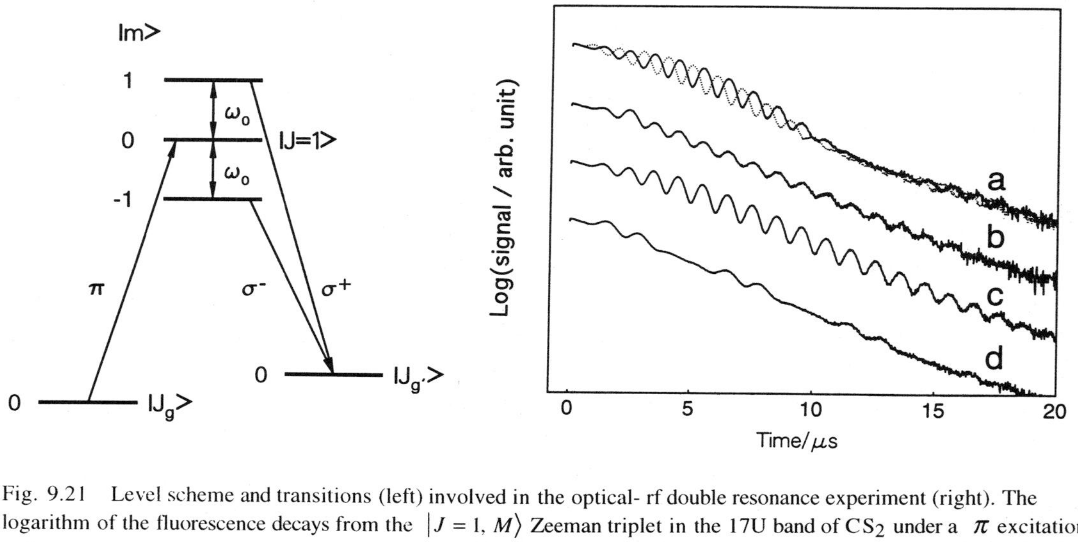

Fig. 9.21 Level scheme and transitions (left) involved in the optical- rf double resonance experiment (right). The logarithm of the fluorescence decays from the $|J = 1, M\rangle$ Zeeman triplet in the 17U band of CS_2 under a π excitation and σ detection configuration is shown in the presence of a continuous rf field (a), of a 2 μs rf pulse (b), of a 6 μs pulse (c), and of an rf field detuned from the resonance frequency by δ = -200 kHz (d). The decay represented by a dotted line in (a) resulted from a 90° phase change of the rf field.

Exploiting the entire dynamical range of the measurement over 7 lifetimes $(\tau = 2.7\mu s)$ and representing the results in a logarithmic plot (Fig. 9.21a, solid line) a periodic modulation of the beat amplitude is revealed. A cyclic transfer of population occurs between the $|J=1, M=0\rangle$ state and the $|J=1, M=\pm 1\rangle$ states during which the $\Delta M = 2$ coherence is built up and transformed back into population difference. The observed modulation period is half the Rabi period, $T_R^A = T_R/2 = \pi\hbar/g_J\mu_B B_1^0$ [41]. The maximum beat amplitude is the same as that obtained by π polarized optical excitation in the absence of the rf field. This result reveals that the system reached the state $|\psi\rangle = \frac{1}{\sqrt{2}}(|J=1, M=+1\rangle + |J=1, M=-1\rangle)$, and implies that all the $|J=1, M=0\rangle$ population has been transferred into the $\Delta M = 2$ coherence when the maximum modulation amplitude is reached. In the same manner an rf field converts population difference into coherence, it can transfer optically created coherence into population difference [41].

An rf pulse of properly tailored length and amplitude can bring the system into any state reached by the driven evolution and the signal during the free evolution after the pulse, the FID, can then be used to characterize this state. This procedure, which is the basis of FT-NMR spectroscopy, is illustrated in Fig. 9.21b and c. Under the present conditions a $T_R^A/2 \approx 6\mu s$ rf pulse transfers the system completely to the $|\psi\rangle = \frac{1}{\sqrt{2}}(|J=1, M=+1\rangle + |J=1, M=-1\rangle)$ state. Furthermore, a detuning $\delta = \omega_0 - \omega_L$ reduces the Rabi period relative to the $\delta = 0$ case according to the generalized Rabi frequency $\omega_e = \left(\delta^2 + (g_J\mu_B B_1^0/\hbar)^2\right)^{1/2}$; however, the detuning also reduces the efficiency of population transfer and consequently the maximum modulation amplitude (Fig. 9.21d).

In the description of the above experiments we treated the rf field classically *i.e.* as a time-dependent perturbation. In an alternative quantum mechanical treatment the rf field is quantized and its coherence is described in terms of coherent "many photon states". The interaction of these states with the molecular states gives rise to eigenstates which are called dressed states [72]. In this picture the eigenstates are excited coherently in the same way as discussed in section 9.2 and the observed dynamics - the Rabi cycling - is manifested by quantum beats among the dressed states [31].

9.5 Concluding remarks

In this chapter we have introduced the principles of quantum beat spectroscopy and have discussed a variety of applications. We have addressed quantum beat spectroscopy performed on a nano- and microsecond time scale with the benefit of a long observation time and in turn with a high frequency resolution. This resolution is beyond the Doppler limit allowing us to measure even smallest energy splittings such as those in magnetic hyperfine and nuclear quadrupole structures as well as Landé *g* factors and electric dipole moments at very low fields. Equally rewarding are the quantum beat measurements of parameters characterizing the dynamics of energy flow. The coupling matrix elements and state densities obtained in this manner are essential for an understanding of non-radiative processes in molecules.

Conventional frequency domain spectroscopy with nanosecond laser pulses is performed in many laboratories. The extra effort that has to be added to conventional laser-induced fluorescence experiments in order to also exploit the time domain is moderate, indeed, but the benefit is considerable: spectroscopy in the frequency and that in the time domain are complementary with the latter revealing the information which is concealed in the former by Doppler broadening.

In the near future the application of quantum beat spectroscopy to radical species will be a fruitful field perhaps even more so than that of the stable closed-shell molecules discussed above. Recently we have demonstrated that the particle density and the rotational temperature required for a quantum beat experiment can be achieved for transient radicals and their complexes. As an example Fig. 9.22 shows the hf quantum beats of a rovibronic level of the Ar·OD complex [40]. The hyperfine constants, the Landé *g* factors and the dipole moments determined in such experiments will contribute significantly to the characterization of the electronic structure. Moreover, the dynamics of polyatomic radicals is promising but as yet poorly studied by time domain spectroscopy. In particular, interesting new effects such as the influence of magnetic interactions on radiationless processes in paramagnetic species can be assessed.

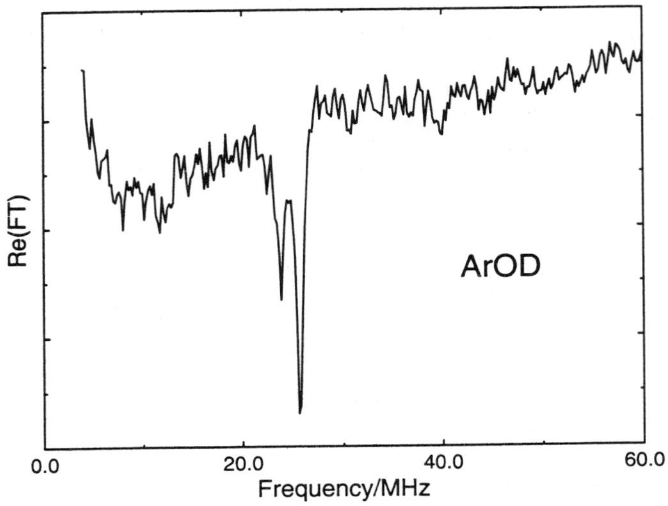

Fig. 9.22 Hyperfine quantum beats observed in the $N = 1$ rotational level of ArOD upon excitation of the $(0, 0, 3)A^2\Sigma^+ - X^2\Pi$ transition [40]

In our account we have not addressed coherence phenomena which occur on the picosecond and femtosecond time scale. Coherent excitation of rotational levels with picosecond pulses gives rise to quantum beats which - in the ensemble average - lead to a transient signal known as rotational recurrences. Such recurrences were successfully utilized for the determination of rotational constants from which the geometry of molecules and complexes could be obtained. This research topic has matured in the last decade and several review articles give an excellent account of its status [33,61]. The large Fourier bandwidth of femtosecond pulses allows coherent excitation of levels with even larger energy separations. Due to this feature molecules can be prepared in vibrational superposition states or vibrational wave packets and their time evolution can be probed with a second femtosecond pulse. As such experiments opened the possibility to observe bond fission in real time, the term femtochemistry was coined for this rapidly developing field within chemical reaction dynamics [73].

Acknowledgement: We thank the Schweizerischer Nationalfonds for the continuous support of our quantum beat work.

References

1. R. N. Ernst, G. Bodenhausen and A. Wokaun, *Principles of Nuclear Nagnetic Resonance in One and Two Dimensisions*, Clarendon Press, Oxford, (1987).
2. E. B. Alexandrov, *Opt. Spectrosc.* **17**, 957 (1964).
3. J. N. Dodd, R. D. Kaul and D. M. Warrington, *Proc.Phys.Soc. (London)* **84**, 176 (1964).
4. E. B. Alexandrov, M. P. Chaika and G. I. Khvostanko, *Interference of Atomic States*, Springer, Berlin (1993).
5. J. N. Dodd and G. W. Series, In *Progress in Atomic Spectroscopy* (W. Hanle and H. Kleinpoppen, eds.), Plenum, New York (1978).
6. S. Haroche, In *High-Resolution Laser Spectroscopy* (K. Shimoda, ed.), Springer Verlag, Berlin (1976).
7. J. Chaiken, T. Benson, M. Gurnick and J. D. McDonald, *Chem. Phys. Lett.* **61**, 195 (1979); J. Chaiken, M. Gurnick and J.D. McDonald, *J. Chem. Phys.* **74**, 106 (1981); *ibid.* **74**, 117 (1981); J. Chaiken and J.D. McDonald, *ibid.* **77**, 669 (1982).
8. W. Henke, H. L. Selzle, T. R. Hays and E. W. Schlag, *Z. Naturf. Teil A* **35**, 1271 (1980); W. Henke, H. L. Selzle, T. R. Hays, S. H. Lin and E. W. Schlag, *Chem. Phys. Lett.* **77**,448 (1981).

9. W. R. Lambert, P. M. Felker and H. A. Zewail, *J.Chem.Phys.* **75**, 5958 (1981).
10. S. Okajima, H. Saigusa and E. C. Lim, *J. Chem. Phys.* **76**, 2096 (1982); H. Saigusa and E. C. Lim, *ibid.* **78**, 91 (1983).
11. B. J. v. d. Meer, H. T. Jonkman, G. M. t. Horst and J. Kommandeur, *J. Chem. Phys.* **76**, 2099 (1982).
12. W. Sharfin, M. Ivanco and S. C. Wallace, *J. Chem. Phys.* **76**, 2095 (1982); M. Ivanco, J. Hager, W. Sharfin and S. C. Wallace, *J. Chem. Phys.* **78**, 6531 (1983); H. Watanabe, S. Tsuchiya and S. Koda. *Faraday Disc, Chem. Soc.* **75**, 365 (1983).
13. J. Kommandeur, W. A. Majewski, W. L. Meerts and D. W. Pratt, *Ann. Rev. Phys. Chem.* **38**, 433 (1987).
14. H. Stafast, H. Bitto and J. R. Huber, *J. Chem. Phys.* **79**, 3660 (1983).
15. H. Bitto, H. Stafast, P. Russegger and J. R. Huber, *Chem. Phys.* **84**, 249 (1984).
16. P. J. Brucat and R. N. Zare, *J. Chem. Phys.* **78**, 100 (1983); *ibid.* **81**, 2562 (1984); *Mol. Phys.* **55**, 277 (1985).
17. J. R. Huber, *Comments At. Mol. Phys.* **23**, 271 (1990).
18. H. Bitto and J. R. Huber, *Opt. Commun.* **80**, 184 (1990); E. Hack and J. R. Huber, *Int. Rev. Phys. Chem.* **10**, 287 (1991); H. Bitto and J. R. Huber, *Acc. Chem. Res.* **25**, 65 (1992).
19. J. Mühlbach, M. Dubs, H. Bitto and J. R. Huber, *Chem. Phys. Lett.* **111**, 288 (1984).
20. M. Dubs, J. Mühlbach, H. Bitto, P. Schmidt and J. R. Huber, *J. Chem. Phys.* **83**, 3755 (1985).
21. M. Dubs, P. Schmidt and J. R. Huber, *J. Chem. Phys.* **85**, 6335 (1986).
22. H. Bitto, M. P. Docker, P. Schmidt and J. R. Huber, *J. Chem. Phys.* **92**, 187 (1990).
23. M. Dubs, J. Mühlbach and J. R. Huber, *J. Chem. Phys.* **85**, 1288 (1986).
24. P. Schmidt, H. Bitto and J. R. Huber, *J. Chem. Phys.* **88**, 696 (1987).
25. P. Schmidt, H. Bitto and J. R. Huber, *Z. Phys. D* **7**, 77 (1987).
26. J. Mühlbach and J. R. Huber, *J. Chem. Phys.* **85**, 4411 (1986).
27. H. Bitto S. Derler and J. R. Huber, *Chem. Phys. Lett.* **162**, 291 (1989).
28. A. Corney, *Atomic and Laser Spectroscopy,* Clarendon Press, Oxford (1979).
29. R. M. Herman, H. Grotch, R. Kornblith and J. H. Eberly, *Phys. Rev. A* **11**, 1389 (1975).
30. W. W. Chow, M. O. Scully and J. O. Stoner, *Phys. Rev. A* **11**, 1380 (1975).
31. A. Levinger, H. Bitto and J. R. Huber, *Opt. Commun.* **103**, 381 (1993).
32. T. Hellmuth, H. Walther, A. Zajonic and W. Schleich, *Phys. Rev. A* **35**, 2532 (1987).
33. P. M. Felker and A. H. Zewail, In *Femtosecond Chemistry* (J. Manz and L. Wöste, eds.), VCH, Weinheim (1995).
34. U. Fano and J. H. Macek, *Rev. Mod. Phys.* **45**, 553 (1973).
35. R. N. Zare, *Angular Momentum,* Wiley-Interscience, New York (1988).
36. U. Fano, *Rev. Mod. Phys.* **29**, 74 (1957).
37. K. Blum, *Density Matrix Theory and Applications,* Plenum Press, New York (1981).
38. M. P. Silverman, S. Haroche and M. Gross, *Phys. Rev. A* **18**, 1507 (1978).
39. H. J. Andrä, In *Progress in Atomic Spectroscopy, B* (W. Hanle and H. Kleinpoppen, ed.), Plenum Press, New York (1979).
40. I. M. Povey, R. T. Carter, H. Bitto and J. R. Huber, *Chem. Phys. Lett.* **248**, 470 (1996).
41. H. Bitto, A. Levinger and J. R. Huber, *Z. Phys. D.* **28**, 303 (1993).
42. E. Hack, H. Bitto and J. R. Huber, *Z. Phys. D* **18**, 33 (1991).
43. H. Bitto, A. Ruzicic and J. R. Huber, *Chem. Phys.* **189**, 713 (1994).
44. P. R. Willmott, H. Bitto and J. R. Huber, *Chem. Phys. Lett* **188**, 369 (1992).
45. S. Schenk, R. C. Hilborn and H. Metcalf, *Phys. Rev. Lett.* **31**, 189 (1973).
46. H. Figger and H. Walther, *Z. Phys.* **267**, 1 (1974).
47. T. Walther, H. Bitto and J. R. Huber, *Chem. Phys. Lett.* **209**, 455 (1993).
48. D. T. Cramb, H. Bitto and J. R. Huber, *J. Chem. Phys.* **96**, 8761 (1992).
49. G. W. Lodge, J. J. Tiee and F. B. Wampler, *J. Chem. Phys.* **84**, 3624 (1986).
50. C. H. Townes and A. L. Schawlow, *Microwave Spectroscopy,* Dover, New York (1955).
51. A. Hese, A. Renn and H. S. Schweda, *Opt. Commun.* **20**, 385 (1977).
52. M. Brieger, A. Renn, A. Hese and A. Sodeik, *Chem. Phys. Lett.* **76**, 465 (1980); M. Brieger, A. Renn, A Sodeik and A. Hese, *Chem. Phys.* **75**, 1 (1983); H. Büsener, F. Heinrich and A. Hese *ibid.* **112** 139 (1987).
53. P. H. Vaccaro, A. Zabludoff, M. E. Carrera-Patiño, J. L. Kinsey and R. W. Field, *J. Chem. Phys.* **90**, 4150 (1989).
54. N. Ohta and T. Tanaka, *J. Chem. Phys.* **99**, 3312 (1993).
55. N. Ochi and S. Tsuchiya, *Chem. Phys.* **152**, 319 (1991).
56. M. A. Mason and K. K. Lehmann, *J. Chem. Phys.* **98**, 5184 (1993).
57. P. R. Willmott and H. Bitto, *Chem. Phys. Lett.* **207**, 93 (1993); H. Bitto, *Chem. Phys.* **186**, 105 (1994).
58. T. A. Cool and N. Hemmi, *J. Chem. Phys.* **103**, 3357 (1995).
59. N. Ochi, H. Watanabe, S. Tsuchiya and S. Koda, *Chem. Phys.* **113**, 271 (1987).
60. D. L. Warnaar and S. J. Silvers, *Chem. Phys.* **180**, 89 (1994).
61. P. M. Felker and A. H. Zewail, In *Advances in Chemical Physics* , Wiley, New York (1988).
62. M. Bixon and J. Jortner, *J. Chem. Phys.* **48**, 715 (1968); *ibid.* **50**, 3284 (1969); *ibid.* **50** 4061 (1969); J. Jortner and S. Mukamel, In *The World of Quantum Chemistry* (R. Daudel, ed.), Reidel, Dortrecht (1974).
63. A. Tramer and R. Voltz, In *Excited States* (E. C. Lim, ed.), Academic, New York (1979).

64. J. M. Delory and C. Tric, *Chem. Phys.* **3**, 54 (1974).
65. T. Uzer, *Phys. Rep.* **199**, 73 (1991).
66. H. Bitto, P. R. Willmott and J. R. Huber, *J. Chem. Phys.* **95**, 4765 (1991).
67. S. Derler, H. Bitto and J. R. Huber, *Chem. Phys.* **169**, 275 (1993).
68. R. T. Carter, H. Bitto and J. R. Huber, *J. Chem. Phys.* **102**, 5890 (1995).
69. H. Bitto and P. R. Willmott, *Chem. Phys.* **165**, 113 (1992).
70. L. Leviandier, M. Lombardi, R. Jost and J. P. Pique, *Phys. Rev. Lett.* **56**, 2449 (1986).
71. R. T. Carter, T. Walther, H. Bitto and J. R. Huber, *Chem. Phys. Lett.* **240**, 79 (1995).
72. C. Cohen-Tannoudji, J. Dupont-Roc and G. Grynberg, *Atom-Photon Interactions,* Wiley, New York (1992).
73. A. H. Zewail, In *Femtosecond Chemistry* (J. Manz and L. Wöste, eds.), VCH, Weinheim (1995).

10 Sum Frequency Generation (SFG) Spectroscopy

Y. R. Shen
Department of Physics, University of California and Materials Sciences Division, Lawrence Berkeley National Laboratory, Berkeley, CA 94720 U.S.A.

10.1 Introduction

Spectroscopic studies are essential for real understanding of surfaces and interfaces. Among the existing surface spectroscopic techniques, those employing optics are particularly attractive. Unlike the others, they can be used for in-situ remote sensing of surfaces even in hostile environments and can be applied to any interfaces accessible by light. Linear optical spectroscopy has been well developed, but unfortunately it lacks the intrinsic surface specificity needed for most surface studies. Nonlinear optical spectroscopies, on the other hand, can be highly surface-specific by symmetry if the nonlinear optical process involved is of even order [1]. Second-order nonlinear processes such as second harmonic generation (SHG), sum-frequency generation (SFG), and difference frequency generation (DFG) are naturally more appealing because of their simplicity and sensitivity. In this chapter, we shall focus our discussion on SFG spectroscopy [2]. SHG and DFG can be regarded as special cases of SFG.

Optical SFG describes a process in which two input laser beams at frequencies ω_1 and ω_2 interact in a medium and generate an output at the sum frequency $\omega = \omega_1 + \omega_2$. Being second order, it is forbidden in media with inversion symmetry. At a surface or interface, however, the inversion symmetry is necessarily broken and SFG becomes allowed. This then makes SFG highly surface-specific for interfaces between two centrosymmetric media. More generally, even for media without inversion symmetry, surfaces and bulks often have different symmetries. With proper input and output beam polarizations, one can suppress the bulk contribution even if it is present [3].

Spectroscopic information can be obtained from resonant enhancement of SFG if either $\omega_1(\omega_2)$ or $\omega_1 + \omega_2$ or both is tuned over resonances (Fig. 10.1). Using pulsed lasers, SFG also has the sensitivity to detect a submonolayer of atoms or molecules. These two characteristics combined make SFG a most powerful and versatile surface spectroscopic tool. Because of the intrinsic ability of SFG to discriminate against bulk contribution, the measurement of a surface spectrum is now possible even in the presence of a strong overlapping bulk spectrum. This allows surface spectroscopic studies of many hitherto unexplored systems including pure liquids and solids. As a second-order effect, SFG is sensitive to the polar orientation of molecules or groups of atoms in molecules. Since the process is coherent and highly directional, it is capable of *in-situ,* remote sensing measurements in hostile environments. With ultrashort laser pulses, SFG can also be used to study surface dynamics with a subpicosecond time resolution. These unique features of SFG have led to many unique applications, opening the doors to some very exciting new areas of research in surface science.

SFG surface spectroscopy was developed as a natural extension of SHG as a surface analytical tool [1,2]. Heinz *et al.* first demonstrated that SHG can be used for spectroscopic studies of surface monolayers [4]. With two input frequencies that can both be varied, SFG is clearly more versatile than SHG. In particular, for surface vibrational spectroscopy, SHG is not applicable because the output appears in the infrared where photodetectors are generally not sufficiently sensitive for detection of a singlemonolayer. The situation is different with SFG. Having ω_1 tunable in the infrared and ω_2 in the visible, the SF output appears in the visible and can be easily detected by a photomultiplier [5].

SFG surface spectroscopy has been applied successfully to different interfacial systems, many of

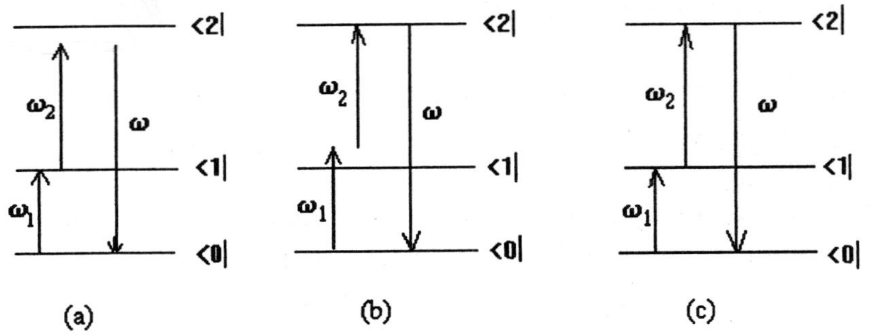

Fig. 10.1 Schematic diagrams of resonant SFG processes: (a) ω_1 at resonance; (b) $\omega_1 + \omega_2$ at resonance; (c) both ω_1 and $\omega_1 + \omega_2 = \omega$ at resonance.

which cannot be probed by other means. These include studies of free liquid interfaces, solid/solid, liquid/solid, and liquid/liquid buried interfaces, surfactant conformations, chemical vapor depositions, surface catalysis under real atmosphere, ultrafast surface dynamics, and others. In this chapter, we give a brief description of the basic theory underlying the technique and the practical aspects of the experimental arrangement. We shall discuss, with selective examples, a range of applications and show how information about an interface can be deduced from a surface SFG spectrum.

10.2 Basic Theory

The theory of surface SFG has been worked out in detail and can be found in the literature [6]. Here, because of space limitation, we can only give a sketch of the theory. We shall emphasize the physical understanding behind the theoretical derivations in our discussion.

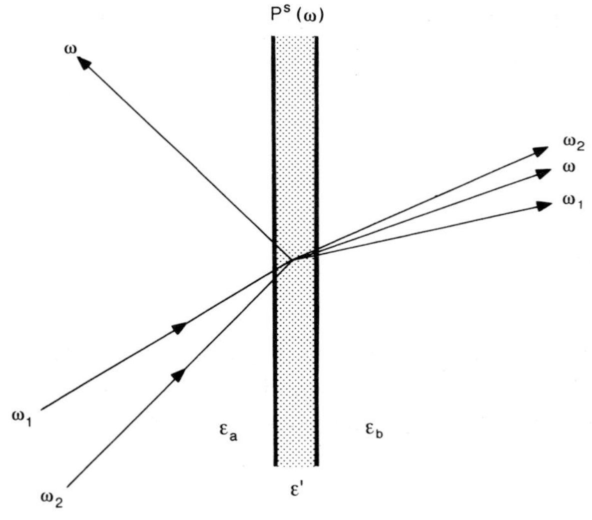

Fig. 10.2 Geometry of SFG from an interface in the reflected direction. The polarization sheet $\vec{P}^S(\omega = \omega_1 + \omega_2)$ is imbedded in a thin layer with dielectric constant ε'.

As is well known, the source of an electromagnetic wave is generally the polarization $\vec{P}$ (the oscillating dipoles per unit volume) [1]. In the case of SFG, $\vec{P}$ is nonlinearly induced in a medium by mixing of two input fields at ω_1 and ω_2 and can be described by

$$\vec{P}^{(2)}(\omega_1+\omega_2)=\vec{\chi}^{(2)}(\omega_1+\omega_2):\vec{E}(\omega_1)\vec{E}(\omega_2), \tag{10.1}$$

where $\vec{\chi}^{(2)}$ is the second order electric susceptibility and : denotes tensor product. Solution of the Maxwell wave equation with $\vec{P}^{(2)}(\omega_1+\omega_2)$ as the source term and with proper boundary conditions then leads to an expression for the SF output.

For a better understanding of SFG from an interface, we can divide the calculation into several logical steps. First, we consider SFG from a polarization sheet, given by

$$\begin{aligned}\vec{P}^{\mathrm{S}}(\omega=\omega_1+\omega_2)&=\vec{P}^{\mathrm{S}}\delta(z)\exp(\mathrm{i}\vec{k}\cdot\vec{r}-\mathrm{i}\omega t)\\&=\vec{\chi}_s^{(2)}:\vec{E}_L(\omega_1)\vec{E}_L(\omega_2),\end{aligned} \tag{10.2}$$

embedded between two semi-infinite dielectric media of dielectric constants ε_{a} (for $z<0$) and ε_{b} (for $z>0$), respectively. The pump fields in the polarization sheet are denoted by $\vec{E}_L(\omega_i)$. The geometry is sketched in Fig. 10.2, with the linear dielectric constant of the polarization sheet labelled as ε'. The solution of SFG from such a three-layer system is easily obtained. We find for pump waves incident from medium a, the SF output field reflected back into medium a is given by

$$\begin{aligned}E_p(\omega)&=\frac{\mathrm{i}4\pi k_{\mathrm{a}}}{\varepsilon_{\mathrm{b}}k_{\mathrm{a}z}+\varepsilon_{\mathrm{a}}k_{\mathrm{b}z}}\left[k_{\mathrm{b}z}P_x^{\mathrm{S}}+\varepsilon_{\mathrm{b}}k_x(P_z^{\mathrm{S}}/\varepsilon')\right]\exp(\mathrm{i}\vec{k}_{\mathrm{a}}\cdot\vec{r}-\mathrm{i}\omega t)\\E_s(\omega)&=\frac{\mathrm{i}4\pi k_{\mathrm{a}}}{k_{\mathrm{a}z}+k_{\mathrm{b}z}}\left[\frac{k_{\mathrm{a}}}{\varepsilon_{\mathrm{a}}}P_y^{\mathrm{S}}\right]\exp(\mathrm{i}\vec{k}_{\mathrm{a}}\cdot\vec{r}-\mathrm{i}\omega t)\end{aligned} \tag{10.3}$$

where $\vec{k}_{\mathrm{a}}$ and $\vec{k}_{\mathrm{b}}$ are SF wave vectors in media a and b, respectively, and the subscripts p and s denote the $\hat{p}$- and $\hat{s}$-polarized components of the field, i.e., polarization parallel and perpendicular to the plane of incidence, respectively. Expressions for SF outputs from other beam geometries can also be found.

In the absence of boundary surfaces at $z=0$, i.e., when $\varepsilon_{\mathrm{a}}=\varepsilon'=\varepsilon_{\mathrm{b}}=\varepsilon$, Eq. (10.3) reduces to

$$\begin{aligned}E_{po}(\omega)&=\frac{\mathrm{i}2\pi k}{\varepsilon k_z}\left(k_zP_x^{\mathrm{S}}+k_xP_z^{\mathrm{S}}\right)\exp(\mathrm{i}\vec{k}\cdot\vec{r}-\mathrm{i}\omega t)\\E_{so}(\omega)&=\frac{\mathrm{i}2\pi k^2}{\varepsilon k_z}P_y^{\mathrm{S}}\exp(\mathrm{i}\vec{k}\cdot\vec{r}-\mathrm{i}\omega t).\end{aligned} \tag{10.4}$$

This is a well-known result that can be derived from the simple theory of radiation for a thin sheet of dipole oscillators embedded in an infinite, uniform dielectric medium. Comparison of the fields in Eqs. (10.3) and (10.4) yields the ratios for the three field components along $\hat{x}$, $\hat{y}$ and $\hat{z}$.

$$\begin{aligned}(E_p/E_{po})_x&\equiv L_{xx}=\frac{2\varepsilon_{\mathrm{a}}k_{\mathrm{b}z}}{\varepsilon_b k_{\mathrm{a}z}+\varepsilon_{\mathrm{a}}k_{\mathrm{b}z}}\\(E_p/E_{po})_z&\equiv L_{zz}=\frac{2\varepsilon_{\mathrm{a}}k_{\mathrm{a}z}(\varepsilon_{\mathrm{b}}/\varepsilon')}{\varepsilon_b k_{\mathrm{a}z}+\varepsilon_{\mathrm{a}}k_{\mathrm{b}z}}\\(E_s/E_{so})_y&\equiv L_{yy}=\frac{2k_{\mathrm{a}z}}{k_{\mathrm{a}z}+k_{\mathrm{b}z}}\end{aligned} \tag{10.5}$$

We note that these L factors are identical to the transmission Fresnel coefficients relating the propagating field in medium a to the field in the polarization sheet. Physically, $\vec{L} = (L_{xx}, L_{yy}, L_{zz})$ acts as a macroscopic local-field correction that accounts for the effect of boundary surfaces in coupling the radiation field in or out of the polarization sheet. We can write

$$\vec{E}(\omega) = \vec{L}(\omega) \cdot \vec{E}_0(\omega) \tag{10.6}$$

with $\vec{E}_0$ given in Eq. (10.4).

Now the pump fields inside the polarization sheet, $\vec{E}_L(\omega_i)$, must also be related to the pump field incident from medium a (Fig. 10.1) by a similar Fresnel coefficient, i.e., $\vec{E}_L(\omega_i) = \vec{L}(\omega_i) \cdot \vec{E}_\mathrm{a}(\omega_i)$. Equation (10.2) becomes

$$\vec{P}^\mathrm{s}(\omega) = \vec{\chi}_\mathrm{s}^{(2)}(\omega = \omega_1 + \omega_2) : [\vec{L}(\omega_1) \cdot \vec{E}(\omega_1)][\vec{L}(\omega_2) \cdot \vec{E}(\omega_2)] \tag{10.7}$$

From Eqs. (10.4), (10.6), and (10.7), we can then immediately find the SF output field

$$\begin{aligned} E_p(\omega) &= \mathrm{i}(2\pi\omega / c)[L_{xx}(\omega)\chi_{s,xjk}^{(2)} L_{jj}(\omega_1) L_{kk}(\omega_2) \\ &\quad + (k_x(\omega) / k_{az}(\omega)) L_{zz}(\omega)\chi_{s,zjk}^{(2)} L_{jj}(\omega_1) L_{kk}(\omega_2)] E_{aj}(\omega_1) E_{ak}(\omega_2) \\ E_s(\omega) &= \mathrm{i}(2\pi\omega / c)(k_a(\omega) / k_{az}(\omega)) L_{yy}(\omega)\chi_{s,yjk}^{(2)} L_{jj}(\omega_1) L_{kk}(\omega_2) E_{aj}(\omega_1) E_{ak}(\omega_2). \end{aligned} \tag{10.8}$$

The above discussion shows that we can obtain the expression without derivation for the SF output for any beam geometry as long as the expression for $\vec{E}_0$ in Eq. (10.4) and the proper Fresnel coefficients are known.

The SF output intensity can be readily calculated knowing $\vec{E}(\omega)$. For the case given by Eq. (10.8) (Fig. 10.1), we have

$$\begin{aligned} I(\omega) &= c\sqrt{\varepsilon_\mathrm{a}(\omega)} \left|\vec{E}_\mathrm{a}(\omega)\right|^2 / 2\pi \\ &= \frac{8\pi^3\omega^2 \sec^2\theta_\omega}{c^3[\varepsilon_\mathrm{a}(\omega)\varepsilon_\mathrm{a}(\omega_1)\varepsilon_\mathrm{a}(\omega_2)]^{1/2}} \left|\vec{e}'^{\dagger}(\omega) \cdot \vec{\chi}_\mathrm{s}^{(2)} : \vec{e}'(\omega_1)\vec{e}'(\omega_2)\right|^2 I_\mathrm{a}(\omega_1) I_\mathrm{a}(\omega_2) \ . \end{aligned} \tag{10.9}$$

Here, θ_ω is the reflection angle of the SF output, $\vec{e}'(\Omega) \equiv \vec{L}(\Omega) \cdot \hat{e}(\Omega)$ with $\hat{e}(\Omega)$ being a unit vector describing the polarization of the field at Ω, which stands for a general variable denoting ω, ω_1 or ω_2, and $I_\mathrm{a}(\omega_i)$ is the input laser intensity. If the input lasers have a pulsewidth T and an overlapping beam cross section A at the sample, then the SF output, in terms of photons per pulse, can be written as

$$\begin{aligned} S(\omega) &= I(\omega) AT / \hbar\omega \\ &= \frac{8\pi^3\omega \sec^2\theta_\omega}{\hbar c^3[\varepsilon_\mathrm{a}(\omega)\varepsilon_\mathrm{a}(\omega_1)\varepsilon_\mathrm{a}(\omega_2)]^{1/2}} \left|\vec{e}'^{\dagger}(\omega) \cdot \vec{\chi}_\mathrm{s}^{(2)} : \vec{e}'(\omega_1)\vec{e}'(\omega_2)\right|^2 I_\mathrm{a}(\omega_1) I_\mathrm{a}(\omega_2) AT. \end{aligned} \tag{10.10}$$

So far, we have assumed that the induced polarization sheet at the boundary surface is the only source for the SFG. More generally, however, nonlinear polarizations induced in the bulk of the two media may also contribute to the SFG. It is possible to combine the surface and bulk contributions in such a way that they can be described by an effective polarization sheet with a surface polarization $\vec{P}_\mathrm{eff}^\mathrm{s}$, or an effective surface nonlinear susceptibility $\vec{\chi}^{(2)}_{s,\text{eff}}$ at the boundary [6]. Then the results of Eqs. (10.8-10) are still valid if $\vec{P}^{S}$ and $\vec{\chi}^{(2)}_{s}$ are replaced by $\vec{P}^{s}_{\text{eff}}$ and $\vec{\chi}^{(2)}_{s,\text{eff}}$. Consider, for example, the case where the bulk nonlinear polarization $\vec{P}^{B}(\omega=\omega_1+\omega_2)$ is negligible in medium a, but non-negligible in medium b. This is a problem that has been worked out by a number of researchers [7]. The solution is described by Eqs. (10.8-10), but $\vec{\chi}^{(2)}_{S}$ in the equations should be replaced by

$$(\vec{\chi}^{(2)}_{s,\text{eff}})_{ijk} = \chi^{(2)}_{s,ijk} + \chi^{(2)}_{B,ijk} / \left[k_{bz}(\omega) + k_{bz}(\omega_1) + k_{bz}(\omega_2)\right] F_i(\omega) F_j(\omega_1) F_k(\omega_2) \tag{10.11}$$

where $\vec{\chi}^{(2)}_{B}$ is the bulk nonlinear susceptibility of medium b, defined by $\vec{P}^{B}(\omega)=\vec{\chi}^{(2)}_{B} : \vec{E}_b(\omega_1)\vec{E}_b(\omega_2)$, $F_j(\Omega) \equiv \varepsilon_b(\Omega)/\varepsilon'$ for $j = z$, and $F_j(\Omega) = 1$ for $j = x, y$.

From Eq. (10.11), we notice that for SFG to be surface-specific, we must have $|\chi^{(2)}_{S}| >> |\chi^{(2)}_{B}/2k_b(\omega)|$. This inequality generally holds only if the bulk has inversion symmetry since otherwise one would find $|\chi^{(2)}_{S}| \sim |\chi^{(2)}_{B} a|$, with a being the monolayer thickness. If the surface is a simple termination of the bulk structure and both $\chi^{(2)}_{s}$ and $\chi^{(2)}_{B}$ arise from electric quadrupole contributions due to field gradients, we would find $|\chi^{(2)}_{S}| \sim |\chi^{(2)}_{B}/k|$ because the fields should vary on the scale of $1/k$ in the bulk but on the scale of the surface layer thickness at the surface. However, $\chi^{(2)}_{S}$ could be dominated by an electric-dipole contribution if the surface consists of a polar layer. In such cases, the surface contribution to SFG could easily dominate over that of the bulk.

A crude estimate of the signal strength of surface SFG is in order. For a polar monolayer of molecules at an interface, the resonant $\chi^{(2)}_{S}$ is typically 10^{-14} - 10^{-16} esu. If we take $|\chi^{(2)}_{S}| \sim 10^{-15}$ esu, $I_1(\omega_1) \sim I_2(\omega_2) \sim 10^{10}$ W/cm^2, $A \sim 10^{-3}$ cm^2, $T = 10$ ps, and $\theta_{\omega} \sim 45°$, Eq. (10.10) predicts a signal of $\sim 10^4$ photon/pulse. Such a signal can easily be detected.

In the above discussion, we have assumed a three-layer model (Fig. 10.1). This is of course a simplified model. In reality, the interfacial layer should be the surface-bulk transition layer in which the structure of the medium and the fields vary continuously [6]. The surface nonlinear polarization must be an integral of the bulk nonlinear polarization across the interfacial layer.

$$\begin{aligned} P^{s}_{i} &= \int_{I} P^{B}_{i}(z)\,dz \qquad \text{for } i = x, y \\ P^{s}_{z}/\varepsilon' &= \int_{I} [P^{B}_{z}(z)/\varepsilon(z)]\,dz \end{aligned} \tag{10.12}$$

where I denotes the interfacial layer. For the z component, we have considered P^{s}_{z}/ε' instead of P^{s}_{z} since the former is the effective radiation source, as can be seen in Eq. (10.3). Because of the rapid field variation across the interfacial layer, the multipole expansion of the polarization in this region is also important. We have [1]

$$\vec{P}^{B}(\omega)=\vec{P}^{(2)}_{D}(\omega) - \nabla\cdot\vec{Q}^{(2)}(\omega) + i\frac{c}{2\omega}\nabla\times\vec{M}(\omega) + \cdots \tag{10.13}$$

where $\vec{P}$, $\vec{Q}$, and $\vec{M}$ denote electric-dipole polarization, electric-quadrupole polarization, and magnetization, respectively. As quadratic functions of the input fields, these multipoles can be expressed as

$$\vec{P}_{\rm D}^{(2)}(\omega)=\vec{\chi}_{\rm D}^{(2)}:\vec{E}(\omega_1)\vec{E}(\omega_2)+\vec{\chi}_{\rm DQ}^{(2)}:\nabla[\vec{E}(\omega_1)\vec{E}(\omega_2)]+\vec{\chi}_{\rm DM}^{(2)}:\nabla\times[\vec{E}(\omega_1)\vec{E}(\omega_2)]+\cdots$$
$$\vec{Q}^{(2)}(\omega)=\vec{\chi}_{\rm Q}^{(2)}:\vec{E}(\omega_1)\vec{E}(\omega_2) \qquad (10.14)$$
$$\vec{M}^{(2)}(\omega)=\vec{\chi}_{\rm M}^{(2)}:\vec{E}(\omega_1)\vec{E}(\omega_2)$$

Inserting Eqs. (10.13) and (10.14) into Eq. (10.12) makes the expression of $\vec{P}^{\rm S}$ very complicated. We shall not discuss the complications here, but simply refer the readers to Ref. [6]. In the following discussion, we shall assume that the surface polarization is dominated by the electric-dipole term $\vec{P}_{\rm D}^{(2)}=\vec{\chi}_{\rm D}^{(2)}:\vec{E}(\omega_1)\vec{E}(\omega_2)$ in $\vec{P}^{\rm B}(z,\omega)$. Note that in media with inversion symmetry, $\vec{\chi}_{\rm D}^{(2)}$ vanishes in the bulk, but can become significant at the interface.

We now turn our attention to $\vec{\chi}_{\rm s}^{(2)}$ or $\vec{\chi}_{\rm D}^{(2)}$. The surface nonlinear optical susceptibility, $\vec{\chi}_{\rm s}^{(2)}$, of an interface reflects the intrinsic property of the interface and is the quantity we would like to measure. As seen in Eqs. (10.10) and (10.11), this is what can be achieved by SFG if the bulk contribution is negligible. Although the output intensity measurements yield only absolute values of $\chi_{{\rm s},ijk}^{(2)}$, output interference measurements can determine phases of $\chi_{{\rm s},ijk}^{(2)}$. To see what information $\chi_{{\rm s},ijk}^{(2)}$ may contain, we show here explicitly the microscopic expression of $\chi_{{\rm s},ijk}^{(2)}$ under the electric-dipole approximation.

$$\begin{aligned}\chi_{{\rm s},ijk}^{(2)} &= \int_{\rm I}\chi_{{\rm D},ijk}^{(2)}(z)dz\\ &= \int_{\rm I}dz\int d\Omega\, f(\Omega)\Bigg\{\sum_{\rm g,n,n'}\left(-\frac{e^3}{\hbar^2}\right)\Bigg[\frac{\langle {\rm g}|r_i|{\rm n'}\rangle\langle {\rm n'}|r_j|{\rm n}\rangle\langle {\rm n}|r_k|{\rm g}\rangle}{(\omega-\omega_{\rm n'g}+i\Gamma_{\rm n'g})(\omega_2-\omega_{\rm ng}+i\Gamma_{\rm ng})}\\ &\quad+\frac{\langle {\rm g}|r_i|{\rm n'}\rangle\langle {\rm n'}|r_k|{\rm n}\rangle\langle {\rm n}|r_j|{\rm g}\rangle}{(\omega-\omega_{\rm n'g}+i\Gamma_{\rm n'g})(\omega_1-\omega_{ng}+i\Gamma_{\rm ng})}+6\text{ other terms}\Bigg]\rho_{gg}^{o}\Bigg\}\end{aligned} \qquad (10.15)$$

where $\rho_{\rm gg}^0$ is the population in the ⟨g| state, $\omega_{\rm ng}$ and Γ_{ng} are the transition frequency from ⟨g| to ⟨n| and the associated damping constant, and f(Ω) denotes the distribution function of a set of parameters, Ω, with $\int f(\Omega)d\Omega=N(z)=$ density of molecules at z. We have assumed here that the interfacial layer is composed of a set of localized, noninteracting molecules. Extension to the more general case is possible, but not essential for our discussion.

The expression in Eq. (10.15) shows that $\chi_{{\rm s},ijk}^{(2)}$ is resonantly enhanced if ω_1, or ω_2, or $\omega_1+\omega_2$ approaches a resonance and the resonant enhancement of SFG naturally provides the spectroscopic information. We note that even at a resonance, the nonresonant terms in $\chi_{{\rm s},ijk}^{(2)}$ are not negligible. We can write

$$\vec{\chi}_{\rm s}^{(2)}=\vec{\chi}_{\rm NR}^{(2)}+\sum_{\rm q}\frac{\vec{A}_{\rm q}}{\omega-\omega_{\rm q}+i\Gamma_{\rm q}} \qquad (10.16a)$$

and according to Eq. (10.10), the SF output is

$$S\propto\left|\vec{e}'^{\dagger}(\omega)\cdot\vec{\chi}_{\rm s}^{(2)}\cdot\vec{e}'(\omega_1)\vec{e}'(\omega_2)\right|^2 \qquad (10.16b)$$

where q refers to individual modes near resonance. Interference between the resonant and nonresonant terms causes the SFG spectral profile to be different from those observed in conventional spectroscopy. Similar to coherent anti-Stokes Raman spectroscopy, the profile of a resonance can appear in various different forms depending on the interference (Fig. 10.3). In order to deduce the parameters characterizing the resonances, Eq. (10.16) must be used to fit the observed spectrum. In some cases, however, $\chi^{(2)}_{NR}$ is negligibly small and resonances are far apart; well defined resonant peaks can then be expected in the SFG spectrum.

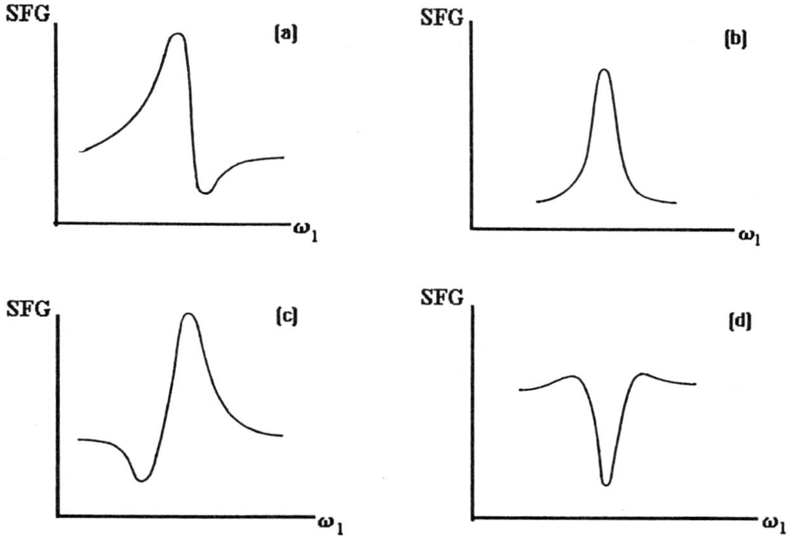

Fig. 10.3 Possible spectral profiles of an SFG resonance with the signal proportional to $\left|\chi^{(2)}_{NR} + A/(\omega_1 - \omega_0 + i\Gamma)\right|^2$, where $\chi^{(2)}_{NR}$ is complex in general and A is real. (a) $\chi^{(2)}_{NR}$ is real and has the same sign as A. (b) $\left|\chi^{(2)}_{NR}\right| << |A/\Gamma|$. (c) $\chi^{(2)}_{NR}$ is real with a sign opposite to A. (d) $\chi^{(2)}_{NR}$ is imaginary and opposite in sign compared to $A/i\Gamma$ with $\left|\chi^{(2)}_{NR}\right| > |A/\Gamma|$.

We conclude the theoretical discussion here by briefly discussing some other characteristic features of surface SFG spectroscopy. The technique is often used to probe low lying surface resonances. From Fig. 10.1(a) and Eq. (10.15), one may recognize that $\vec{A}_q$ in Eq. (10.16) is proportional to the matrix element product of infrared and Raman transitions from ⟨g| to ⟨n|. In fact, we can write $\vec{A}_q = \vec{M}_q \vec{\mu}_q$ with $\vec{M}_q$ and $\vec{\mu}_q$ being the Raman and infrared transition probability amplitudes, respectively. For $\vec{A}_q$ to be non-vanishing, the q mode must be both Raman and infrared active. This is an important selection rule for SFG spectroscopy. We also see in Eq. (10.15) that $\vec{\chi}_s$ depends on the distribution function $f(\Omega)$, where Ω may denote the orientation and arrangement of a molecule or group of atoms in a molecule. Being a second-order process, SFG is very sensitive to polar orientations, with $\chi^{(2)}_s$ vanishing if the average orientation is nonpolar. Finally, the surface symmetry is directly reflected in the symmetry of $\chi^{(2)}_s$. The symmetry relations among $\chi^{(2)}_{s,ijk}$ lead to the vanishing of some $\chi^{(2)}_{s,ijk}$ elements and inter-dependences of others. The nonvanishing, independent $\chi^{(2)}_{s,ijk}$ elements can be determined by SFG measurements using different input/output polarization combinations. In many cases, hyperpolarizabilities $\alpha^{(2)}_{\xi,\eta,\zeta}$ of a molecule or group of atoms in a molecule at an interface are the quantities of interest. If intermolecular interaction is negligible, then $\chi^{(2)}_{s,ijk}$ and $\alpha^{(2)}_{\xi,\eta,\zeta}$ are related by a coordinate transformation linking the lab coordinates $\hat{x}, \hat{y}, \hat{z}$ and the molecular coordinates $\hat{X}, \hat{Y}, \hat{Z}$.

$$\chi^{(2)}_{s,ijk} = N_s \langle (\hat{i} \cdot \hat{\xi})(\hat{j} \cdot \hat{\eta})(\hat{k} \cdot \hat{\zeta}) \rangle \alpha^{(2)}_{\xi\eta\zeta} \qquad (10.17)$$

Here, N_S is the surface density of molecules in the interfacial layer, and the angular brackets denote the orientational average or $\int d\Omega\, f(\Omega)$ in Eq. (10.15).

10.3 Experimental Arrangement

The experimental arrangement for SFG spectroscopic measurements is usually quite simple except for the laser system. A typical setup is sketched in Fig. 10.4. The two input laser beams should be well overlapped on the sample. Since the SF output is highly directional, spatial filtering can be used to effectively discriminate against the background noise resulting from laser scattering and induced fluorescence. Spectral filtering can be achieved by placing a spectrometer or a set of interference filters in the detection arm to allow further elimination of the unwanted background light. The SF signal is finally detected by a photomultiplier connected to a gated integrator and computer system. For adjustment of input and output beam polarizations, polarizers are inserted in the optical paths before and after the sample.

The important and expensive part of the experimental setup is the laser system. For SFG spectroscopy, a tunable laser is needed. A short-pulse laser is preferred because according to Eq. (10.10), the optimized signal is inversely proportional to the pulse duration if $I(\omega_i)T$ is the quantity that is limited by laser damage of the sample, as is usually the case. A widely tunable, high-energy, short-pulse laser is difficult to find. Fortunately, recent advances in optical parametric oscillators and amplifiers have changed the scene [8]. Take the picosecond Nd:YAG-laser-pumped optical parametric generator and amplifier as an example [9]. Together with sum and difference frequency generators, coherent pulses tunable from 200 nm to 20 μm [10] with tens to hundreds of μJ per pulse can be obtained from a small table-top system (Fig. 10.5). Such a tunable source is fairly ideal for surface SFG spectroscopy as it can cover a wide range of electronic and vibrational transitions. Nanosecond optical parametric oscillators with sum and difference frequency generators are commercially available. They can also be used for SFG spectroscopy. Their higher energy per pulse (tens of mJ per pulse) partly offsets the loss of signal due to the longer pulse. Stimulated Raman scattering in gases has also been used to generate tunable nanosecond pulses in the infrared, and has been employed for SFG spectroscopy [11]. More recently, free electron lasers have also become a powerful tunable source for SFG spectroscopy as they can be tuned, in principle, throughout the mid- and far-infrared [12]. In the visible range, dye lasers can still be used, but are likely to be replaced by optical parametric generators.

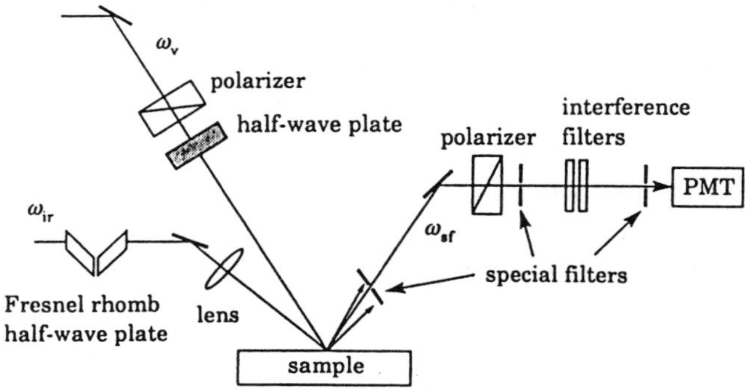

Fig. 10.4 A typical experimental setup for SFG.

As we mentioned earlier, the maximum laser flux allowed on a sample is determined by the laser damage threshold. To optimize the SF signal, the pump area on the sample surface should be as small as possible. Given the input laser powers, the signal is inversely proportional to the pump area. Scanning the tunable source may cause the overlap of the input beams on the sample surface to change. Therefore, in the measurement, normalization of the observed spectrum against a known reference is important.

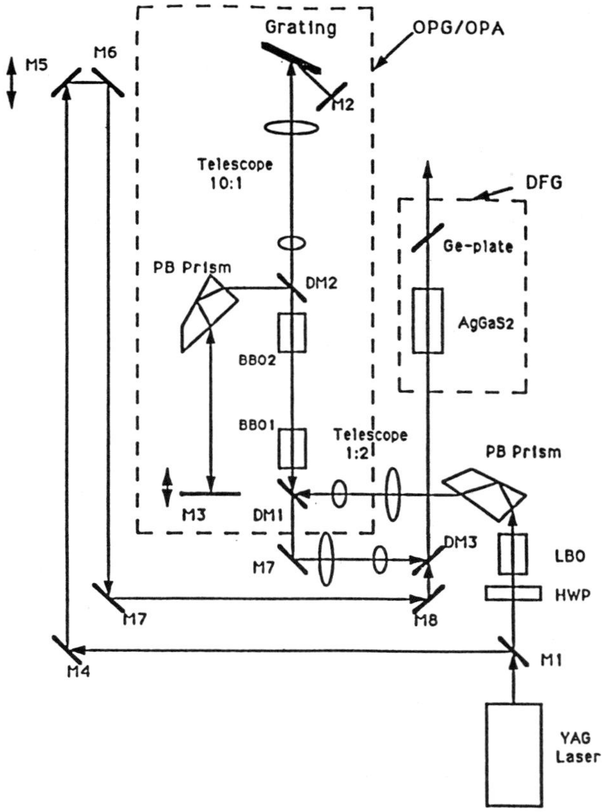

Fig. 10.5 Schematic of an optical parametric generator and amplifier (OPG/OPA) system using BBO crystals pumped at 532 nm. The system generates tunable picosecond pulses from 650 nm to 2.5 μm and has an output bandwidth near the transform limit. The output wavelength can be extended to the UV by SFG and mid-IR by difference frequency generation (DFG) in a nonlinear crystal. Shown in the figure is a DFG stage using an $AgGaS_2$ crystal to extend the output to ~ 10 μm. (After J. Y. Zhang, J. Y. Huang, Y. R. Shen, and C. Chen, *J. Opt. Soc. Am.* **B10**, 1758 (1993).)

10.4 Applications: Selective Examples

The main advantage of SFG as a surface spectroscopic technique is its versatility. It can be applied to all interfaces accessible by light. There already exist in the literature many reports on successful applications of SFG spectroscopy to a wide variety of interfacial problems. Here we only have space to discuss a selective few. We shall focus on those that are unique with SFG.

10.4.1 Surface Specificity

We first show an example illustrating the surface specificity of SFG spectroscopy. Figure 10.6 depicts the surface vibrational spectra in the CH stretch region for three interfaces: hexadecane/silica, hexadecane/octadecyltrichloride(OTS) monolayer/silica, and CCl_4/OTS/silica [13]. The (p, p, p) (SF output, visible input, and infrared input are all p-polarized) polarization combination was used. It is seen that for the hexadecane/silica interface, the SFG spectrum has hardly any detectable features even though the infrared spectrum of liquid hexadecane is known to have very strong absorption bands in this region arising from the large number of CH_2 modes in molecule. For the hexadecane/OTS/silica interface, however, a clear SFG spectrum is observed. The three well resolved peaks at 2870, 2935, and 2965 cm^{-1} can be identified as the symmetric stretch of CH_3, Fermi resonance between symmetric stretch and bending overtone of CH_3, and antisymmetric stretch of CH_3, respectively. This spectrum obviously originates from the OTS monolayer adsorbed at the interface. No CH_2 stretch modes are clearly present, indicating that the alkyl chains of the OTS molecules have few trans-gauche defects since, for a straight chain, the CH_2 modes must be either Raman or infrared active, but not both [14]. That the observed SFG spectrum for the hexadecane/OTS/silica interface must arise from the OTS monolayer is further confirmed by the observed spectrum for the CCl_4/OTS/silica interface. As seen in Fig. 10.6, the spectrum remains basically unchanged with CCl_4 replacing hexadecane.

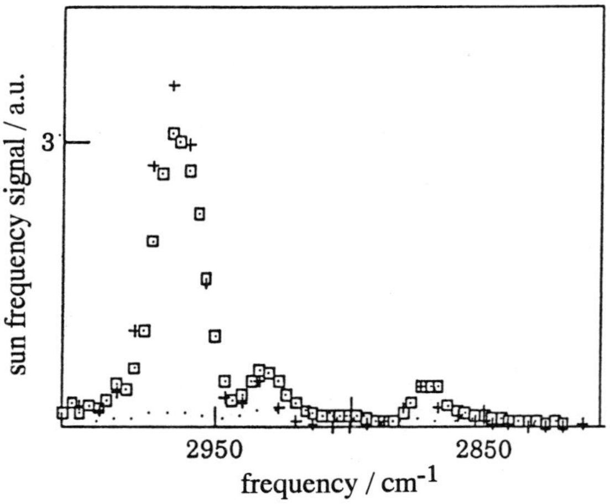

Fig. 10.6 SFG spectra at different interfaces in the (*p*-out, *p*-vis, *p*-infrared) polarization combination. Dots: silica-hexadecane interface. Dotted squares: silica-OTS-CCl_4 interfaces. Crosses: silica-OTS-hexadecane interface. (After Ref. [13].)

The above example clearly illustrates the surface-specific nature of SFG. The polar orientation of the OTS molecules at the interface leads to an electric-dipole-dominated surface nonlinear susceptibility whose contribution to SFG overwhelms the electric-quadrupole bulk contribution. We shall see later many other examples demonstrating the surface specificity of SFG spectroscopy.

10.4.2 Chain Conformation and Orientation

As seen in the above example, the unusual selection rule of SFG spectroscopy suppresses the CH_2 modes of a straight alkyl chain, permitting the CH_3 modes to be readily observable, in contrast to IR and Raman spectroscopies. Trans-gauche defects in the chains break the symmetry and make the CH_2 modes allowed. The appearance of CH_2 peaks in the SFG spectrum is therefore an indication of the existence of defects in the alkyl chains of the molecular monolayer. Because defects disturb the orientational order of the terminal CH_3 group on the chains, the CH_3 peaks are expected to decrease accordingly. Thus, the SFG spectrum can allow an evaluation of the chain conformation of a surface monolayer, which is of great importance in surfactant chemistry, biology, and other areas of science and technology. Figure 10.7 provides an example [14]. The SFG spectra in the CH stretch region were taken for monolayers of pentadecanoic acid (PDA) adsorbed on water at three different surface densities: 47, 34, and 22 Å^2/molecule. The right and left panels correspond to two different input/output polarization combinations, (s, s, p) and (s, p, s), respectively. In the case of 22 Å^2/molecule, the main peaks can all be assigned to CH_3 stretch modes as were the peaks in Fig. 10.6. The small peaks at 2850 cm^{-1} in the (s, s, p) spectrum and at 2890 cm^{-1} in the (s, p, s) spectrum can be identified with the symmetric and antisymmetric CH_2 stretches. The weakness of the CH_2 peaks indicates that the alkyl chains of this densely packed PDA monolayer have hardly any defects. With the surface density of PDA reduced, the CH_3 peak intensities drop accordingly, but the CH_2 peaks become appreciably stronger. This indicates that the chains must have developed more defects as the chain density decreases. Unfortunately, a quantitative calculation is not yet available to relate

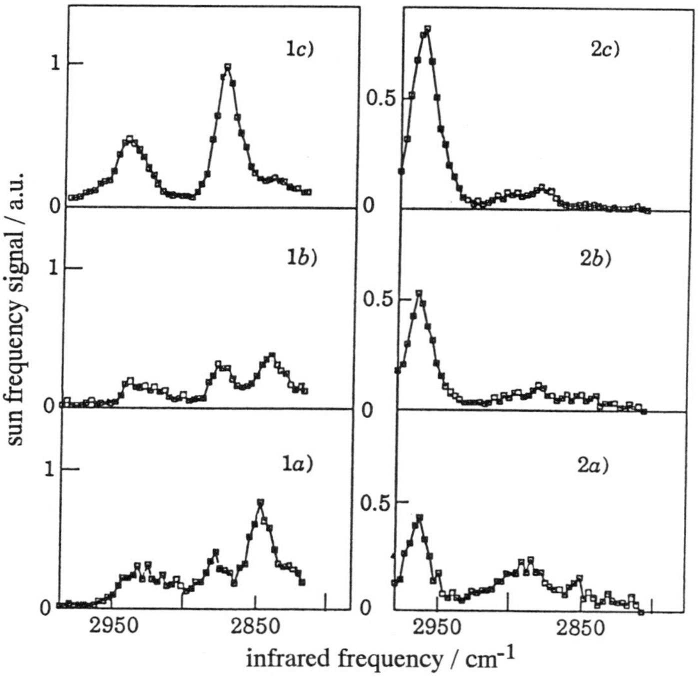

Fig. 10.7 SFG spectra of PDA at different surface coverages normalized per molecule. 1a)-1c) were taken with the *s*-visible, *p*-infrared polarization combination. 2a)-2c) were taken with *p*-visible, *s*-infrared: a) 47 Å^2/molecule, b) 34 Å^2/molecule, c) 22 Å^2/molecule. (After Ref. [14].)

the SFG spectrum to the average number of defects on a chain, thus precluding a quantitative evaluation of the chain conformation.

In the case of straightened alkyl chains, the average chain orientation can be inferred from the average orientation of the terminal CH_3 groups (14). The latter can be deduced from the polarization-dependent SFG spectra of the CH_3 modes. Consider the case where the chain distribution is azimuthally isotropic. Equation (10.17) gives, for the symmetric CH_3 stretch [15,16]

$$\begin{aligned} \left(\chi^{(2)}_{yyz}\right)_s &= \alpha_s[(r+1)\langle\cos\theta\rangle + (r-1)\langle\cos^3\theta\rangle] \\ \left(\chi^{(2)}_{yzy}\right)_s &= \alpha_s(1-r)(\langle\cos\theta\rangle - \langle\cos^3\theta\rangle), \end{aligned} \tag{10.18a}$$

and for the antisymmetric CH_3 stretch

$$\begin{aligned} \left(\chi^{(2)}_{yyz}\right)_d &= -\alpha_d(\langle\cos\theta\rangle - \langle\cos^3\theta\rangle)/2 \\ \left(\chi^{(2)}_{yzy}\right)_d &= \alpha_d\langle\cos^3\theta\rangle/2 \end{aligned} \tag{10.18b}$$

with $\alpha_s \equiv N_s(\alpha^{(2)}_{zzz})_s$, $r \equiv (\alpha^{(2)}_{xxz}/\alpha^{(2)}_{zzz})_s$, and $\alpha_d \equiv N_s(\alpha^{(2)}_{zxx})_d$. Here, $\hat{z}$ is along the surface normal, $\hat{Z}$ is along the symmetry axis of the CH_3 group, θ is the angle between $\hat{z}$ and $\hat{Z}$, and $\vec{\alpha}^{(2)}$ is the hyperpolarizability of the CH_3 group with $\alpha^{(2)}_{xxz}$, $\alpha^{(2)}_{zxx}$, and $\alpha^{(2)}_{zzz}$ being the only independent, nonvanishing elements. Experimentally, $\chi^{(2)}_{yyz}$ and $\chi^{(2)}_{yzy}$ can be deduced from the SFG spectra with (*s, s, p*) and (*s, p, s*) polarization combinations, respectively. From the ratio of $\chi^{(2)}_{yyz}/\chi^{(2)}_{yzy}$, the angle θ can be determined, assuming that the orientational distribution of θ is sharp. Application of this analysis to the spectra in Fig. 10.7 for the densely packed PDA monolayer finds that the CH_3 groups of the alkyl chains are tilted ~35° away from the surface normal, implying that the chains are oriented essentially along the surface normal [14]. Wolfrum *et al.* used this method to determine the chain orientations of fatty alcohols and found excellent agreement with the X-ray results [15].

Long-chain surfactant monolayers adsorbed at air/liquid, liquid/liquid, and liquid/solid interfaces have been studied by a number of groups using SFG spectroscopy [14-19]. Chain conformations and orientations in response to environmental changes are information needed for the understanding of interfacial properties of such systems. In some cases, information on the spatial arrangement of the chains at an interface can also be deduced from the measurements [19].

10.4.3 Pure Liquid Interfaces

SFG can be used to study molecular monolayers adsorbed at interfaces, but is not the only spectroscopic technique for such studies. Linear optical spectroscopy can be equally effective although the selection rules are different. This is not the case if one is interested in surface spectroscopy of pure liquids. Here, for lack of surface specificity, linear optical techniques are unable to yield a surface spectrum not dominated by bulk contributions. Raman spectroscopy has the same difficulty. SFG appears to be the only spectroscopic method that can be adapted for liquid interfacial studies.

Liquid interfaces are certainly not less important than solid surfaces, which have been extensively investigated, but they have hardly been explored experimentally at the molecular level owing to the lack of experimental tools. Unlike solids, molecules in liquids are still free to move. They can orient and arrange themselves at the surface to minimize the free energy of the system, resulting in specific surface physical and chemical properties. We expect that molecular orientation and arrangement at a pure liquid surface can be deduced from the surfacevibrational spectrum obtained by SFG. For illustration, we use water as an example [20].

Water is the most important liquid on earth. Many theoretical calculations have been carried out to predict water structures at various interfaces [21]. The results however are rather confusing. Surface vibrational spectra, if available, can provide a meaningful check on the theory. Figure 10.8 shows the SFG spectra of the air/water interface in the OH stretch region [20]. The (s, p, s) spectrum is very weak, with

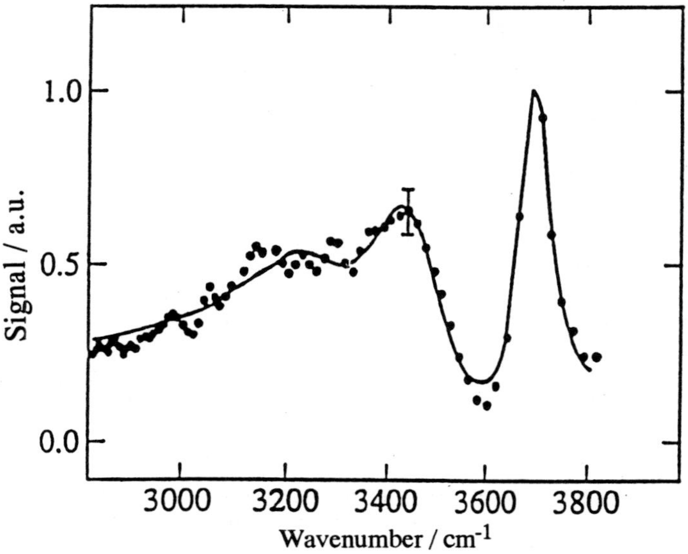

Fig. 10.8 SFG spectrum from the air/water interface with the (ssp) polarization combination. (After Ref. [20].)

hardly discernible features. The (s, s, p) spectrum exhibits three prominent peaks at 3680, 3400, and 3200 cm^{-1}; they can be identified with OH stretches associated with dangling OH bonds, OH with H bonded to a neighbor in a disordered structure, and OH with H bonded to a neighbor in a well ordered structure like ice, respectively. The one at 3680 cm^{-1} is most significant. Its presence indicates that, first,the spectrum must come from the water monolayer at the surface since no dangling bonds would exist in the sublayers, and second, at least part of the surface water molecules must be oriented with one OH bond pointing out of the liquid. Quantitative measurements showed that about 25% of the water molecules in the surface monolayer have contributed one dangling bond per molecule. This is exactly what one would find on a hexagonal ice surface. Therefore, the result suggests that the surface structure of water at the air/water interface is ice-like. However, unlike ice, the tetrahedral hydrogen-bonding network is significantly disordered at the water surface, as evidenced by the strong peak at 3400 cm^{-1}.

The free OH peak at 3680 cm^{-1} is completely suppressed when a monolayer of surfactant molecules adsorbs at the air/water interface [20]. The adsorbed molecules must have either terminated the OH dangling bonds or rearranged the surface water molecules to eliminate the dangling bonds. We can use the surface vibrational spectrum to understand hydrophobicity and hydrophilicity at the molecular level [20]. It was found that, against a hydrophilic surface, the water surface spectrum has the dangling OH peak at 3680 cm^{-1} completely suppressed. Clearly, hydrophilic interaction must have terminated the dangling OH bonds. On the other hand, for a water/hydrophobic interface, the 3680 cm^{-1} peak is as pronounced as ever, indicating that the dangling OH bonds are unaffected by the presence of the hydrophobic surface. Thus the free OH peak can be regarded as a signature of the hydrophobic interface, signifying the lack of bonding interactions between water and the non-wetting surface. Two examples are presented in Fig. 10.9. In both cases, the strength of the free OH peak also provides valuable information. As for the air/water interface, it indicates that the water interfacial structure is ice-like. In the bonded OH region (3000-3600 cm^{-1}), the spectra of the two cases look different. For the water/hydrophobic solid surface, the spectrum closely resembles that of ice with the peak at 3400 cm^{-1} very much suppressed. This suggests that the interfacial ice-like structure of water is highly ordered. This result can be expected if we realize that the water surface layer, being against a solid wall, is not so flexible and cannot allow much dis-order in its hydrogen-bonding network. For the flexible water/oil interface, the spectrum resembles that of the free

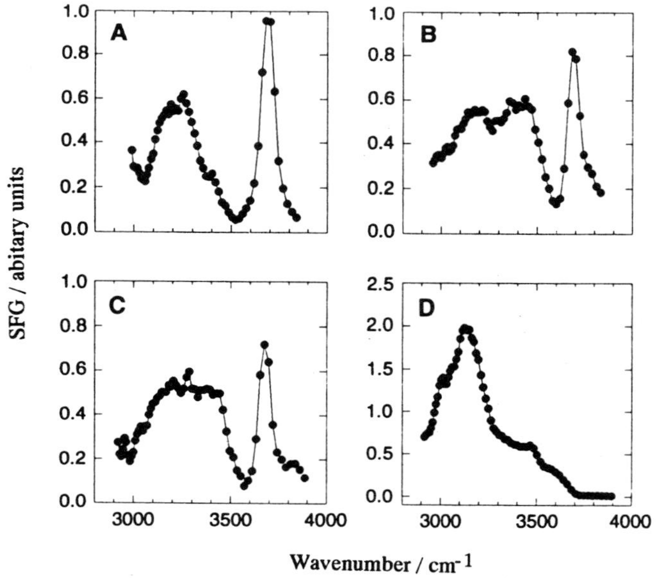

Fig. 10.9 SFG spectra from (A) the quartz-OTS-water interface, (B) the air-water interface, (C) the hexane-water interface, and (D) the quartz-ice interface. The SF output, visible input, and IR input were *s*-, *s*-, and p-polarized, respectively. (After Ref. [20].)

the free water surface, as one would expect.

The surface structure of water at a charged water/solid interface and its reponse to the bulk pH value and salt concentration have also been studied by SFG vibrational spectroscopy [20]. The result shows that the surface water structure varies significantly with surface charges. This should be an important piece of information in the search for an understanding of electrochemistry. SFG is currently the most viable spectroscopic method for in-situ studies of electrochemical reactions at surfaces. It has already been used in several cases to study the nonlinear response of electrode surface and to identify surface species that appear and disappear in an electrochemical process [22].

SFG spectra for other pure liquid interfaces such as alcohols have also been reported [23]. They have provided useful information about the surface structures of those liquids. It is hoped that such work will eventually lead to some general understanding of liquid surfaces and interfaces. The difficulty in the study often lies in our inability to assign and analyze the observed spectral features. Much work is needed before we can relate a surface spectrum unambiguously to a surface structure.

10.4.4 Phase Transitions of Surface Monolayers

Surface monolayers of pure liquids or adsorbed monolayers on liquids can undergo various phase transitions. They can be probed by surface tension measurements, ellipsometry, and X-ray reflection and diffraction, but none of these techniques can provide as detailed information about the nature of the phase transitions as surface vibrational spectroscopy. We consider two cases here [24]. The first one is a monolayer of long-chain alcohol ($C_{12}H_{26}O$) floating on water. It was found that the monolayer undergoes a freezing transition at 39°C, which is 17°C higher than the bulk alcohol freezing temperature. Ellipsometry measurements suggested that the transition is characterized by a sudden change in layer thickness [25]. This is, however, not supported by the SFG spectra presented in Fig. 10.10 [24]. It is seen that the spectra before and after the freezing transition are very much the same except that the former is 40% weaker than the latter and has a somewhat more pronounced CH_2 mode at 2840 cm^{-1}. The other peaks can all be identi-

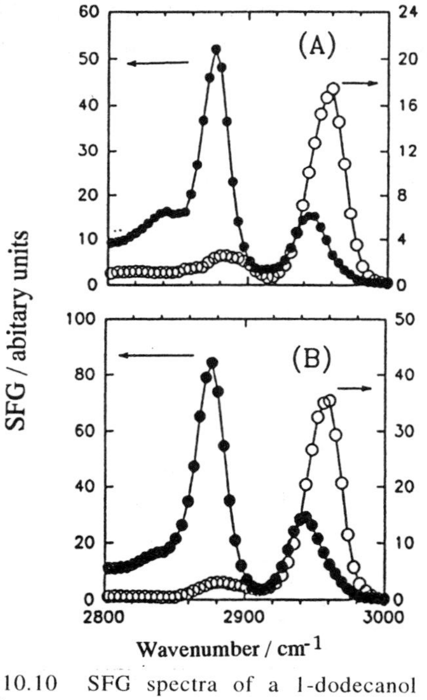

Fig. 10.10 SFG spectra of a 1-dodecanol monolayer on water surface at (A) 43°C and (B) 27°C; solid circles are for the *ssp* polarization combination and open circles are for the *sps* polarization combination. (After Ref. [24].)

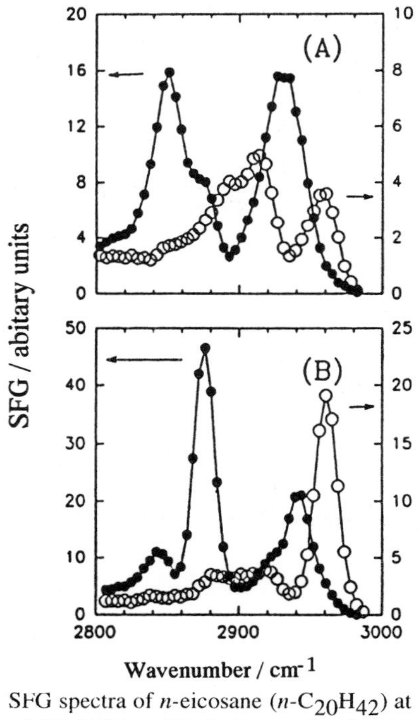

Fig. 10.11 SFG spectra of *n*-eicosane (n-$C_{20}H_{42}$) at (A) 40°C and (B) 37°C; solid circles are for the *ssp* polarization combination and open circles are for the *sps* polarization combination. (After Ref. [24].)

fied as various CH_3 stretch modes, as they were for the spectrum in Fig. 10.6 or 10.7. These spectra clearly indicate that both before and after the transition, the alcohol molecules are oriented perpendicular to the surface with nearly straight chains. The surface freezing transition is characterized by a simple density increase together with a slight reduction of defects on the hydrocarbon chains.

The second case occurs in pure n-alkane (C_nH_{2n+2}) liquids. It was found that for $n \geq 16$, the liquid would experience a surface monolayer freezing transition at a temperature several degrees higher than the bulk freezing temperature [26], but the nature of the transition was not completely clear. Again, the SFG spectra can provide some clues [24]. As shown in Fig. 10.11, the spectra before and after the surface monolayer transition are significantly different. Using the spectra in Fig. 10.7 as references, we notice that the spectra before the transition describe a surface monolayer of vertically oriented chains with an appreciable number of defects. After the transition, the monolayer becomes much better ordered, with the chains in an essentially all-trans configuration. Therefore, the surface freezing transition is mainly characterized by a change in the chain configuration.

In a study of binary mixtures of CH_3CN and water using surface SFG spectroscopy, Zhang *et al.* [27] found that there are sudden changes in the stretch vibrational frequency of CN and the orientation of the symmetry axis of CH_3CN at the air/liquid interface when the bulk molar concentration of CH_3CN exceeds 0.07. At this bulk composition, CH_3CN presumably has reached a full monolayer at the interface. They attributed the sudden changes to an interfacial structural phase transition in which the chemical environment is drastically modified. However, this conclusion is yet to be corroborated by other experimental tests, including the question whether the changes are indeed so sudden that they can legitimately be called a phase transition.

10.4.5 Surface Catalysis

Understanding surface catalytic reactions has long been a goal of modern surface science. Many studies of surface reactions on well defined substrate surfaces in ultrahigh vacuum (UHV) have been carried out in the past several decades. They were designed to search for a microscopic understanding of catalytic reactions. However, surface reactions in UHV could be characteristically different from those in a real gas

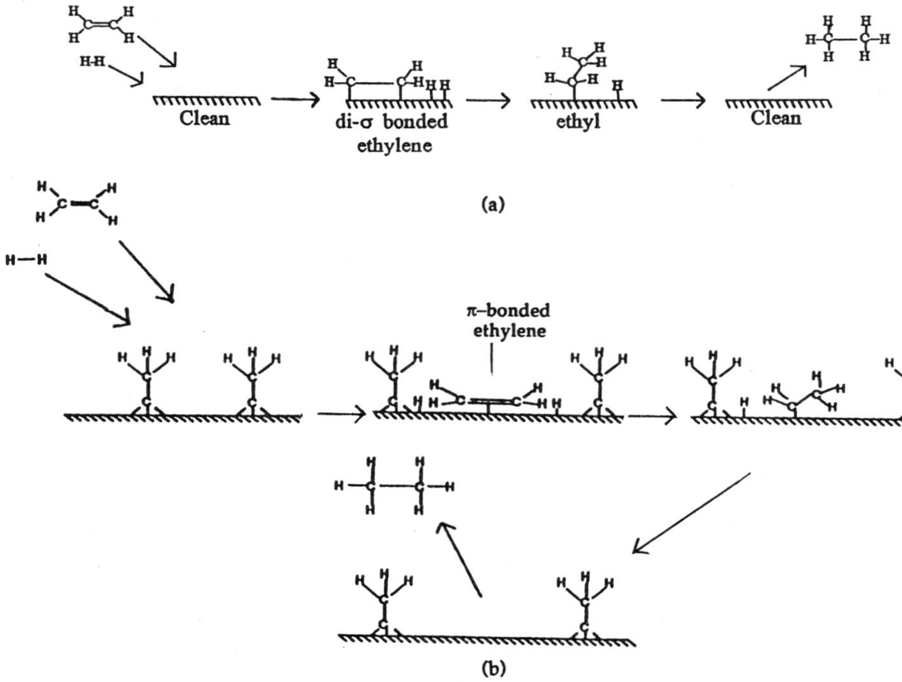

Fig. 10.12 (a) Horiuti - Polanyi mechanism for ethylene hydrogenation on platinum.
(b) New mechanism for ethylene hydrogenation on platinum.

atmosphere. To monitor catalytic reactions in practical circumstances, one would then need *in-situ* surface probes that can also be operative in real atmospheres. Unfortunately this requirement cannot be met by the conventional surface analytical tools. Now SFG spectroscopy provides a great opportunity for such studies.

We use hydrogenation of ethylene (C_2H_4) on Pt(111) into ethane (C_2H_6) as an example [28]. As proposed by Horiuti and Polanyi in 1934 [29], the process was widely believed to proceed in the steps described in Fig. 10.12(a). The C_2H_4 molecules are supposed to be di-σ bonded to the Pt(111) surface and hydrogenated first to ethyl (C_2H_5) and then to C_2H_6 by the hydrogen atoms adsorbed on the surface. Without sufficient H on the surface, C_2H_4 tends to dehydrogenate to ethylidyne above 240 K. SFG spec-

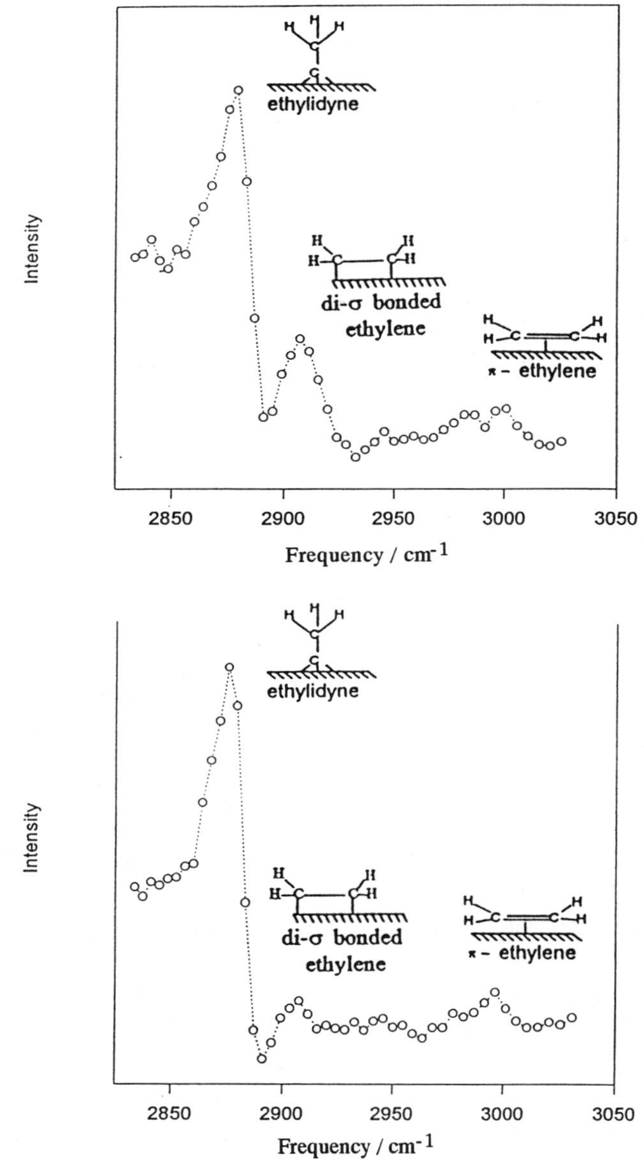

Fig. 10.13 (a) SFG spectrum of the Pt(111) surface during ethylene hydrogenation with 100 Torr H_2, 35 Torr C_2H_4, and 615 Torr He at 295 K. (b) The SFG spectrum under the same conditions as (a), but on a surface which was precovered in UHV with 0.25 ML of ethylidyne. (After Ref. [28].)

troscopy in the CH stretch region can be used to identify these various species on the surface: ethylidyne is characterized by a CH_3 symmetric stretch peak at 2884 cm^{-1}, di-σ bonded ethylene by a CH_2 symmetric stretch peak at 2904 cm^{-1}, and ethyl by peaks at 2860, 2920, and 2970 cm^{-1}. It is also possible to have ethylene π-bonded on Pt(111). The spectral feature is a weak CH_2 symmetric stretch mode at ~3000 cm^{-1}.

We can record the surface SFG spectrum *in-situ* during a reaction process in a real atmosphere. Figure 10.13 displays the spectra for two related cases [28]. One was taken during ethylene hydrogenation at 295 K with 100 Torr H_2, 35 Torr ethylene, and 625 Torr He (1 Torr=133.322 Pa); the process started with a bare, clean Pt(111) crystal. The other was taken under the same conditions, but the Pt(111) surface was initially covered with ethylidyne. Gas chromatography was used to monitor the ethane production. It revealed a production rate of 11 ± 1 ethane molecules (from ethylene) per surface Pt atom per second in the first case and 12 ± 1 in the second case. This means that the production rates of the two cases are practically the same. The two spectra in Fig. 10.13 are also about the same. They show that the surface is mostly covered by ethylidyne (0.15 ML in one case and 0.21 ML in the other, with 0.25 ML being the saturation coverage), and partly covered by di-σ- and π-bonded ethylene. While the π-bonded ethylene coverages in the two cases are the same, the di-σ bonded ethylene coverages differ by a factor of 4. Since the production rates of ethane in the two cases are essentially the same, this result indicates that di-σ bonded ethylene cannot be the important intermediate species in this hydrogenation process. Ethylidyne is known to be just a spectator in the process [30]. Therefore π-bonded ethylene must be the intermediate species truly responsible. The relevant steps for the hydrogenation process are described in Fig. 10.12(b). We notice that the π-bonded ethylene is not the primary adsorbed species on Pt(111) and is only weakly bonded to the surface. This is presumably the reason why ethylene hydrogenation is not sensitive to the Pt surface structure [30].

Catalytic reactions can also take place at liquid/solid interfaces. Clearly, SFG spectroscopy again will be a unique probe for this type of reaction.

10.4.6 Surface States of Buried Interfaces

SFG spectroscopy is an effective means to study buried interfaces. Liquid/solidinterfaces are good examples. The intrinsic surface specificity of SFG permits spectroscopic measurements of the interfacial layers even in the presence of strong bulk absorption in the same spectral region.

Here we focus our attention on solid/solid interfaces. Such interfaces are of great importance for modern electronics, but few techniques are available for their studies. SFG is currently the only spectroscopic technique applicable. Heinz *et al.* first used it to probe surface states of a CaF_2/Si(111) interface [31], which is a prototype insulator/semiconductor interface. Figure 10.14 shows the spectra obtained by SHG ($\omega_1 + \omega_1 \rightarrow 2\omega_1$) and SFG ($\omega_1 + \omega_2 \rightarrow \omega$) with ω_1 scanned in the 2.2 - 2.5 eV range. The spectral features arise from one-photon resonances of ω_1 with electronic transitions. The main peak at 2.4 eV can be identified as due to transitions from the occupied to the empty interface states. It measures the band gap of the interfacial electronic structure, and is very different from the band gaps of the bulk Si (1.1 eV) and CaF_2 (12.1 eV). The peak at 2.26 eV is believed to be associated with the n = 1 surface exciton, but the one at 2.32 eV has not been identified.

Similar experiments were carried out by Daum *et al.* on SiO_2/Si interfaces [32]. They observed a strong two-photon resonance at 3.3 eV and found that it arises from direct band gap transitions between valence and conduction band states associated with a few monolayers of strained Si at the interface. The same resonance was observed at clean Si(100)-(2x1) and Si(111)-(7x7) surfaces where strained Si sublayers also exist due to surface reconstructions.

Yodh and coworkers have also used SHG and SFG spectroscopy to study interface states of ZnSe/GaAs(001) heterojunctions [33]. In this case, the semiconductor bulk has no inversion symmetry which makes SHG and SFG non-surface-specific in general. However, as we mentioned earlier, the symmetries of the interface and bulk are still intrinsically different, and selective input/output beam polarizations can render SHG and SFG surface-specific [3]. Using this scheme, Yeganeh *et al.* found a two-photon resonance at 2.72 eV that can be assigned to a transition between the valence states of ZnSe and a quantum-well state on the GaAs side at the buried interface [33]. The interfacial quantum well was created by interdiffusion of Zn into GaAs and Ga into ZnSe during the sample growth. This resonance is sensitive to a variety of structural phenomena; any process that modifies the band profile near the hetero-junction, such as lattice strain and interfacial reconstruction, can affect the strength of the resonance.

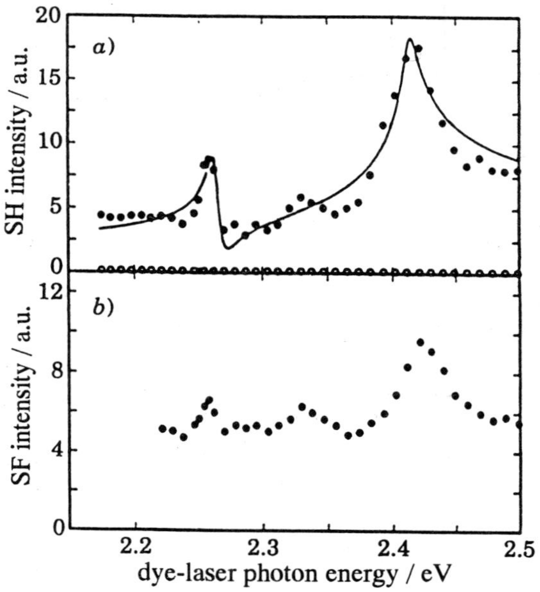

Fig. 10.14 Resonant three-wave mixing signals associated with ${}^{s}\chi^{(2)}_{zzz}$ as a function of the energy of a photon from the tunable dye laser. (a) Results for the SHG process and b) the case of SFG involving mixing the output of the dye laser with a Nd:YAG laser at 1.17 eV. The filled symbols refer to signal from the CaF_2/Si(111) sample, the open symbols to a Si(111) surface covered by the native oxide. The solid curve in a) is a fit to theory. (After Ref. [31].)

More recently, with the same technique, the same research group has studied interface states of buried metal/GaAs junctions [34] and found that these states are sensitive to preparation of the interface. For the As-rich, Au/GaAs (n-type) interface, a one-photon resonance at 0.715 eV arising from transitions from an interface state to the conduction band states was observed. The same resonance was found in As-capped GaAs (n-type) samples. For Ga-rich Au/GaAs (n-type) interface, the spectrum exhibited two such resonances. However, no such resonance was observed in Au/GaAs (p-type) systems. These interface states have been identified as As and Ga p-like defect states associated with As and Ga atomic displacements just below the buried interface.

10.4.7 Ultrafast Surface Dynamics

Conventional surface probes are limited in time response to the millisecond range. Therefore, ultrafast surface dynamics is a field whose exploration had to await the arrival of ultrashort laser pulses. Surface dynamics can now be studied with picosecond or subpicosecond time resolution, and SHG and SFG spectroscopy are best suited for this type of work, namely, studies of transient excitations and relaxations of surface species or surface states and detection of intermediate species in surface reactions. We consider, as examples, measurements of longitudinal and transverse relaxations of surface molecular vibrations using SFG.

Longitudinal relaxation here refers to relaxation of population distribution back to thermal equilibrium. Vibrational energy relaxation of adsorbates on substrates is often intimately connected to surface excitations

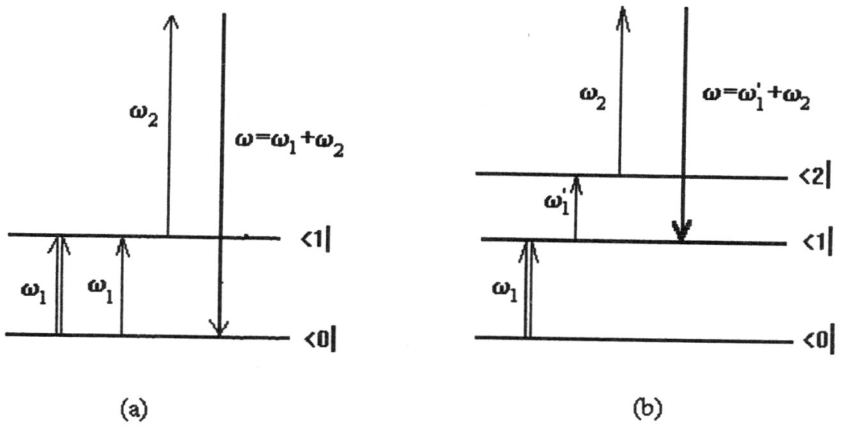

Fig. 10.15 Schematics describing the pump/probe processes using resonant SFG to probe (a) the population recovery and (b) the population decay from the $v = 1$ state after pumping of the $v = 0$ to $v = 1$ transition (indicated by the heavy arrows).

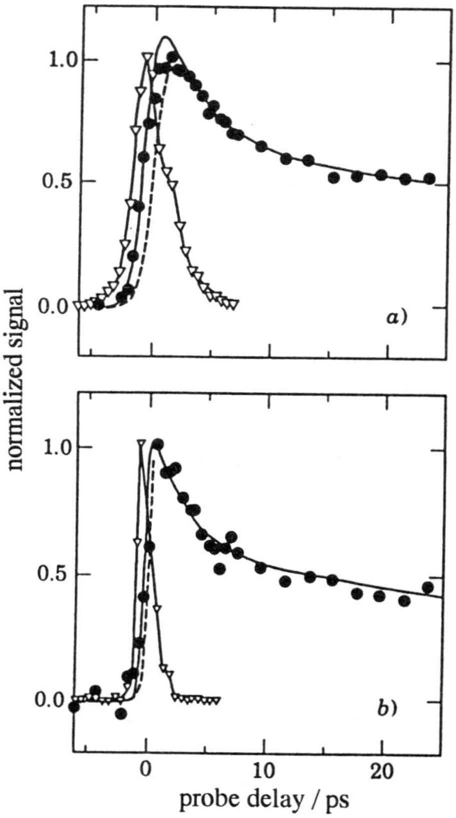

Fig. 10.16 Transient vibrationally resonant SFG signals at 300 K in experiments carried out with pulse widths (FWHM) a) $\tau_p = 2.5$ ps or b) $\tau_p = 1.0$ ps and the corresponding visible-infrared cross-correlations. Solid points, normalized transient SFG signals $1 - [S(\tau_d)/S_o]^{1/2}$; solid lines, fits to transient SFG signals according to a three-level model; lines with inverted triangles, visible-infrared cross-correlation. (After Ref. [35].)

and reactions. Ultrafast infrared spectroscopy has been developed for such studies, but the poor signal-to-noise ratio usually limits the investigations to samples with high surface areas, such as powder or porous semiconductors.materials. SFG spectroscopy permits studies of well-defined crystalline surfaces of metals and semiconductors.

Harris *et al.* first used the technique to study longitudinal relaxation of the symmetric CH stretch vibration of methyl thiolate (CH_3S) on Ag(111) [35]. As sketched in Fig. 10.15, the picosecond pump and probe pulses are at the same frequency and resonant with the $v = 0$ to $v = 1$ transition. The pump pulse excites some population from $v = 0$ to $v = 1$ and the time-delayed probe pulse probes via SFG the relaxation back to equilibrium of the population difference between $v = 0$ and $v = 1$. Figure 10.16 depicts the experimental results that exhibit a bi-exponential relaxation. This suggests that an intermediate state is involved in the relaxation from $v = 1$ to $v = 0$. The population difference recovers in a time less than 100 ps, which seems to be dominated by intramolecular relaxation rather than energy transfer to electron-hole pair excitations in the Ag metal substrate. This is believed to be generally true for sufficiently large polyatomic molecules if the vibrational excitation is localized on an atomic group situated away from the metal.

Harris and coworkers also studied the CO stretch on Cu(100) using SFG [36]. For such small molecules, electron-hole pair excitations become the dominant mechanism for longitudinal vibrational relaxation. A similar result was obtained by Beckerle *et al.* studying CO on Pt(111) [37]. As a further confirmation of this picture, Peremans *et al.* measured the CO vibrational relaxation at a Pt(100) electrochemical interface and found that it was not affected by the interfacial electric field or environment [38].

Vibrational relaxations of adsorbates on semiconductor surfaces have also been studied, namely, Si-H and C-H stretches for H on Si and diamond, respectively [39,40]. As mentioned above, the population recovery time depends on the possible existence of intermediate states in the relaxation pathway. It is therefore generally not the same as the longitudinal lifetime of the excited state. To find the latter, a different scheme must be used. As described in Fig. 10.15, this can be achieved by time-delayed probing via SFG of the $v = 1$ to $v = 2$ transition after population is pumped from $v = 0$ to $v = 1$. The time-dependent SFG signal is directly proportional to the population in the $v = 1$ state and provides a direct measurement of the $v = 1$ lifetime. This is shown in Fig. 10.17 for the Si-H vibration of H/Si(111)[39]. The measured relaxation time of $T_1 = 800$ ps at room temperature is long compared to what one would normally expect for chemisorbed molecules on semiconductors and metals. This is because the H-Si stretch vibration is effectively decoupled from the electronic excitations of Si as well as the H-Si bending vibration and surface Si phonons. The main relaxation pathway is believed to be the four-phonon energy transfer to the H-Si bending and surface Si phonon modes. A similar situation occurs for H on diamond C(111)[40],

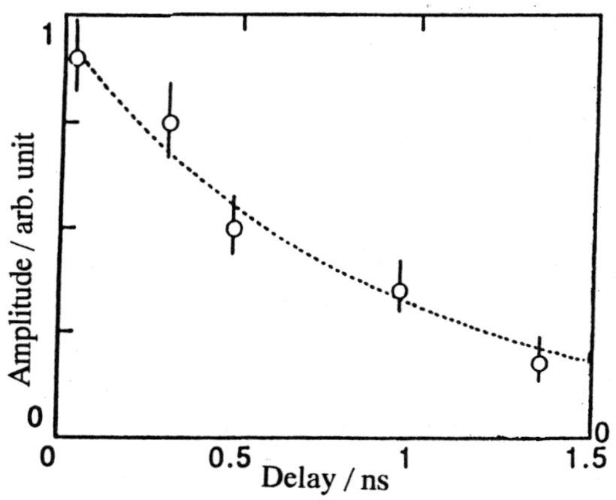

Fig. 10.17 SFG amplitude as a function of time delay in the pump/probe experiment on the H-Si stretch vibration of H/Si(111) using the scheme described in Fig. 15 (b). The dotted line is a theoretical fit giving a relaxation time of $T_1 = 0.9$ ns. (After Ref. [39].)

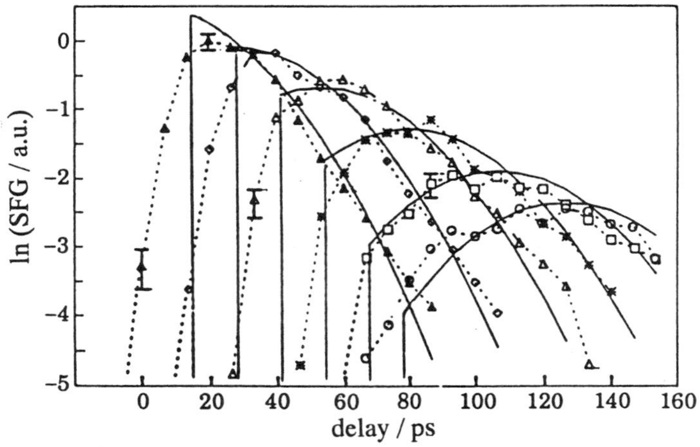

Fig. 10.18 Logarithmic plot of the surface SFG echo signal from the stretch vibration of H/Si(111) as a function of time delay of the visible probe. The data points connected by dashed lines are shown for several delays between the two infrared pulses: 15 ps, ▲; 28 ps, ◊; 42 ps, Δ; 55 ps, *; 68 ps, □; 78 ps, o. The solid lines are theoretical fits. (After Ref. [41].)

although the lifetime of the CH stretch is significantly shorter because of the more efficient coupling in relaxation.

Transverse (or dephasing) relaxation of a surface excitation can also be measured by SFG [41,42]. It is known that, for an inhomogeneously broadened transition, dephasing of an excitation can only be monitored by coherent transient processes such as photon echoes. However to observe photon echoes from surface vibrations is difficult because of the limited sensitivity of infrared detectors. With SFG it becomes possible. Instead of waiting for the rephased oscillating dipoles to radiate the infrared photon echo, we can wave-mix the oscillating dipoles with an incoming visible laser pulse and produce an SFG signal, which is effectively an up-converted photon echo. Guyot-Sionnest used this scheme to observe photon echoes from the surface H-Si stretch vibration of H/Si(111). A dephasing time of 85 psec at 120 K was deduced from the measurement (Fig. 10.18); it corresponds to a homogeneous linewidth of 0.12 cm^{-1}. In comparison, the observed spectral linewidth is 0.3 cm^{-1}, which is obviously dominated by inhomogeneous broadening. Quadratic anharmonic coupling of the H-Si stretch vibration with Si surface phonons at ~ 190 cm^{-1} is believed to be mainly responsible for the dephasing.

10.5 Conclusion

SFG is a most powerful and versatile spectroscopic technique for surface and interface studies. In many cases, it is the only technique available. A number of examples have been given in this article for illustration. As a probe that can be applied to any interface accessible by light, we should be able to find many more applications. Examples are interfacial biological systems, chemical vapor deposition, plasma etching, corrosion, reactions at liquid/solid interfaces, and surface dynamics at liquid interfaces. All of these are problems that can hardly be studied in-situ by other means.

As one would expect, SFG is also not a technique without difficulties. First, tunable laser sources are still limited. Although infrared tunability can be extended to ~ 20 μm in a table-top setup, the energy per pulse is low (~ 10 μJ/pulse) toward the long wavelength end, making the SFG sensitivity poor. The free electron laser allows wavelength coverage over the entire IR range, but it is still not a facility easily available to all researchers. Second, the interpretation of an observed SFG spectrum may not be straightforward. Like all new spectroscopies, assignment of spectral features is a delicate matter, especially when the line profiles are not the same as those appearing in the usual absorption or emission spectrum. In general, it is also necessary to be sure that the bulk contribution to the SFG spectrum is negligible. Otherwise, to obtain the surface or interfacial spectrum, the bulk contribution must be measured separately and subtracted from the observed spectrum. Finally, the attainable signal strength of surface SFG is still rather low and makes spectroscopic measurements time consuming and difficult for routine applications.

We also note that in most surface science problems, a single probe cannot provide the complete answer. SFG spectroscopy is certainly not an exception. Complementary information obtained by other methods is often needed for a full appreciation of the SFG result. However, SFG spectroscopy does have the capability to yield some unique information about surfaces and interfaces. Judging from the growing interest in this area, it is likely to become a common tool in surface science laboratories in the near future.

This work was supported by the Director, Office of Energy Research, Office of Basic Energy Sciences, Materials Sciences Division, of the U.S. Department of Energy under Contract No. DE-AC03-76SF00098.

References

1. See for example, Y. R. Shen. *The Principles of Nonlinear Optics* , Chap. 2 and 25, J. Wiley, New York (1984).
2. Recent reviews on the topic can be found in: Y. R. Shen. In *Frontiers in Laser Spectroscopy, Proc. Int. School of Physics "Enrico Fermi," Course CXX*, (T. Hänsch and M. Inguscio, ed.), p139, N. Holland, Amsterdam(1994); J. Y. Huang and Y. R. Shen. In *Laser Spectroscopy and Photochemistry on Metal Surfaces* (H. L. Dai and W. Ho, ed.) World Scientific, Singapore, to be published.
3. T. Stehlin, M. Feller, P. Guyot-Sionnest and Y. R. Shen. *Optics Lett.* **13**, 389 (1988).
4. T. F. Heinz, C. K. Chen, D. Ricard and Y. R. Shen. *Phys. Rev. Lett.* **48**, 478 (1982).
5. X. D. Zhu, H. Suhr and Y. R. Shen. *Phys. Rev.* **B35**, 3047 (1987).
6. P. Guyot-Sionnest, W. Chen and Y. R. Shen. *Phys. Rev.* **B33**, 8254 (1986); P. Guyot-Sionnest and Y. R. Shen. *Phys. Rev.* **B35**, 4420 (1987); *Phys. Rev.* **A38**, 7985 (1989) and references therein.
7. See, for example, C. C. Wang. *Phys. Rev.* **178**, 1457 (1969).
8. C. L. Tang and L. K. Chang. *Fundamentals of Optical Processes and Oscillators* in *Laser Science and Technology*, (V. S. Letokhov *et al.* ed.) Vol. 20, Harwood Acad., Switzerland, 1995; J. Y. Zhang, J. Y. Huang and Y. R. Shen. *Optical Parametric Generation and Amplification,* in *Laser Science and Technology* (V. S. Letokhov *et al.* ed.), Vol. 19, Harwood Acad., Switzerland (1995); see also, Special Issues on Optical Parametric Oscillators and Amplifiers, *J. Opt. Soc. Am.* **B10**, Nos. 9 and 11 (1993).
9. J. Y. Zhang, J. Y. Huang, Y. R. Shen and C. Chen. *J. Opt. Soc. Am.* **B10**, 1758 (1993).
10. Extension to 20 µm by difference frequency generation in CdSe has recently been achieved by A. Dhirani and P. Guyot-Sionnest. *Optics Lett.* **20**, 1104 (1995).
11. A. L. Harris, C. E. D. Chidsey, N. J. Levinos, and D. N. Loiacono. *Chem. Phys. Lett.* **141**, 350 (1987); S. R. Hatch, R. S. Polizzotti, S. Dougal and P. Rabinowitz. *Chem. Phys. Lett.* **196**, 97 (1992).
12. A. Peremans *et al. J. Electron Spectrosc. and Relat. Phenom.* **64/65**, 391 (1993); A. Peremans *et al. Nucl. Instr. and Meth.* **A331**, 146 (1994); E. R. Eliel *et al. Proc. 15th Int. Conf. on Coherent and Nonlinear Optics*, St. Petersburg, Russia (June, 1995).
13. P. Guyot-Sionnest, R. Superfine, J. H. Hunt and Y R Shen. *Chem. Phys. Lett.* **144**, 1 (1988).
14. P. Guyot-Sionnest, J. H. Hunt, and Y. R. Shen. *Phys. Rev. Lett.* **59**, 1597 (1987).
15. K. Wolfrum, J. Lobau and A. Laubereau. *Appl. Phys.* **A59**, 605 (1994).
16. C. Hirose, N. Akamatsu and K. Domen. *J. Chem. Phys.* **96**, 997 (1992); *Appl. Spec.* **46**, 1051 (1992).
17. T. H. Ong, P. B. Davis and C. D. Bain. *Langmuir* **9**, 1836 (1993); R. N. Ward, P. B. Davis and C. D. Bain. *J. Phys. Chem.* **97**, 7141 (1993); C. D. Bain, P. B. Davis and R. N. Ward. *Langmuir* **10**, 2060 (1994); R. N. Ward, D. C. Duffy, P. B. Davis and C. D. Bain. *J. Phys. Chem.* **98**, 8536 (1994).
18. M. C. Messmer, J. C. Conboy and G. L. Richmond. *J. Am. Chem. Soc.* **117**, 8039 (1995).
19. M. S. Yeganeh, S. M. Dougal, R. S. Polizzotti and P. Rabinowitz. *Phys. Rev. Lett.* **74**, 1811 (1995).
20. Q. Du, R. Superfine, E. Freysz and Y. R. Shen. *Phys. Rev. Lett.* **70**, 2313 (1993); Q. Du, E. Freysz and Y. R. Shen. *Phys. Rev. Lett.* **72**, 238 (1994); *Science* **264**, 826 (1994).
21. See, for example, a recent article on the subject by I. Benjamin. *Phys. Rev. Lett.* **73**, 2083 (1994), and references therein.
22. P. Guyot-Sionnest, A. Tadjeddine and A. Liebsch. *Phys. Rev. Lett.* **64**, 1678 (1990); P. Guyot-Sionnest and A. Tadjeddine. Chem. Phys. Lett. **172**, 5 (1990); A. Tadjeddine and P. Guyot-Sionnest. *Electrochemica Acta* **36**, 1849 (1991).
23. C. D. Stanners, Q. Du, R. P. Chin, P. Cremer, G. A. Somorjai and Y. R. Shen. *Chem. Phys. Lett.* **232**, 407 (1995).
24. G. A. Sefler, Q. Du, P. B. Miranda and Y. R. Shen. *Chem. Phys. Lett.* **235**, 347 (1995).

25. B. Berge and A. Renault. *Europhys. Lett.* **21**, 773 (1993).
26. X. Z. Wu, E. B. Sirota, S. K. Sinha, B. M. Ocko and M. Deutch. *Phys. Rev. Lett.* **70**, 958 (1993); X. Z. Wu, B. M. Ocko, E. B. Sirota, S. K. Sinha and M. Deutch. *Physica* **A200**, 751 (1993); X. Z. Wu, B. M. Ocko, E. B. Sirota, S. K. Sinha, M. Deutch, B. H. Cao and M. W. Kim. *Science* **261**, 1018 (1993).
27. D. Zhang, J. H. Gutow, K. B. Eisenthal and T. F. Heinz. J. *Chem. Phys.* **98**, 5099 (1993).
28. P. S. Cremer, X. Su, Y. R. Shen and G. A. Somorjai. *J. Chem. Phys.* (to be published).
29. I. Horiuti and M. Polanyi. *Trans. Faraday Soc.* **30**, 1164 (1934).
30. S. Davis, F. Zaera, B. Gordon and G. A. Somorjai. *J. Catalysis* **92**, 250 (1985); T. Beebe and J. Yates. *J. Am. Chem. Soc.* **108**, 663 (1986); J. Rekoske, R. Cortright, S. Goddard, S. Sharma and J. Dumesic. *J. Phys. Chem.* **96**, 1880 (1992).
31. T. F. Heinz, F. J. Himpsel, E. Polange and E. Burstein. *Phys. Rev. Lett.* **63**, 644 (1989).
32. W. Daum, H. J. Krause, U. Reichel and H. Ibach. *Phys. Rev. Lett.* **71**, 1234 (1993).
33. M. S. Yeganeh, J. Qi, A. G. Yodh and M. C. Tamargo. *Phys. Rev. Lett.* **68**, 3761 (1992); **69**, 3579 (1992).
34. J. Qi, W. Angerer, M. S. Yeganeh, A. G. Yodh and W. M. Theis. *Phys. Rev. Lett.* **75**, 3174 (1995).
35. A. L. Harris, L. Rothberg, L. H. Dubois, N. J. Levinos and L. Dahr. *Phys. Rev. Lett.* **64**, 2086 (1990); A. L. Harris, L. Rothberg, L. Dahr, N. J. Levinos and L. H. Dubois. *J. Chem. Phys.* **94**, 2438 (1991).
36. A. L. Harris, N. J. Levinos, L. Rothberg, L. H. Dubois, L. Dahr, S. F. Shane and M. Morin. *J. Electron Spectrosc. and Relat. Phenom.* **54/55**, 5 (1990); M. Morin, N. J. Levinos and A. L. Harris. *J. Chem. Phys.* **96**, 3950 (1992).
37. J. D. Beckerle, M. P. Casassa, R. R. Cavanagh, E. J. Heilweil and J. C. Stephenson. *Phys. Rev. Lett.* **64**, 2090 (1990); J. D. Beckerle, M. P. Casassa, E. J. Heilweil, R. R. Cavanagh and J. C. Stephenson. *J. Electron Spectrosc. and Relat. Phenom.* **54/55**, 17 (1990).
38. A. Peremans, A. Tadjeddine, and P. Guyot-Sionnest (to be published).
39. P. Guyot-Sionnest. *Phys. Rev. Lett.* **67**, 2323 (1991).
40. R. P. Chin, X. Blase, Y. R. Shen, and S. G. Louie. Eur. Phys. Lett. **30**, 399 (1995).
41. X. D. Zhu and Y. R. Shen. *Appl. Phys.* **B50**, 535 (1990).
42. P. Guyot-Sionnest. *Phys. Rev. Lett.* **66**, 1489 (1991).

Index

Please note: all page numbers in **bold** refer to tables and all those in *italic* to figures.